APS

Advances in Pharmacological Sciences

Thermal Balance in Health and Disease

Recent Basic Research and Clinical Progress

Edited by
E. Zeisberger
E. Schönbaum
P. Lomax

Birkhäuser Verlag
Basel · Boston · Berlin

Editors:

Professor Dr. Eugen Zeisberger
Physiologisches Institut
Klinikum der Justus Liebig Universität
Aulweg 129
D-35392 Giessen
Germany

Professor Dr. Peter Lomax
Department of Pharmacology
UCLA School of Medicine
University of California
Los Angeles CA 90024-1735
USA

Dr. Eduard Schönbaum*
Department of Medical Pharmacology
Leiden Amsterdam Centre for Drug Research
State University of Leiden
Leiden
The Netherlands

* *Correspondence address:*
Peelkensweg 4
5428 NM Venhorst, N.BR.
The Netherlands

A CIP catalogue record for this book is available from the Library of Congress, Washington D.C., USA

Deutsche Bibliothek Cataloging-in-Publication Data
Thermal balance in health and disease: recent basic research and clinical progress / ed. by E. Zeisberger ... – Basel ; Boston ; Berlin : Birkhäuser, 1994
(Advances in pharmacological sciences)
ISBN 3-7643-5054-7 (Basel ...)
ISBN 0-8176-5054-7 (Boston)
NE: Zeisberger, Eugen [Hrsg.]

Camera-ready copy prepared by the editor and authors
Printed on acid-free paper produced from chlorine-free pulp
Printed in Germany

ISBN 3-7643-5054-7
ISBN 0-8176-5054-7

9 8 7 6 5 4 3 2 1

CONTENTS

Section III: ADAPTATION

Section V:
FEVER AND ANTIPYRESIS

Section VI: PHARMACOLOGICAL ASPECTS

PREFACE

The German land of Hesse (Bundesland Hessen) is well known for its excellent wines (Rheinhessen), for its financial centre and for its airport (Frankfurt/Main). It is, unjustifiably, far less well known for its importance to thermoregulation. Within the small area of Bad Nauheim, Giessen and Marburg (in alphabetical order) a greater concentration of students of thermoregulation has been trained or worked or is working than probably in any other single area of Europe, if not the world. It is thus most appropriate to hold the Ninth International Symposium on the Pharmacology of Thermoregulation in Giessen, from August 7 to 12, 1994.

The community of students of thermoregulation was extremely grateful that Professor Kurt Brück and the Justus-Liebig-University of Giessen had extended this invitation. The same community was even more saddened that Kurt Brück died on April 27, 1992 and thus could not participate in this event that had been planned to honour his achievements. As a small token of friendship and affection, the many friends of Kurt Brück from all over the world who will participate in this event dedicate this symposium to the memory of Kurt Brück. Another staunch friend, researcher of temperature regulation and supporter of this series of symposia died in October 1993: Professor Wilhelm Siegmund Feldberg. We feel fortunate to be able to start this volume with two brief informal and personal accounts of professors Brück and Feldberg.

It is interesting to have a quick look at the course of development of this series of symposia. They were started in order to bring two disciplines, temperature regulation and pharmacology, together - hoping that some cross-fertilisation would ensue: 25 years ago one felt that this would be highly desirable. Not just the fact of their survival, but the way in which these symposia found a place within and between similar meetings and the steadily growing collaboration(s) by students of this multidisciplinary area suggests that the endeavour and efforts were worthwhile. It is with some pride that we can say that our clinical colleagues' interest has increased steadily as has their active participation, witness this book.

A great number of people have given generously of their precious time and wisdom. The first to be congratulated are, of course, the authors of the more than seventy chapters of this book and the even greater number of those responsible for abstracts and posters on display during the meeting. However, they would not have been able to present anything if members of the scientific staff and the support staff of the Justus-Liebig-University in Giessen had not been such enthusiastic helpers. The financial support of the Deutsche Forschungsgemeinschaft, the Hessisches Ministerium für Wissenschaft und Kunst, the Giessener Hochschulgesellschaft and the Thermal Commission of the IUPS made the meeting possible. The organizers have done all that is possible to put the generous support to good use.

After more than twenty years of near perfect collaboration with Dr. Thomas Karger and Mr. Rolf Steinebrunner of S. Karger, Publishers, Basle, Switzerland, who published the first eight volumes, this ninth volume is the first to be published by Birkhäuser Verlag, also from Basle. The collaboration with Dr. J.C.F. Habicht who appears to be equally at home in physiology and pharmacology as in the art and science of publishing was excellent. Dr. Habicht has made it possible for this volume to be (most likely) available in Giessen at the time of the meeting.

March 1994
The editors.

Thermal Balance in Health and Disease
Advances in Pharmacological Sciences

KURT BRÜCK-HIS LIFE AND CONTRIBUTION TO THERMAL PHYSIOLOGY

E. Zeisberger

Physiologisches Institut, Klinikum der Justus Liebig Universität Giessen, Aulweg 129, 35392 Giessen, Germany

Kurt Brück
1925–1992

Life

Kurt Brück, the director of the Physiological Institute at the Medical Faculty of the Justus-Liebig-University, died on April 27,1992 at the age of 66 after a distinguished career in paediatrics and physiology. Born in Köln in 1925, Kurt Brück grew up in Mannheim and studied medicine at Heidelberg and Frankfurt. After obtaining his doctorate he worked as assistant professor at the Physiological Institute in Heidelberg under Prof. Dr. Hans Schaefer (1951-53) and thereafter as paediatrician at the University Children´s Hospital in Hamburg under Prof. Dr. Karl-Heinz Schäfer (1953-55). From 1956 he was assistant professor at the Physiological Institute in Marburg under Prof. Dr. Herbert Hensel, where he continued his studies on neonates. After a one year stay in the Department of Pediatrics of the University of California at Los Angeles, he returned to Marburg, where he assumed in 1959 the position of professor of physiology. His research concerned the development of thermoregulation in the newborn, in which he combined pioneer studies on human neonates with experimental studies on small animals. This integration of basic and clinical research was quite modern at that time and is still widely regarded. His contributions to textbooks of paediatrics are cornerstones of today´s knowledge

in this field and have been reedited several times, as recently as 1991 (1). From that time dates his contribution on Homeothermic Organisms, which he wrote together with H. Hensel and P. Raths in the classical book Temperature and Life (2). His research papers from that time, concerning the developmental and thermoadaptive changes in the capacity for nonshivering thermogenesis and the integration of nonshivering thermogenesis in the central control of thermoregulation, which he published with his co-workers, have inspired many research groups all over the world.
In 1970 Brück moved to Giessen, where he assumed the leadership of the newly founded 2nd Department of the Physiological Institute, which was at that time headed by Prof. Dr. Rudolf Thauer of the Kerckhoff-Institute of the Max-Planck-Society in Bad Nauheim. The next few years in Giessen were occupied in organizing the work in the new institute and by creating new concepts for teaching. From 1973 until his death he was the director of the Physiological Institute in Giessen. Under his leadership he integrated several groups of investigators interested in various aspects of environmental physiology and of thermal biology, and regular discussions among them promoted the development of new thermoregulatory concepts. His favorite topic was thermal adaptation which he tried to investigate both by a biocybernetical approach based, on experiments in man, as well as by a neurological and neuroendocrinological approach, based on experiments in animals. From this research Brück formed some basic concepts on adaptive changes in thermoregulation and on modifications in thermoafferent networks, which he published together with his co-workers in recent reviews (3,4). He also contributed chapters on Heat Balance and the Regulation of Body Temperature and on Functions of the Endocrine System for the textbook Human Physiology (5,6) which is very popular in Germany. Brück devoted much time to serving his University, the scientific community, and national and international needs. He was member of the German Physiological Society (1981-1982 Chairman), member of Thermal Commission of IUPS (1971-1977), member of Special Research Programs of Deutsche Forschungsgemeinschaft: SFB 122 - Adaptation and Rehabilitation (1971-1979), SPP - Thermoregulation and -adaptation (1979-1984), from 1985 a honorary member of Hungarian Physiological Society, and in 1988 he was among the founders of the Institute for Psychobiology and Behavioral Medicine at JLU Giessen. He supported several scientific journals as both a referee and as a source of advice on issues of policy regarding thermal physiology.

Research

Brück was a devoted researcher who lived with and for his science. His scientific career can be divided into 3 periods:
1. Neonatal thermoregulation
2. Animal studies in thermal adaptation
3. Thermoregulatory studies in adult men

1. Period

Brück was one of the few people I know who left the clinic to work in the less lucrative theoretical Institute. In the first period of his career, in Marburg, he cooperated closely with clinicians. Together with his wife, Monica, and Horst Lemtis from the Department of Obstetrics and Gynecology he investigated the thermoregulatory changes of skin blood flow in newborn and premature babies. Later he included measurements of energy metabolism by indirect calorimetry in order to evaluate the thermoregulatory ability of newborn and premature infants. The results of about 15 papers from this period influenced not only thermal physiologists but all of his colleagues in pediatric medicine. One of his papers (7) found its way into the list of the renowned Citation Classics. His contributions of this time period are still highly appreciated. In the condolence letter on Brück's death, Volker von Loewenick, a former chairman of the German Society for Perinatal Medicine, wrote that this society is mourning one of its most profiled members who taught them in the sixties so much about the thermoregulation of newborn infants. He was a pioneer researcher at a time when it was generally believed that newborns, and even more so premature infants, were unable to regulate their temperatures. We owe him much for his impetus for progress and advancement in clinical research in this field.

2.Period

I first met Kurt Brück at a conference in 1966 just at the beginning of the second period of his career. We soon realized that we had similar interests. Both of us were investigating the extent of nonshivering thermogenesis, though by different methods. To compare these methods, Brück invited me to Marburg. My one-year stay in Marburg was occupied with intense work. Fig. 1 demonstrates the first result of this cooperation.

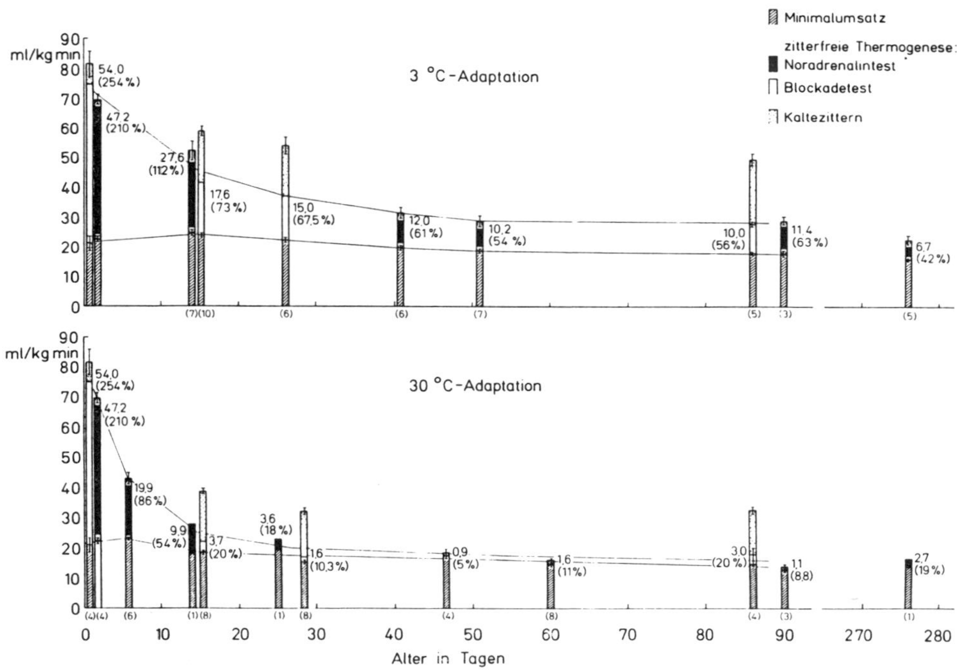

Fig. 1: The extent of nonshivering thermogenesis (NST) in guinea pig in dependence of age (in days) and adaptation to 3°C (top) or 30°C (below), determined by two different methods. Shaded columns denote the resting oxygen uptake, black columns the additional increase in oxygen uptake evoked by an i.m. noradrenaline injection (the noradrenaline test). White columns denote the extent of NST evaluated from the difference between oxygen uptake in cold and oxygen uptake in cold reduced by an application of an adrenergic β-blocker (the blocking test). Columns denoted by dots represent the increase of oxygen uptake due to shivering. The numbers beside the columns give the extent of NST in ml O_2/kg•min, or in % (the resting metabolic rate being set for 100%). The numbers below the columns indicate the number of animals. Reprinted from (8) with permission.

We found that both of the methods employed to determine the extent of nonshivering thermogenesis yielded comparable results and we described, for the first time, this extent and its dependence on age and ambient temperature in guinea pigs (8). The Fig. 3 may be familiar to many readers because it has been reprinted in several publications and in slightly modified form also in several reviews. Another result of this cooperation was a comparative study (9) showing that some species completely lack nonshivering thermogenesis. Together with Wolf Wünnenberg I contributed to this study in the miniature pig and we often remembered the nights spent in the Institute waiting for a parturition and the exciting work with the newborn pigs. At that time I learned of another important finding of Brück and Wünnenberg (10), illustrated in Fig. 2.

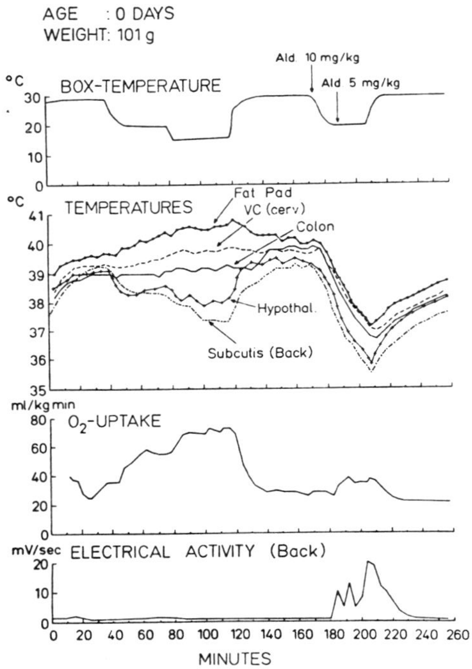

It demonstrates an experiment in a newborn guinea pig. It is interesting in several aspects. First, it demonstrates that the newborn guinea pig has a well-functioning thermoregulatory system. The animal responds to external cooling by an increase in oxygen uptake accompanied by an increase in temperature of brown adipose tissue and in the cervical vertebral canal, but during this time there is no increase in electrical muscle activity, i.e., no shivering. Further it demonstrates that this nonshivering thermogenesis in the brown adipose tissue is controlled by sympathetic nerves via adrenergic ß-receptors, because it can be inhibited by the ß-blocker pronethelol. Third, after this blockade and the decrease in body temperature shivering was elicited. Brück deduced that the shivering activity was suppressed as long as the interscapular brown adipose tissue produced enough heat to keep the spinal cord warm. Together with Wünnenberg he demonstrated later the existence of thermoregulatory structures in the spinal cord and described their projections to the brain (11,12).

Fig. 2: Experiment in newborn guinea pig. The first panel depicts changes in chamber temperature, the second different body temperatures(Fettkörper = interscapular brown adipose tissue, VK = cervical part of the vertebral canal, Subkutis am Rücken = subcutaneous temperature in the area of the os sacrum), the third the changes in oxygen uptake and the fourth the changes of the electrical muscle activity at the back. Ald. = intraperitoneal administration of Alderline (β-receptor blocker). For further explanation see text. Reprinted from (10) with permission.

At the same time we were kept busy by two other findings. Brück and Wünnenberg decided to evaluate thermoregulatory responses more quantitatively and described the threshold temperatures for the onset of shivering and found that these were lowered in cold adapted animals (13). Inspired by a paper by Feldberg and Myers (14) we started to test central effects of noradrenaline. Unfortunately, we had to interrupt these experiments because of my return to Prague. What an exciting time, the years 1966 and 1967! What an accumulation of findings, new techniques and ideas! We were very busy, but lucky and happy. Kurt Brück, too, often remembered this time of his life as the most enjoyable. At that time I did not realize that our cooperation would soon be continued and last for another 25 years. The political developments, the reinstatement of the dogmatic communist regime in 1968, forced my family to emigrate from Czechoslovakia. Of course, I decided to go to Marburg. Kurt Brück was able to offer me a position, and thus I was able to continue the research I had interrupted a year earlier. I will never forget the sympathy and help our family received in those days from the families Brück and Wünnenberg. The cordial atmosphere in the Institute helped us soon to feel at home in Germany. The paper on central effects of noradrenaline on thermoregulation in guinea pigs was published in due course (15), and all the foregoing findings could be unified in a concept of meshed control of two effector systems, non-shivering and shivering thermogenesis (16).

3. Period

At that time the third period of Brück's career had already begun. I participated, but not as a researcher but as a subject. Before Brück started the series of experiments in adult men, he tested the experimental procedure on himself, and afterwards he asked for our help as subjects. Fig. 3 shows the main results of this study.

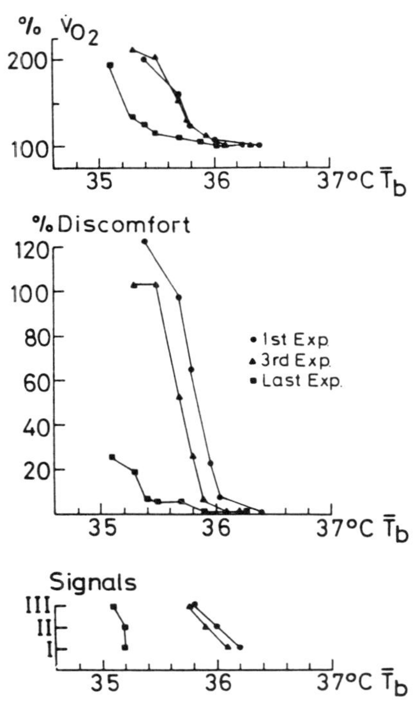

It demonstrates that in man the onset of shivering can be shifted to lower mean body temperatures within a week of daily (1 hour) exposure to cold (+5° C to -5° C). Not only did oxygen uptake start to increase at a lower mean body temperature, but also cold sensation, expressed as percentages of discomfort, was shifted downwards. The paper with these results appeared later (17), because Brück added another series of subjects, but the studies I participated in were begun in Marburg 1969. At that time Brück was appointed head of the Physiological Department in Giessen. He invited Wünnenberg and me to come with him to Giessen and found positions for us. Thus we continued our cooperation in Giessen. In 1973 the Department moved into the new building of the Physiological Institute, where Brück continued to be director until his death.

In Giessen we worked first on ideas we brought from Marburg, but soon the group began to differentiate. Wünnenberg specialized on electrophysiological methods, and investigated central thermosensitivity in hibernators. In 1976 he was appointed head of the Zoological Institute in Kiel and left Giessen. Gerhard Heldmaier came to Giessen, but stayed only short time at the Institute before he got an appointment in Frankfurt and later in Marburg.

Fig. 3: Example of changes in oxygen uptake due to shivering in an adult men during a week of adaptation to daily exposures to temperatures between +5° C and -5° C for about 1 hour. The lower two panels denote the onset and increase of subjective discomfort , measured by two different methods. Reprinted from (17) with permission.

In 1977 Claus Jessen joined us with his research group. One year later I was appointed Professor at Giessen and formed my own research group. We continued the work on central transmitters and became increasingly interested in interactions between peripheral and central transmitters at the hypothalamic level. We used the febrile response and its suppression by endogenous antipyretics as a model for interactions between the immune and central nervous systems (18). Kurt Brück favoured this orientation, being a member of the pathophysiological section of the Physiological Society, founded by his former colleague and friend Professor Dr. E. Betz from Tübingen. Unfortunately, Brück was too occupied to participate in our investigations of fever.

Kurt continued, with his own group of co-workers, the investigations in men. These had different aspects, ranging from exercise and sport physiology, endocrinology, nutrition, psychobiology to effects of neurosurgery. Another aspect of Brück's later work was electrophysiological studies on neuronal plasticity in animals in cooperation with Peter Hinckel. It is impossible to mention all of Brück's publications (more than 200), but I hope I have been able to illustrate at least a small period of his life. Although we did not work together in the final part of his career, we had good and friendly contacts. I remember the daily walk to lunch and our discussions with him, Jessen and other colleagues at tea-time, where we may have had different opinions, but were always able to reach a reasonable compromise.

Brück contributed to many conferences. The last we attended together was the Regional Meeting of the International Union of Physiological Sciences in Prague (1991). Brück was very pleased that both of us could help to organize and chair two of the four round-table discussions of the Regional Thermoregulatory Group. After our return home, he asked me whether it would be possible to organize a similar meeting in Giessen, perhaps the next Pharmacology of Thermoregulation Symposium. At the meeting in Kananaskis a month later I made this proposal and received an invitation to undertake the responsibility for the organization. Brück was very pleased about this decision and promised his help, but unfortunately a few months later he fell seriously ill and died in april 1992. We dedicate this volume to his memory.

Brück was not only an internationally recognized scientist and devoted teacher, he was also a man who had a decisive influence on his colleagues and pupils. He covered his field of science with competence and a deep sense of fascination with living phenomena. His extreme kindness, modesty and a constant desire to share his broad knowledge with others were highly appreciated by all who knew him. His loss will be particularly felt by the members of the Physiological Institute in Giessen, for whom he created a stimulating environment for so many years, but he will be missed also by the international community of researchers in the field of thermoregulation.

References

1. Brück, K (1992): Neonatal thermal regulation. In: Fetal and Neonatal Physiology. Polin, RA, Fox, WW eds. Philadelphia : W.B. Saunders Comp., 488-515.

2. Hensel, H, Brück, K, Raths, P (1973): Homeothermic organisms. In: Temperature and Life. Precht, H, Christophersen, J, Hensel, H, Larcher, W eds. Berlin-Heidelberg-New York: Springer, 503-732.

3. Brück, K, Hinckel, P (1990): Thermoafferent networks and their adaptive modifications. In: Thermoregulation: Physiology and Biochemistry. Schönbaum, E, Lomax, P eds. New York: Pergamon Press, 129-152.

4. Brück, K, Zeisberger, E (1990): Adaptive changes in thermoregulation and their neuropharmacological basis. In: Thermoregulation: Physiology and Biochemistry. Schönbaum, E, Lomax, P eds. New York: Pergamon Press, 255-307.

5. Brück, K (1983): Heat balance and the regulation of body temperature. In: Human Physiology. Schmidt, RF, Thews, G eds. Berlin-Heidelberg-New York: Springer, 531-547.

6. Brück, K (1983): Functions of the endocrine system. In: Human Physiology. Schmidt, RF, Thews, G eds. Berlin-Heidelberg-New York: Springer, 658-687.

7. Brück, K (1961): Temperature regulation of the newborn infant. Biol. Neonat. (Basel) 3, 65-119.

8. Zeisberger, E, Brück, K, Wünnenberg, W, Wietasch, C (1967): Das Ausmaß der zitterfreien Thermogenese des Meerschweinchens in Abhängigkeit vom Lebensalter. Pflügers Archiv 296, 276-288.

9. Brück, K, Wünnenberg, W, Zeisberger, E (1969): Comparison of cold-adaptive metabolic modifications in different species, with special reference to the miniature pig. Fed. Proc. 28, 1035-1041.

10. Brück, K, Wünnenberg, W (1966): Beziehung zwischen Thermogenese im braunen Fettgewebe, Temperatur im cervicalen Anteil des Vertebralkanals und Kältezittern. Pflügers Archiv 290, 167-183.

11. Wünnenberg, W, Brück, K (1968): Zur Funktionsweise thermoreceptiver Strukturen im Cervicalmark des Meerschweinchens. Pflügers Archiv 299, 1-10.

12. Wünnenberg, W, Brück K (1970): Studies on the ascending pathways from the thermosensitive region of the spinal cord. Pflügers Archiv 321, 233-241.

13. Brück, K, Wünnenberg, W (1967): Eine kälteadaptative Modifikation: Senkung der Schwellentemperatur für Kältezittern. Pflügers Archiv 293, 226-235.

14. Feldberg, W, Myers, RD (1964): Effects on temperature of amines injected into the cerebral ventricles. A new concept of temperature regulation. J Physiol (London) 173, 226-237.

15. Zeisberger, E, Brück, K (1971): Central effects of noradrenaline on the control of body temperature in the guinea pig. Pflügers Archiv 322, 152-166.

16. Brück, K, Wünnenberg, W, Zeisberger, E (1971): Integration of non-shivering thermogenesis in the thermoregulatory system. In: Nonshivering Thermogenesis. Jansky, L ed. Prague: Academia, 278-305.

17. Brück, K, Baum, E, Schwennicke, P (1976): Cold-adaptive modifications in man induced by repeated short-term cold-exposures and during a 10-day and -night cold-exposure. Pflügers Archiv 363, 125-133.

18. Zeisberger, E, Roth, J (1994): Central regulation of adaptive responses to heat and cold. In: APS Handbook of Physiology, Section 4, "Adaptation to the Environment". Fregly, MJ, Blatteis, CM eds. Bethesda: Oxford University Press, in press.

Thermal Balance in Health and Disease
Advances in Pharmacological Sciences

W.S. FELDBERG - HIS CONTRIBUTION TO THE PHARMACOLOGY OF THERMOREGULATION

A.S. Milton

Professor of Pharmacology in the University of Aberdeen, Scotland

Wilhelm Siegmund Feldberg
1900–1993

In the beginning

Wilhelm Siegmund Feldberg was born in Hamburg on the 19th November 1900. He graduated in medicine in Berlin in 1925 and then came to England where he worked as a visiting scientist with J.N. Langley and Joseph Barcroft in Cambridge and with Henry Dale at the National Institute for Medical Research in Hampstead. Thus began his scientific career in medical research, a career which he pursued for the rest of his life. On returning to Germany in 1928, he worked with E. Schilf on the role of histamine in allergic reactions, which research lead to our understanding of the pathology of asthma; together they published a book in 1930 on histamine, which they dedicated to Dale.(13) In an essay which he wrote in 1977 (4) on the early history of acetylcholine transmission, Feldberg recounts how he came back to work with Dale in the 1930's, how one day in 1933 shortly after Hitler came to power, the director of the Institute in Berlin where he was working sent for him and informed him that he had been dismissed, must leave the Institute at the latest by midnight that day, and was not allowed to enter it anymore. It is a measure of the man that he continued to carry out experiments by working during the night when no-one was around and knew that he was there. Some weeks later he met a representative of the Rockefeller Foundation whom he asked whether there was anything that the Rockefeller Foundation could do to get him out of Germany. Professor Feldberg recounted this episode saying that the representative although being sympathetic told him that there were so many scientists wishing to leave Germany that there was very little that he could do. He then asked whether Feldberg could help him, as he was looking for a person called Feldberg as he had a message for Feldberg from Sir Henry Dale in England who had told him that he had to get Feldberg to England. Feldberg told him that he was that very person, and that is how it happened that on the 7th July 1933 he landed at Harwich with his wife Katherine and their two young children. Feldberg brought with him to Dale two techniques which were the keys which resulted in the first direct experimental evidence for the role of acetylcholine, both in ganglionic and neuromuscular transmission. These two keys were eserine which prevented the breakdown of acetylcholine by cholinesterase, and the leech muscle for the estimation of acetylcholine, the most sensitive method available at that time. In the three years in which Feldberg worked with Dale, his name appears on twenty four separate publications in the Journal of Physiology, this research in collaboration with Dale, J.H. Gaddum, Marthe Vogt and G.L. Brown provided the final proof for chemical transmission. With no permanent position available in the U.K., Professor Feldberg and his family departed on the 15th April 1936 to Australia where he worked in Melbourne for two years, continuing his research into the release of histamine, studying the effects of venoms and bacterial toxins, before returning to Cambridge in 1938 as University Reader in Physiology. He stayed in Cambridge until 1949 when he was appointed head of the Division of Physiology and Pharmacology at the National Institute of Medical Research which by that time had moved from Hampstead to Mill Hill. When he reached retirement age he stayed on at Mill Hill firstly as Head of Neuropharmacology and then as an honorary research worker, finally retiring in 1990.

The inner surface of the brain

It was in the early 1950's that Feldberg started on his research on the inner surface of the brain where he devised a series of cannulae which could be implanted into the cerebral ventricular system to allow the direct administration of drugs into the brain. An account of this research is published in a book which he wrote in 1960.(3) A part of this initial research was on tremor, particularly that produced by the injection of tubocurarine into the ventricular system. In a paper written with Laurence Malcolm in 1959 (6), he observed that the tremor produced by tubocurarine could be abolished by adrenaline and he also mentions in this paper that he makes no distinction between tremor and shivering. At no time in these early experiments did he record body temperature, yet these techniques were the trigger for so much of what was to come.

Monoamines and temperature regulation

In 1963, he was joined by R.D. Myers from Purdue University in Indiana. Myers and Villablanca had carried out experiments in which they had injected bacterial pyrogens into cerebral ventricles and showed that the animals developed fever (Villablanca & Myers, 1975) (18). Myers persuaded Feldberg that he should monitor body temperature during their icv experiments. As Feldberg said many times later, though perhaps jokingly, this was one of the greatest mistakes of his life. Within a very short time, Feldberg and Myers had shown that in the conscious cat, i.c.v. injection of 5-hydroxytryptamine produced a rise in deep body temperature, whereas adrenaline and noradrenaline produced a fall. In 1964 (8), they published a paper in the Journal of Physiology entitled 'Effects on temperature of amines injected the cerebral ventricles, a new concept of temperature regulation'.(see Fig 1) This paper opened a new chapter in the physiology and pharmacology of temperature regulation which has continued to this day. As does all good research it produced controversy, particularly when it was shown by other research workers that in the rabbit and sheep the effects of the amines were reversed with 5-HT producing a fall in deep body temperature whereas noradrenaline produced a rise. As well as studying the effects of the monoamines on body temperature, Feldberg and Myers also measured the output of the monoamines.

This research on body temperature and the monoamines was continued by Feldberg for many years with a variety of distinguished visitors to his laboratory. In the paper in 1964, Feldberg and Myers state "a fall in temperature may therefore be visualized as being due to an augmentation in the release of the catecholamines together with inhibition of the release of 5-HT, or solely to an inhibition of the release of the catecholamines. And a rise in temperature would be due to the augmentation in the release of 5-HT together with inhibition of the release of adrenaline and noradrenaline, or solely to an inhibition of the release of 5-HT. The rise in temperature produced by bacterial pyrogen i.e.. the fever of infections and diseases, may be brought about in one of these way. In addition, pyrogens may render the hypothalamus more sensitive to the action of 5-HT, and there is a further possibility that pyrogens do not act solely through the release of hypothalamic amines, but mimic the effect of 5-HT."

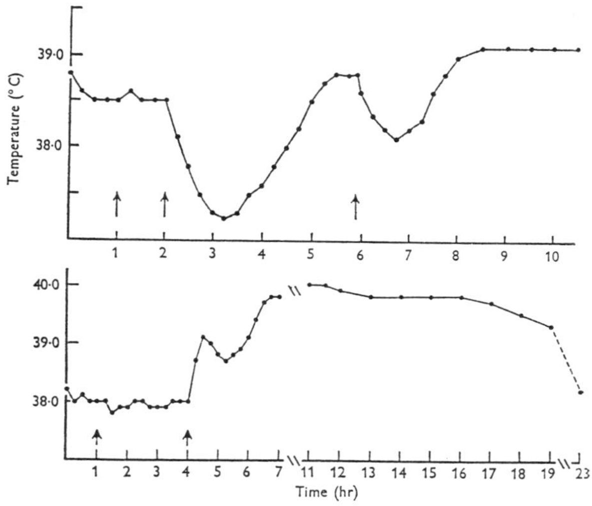

Fig.1: Record of rectal temperatures of an unanaesthetized cat. The arrows indicate injections into the cerebral ventricles. Upper trace: 0.1 ml. 0.9% NaCl (1st arrow), 50 µg adrenaline (2nd arrow) and 50 µg noradrenaline (3rd arrow). Lower trace: 100µg creatinine sulphate (1st arrow and of 200 µg 5-HT (2nd arrow). (J.Physiol.1964)

Prostaglandins and fever

It was during the late 60s that I became interested in temperature regulation and in particular the antipyretic actions of the aspirin-like drugs. I had previously been interested in the i.c.v. effects of nicotine. I had wondered whether the tremor produced by nicotine was in fact a shivering response. I was joined by Sabine Wendlandt, an East German who had recently graduated in Psychology from McGill University in Canada, and we decided to investigate Feldberg's monoaminergic theory of fever and attempt to fit in to this concept the mode of action of the antipyretic agents. In our first experiments, we showed in 1968 (15) that the fever produced by the i.c.v. injection of pyrogen was abolished by the antipyretic agent 4-acetamidophenol (paracetamol, acetaminophen). We also showed that during fever a biologically active substance was present in cerebral spinal fluid. However, when we attempted to assay this on the rat fundus strip, the response of the tissue was not blocked by the 5-HT blocking drug brom-LSD. On the off-chance that the substance might be a prostaglandin (such substances also contract the rat fundus strip), we injected prostaglandin E_1 into the cerebral ventricles and found that it produced a fever identical to pyrogen fever other than it had a shorter duration of action. Sabine Wendlandt and I (16) then showed that the prostaglandin fever could not be abolished or prevented by an antipyretic drug. These observations led us to suggest that a prostaglandin was the central mediator for fever, that pyrogens caused the release of these substances and that antipyretic drugs somehow interfered with this release. As was subsequently shown by Vane in 1972,(17) the non-steroidal antipyretic agents actually work by inhibiting the synthesis of prostaglandins.

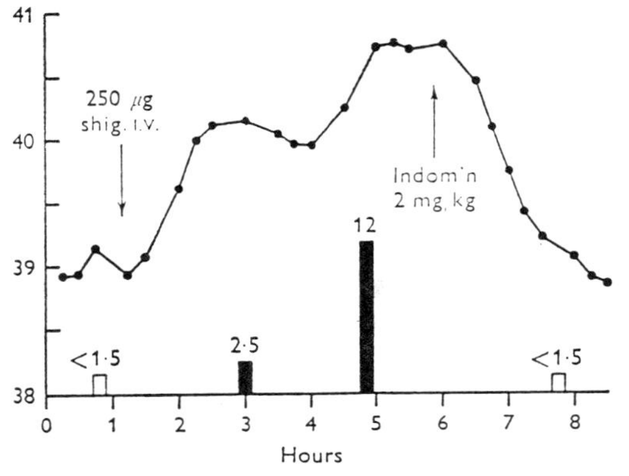

Fig. 2: Record of rectal temperature of an unanaesthetized cat. The height of the columns and the values above refer to PGE_1-like activity in ng/ml of cisternal c.s.f., the position of the columns refers to the time but not to the duration of c.s.f collection. The first arrow indicates an injection of 250 µg pyrogen, the second arrow an injection of indomethacin 2mg/kg (Indom'n). (J. Physiol. 1973)

Interestingly Feldberg and Myers in 1966 (9) had found, in one of their perfusion experiments, that when they perfused the anterior horn of the cat's ventricle a biologically active substance which was not 5-HT was released and they suggested that this might be either a prostaglandin or irin. They did not follow this work up at that time.

Sabine Wendlandt and I were working in the Pharmacology Department of the School of Pharmacy of London University, only a few miles away from the National Institute of Medical Research at Mill Hill and Professor Feldberg kindly invited the two of us to spend one day a week in his laboratory, collaborating with him into the role of prostaglandin. In 1971 he and Saxena (12) had recently confirmed our experiments on PGE_1 and showed that an infusion of PGE_1 into the ventricles could produce a sustained fever They also were able to find a biologically active substance in cerebro-spinal fluid collected from the third cerebral ventricle. Saxena had returned to India by the time Sabine Wendlandt and I joined Professor Feldberg, however, we were joined by another scientist from India, K.P. Gupta. We developed a method for perfusing the cerebral ventricle system from the lateral ventricle to the cisterna magna and collected the perfusate. We were able to show that in control situations no prostaglandin could be detected in the perfusing fluid whereas during fever the levels were elevated. When we administered a variety of antipyretic agents, not only did the fever reduce, but also the levels of prostaglandins.(Fig 2) Our initial results on the prostaglandins and their release during pyrogen fever were presented by Professor Feldberg himself at the first Thermoregulation Symposium held in San Francisco in 1972.(Feldberg & Milton 1973) (7) The research was published in full in a paper in 1973 (Feldberg, Gupta, Milton and Wendlandt) (5)

The 'Set-point" and the Role of Calcium Ions

In experiments in which the ventricles were perfused with artificial CSF containing different proportion of calcium and sodium ions, Feldberg, Myers and Veale in 1970 (10) observed that in the absence of calcium ions, the cats developed an exaggerated febrile response associated with a rapid rise in deep body temperature accompanied by a vigorous shivering.The authors suggest that calcium ions are responsible for maintaining the 'set-point' and that any reduction in the availability of Calcium ions would result in a resetting of the 'set-point' and a rise in body temperature. They suggested that pyrogens might act by this mechanism. Once again Feldberg had initiated a idea which was to produce a flurry of activity in the following years, particularly by Myers and his colleagues. Later experiments by Dey, Feldberg, Gupta, Milton and Wendlandt in 1974 (2) showed this sodium fever was completely unrelated to the bacterial pyrogen fever in that it was not abolished by antipyretic drugs, nor was there any evidence of the release of prostaglandins during the hypothermia.(Fig 3).

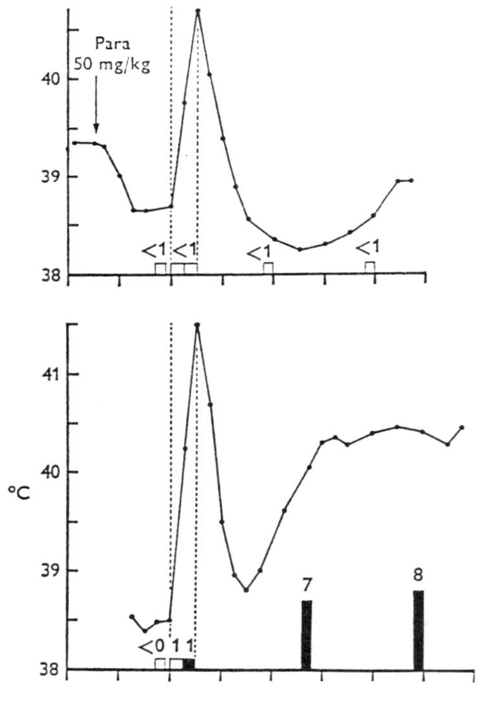

Fig. 3. Two records of rectal temperature obtained from the same cat. The interval between the interrupted vertical lines indicates half hour perfusion at a rate of 0.1ml./min, of the cerebral ventricles with calcium free artificial c.s.f. The height of the vertical columns and the values above refer to PGE_2-like activity in ng/ml. of cisternal effluent or cisternal c.s.f. The arrow in the top record indicates an intra-peritoneal injection of paracetamol 50mg/kg. (Para). (J. Physiol.1974)

In the paper by Feldberg et al (11), they observed a long lasting rise in body temperature following the perfusion, they attributed this to an unknown pyrogen. This is also illustrated in Fig.3. where it is shown to be completely prevented by paracetamol. This is what is now referred to as a 'non-specific' fever which is caused by perturbation of the ventricular system, resulting in a local inflammatory response resulting in the release of PGE_2.

The later years

In 1973, I left for Aberdeen and Professor Feldberg continued his research at Mill Hill. The last occasion that he attended a Thermoregulation Symposium was in 1976 at Banff where he presented a paper on prostaglandin release during shivering produced by anaesthesia. (Dashwood & Feldberg, (1977) (1). Professor Feldberg's last published paper on temperature regulation was presented to a meeting of the Physiological Society in Galway in 1984 when he reported with Scott on fever produced in cats by i.c.v. injection of interferons.(14).During the last few years of his research career, he returned to a previous interest in the control of blood sugar, in particular its central control , and he continued to publish a series of papers on this subject. In 1988 at the meeting of the Physiological Society in Cambridge, his final paper (with D.A. Pyke and W.A. Stubbs) was presented to the Society, "Hyper- and hypoglycaemia produced in anaesthetized cats by phentolamine".(11) At that time, he was approaching his 88th birthday.

Memories

All of those who worked with Professor Feldberg will have their own individual memories of him, and their own collection of stories. Two stick in my mind. The first was the occasion when he was invited by the British Pharmacological Society to be the guest of honour at an official dinner of the Society at a meeting in London. I was in his office when he turned down this invitation and he turned to me and said that he was taking his grand-daughters to see 'Così fan tutte' at Covent Garden the same evening and that Mozart was far more important. The second occasion was when he came to Aberdeen to receive an honorary degree from the University. He was accompanied by Kim O'Rourke, his faithful secretary for so many years and who was soon to become the second Mrs Feldberg. Indeed they first publicly told of their engagement while in Aberdeen. They were staying at the Station Hotel and I was commanded to have breakfast with them as Professor Feldberg maintained that the Station Hotel served the best kippers in Britain. Who can forget visiting 'Lavenham' with the collection of signed Toulouse-Lautrec lithographs encircling the drawing room walls, which he and Katherine had purchased in Paris in happier times and then brought to England. The first time I met Professor Feldberg was at a meeting of the Physiological Society at Mill Hill in 1957, I was a young graduate student at Oxford and I remember being introduced to him and Sir Henry Dale by my supervisor J.H. Burn. The last occasion was at the summer meeting of the Physiological Society in Cambridge, where he presented his final paper to the Society. We sat next to each other at the official dinner and I was able to introduce him to my son Nathaniel (N.G.N. Milton) who was attending his first Physiological Society meeting.

On 23rd October 1993, Professor Feldberg died at his home 'Lavenham', as he approached his ninety third birthday; and so passed into legend one of the great scientists of our time. In the words of Hamlet "I shall not look upon his like again"

"Perhaps it was that I had brought with me a key that would open the doors"

Ipse dixit

References

1. Dashwood, M.R. and Feldberg, W. (1977) Endotoxin fever, prostaglandin and anaesthesia. In: Drugs, Biogenic amines and body temperature. Eds. Cooper, K.E., Lomax, P. and Schönbaum, E. pp. 145-152. Basel, Karger.

2. Dey, P.K., Feldberg, W., Gupta, K.P., Milton, A.S. and Wendlandt, S. (1974) Further studies on the role of prostaglandin in fever. J. Physiol. (Lond) 241, 629-646.

3. Feldberg, W. (1963) A pharmacological approach to the brain from its inner and outer surface. London, Edward Arnold (Publishers) Ltd.

4. Feldberg, W. (1977) The early history of synaptic and neuromuscular transmission by acetylcholine: reminiscences of an eye witness. In: The Pursuit of Nature. pp. 65-83. Cambridge, Cambridge University Press.

5 Feldberg, W., Gupta, K.P., Milton, A.S. and Wendlandt, S. (1973) Effect of pyrogen and antipyretics on prostaglandin activity in cisternal CSF on unanaesthetized cat. J. Physiol. (Lond) 234, 279-303.

6. Feldberg, W. and Malcolm J.L. (1959) Experiments on the site of action of tubocurarine when applied via the cerebral ventricles. J. Physiol. (Lond) 149, 68-77.

7. Feldberg, W. and Milton, A.S. (1973) Prostaglandin fever. in: The Pharmacology of Thermoregulation. Eds. Lomax. P. and Schönbaum, E. pp 302-310. Basel, Karger.

8. Feldberg, W. and Myers, R.D. (1964) Effects on temperature of amines injected into the cerebral ventricles. A new concept of temperature regulation. J. Physiol. (Lond) 173, 226-237.

9. Feldberg, W. and Myers, R.D. (1966) Appearance of 5-hydroxytryptamine and an unidentified pharmacologically active lipid acid in effluent from perfused cerebral ventricles. J. Physiol. (Lond) 184, 837-855.

10. Feldberg, W., Myers, R.D. and Veale, W.L. (1970) Perfusion from cerebral ventricle to cisterna magna in the unanaesthetized cat. Effect of calcium on body temperature. J. Physiol. (Lond) 207, 403-416

11. Feldberg, W., Pyke, D.A. and Stubbs, W.A. (1988) Hyper- and hypoglycaemia produced in anaesthetized cats by phentolamine. J. Physiol. (Lond) 407, 106p.

12. Feldberg, W. and Saxena, P.N. (1971) Fever produced by prostaglandin E_1. J. Physiol. (Lond) 217, 547-556.

13. Feldberg, W and Schilf E. (1930) Histamin. Berlin. Julius Springer.

14. Feldberg, W. and Scott, J.M. (1984) Interferon fever in cats. J. Physiol. (Lond) 355, 28P.

15. Milton, A.S. and Wendlandt, S. (1968) The effect of 4-acetamidophenol in reducing fever produced by the intracerebral injection of 5-hydrotryptamine and pyrogen in the conscious cat. Brit. J. Pharmacol. 34, 215P.

16. Milton, A.S. and Wendlandt, S. (1970) A possible role for prostaglandin E_1 as a modulator for temperature regulation in the central nervous system in the cat. J. Physiol. (Lond) 207, 76-77P.

17. Vane, J.R. (1971) Inhibition of prostaglandin synthesis as a mechanism of action for aspirin-like drugs. Nature New Biol. 23, 232-235.

18. Villablanca, J. and Myers, R.D. (1965) Fever produced by microinjection of typhoid vaccine into hypothalamus of cats. Am. J. Physiol. 208, 703-707.

(Figures and quotation from the Journal of Physiology: with permission.)

Section I:

THERMOSENSITIVITY AND NEURONAL CONNECTIVITY

Thermal Balance in Health and Disease
Advances in Pharmacological Sciences

CELLULAR AND SYNAPTIC MECHANISMS OF THERMOSENSITIVITY IN HYPOTHALAMIC NEURONS

J.A. Boulant

Department of Physiology, College of Medicine, The Ohio State University, 1645 Neil Avenue, Columbus, Ohio 43210 (USA)

The preoptic-anterior hypothalamus (PO/AH) and nearby circumventricular and septal regions are known to have important roles in thermoregulation (reviewed in 1-3). Thermode studies show that PO/AH warming evokes heat loss responses, while PO/AH cooling evokes heat retention and heat production responses. Thermoregulatory responses are elicited also after various neuroactive substances are microinjected into these areas. When pyrogens or drugs alter body temperature, it is often presumed that these substances are acting on the neurons in and near the rostral hypothalamus. The purpose of this paper is to review basic characteristics of the cells that form this hypothalamic network. These cellular properties and synaptic connections should provide clues to understanding mechanisms by which drugs and pyrogens affect neuronal activity to alter thermoregulatory responses.

Criteria for Neuronal Types

Electrophysiological studies have shown that the rostral hypothalamus contains both temperature sensitive and temperature insensitive neurons (1-3). Fig. 1 presents examples of three basic types of neurons in the PO/AH (4). When this region is warmed, firing rates increase in warm sensitive neurons, and firing rates decrease in cold sensitive neurons. The majority of preoptic neurons, however, are considered to be temperature insensitive, showing little or no change in their firing rates when the hypothalamus is warmed or cooled. PO/AH electrophysiological studies often report that more than 60% of the neurons are temperature insensitive, 30% are warm sensitive and less than 10% are cold sensitive (1-3).

Generally, the criteria for neuronal thermosensitivity is based on the thermal coefficient or the slope determined when firing rate (impulses/sec) is plotted as a function of preoptic temperature (T_{po}) (1-3). As illustrated in Fig.2A, if a neuron has a positive thermal coefficient of at least +0.8 impulses/sec/°C, it is classified as warm sensitive.

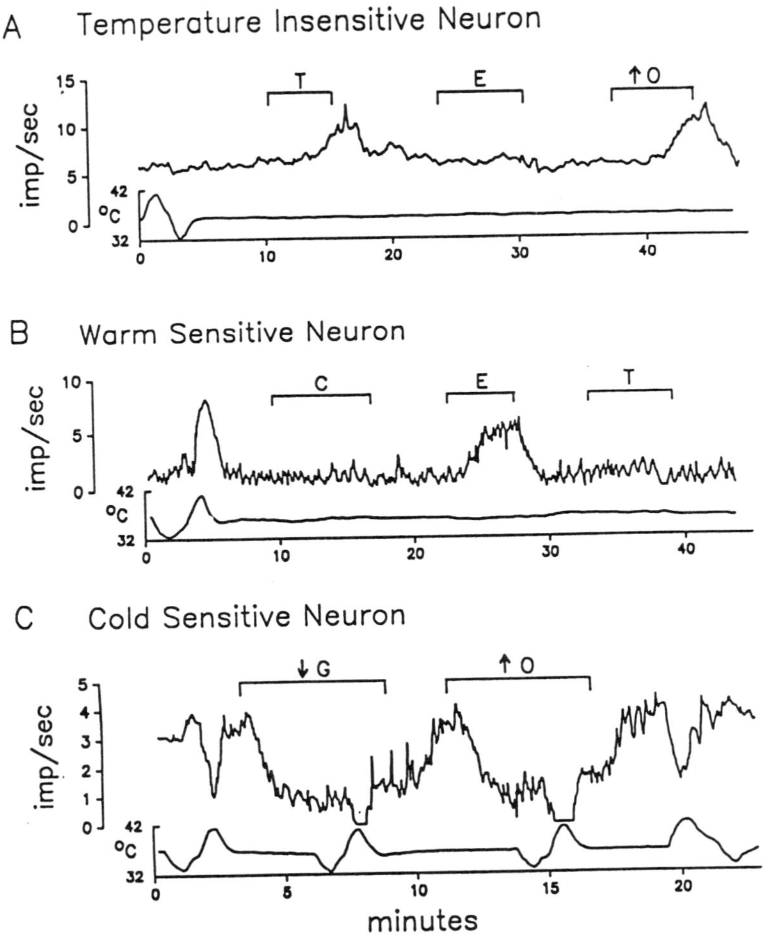

Fig. 1. Firing rates (imp/sec) of: (A) a temperature insensitive neuron, (B) a warm sensitive neuron, and (C) a cold sensitive neuron recorded in rat preoptic tissue slices. Each record shows neuronal firing rates during changes in temperatures (°C) and changes in perfusion media. Normal perfusion medium is 300 mosmol and contains 10 mM glucose. Upper bars indicate times of perfusion with experimental media containing: (T) 30 pg/ml testosterone; (E) 30 pg/ml estradiol; (C) 0.002% ethanol control; (G, low glucose) 1 mM glucose; and (O, hyperosmotic) 309 mosmol. [From Boulant & Silva (4)]

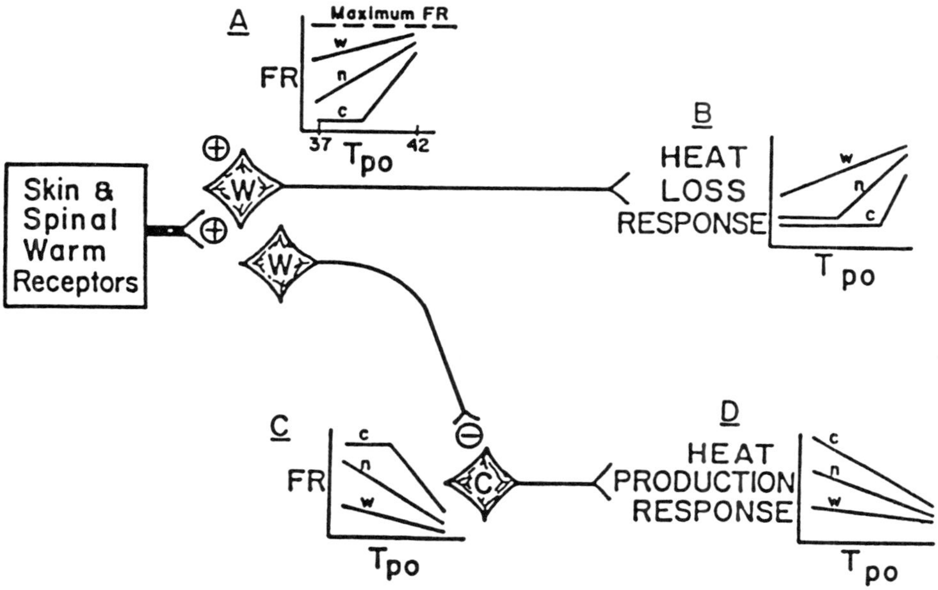

Fig. 2. Neuronal firing rate (FR) responses (A & C) and whole-body thermoregulatory responses (B & D) to changes in preoptic temperature, T_{po}. (A) warm sensitive neuron's FR plotted as a function of T_{po}. Each curve shows the neuron's preoptic thermosensitivity at a different skin or spinal temperature; i.e., w = warm, n = neutral, and c = cold. (B) heat loss response as a function of T_{po}. (C) cold sensitive neuron's firing rate as a function of T_{po}. (D) heat production response as a function of T_{po}. The (+) indicates excitatory synaptic input and the (-) indicates inhibitory synaptic input. [From Boulant & Gonzalez (8)]

If a neuron has a negative thermal coefficient (Fig. 2C) of at least -0.6 impulses/sec/°C, it is classified as cold sensitive. Neurons with lower absolute slopes are classified as temperature insensitive; but among these, distinctions are sometimes made between moderate-slope and low-slope sub-groups of temperature insensitive neurons. For example, neurons with absolute thermal coefficients less than 0.1 or 0.2 impulses/sec/°C are considered to be low-slope temperature insensitive neurons. Compared with higher thermal coefficient neurons, these low-slope neurons are more sensitive to ouabain blockade of the Na-K pump (5) but less sensitive to cAMP-enhancing drugs (6).

Integration of Thermal Information

One of the original reasons for establishing the criteria of +0.8 impulses/sec/°C for warm-sensitivity and -0.6 impulses/sec/°C for cold-sensitivity was that this made distinctions between hypothalamic neurons that received afferent input from skin and spinal thermoreceptors (7). In addition to sensing temperature, preoptic thermosensitive neurons also have the ability to integrate internal thermal information with peripheral thermal information. *In vivo* experiments find very few temperature insensitive neurons that are affected by changes in skin or spinal cord temperature. However, if the above criteria for thermo-sensitivity are used, the majority of preoptic warm and cold sensitive neurons are affected by either skin or spinal temperature (7). Fig. 2 is a simple model illustrating how the integration of central and peripheral thermal information may occur in warm and cold sensitive neurons (8). Some preoptic warm sensitive neurons may control heat loss responses (e.g., panting, sweating). These neurons may also receive afferent input from multisynaptic projections of cutaneous or deep-body thermoreceptors. Preoptic neurons often display the same type of thermosensitivity for both hypothalamic and extra-hypothalamic temperatures (7).

As illustrated in Fig. 2A, if a neuron is excited by increases in hypothalamic temperature, it is usually also excited by increases in an extra-hypothalamic temperature (e.g, skin temperature) (7). This figure shows the neuron's firing rate and preoptic thermosensitivity at 3 different skin temperatures: warm (w), neutral (n) and cold (c). Skin warming increases the neuron's firing rate, but it also slightly decreases the neuron's sensitivity to its own preoptic temperature. Conversely, skin cooling decreases the neuron's firing rate, but it also increases the neuron's thermosensitivity. Since similar responses have been observed in heat loss responses (Fig. 2B) (8), there is reason to believe that some preoptic warm sensitive neurons act as integrators of central and peripheral thermal information and play a role in controlling these thermoregulatory responses.

The firing rate of a preoptic cold sensitive neuron is illustrated in Fig. 2C. Using hypothalamic tissue slices, some electrophysiological studies suggest that these neurons are not inherently thermosensitive, but rather, derive their cold-sensitivity from the synaptic inhibition of nearby warm-sensitive neurons. When tissue slices are exposed to perfusion media that block all synaptic activity, cold-sensitivity is usually lost in PO/AH neurons; however, cold-sensitivity is sometimes retained in other regions (e.g., posterior hypothalamus). Intracellular recordings support the hypothesis that preoptic neuronal cold-sensitivity is synaptically-derived, since the activity of cold sensitive neurons is primarily dependent on inhibitory and excitatory postsynaptic potentials (IPSPs and EPSPs) generated by nearby thermosensitive neurons (9).

Only a small number of PO/AH neurons are cold sensitive, but a very high proportion of these neurons respond to changes in skin and spinal cord temperatures (7). As suggested in Fig. 2C, this may indicate that cold sensitive neurons receive synaptic input from a select group of highly responsive warm sensitive neurons. It may also indicate that cold sensitive neurons receive direct synaptic input from peripheral afferent pathways. During changes in skin or spinal temperature, Fig. 2C shows that the responses of many cold sensitive neurons are the mirror-image of responses of warm sensitive neurons (further supporting the concept that cold sensitive neurons are synaptically inhibited by warm sensitive neurons). Peripheral cooling, for example, not only increases the firing rates of preoptic cold sensitive neurons, but it also increases their preoptic cold-sensitivity (i.e., thermal coefficient). This would be expected if the neuron is inhibited by a warm sensitive neuron which decreased its firing rate and increased its thermosensitivity during peripheral cooling (Fig. 2A). Since similar responses have been observed in whole-body heat production responses (Fig. 2D)(8), there is reason to hypothesize that Fig. 2 illustrates some basic properties of neuronal networks controlling body temperature.

Integration of Thermal and Nonthermal Information

The fact that a preoptic neuron is warm sensitive or cold sensitive does not necessarily mean that the neuron has a role in thermoregulation, nor does it mean that the neuron is not sensitive to endogenous factors associated with other regulatory systems. This is illustrated by the three examples in Fig. 1. The temperature insensitive neuron (Fig. 1A) is unaffected by a perfusion medium containing estradiol, but it is excited by perfusion media containing either testosterone or a slight increase in osmotic pressure. The warm sensitive neuron (Fig. 1B) is excited by estradiol but unaffected by testosterone; and the cold sensitive neuron (Fig. 1C) is inhibited by perfusion media containing either reduced glucose concentrations or elevated osmotic pressure (4).

Of the preoptic neurons tested, low glucose and hyperosmotic media affected 44% of the temperature sensitive neurons and 27% of the temperature insensitive neurons (10). As Fig. 1 illustrates, these two media often have the same effect on the same neuron; and warm sensitive neurons tend to be excited, while cold sensitive neurons are inhibited. Media containing estradiol or testosterone affected 52% of the temperature sensitive neurons and 35% of the temperature insensitive neurons (11). As Fig. 1 suggests, nearly every responsive neuron is affected by only one of the two steroids, and again, warm and cold sensitive neurons show opposite responses. These studies support the possibility that there are broad interactions between hypothalamic neuronal networks controlling different homeostatic systems. These interactions allow nonthermal endogenous factors to easily affect thermoregulation, just as changes in body temperature can easily affect all other regulatory systems.

Cellular Mechanisms of Neuronal Warm-sensitivity

As mentioned above, there is evidence to suggest that PO/AH cold sensitive neurons are not inherently thermosensitive but, instead, are synaptically driven by other thermosensitive neurons. Most PO/AH cold sensitive neurons loose their thermosensitivity during synaptic blockade (12,13), and intracellular recordings indicate that thermal changes in the frequencies of IPSPs and EPSPs are primary determinants of cold-sensitivity (9). On the other hand, PO/AH warm sensitive neurons are inherently thermosensitive since these neurons retain their thermosensitivity during synaptic blockade(12,13). Warm sensitive neurons also display many postsynaptic potentials, but IPSP and EPSP frequencies remain relatively stable during temperature changes (9,14). This observation is very important because it suggests that warm sensitive neurons receive their synaptic input predominantly from nearby PO/AH temperature insensitive neurons. The activity of temperature insensitive neurons, therefore, is probably crucial to the overall activity of the warm sensitive neurons.

Some studies suggest that thermal changes in the resting membrane potential provide the basis for neuronal thermosensitivity (reviewed in 3). These studies imply, for example, that warm sensitive neurons depolarize during warming, thus providing the reason for their increased firing rates with increased temperature. This hypothesis, however, has not been supported by our own intracellular experiments using sharp-tip microelectrodes in rat PO/AH tissue slices (9). These recordings found no consistent changes in resting membrane potential that could account for neuronal thermosensitivity. This phenomenon has been confirmed in recent experiments, using tight-seal whole-cell recordings in warm sensitive preoptic neurons (15). These neurons increase their firing rates during warming, but their resting membrane potentials remain constant.

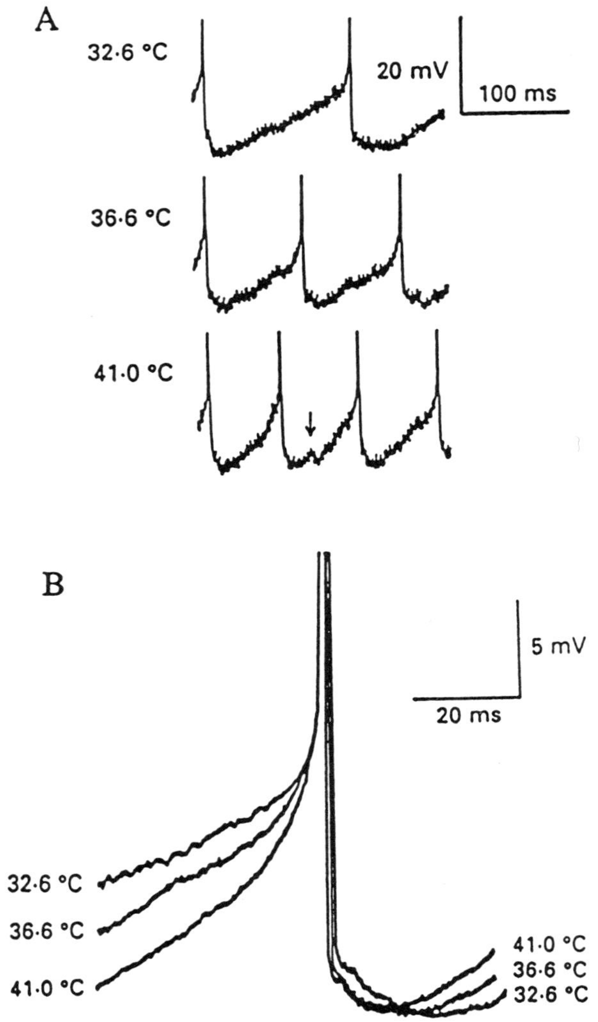

Fig. 3. Intracellular activity of a preoptic warm sensitive neuron recorded at three different temperatures. (A) Records are amplified to show depolarizing prepotentials, and the action potentials are truncated. Resting membrane potential is approximately -60 mV, and this does not change with temperature. The arrow indicates an IPSP. (B) Averaged pre-spike & post-spike activity of the neuron in A, showing that warming increases the rate of rise in the depolarizing prepotential. Each temperature shows the average of 15 action potentials superimposed on threshold. [From Curras, Kelso & Boulant (9)]

Fig. 3 shows an example of a preoptic warm sensitive neuron, recorded intracellularly with a sharp-tip microelectrode (9). As noted in Fig. 3A, warming reduces the time between action potentials (i.e., the interspike interval), which increases the firing rate. Warming does not change the resting membrane potential. What does change, however, is the rate of rise of the depolarizing prepotential preceding each action potential. In warm sensitive neurons, the prepotential's rate of rise is increased during warming, thus producing an increase in firing rate. This is illustrated in Fig. 3B which shows superimposed averaged action potentials at three different temperatures. The ionic conductances that account for the prepotential provide a basis for neuronal warm-sensitivity. Temperature insensitive neurons also display depolarizing prepotentials, but temperature has little or no effect on their rate of rise (9,16). It appears, therefore, that much of the cellular mechanisms both for warm-sensitivity and for temperature-insensitivity reside in ionic conductances that are specific to the different cell types.

Presumably, pyrogens and drugs alter body temperature by changing the firing rates or thermosensitivities of PO/AH neurons. While these actions may occur by changes in resting membrane potential, Fig. 3 shows the importance of the depolarizing prepotential. It is quite possible, therefore, that pyrogens and drugs affect this prepotential which, in turn, affects not only neuronal firing rate but thermosensitivity, as well.

Synaptic Contributions to Neuronal Warm-sensitivity

While it has been suggested that thermal changes in a neuron's input resistance determine differences in neuronal thermosensitivity, our own intracellular studies find that all preoptic neurons show similar changes in resistance during changes in temperature (9,15). Warm sensitive neurons, cold sensitive neurons, and temperature insensitive neurons all decrease their resistances during increases in temperature. Thus, input resistance alone does not provide the key element for distinctions between different types of neuronal thermosensitivity. On the other hand, resistance may be one component that contributes to thermosensitivity, particularly in warm sensitive neurons.

As mentioned above, warm sensitive neurons receive inhibitory and excitatory synaptic input from nearby neurons (9,14). This input comes predominantly from temperature insensitive neurons, since IPSP and EPSP frequencies are generally not affected by temperature changes. Moreover, IPSPs are the most prevalent form of postsynaptic potentials in warm sensitive neurons. If temperature changes the input resistance, it will also change the amplitude of postsynaptic potentials, since (according to Ohm's law: $V = IR$) voltage changes are the product of current changes times the resistance. Even though IPSP frequency remains constant and the ionic current of each IPSP remains constant, if cooling increases the resistance, it will also increase the IPSP amplitude. This increased IPSP amplitude will decrease a neuron's firing rate during cooling; and, in effect, this will increase the neuron's sensitivity to temperature.Thus, thermal effects on IPSPs can enhance a neuron's warm-sensitivity.

Preoptic neurons show thermal changes in IPSP amplitudes that are consistent with the above hypothesis (9). Fig. 4 shows recordings from the same warm sensitive neuron presented in Fig. 3. In comparison with Fig. 3A, Fig. 4A shows that when IPSPs occur, they prolong the interspike interval and decrease the firing rate at each temperature. In addition to the effect of temperature on the depolarizing prepotential, cooling also increases the effectiveness of IPSPs in terms of increasing the interspike interval and decreasing the firing rate. This can be illustrated in this neuron at two different temperatures, by comparing interspike intervals that contain IPSPs with interspike intervals that do not contain IPSPs. For example, at 41.0° the average interspike interval with IPSPs is 98 msec, but the average interspike interval without IPSPs is 74 msec; a difference of 24 msec. At 32.6° the average interspike interval with IPSPs is 189 msec, but the average interspike interval without IPSPs is 142 msec; a difference of 47 msec. Thus, cooling from 41.0° to 32.6° has doubled the effectiveness of IPSPs in terms of lengthening the interspike interval. This cooling has produced a 23 msec increase in the ability of an IPSP to lengthen the interspike interval (i.e. 47 - 24 = 23 msec). Comparable increases in averaged IPSP durations are also shown in Fig. 4B, where cooling from 41.0° to 32.6° produced an increase in IPSP amplitude and a 23 msec increase in IPSP duration.

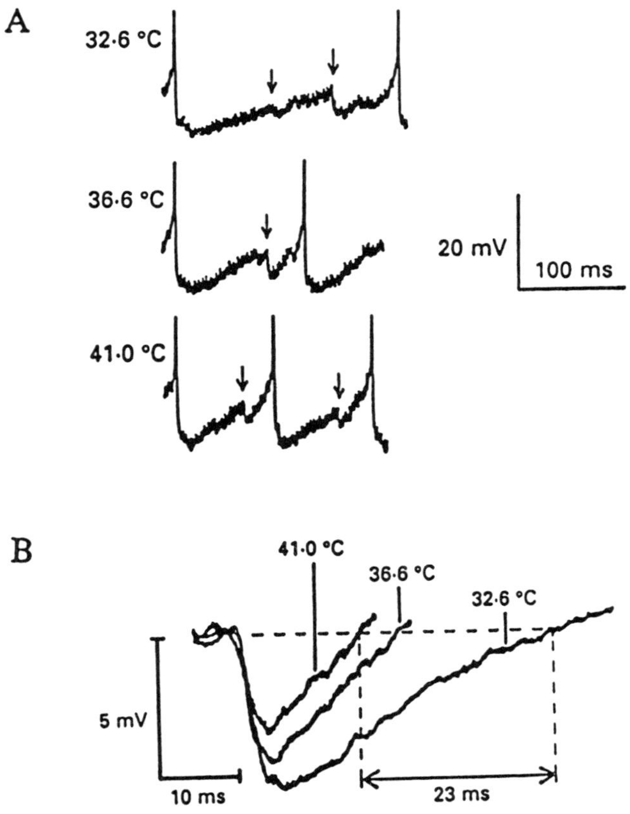

Fig. 4. Intracellular activity of the same preoptic warm sensitive neuron shown in Fig.3. (A) The records are recorded at the same 3 temperatures as in Fig.3, and each record shows at least one IPSP (arrow), which prolongs the interspike interval. (B) Averaged IPSP activity at the 3 temperatures (each n = 16-19), superimposed at the beginning of the hyperpolarization. [From Curras, Kelso & Boulant (9)]

Substantial IPSP activity is observed in both warm sensitive neurons and in temperature insensitive neurons (9,14). Temperature insensitive neurons, however, have very low firing rates compared to warm sensitive neurons (1-3). Therefore, in temperature insensitive neurons, thermally-induced 23 msec changes in IPSP durations will have little or no effect on firing rate responses to temperature. For example, if a temperature insensitive neuron has a spontaneous firing rate of 1 impulse/sec and its interspike interval is 1000 msec, a mere 23 msec increase in this interval will not change the firing rate. On the other hand, if a warm sensitive neuron has a firing rate of 10 impulses/sec, its interspike interval is only 100 msec. A 23 msec increase in this interval will decrease the firing rate by about 2 impulses/sec or 20%. This change in firing rate is not sufficient to solely account for warm-sensitivity, which (as described above) is due primarily to the depolarizing prepotential. Nevertheless, a 20% change in firing rate could significantly contribute to the thermosensitivity of a neuron that is already warm sensitive.

This hypothesis may explain some earlier electrophysiological observations of warm sensitive neurons during perfusions with synaptic blockade media. Most preoptic warm sensitive neurons remain warm sensitive during synaptic blockade; however, usually there are decreases in the thermal coefficients (12,13). This suggests that a neuron can be inherently thermosensitive, but in addition, the thermal effects of local synaptic events can also contribute to the neuron's thermosensitivity.

Implications of Cellular and Synaptic Contributions to Thermosensitivity

The implications of these intracellular studies are important to our view of hypothalamic neuronal networks, and they are equally important to our interpretations of drug and pyrogen actions on hypothalamic neurons. Presently, the most generally-accepted views of hypothalamic neuronal networks originate with a 1965 model proposed by Hammel (17). The basis of this model is that warm sensitive neurons and temperature insensitive neurons send separate, mutually-antagonistic synaptic inputs to thermoregulatory effector neurons. For example, heat loss effector neurons could be excited by warm sensitive neurons and inhibited by temperature insensitive neurons; while heat production effector neurons could be inhibited by warm sensitive neurons and excited by temperature insensitive neurons. Because of this concept, if a drug affects body temperature, we may speculate that it could have opposite effects on warm sensitive and temperature insensitive neurons.

The above intracellular recordings tend to support Hammel's hypothesis regarding cold sensitive neurons, suggesting that (at least in the PO/AH) these neurons are synaptically-driven by nearby thermosensitive neurons (9). On the other hand, intracellular studies present a different view of the synaptic interactions between warm sensitive and temperature insensitive neurons. While warm sensitive preoptic neurons are inherently thermosensitive, they also receive much synaptic input from nearby temperature insensitive neurons. This synaptic input may contribute to both the firing rate and thermosensitivity of warm sensitive neurons. Under these circumstances, therefore, a drug could alter the thermosensitivity of a warm sensitive neuron, not by affecting the neuron directly, but rather, by affecting a temperature insensitive neuron that synaptically inhibits the warm sensitive neuron.

Acknowledgments

The author's work cited here has been supported by National Institutes of Health grant NS-14644 and by the F.A. Hitchcock Professorship for Environmental Physiology. The author wishes to express appreciation to his students and collaborators.

References

1. Boulant, J.A. (1980): Hypothalamic control of thermoregulation: neurophysiological basis. In: Handbook of the Hypothalamus: Behavioral Studies of the Hypothalamus, ed. P.J.Morgane and J. Panksepp. New York: Dekker, vol. 3A, 1-82.

2. Boulant, J.A. and J.B. Dean (1986): Temperature receptors in the central nervous system. Ann. Rev. Physiol. 48, 639-654.

3. Boulant, J.A., M.C. Curras, and J.B. Dean (1989): Neurophysiological aspects of thermoregulation. In: Advances in Comparative and Environmental Physiology, 4, Animal Adaptation to Cold, ed. L.C.H. Wang. Berlin: Springer-Verlag, 117-160.

4. Boulant, J.A. and N.L. Silva (1989): Multisensory hypothalamic neurons may explain interactions among regulatory systems. NIPS 4, 245-248.

5. Curras, M.C. and J.A. Boulant (1989): Effects of ouabain on neuronal thermosensitivity in hypothalamic tissue slices. Am. J. Physiol. 257, R21-R28.

6. Griffin, J.D, M.L. Kaple and J.A. Boulant (1990): Cyclic AMP: a determinant of hypothalamic neuronal thermosensitivity. FASEB Journal 4, A720.

7. Boulant, J.A. and J.D. Hardy (1974): The effect of spinal and skin temperatures on the firing rate and thermosensitivity of preoptic neurons. J. Physiol. 240, 639-660.

8. Boulant, J.A., and R.R. Gonzalez (1977): The effect of skin temperature on the hypothalamic control of heat loss and heat production. Brain Res. 120, 367-372.

9. Curras, M.C., S.R. Kelso and J.A. Boulant (1991): Intracellular analysis of inherent and synaptic activity in hypothalamic thermosensitive neurones in the rat. J. Physiol. 440, 257-271.

10. Silva, N.L. and J.A. Boulant (1984): Effects of osmotic pressure, glucose, and temperature on neurons in preoptic tissue slices. Am. J. Physiol. 247, R335-R345.

11. Silva, N.L. and J.A. Boulant (1986): Effects of testosterone, estradiol, and temperature on neurons in preoptic tissue slices. Am. J. Physiol. 250, R625-R632.

12. Kelso, S.R. and J.A. Boulant (1982): Effect of synaptic blockade on thermosensitive neurons in hypothalamic tissue slices. Am. J. Physiol. 243, R480-R490.

13. Dean, J.B. and J.A. Boulant (1989): Effects of synaptic blockade on thermosensitive neurons in rat diencephalon in vitro. Am. J. Physiol. 257, R65-R73.

14. Griffin, J.D. and J.A. Boulant (1992): Synaptic interactions of hypothalamic temperature sensitive and insensitive neurons. Soc. Neurosci. Absts. 18, 877.

15. Griffin, J.D., P. Wung Burgoon and J.A. Boulant. (1993): Neuronal thermosensitivity does not depend on thermal changes in resting membrane potential. FASEB Journal 7, A17.

16. Griffin, J.D. and J.A. Boulant (1992): Hypothalamic warm sensitive neurons: role of the temperature dependent pacemaker potential. FASEB Journal 6, A1198.

17. Hammel, H.T. (1965): Neurons and temperature regulation. In: Physiological Controls and Regulations, ed. W.S. Yamamoto and J.R. Brobeck. Philadelphia: Saunders, 71-97.

POSSIBLE IMPLICATIONS OF THE PLASTICITY OF TEMPERATURE-SENSITIVE NEURONS IN THE HYPOTHALAMUS

Fr.-K. Pierau, J. Schenda, M. Konrad and H. Sann

Max-Planck-Institut für physiologische und klinische Forschung, W.G. Kerckhoff-Institut, Parkstr. 1, D-61231 Bad Nauheim

Temperature-sensitive neurons in the preoptic area/anterior hypothalamus (PO/AH) play the key role in most concepts of nervous control of body temperature. In these models neuronal temperature sensitivity is considered to be a fixed property. However, there is increasing evidence that the temperature sensitivity of hypothalamic neurons can be modulated particularly by endogenous substances which affect body temperature upon intra-hypothalamic application and/or are involved in fever reactions or antifebrile regulations (1, 2, 3). In this respect, pyrogens were the first substances reported to change the temperature coefficient (TC) of hypothalamic neurons in rabbits (4). In general, changes of temperature sensitivity would imply that the gain of the system may increase or decrease. However, the observed changes of the TC also transform temperature-insensitive neurons (IS) into warm-sensitive neurons (WS) and vice versa. It is tempting to suggest that this results not only in quantitative changes but leads to a reorganization of the neuronal network responsible for the control of body temperature.

Methods

Brain slices (400 µm) from adult Wistar rats encompassing the PO/AH were perfused with oxygenated artificial cerobrospinal fluid (aCSF) in a recording chamber (volume: 0.5 ml; flow rate: 2.5 ml/min). A computer controlled Peltier assembly allowed different constant temperatures and exact sinoidal temperature changes (maximal amplitude: 6 °C; velocity: 0.02-0.05 °C/s). For short-term temperature adaptation slice temperature was kept constant at 36, 38 or 39 °C for at least 10 min before temperature cycles were applied. Extracellular recordings from single units were obtained with glass coated platinum-iridium wire electrodes and conventional recording equipment. Bombesin , substance P (SP), prostaglandin E_2 (PGE_2) and thyreotropin releasing hormone (TRH) were either applied by bolus injection into the inlet of the perfusion chamber or perfused with the aCFS.

For adaptation to warm and cold ambient temperatures rats were kept in climatic chambers for 3 to 6 weeks at temperatures of 5 C and 30 °C, respectively.

The temperature response curve of each neuron was evaluated by a computer program relating firing rate to slice temperature and fitted to piecewise or rectilinear regression lines (5). The slope of the appropriate regression line indicated the temperature coefficient (TC: imp/s/°C) of the neuron. Warm- and cold-sensitive neurons were defined by a TC of an absolute value ≥ 0.6 with a positive sign for warm sensitivity and a negative sign for cold sensitivity. All other recorded neurons were by this definition temperature-insensitive.

Results and Discussion

Effect of hypo- and hyperthermic endogenous substances on hypothalamic neurons

Bombesin and SP were the first peptides which were compared in respect to their effect on tonic activity and temperature sensitivity on PO/AH neurons of rats *in vitro* (1). Both substances increased the tonic activity of warm-sensitive neurons (WS) and temperature-insensitive neurons (IS) but differed considerably in respect to their effect on the TC. Bombesin increased the TC of most IS leading to a transformation into WS of about 50% of the neurons (Fig. 1); the average TC of IS was significantly increased (mean±S.E.M.: 0.16±0.03 --> 0.54±0.06 imp/s/°C, n=48; $p<0.001$ paired t-test). The effect on temperature sensitivity of WS was quite ambiguous since the TC increased in about 45% but decreased in 35% so that the average TC of the whole population was not significantly changed (1.02±0.09 --> 1.17±0.19 imp/s/°C, n=25). The net effect of bombesin was a recruitment of WS which would increase the proportion of WS in the PO/AH. In contrast, the TC of IS was not affected by SP, but the temperature sensitivity of WS was preferentially decreased (Fig. 1). However, the proportion of IS and WS was not changed by SP. The different effects of bombesin and SP on hypothalamic neurons may be used to interprete the different actions the two substances excert on body temperature upon intrahypothalamic application. Only bombesin reduces body temperature while SP has little or no effect on temperature regulation. Since only bombesin but not SP induced transformation of IS into WS the important mechanism for the hypothermic action of a substance might be the recruitment of WS.

Following this notion one would expect that no recruitment of WS should appear after application of substances which are known to increase body temperature. Consequently, PGE_2 which is suggested to be one of the neuromediators of the febrile response (6) should not produce accumulation of WS. Application of PGE_2 does indeed not induce significant recruitment of WS in the PO/AH (2). Figure 1 shows that the TC of the majority of IS is not changed by PGE_2 but that the TC of WS is predominantly decreased. Recruitment of WS does not occur since the small number of transformed IS is compensated by the WS transformed into IS. In WS the TC is preferentially decreased by PGE_2 leading to a transformation of a number of WS into IS, which is also reflected by the significant decrease of the mean TC (1.08±0.12 --> 0.68±0.12 imp/s/°C; $p<0.05$) (2). This could mean that both the functional loss of WS and the decrease of the gain of the neuronal system may be important central mechanisms to increase body temperature.

Suggesting that the recruitment of WS in the PO/AH is characteristic for an endogenous substance causing hypothermia one would predict from the data in Fig. 1 that TRH belongs to this group, since its predominant effect is the transformation of IS into WS. This is also indicated by the increase of the average TC of IS (0.16±0.09 --> 0.88±0.25 imp/s/°C; $p<0.05$). Surprisingly, TRH appears to do just the opposite in producing hyperthermia upon injection into the hypothalamus (7). However, the TRH effect in rats depends on body and ambient temperatures.

At an ambient temperature of 22 °C TRH induces a short hyperthermia (5 min) which is followed by a long lasting decrease in body temperature (8, 9). We suppose that the increase of the TC of IS corresponds to the hypothermic phase observed *in vivo*, since in our experiments the TC was always estimated 10 min after the start of superfusion. It appears therefore that the action of TRH on PO/AH neurons is in agreement with the hypothesis that recruitment of WS is characteristic for a hypothermic action of a substance.

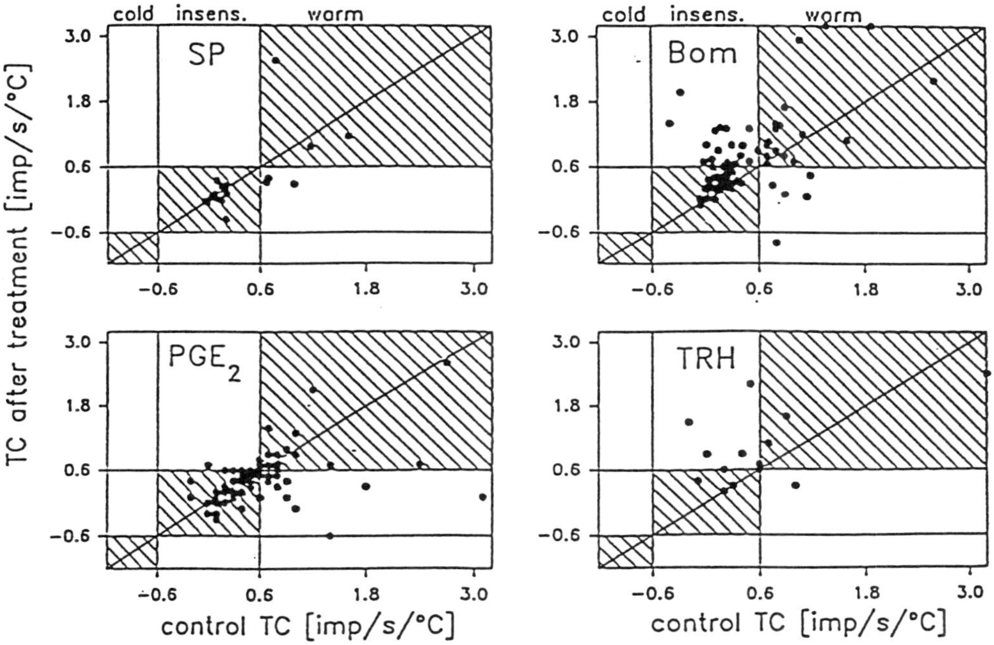

Fig.1: Effects of different endogenous substances on TC of PO/AH neurons. The TC after application of a substance is plotted as function of the TC during the control period. The distance of a dot from the line of identity indicates the degree of change. Dots in clear areas represent neurons in which the TC-changes were big enough to transform this neuron into another category. Vertical lines define the different categories of neurons according to their TC: cold: cold-sensitive; insens: temperature-insensitive; warm: warm-sensitive. SP and Bom: bolus injection, 1µg in 0.1 ml; PGE_2 and TRH: superfusion, 3.5 ng/ml and 10 ng/ml, respectively. Determination of TC 10-15 min after bolus injection or beginning of superfusion.

The effect of various hypo- and hyperthermic substances on the temperature sensitivity of PO/AH neurons suggests that the temperature characteristic of these neurons is not a fixed parameter but can change under different conditions. Provided that the investigated neurons are indeed involved in the control of body temperature one could expect that the response characteristic of the neuronal network (of unknown structure) to which the neurons belong can change under the different demands of temperature regulation, particularly during extensive heat load and elevated temperatures during fever reaction. Therefore we have tested whether changes of the local temperature itself can affect the temperature characteristic of PO/AH neurons. In these experiments the slice temperature was equilibrated at 36, 38 or 39 °C for 10 min before the temperature cycle was applied. In 9 of the 24 investigated WS in which all 3 temperatures could be investigated the TC was stable at the different constant temperatures. However, the TC of 9 neurons increased with increasing temperatures i.e. it was high at 39 °C and low at 36 °C. In the remaining 6 neurons the TC had its highest level at 38 °C. For IS stable TC were observed in 9 neurons (50%), while in 6 neurons the TC was increased with increasing temperatures. In the remaining 3 neurons the TC reached its highest value at 36 °C and its lowest value at 39 °C.

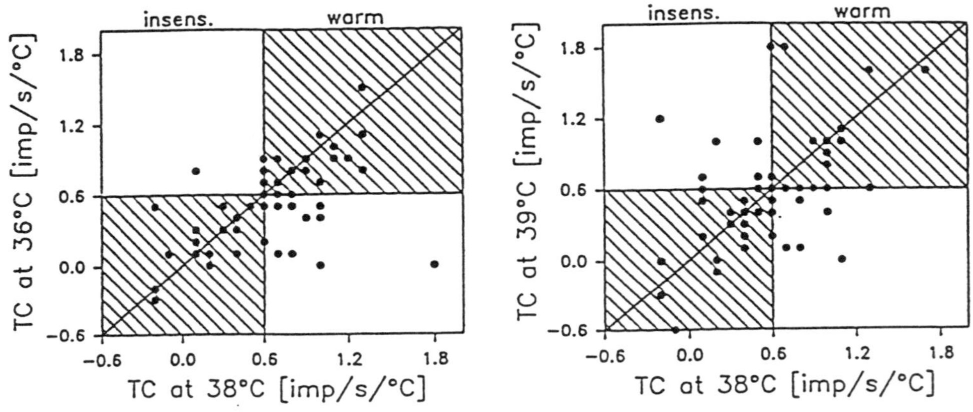

Fig.2: Effect of different constant temperatures (short term adaptation) on TC of rat PO/AH neurons. The TC at 38°C is plotted against the TC at 36°C (left) and 39°C (right), respectively. For details see legend Fig. 1.

The deviations of the TC at various constant temperatures caused transformation of some of the neurons into a different category (Fig. 2). Lowering the temperature from 38 to 36 °C quite consistently decreased the TC of WS leading to a conversion of 11 of the neurons into IS. Only in a small number of IS the TC was significantly changed . The short term adaptation to 39 °C resulted in the transformation of some WS into IS. However, the TC variations were considerably higher in the IS and the transformation of WS to IS was almost balanced by the conversion of IS to WS (Fig. 2). The results suggest that the temperature sensitivity is independent from local temperature changes in about 50% of the PO/AH neurons. But, deviations of the TC at different temperatures occur in the other half of the neurons indicating the plasticity of the hypothalamic neuronal network. In regard to IS, one might speculate, that those neurons which exhibit a stable TC may serve as reference neurons.

Acclimatization and adaptation to different ambient temperatures cause changes in the behavioral and physical responses of homeotherms to temperature load, probably mainly due to central mechanisms of the temperature control system (10). Given the high plasticity of the temperature characteristic of hypothalamic neurons one would expect that adaptation to different ambient temperatures might result in a change of the temperature sensitivity and consequently in the distribution of temperature-sensitive and -insensitive neurons. Table 1 demonstrates indeed that the distribution of rat PO/AH neurons is different after three weeks adaptation to 5 °C or to 30 °C, respectively. In control rats the portion of WS was about 40%, which is in good agreement with the data from the literature (10,11). After adaptation to cold ambient temperatures the number of WS was increased by about 12%. The number of neurons demonstrating a high TC (>1.2) was also increased under these conditions. In contrast, the number of WS was remarkably reduced to 29% in rats adapted to an ambient temperature of 30 °C and the number of neurons demonstrating a TC >1.2 was also decreased. No cold-sensitive neuron was observed in warm adapted rats but in general the number of cold-sensitive neurons was very small under our experimental conditions (1) and no conclusions can be drawn about adaptive changes of cold-sensitive neurons.

Table 1: Effect of adaptation to cold (5°C) and warm (30°C) ambient temperatures on the TC of rat PO/AH neurons. Control data are collected by 4 different investigators from our lab.

	Number of neurons			
	temperature-insensitive	warm-sensitive	cold-sensitive	total
cold-adapted (5 °C)	49 (47%)	55 (52%)	1 (1%)	105
control (22 °C)	233 (59%)	157 (39.5%)	6 (1.5%)	396
warm-adapted (30 °C)	79 (71%)	32 (29%)		111

In conclusion, the neuronal network - of hitherto unknown structure - responsible for the control of body temperature represents a flexible system. Endogenous substances which affect body temperature not only influence the tonic activity of hypothalamic neurons but also change the temperature characteristic of temperature-sensitive and -insensitive neurons. Transformation of IS into WS may be regarded as a common site of action of *hypothermic* substances, leading to a recruitment of WS. On the other hand conversion of WS into IS and the resulting decrease of the portion of WS might be an important central mechanism for the action of *hyperthermic* substances. The output characteristics of the neuronal system appear to depend on local temperatures and may vary under different physiological and pathophysiological conditions. Circadian changes of the temperature sensitivity of hypothalamic neurons resulting in a nocturnal increase of the portion of WS have been recently reported (12). Temperature-insensitive neurons appear not to be a uniform population but may include neurons which become temperature-sensitive under certain conditions and others which remain temperature-insensitive under all circumstances. Both types might serve as reference neurons depending on the actual demands. The distribution of temperature-sensitive and -insensitive neurons appears to change with adaptation to different ambient temperatures indicating that the plasticity of the hypothalamic network may enable adjustment to long term changes of temperature regulation.

References

1. Schmid, HA, Jansky, L, Pierau, Fr-K (1993): Temperature sensitivity of neurons in slices of the rat PO/AH area: effect of bombesin and substance P. Am J Physiol 264, R449-R455.

2. Kulchitsky, V, Sann, H, Pierau, Fr-K (1994): Effects of prostaglandin E2 on the thermosensitivity of anterior and posterior hypothalamic neurones in vitro. In: Integrative and Cellular Aspects of Autonomic Functions. Pleschka, K, Gerstberger, R, eds. London, Paris: John Libbey Eurotext, 37-46.

3. Moravec, J, Pierau, Fr-K (1994): Arginine vasopressin modifies the firing rate and thermosensitivity of neurons in slices of the rat PO/AH area. In: Integrative and Cellular Aspects of Autonomic Functions. Pleschka, K, Gerstberger, R, eds. London, Paris: John Libbey Eurotext Ltd (in press).

4. Cabanac, M, Stolwijk, AJ, Hardy, JD (1968): Effect of temperature and pyrogens on single unit activity in the rabbits brain stem. J Appl Physiol 24, 645-652.

5. Vieth, E (1989): Fitting piecewise linear regression functions to biological responses. J Appl Physiol 67, 390-396.

6. Stitt, JT (1986): Prostaglandin E as the neural mediator of the febrile response. Yale J Biol Med 59, 137-149.

7. Cohn, ML, Cohn, M, Traube, D (1980): Thyrotropin releasing hormone induced hyperthermia in the rat inhibited by lysine acetylsalicylate and indometacin. In: Thermoregulatory Mechanisms and their Therapeutic Implications. Cox, B, Lomax, P, Milton, AS, Schönbaum, E eds. Basel: Karger, 198-201.

8. Lin, MT, Chandra, A, Chern, YF, Tsay, BL (1980): Effects of thyrotropin-releasing hormone (TRH) on thermoregulation in the rat. Experimentia 36, 1077-1078.

9. Hori, T, Yamasaki, M, Asami, T, Koga, H, Kiyohara, T (1988): Responses of anterior hypothalamic-preoptic thermosensitive neurons to thyrotropin releasing hormone and cyclo (his-pro). Neuropharmacology 27, 895-901.

10. Hensel, H, Brück, K, Raths, P (1973): Homeothermic organisms. In: Temperature and Life. Precht, H, Christophersen, J, Hensel, H eds. Berlin, Heidelberg, New York: Springer-Verlag, 503-761.

11. Boulant, JA, Dean, JB (1986): Temperature receptors in the central nervous system. Annu Rev Physiol 48, 639-654.

12. Derambure, PS, Dean, JB, Boulant, JA (1994): Circadian changes in neuronal thermosensitivity in the rat suprachismatic nucleus. In: Integrative and Cellular Aspects of Autonomic Functions. Pleschka, K, Gerstberger, R, eds. London, Paris: John Libbey Eurotext, 269-274.

Thermal Balance in Health and Disease
Advances in Pharmacological Sciences

TEMPERATURE-SENSITIVE COUPLING IN HYPOTHALAMIC AND SPINAL NEURONES OF THE RAT IN VITRO

H. Sann, M. Konrad*, U. Pehl, H.A. Schmid, Fr.-K. Pierau and E. Simon

MPI für physiologische und klinische Forschung, Parkstraße 1, D-61231 Bad Nauheim, F.R.G.
*Present adress: Department of Physiology and Developmental Biology, Charles University, Prague, Czech Republic

There are several regions in the central nervous system which are known to be involved in temperature regulation, including the anterior hypothalamus and the spinal cord, where intrinsic temperature-sensitive neurones have been observed (for review: 1,2). The temperature sensitivity of these neurones is defined by a change in discharge rate to a given change of local temperature (1). We here report a temperature-sensitive neuronal mechanism, which appears to be independent from changes in firing rate. In the course of experiments conducted to characterize the temperature sensitivity of hypothalamic and spinal neurones in rats, occasionally recordings were obtained in which a regular spike was always preceded by one or more small spikes or by a hump in the rising phase (termed here pre-spike). The delay between these coupled events exhibited a marked temperature dependence.

Methods

Male Wistar rats were decapitated and the brain or the lumbar spinal cord were quickly removed. Four 500 µm thick coronal slices were cut from the brain including the preoptic area/anterior hypothalamus (PO/AH) or from the lumbal spinal cord (SC) as described previously (3). After equilibration for at least 2 h, a slice was transferred to a recording chamber and continuously superfused with oxygenated (95% O_2 and 5% CO_2) artificial cerebrospinal fluid containing (mM) NaCl 124.0, KCl 5.0, NaH_2PO_4 1.2, $MgSO_4$ 1.3, $NaHCO_3$ 26.0, $CaCl_2$,1.2 and glucose 10.0. The temperature of the recording chamber was controlled using a Peltier thermoassembly.

Single units were recorded extracellularly with glass-coated platinum-iridium electrodes using conventional techniques. The two spikes were discriminated using a time-window discriminator and a conventional discriminator. To demonstrate the variations in coupling time, the delay time between the positive peak of the pre-spike (or the occurrence of the hump) and that of the large spike were plotted against time as a dot-raster display (Fig. 1). Cross-correlations between the two events were analysed with respect of both events. To evaluate neuronal temperature sensitivity the slice temperature was altered sinusoidally from 38 °C (PO/AH) or 37 °C (SC) by ± 3 °C or ± 4 °C, respectively. In addition, ramp-like temperature changes were performed in SC slices (see Fig. 1 B). The temperature sensitivity of the firing rate was calculated by a special computer program relating the discharge rate of the unit to the average temperature and fitting either one linear or two piecewise regression lines to the data. The slope of the steepest regression line was used as the temperature coefficient (TC) of the unit. Neurones with a TC ≥ 0.6 imp/s/°C were considered as warm-sensitive, those with a TC between -0.6 and 0.6 imp/s/°C were considered as temperature-insensitive (3).

In some of the PO/AH neurones the hypothermic neuropeptide bombesin was injected into the superfusion medium as a bolus (1 µg, 0.1 ml of 10 µg/ml). After bombesin application the temperature sensitivity was again determined, to detect changes in temperature sensitivity induced by the peptide.

Values are expressed as mean ± S.D. and range. For statistical comparisons an independent or a paired t-test were used.

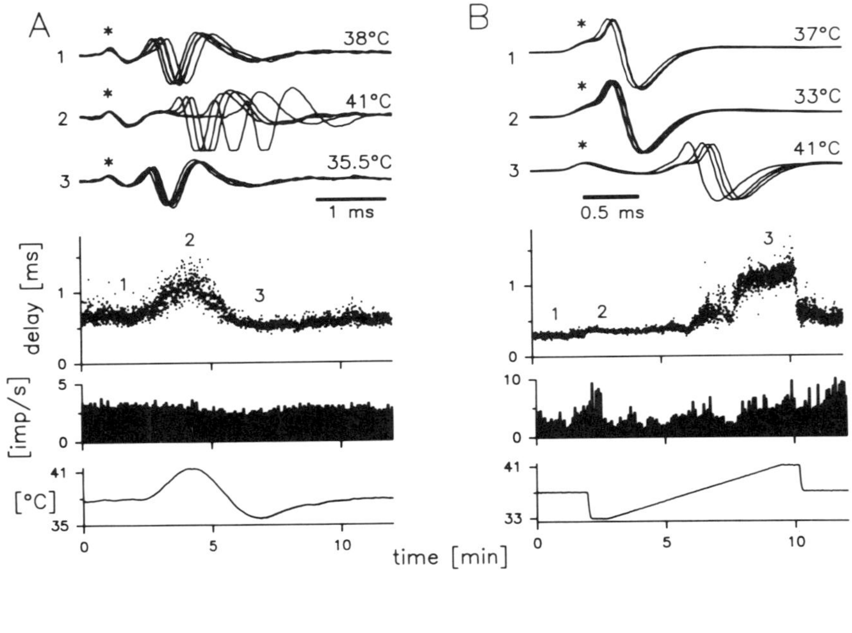

Figure 1: Discharge rate [imp/s] and coupling delay [ms] of one PO/AH (A) and one spinal neurone (B) during temperature stimulation [°C]. The original spike recordings (1 - 3; 5 spikes superimposed in each trace, * labels the pre-spike) in the upper part of A and B were taken at the time indicated in the coupling delay plot. The time between pre-spike and larger spike is plotted in the delay plot.

Results

Temperature-sensitive coupling between a small pre-spike and a regular spike was observed in 7 of 218 PO/AH units (3.3 %) and 4 of 106 SC units (3.8 %). All of these units exhibited an initial hump in the rising phase (Fig. 1 B) or a small pre-spike preceding a regular spike (Fig. 1 A) at basal slice temperature. The effect of temperature changes on the discharge rate and the coupling time for one PO/AH (A) and one SC unit (B) is shown in Fig. 1. As can be seen from the spike recordings, the delay time between the pre-spike and the large spike (Fig 1. A 1) or the hump and the large spike (Fig. 1 B 1) already varied at control temperature. Variations of the interval (jitter) occurred in a random manner and were usually not related to the discharge rate of the neurone. The average delay between the two events in all 11 units was 0.6 ± 0.5 ms (mean ± S.D., range: 0.3 - 1.7 ms , 37-38°C) and showed a jitter of 0.3 ± 0.3 ms (0.03 - 1.2 ms). In all neurones, warming increased the coupling time and its variability (Fig. 1). At 41°C the average delay was significantly increased to 1.4 ± 1.1 ms (0.4 - 3.8 ms, $p < 0.05$) and the jitter was 0.7 ± 0.5 ms (0.13 - 1.9 ms, $p < 0.05$). The temperature-dependent increase in coupling time was usually not correlated with an increase in discharge rate (Fig. 1). Cooling had less pronounced effects on the coupling time; it was unchanged (5 units), slightly increased (3 units) or reduced (2 units) (One SC neurone was not active at 33-34°C). In average, the delay (0.7 ± 0.6 ms, 0.3 - 1.8 ms, n = 10) and the jitter (0.4 ± 0.6 ms, 0.1 - 2 ms) were not significantly different from that found at basal temperature.

For all units the Q_{10} of the temperature/coupling time relationship was greater than 3. In 10 of 11 units the Q_{10} of the coupling delay was considerably larger in the upper temperature range. Thus, between 37/38°C and 41°C the average Q_{10} was 6.6 ± 4.5 (1.4 - 14.6, n = 11), while it was only 2.2 ± 3.9 (0.1 - 13) between 33/35°C and 37/38°C. Temperature-sensitive coupling was observed in both temperature-insensitive (6 PO/AH, 1 SC) and warm-sensitive neurones (1 PO/AH, 3 SC). The mean TC (0.27 ± 0.24 imp/s/°C) of the coupled PO/AH neurones was not significantly different from the rest of the sample (TC: 0.56 ± 0.55 imp/s/°C, n = 211). Similarly, the TC of the SC units did not differ from the rest of the spinal population investigated. In the PO/AH, five coupled units were located in the medial preoptic area, one in the medial preoptic nucleus and one in the periventricular nucleus. Two SC units were found in lamina II, one in lamina IV and one in lamina V.

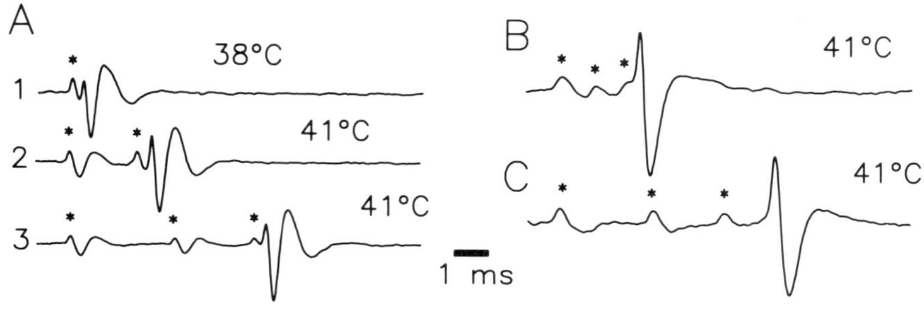

Figure 2: Original spike recordings from one PO/AH (A) unit and two spinal units (B, C) at the indicated temperatures. * indicates the occurrence of the pre-spike. In A, examples for a short (A 2) and a long (A 3) delay time at 41°C are shown.

In 4 recordings (2 PO/AH, 2 SC) two or sometimes even three pre-spikes were observed at 41°C (Fig. 2). The amplitude of the second and third pre-spike was slightly reduced and the intervals between the pre-spikes showed considerable variations. The shortest interval was 1.9 ms in the PO/AH (Fig. 2 A 2) and 1 ms in the SC (Fig. 2 B). A single pre-spike or the last of the repeated pre-spikes was invariably followed by a regular spike, i.e. the coupling factor was 100%. In none of these cases, a small spike was observed following the larger spike.

In one recording, superfusion with synaptic blockade solution (0.2 mM Ca^{2+}, 9 mM Mg^{2+}) reduced and finally completely inhibited the activity of both events (pre-spike and larger spike) with no obvious change in the coupling time.

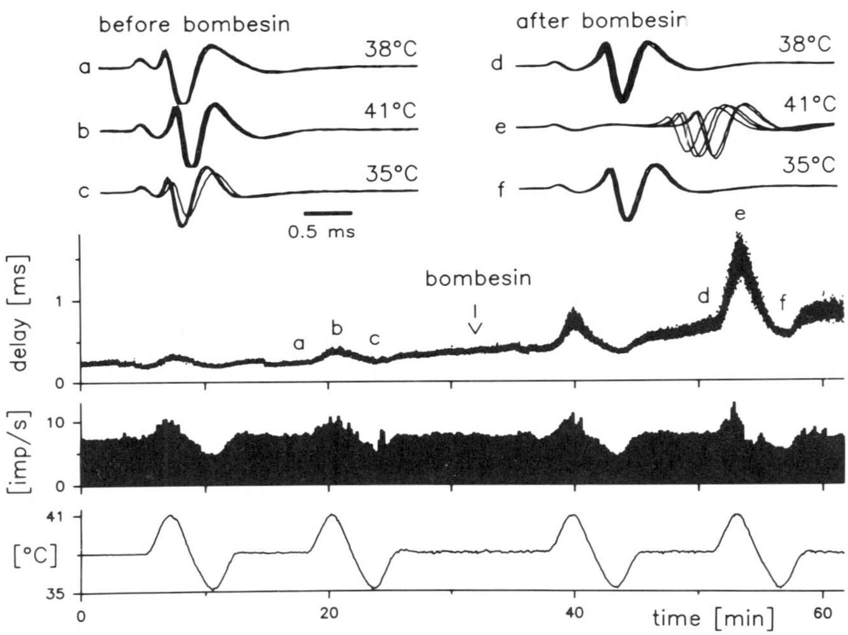

Figure 3: Influence of bombesin (1µg bolus injection) on the coupling delay [ms] and discharge rate [imp/s] in a warm-sensitive PO/AH unit. Original spike recordings (a - f; 5 spikes superimposed in each trace) were taken at times indicated in the coupling delay plot.

Effect of bombesin on the temperature-sensitive coupling

In 5 PO/AH units, which had exhibited temperature-sensitive coupling under control conditions the effects of bombesin were analysed. During bombesin-induced excitation the coupling time was slightly increased in 2 units but was unchanged in the remaining units, demonstrating again that changes in coupling time are not related to changes in discharge rate. Although 4 of 5 units significantly increased their TC after bombesin application (from 0.2 ± 0.24 to 0.42 ± 0.29 imp/s/°C, n = 5, $p < 0.05$) the temperature-sensitive coupling was only slightly increased in one unit.

In 9 recordings from the PO/AH and 4 from the SC demonstrating a pre-spike or an initial hump in the rising phase no or only a very weak temperature dependence of the coupling was noted. In four of these PO/AH units the coupling time became temperature-sensitive after the application of bombesin. The responses of one of these neurones is depicted in Fig. 3. While there was hardly any effect of temperature on the coupling time during the two control temperature stimuli, the temperature dependence of the delay and the variability of the coupling were markedly enhanced after a bolus injection of bombesin. Again, there was no obvious dependence of this effect on the discharge rate or the TC of the unit. In units which exhibited a bombesin-induced temperature sensitivity of the coupling time, the firing rate was increased (3 units) or not altered (one unit) by bombesin. The mean TC of these units was only slightly increased after bombesin application (from 0.63 ± 0.4 to 0.78 ± 0.57 imp/s/°C, n = 4).

Discussion

Although the shape of an extracellular spike can vary according to the position or the properties of the electrode and on the types of electronic filters used, it is very unlikely that the observed pre-spike/spike relationship or its temperature dependence was due to technical artefacts. Considering that both events were picked up with one electrode, a close anatomical relation of the two different sources of the events can be proposed. The observation that the pre-spike always exhibited a lower amplitude than the following spike, might indicate that it originated from a relatively small structure, e.g. a small cell body, an axon or a dendrite. There are two possible explanations for our observations: (I.) the pre-spike and the large spike might arise from two different neurones or (II.) from different compartments of one neurone.

I.) Pre-spike and spike arise from two different neurones

Although the variability of the coupling time (jitter) could suggest chemical synaptic mechanisms, coupling times of less than 1 ms are unlikely for chemical transmission (4). Another possibility could be that two separate neurones are excited by a common input. This appears to be unlikely, since due to the strong jitter one would suggest that the smaller spike should at least sometimes follow the larger spike. In all experiments, however, the small spike preceded the large spike, although small spikes could discharge in bursts of up to 3 spikes with inter-spike-intervals of 1-5 ms (Fig. 2). The latter observation could indicate that a resetting mechanism for the pre-spike by the coupled large spike has to be considered. Since chemical synapses are usually unidirectional, a feedback inhibition can only be obtained via an interneurone.

Such an interneurone would not be necessary for a bidirectional electrical synapse. The coupling time of an electrical synapse is usually short and similar to that found in our study; it might, however, not exhibit a strong jitter. Although electrical coupling via gap junctions is important during development of the mammalian central nervous system, in adult animals electrical synapses with such a high safety factor are not described to our knowledge. Nevertheless, we cannot exclude such a possibility.

II.) Pre-spike and spike arise from different compartments of the same neurone

Different spike initiation points in the initial segment, the soma and also the dendrites of one cell have been described (4, 5). Since the time between an initial segment spike and a soma spike is much shorter than the delay observed in this study, the two events more likely represent a dendritic and a somatic spike. The configuration and size of the coupled events in our experiments are in agreement with the idea, that the recording electrode, positioned close to a dendrite, had picked up a small dendritic spike which triggered a larger soma spike. Although in many neurones dendritic spikes might represent slow Ca^{2+} spikes (5), there are reports indicating that fast, Na^{+}-dependent dendritic spikes can burst with inter-spike intervals as low as 1 ms before triggering a soma spike (5, 6), similar to the presented results. The variation of the coupling time could be due to spatial and temporal summation of IPSPs, EPSPs and dendritic spikes. This could also explain that sometimes up to 3 pre-spikes have to be integrated to evoke a soma spike. Functionally, such dendritic spikes might be important in neurones with large dendritic arborizations to amplify distal synaptic input and thus increase the probability of these synapses in generating a soma spike. Our data suggest that the incidence of larger spikes evoked by pre-spikes is 100%, which is much higher than in fast pyramidal tract neurones (40%) (6).

The temperature sensitivity of the coupling could be at least partially explained by a decrease in membrane input resistance due to local warming as it has been reported in spinal (7,8) and hypothalamic neurones (9). In spinal motoneurones this decrease in input resistance decreased the excitability of a given neurone and consequentely increase the latency of monosynaptically evoked spikes (7,8). The mean Q_{10}, however, of these mechanisms was found to be only 2.1 (8), in comparison to a Q_{10} of 6.6 for the coupling time in this study. This difference is not unexpected since a change in membrane resistance might also affect other parameters of integration in the cell soma such as postsynaptic potentials. In addition, an increase of inhibitory inputs from warm-sensitive neurones during warming might hyperpolarize the soma and thus increase coupling time. The failure of a dendritic spike to evoke a soma spike during hyperpolarization has been demonstrated in other preparations (6, 10).

In conclusion, the observed coupled spikes might either originate from two neurones which are coupled electrically or from a dendrite and the soma of the same neurone. The possible relevance of the observed temperature-sensitive coupling for the neuronal mechanism of temperature regulation remains to be established.

References

1. Boulant, JA, Curras, MC, Dean, JB (1989): Neurophysiological aspects of thermoregulation. In: Advances in Comparative and Environmental Physiology, Vol 4, Wang, LCH ed. Berlin, Heidelberg: Springer, 117-159.

2. Simon, E, Pierau, Fr-K, Taylor, DCM (1986): Central and peripheral thermal control of effectors in homeothermic temperature regulation. Physiol Rev 66, 235-300.

3. Schmid, HA, Pierau, Fr-K (1993): Temperature sensitivity of neurons in slices of the rat PO/AH hypothalamic area: effect of calcium. Am J Physiol 264, R440-R448.

4. Hubbard, JI, Llinas, R, Quastel, DMJ (1969): Electrophysiological analysis of synaptic transmission. London: Edward Arnold (Publishers) LTD, 1-372.

5. Huguenard, JR, Hamill, OP, Prince, DA (1989): Sodium channels in dendrites of rat cortical pyramidal neurons. Proc Natl Acad Sci USA 86, 2473-2477.

6. Deschenes, M (1981): Dendritic spikes induced in fast pyramidal tract neurons by thalamic stimulation. Exp Brain Res 43, 304-308.

7. Klee, MR, Pierau, Fr-K, Faber, DS (1974): Temperature effects on resting potential and spike parameters of cat motoneurons. Exp Brain Res 19, 478-492.

8. Pierau, Fr-K, Klee, MR, Klussmann, FW (1976): Effect of temperature on postsynaptic potentials of cat spinal motoneurones. Brain Res 114, 21-34.

9. Curras, MC, Kelso, SR, Boulant, JA (1991): Intracellular analysis of inherent and synaptic activity in hypothalamic thermosensitive neurones in the rat. J Physiol 440, 257-271.

10. Schwartzkroin, PA (1977): Further characteristics of hippocampal CA1 cells in vitro. Brain Res 128, 53-68.

Thermal Balance in Health and Disease
Advances in Pharmacological Sciences

LAMINA-SPECIFIC EFFECTS OF NITRIC OXIDE ON TEMPERATURE SENSITIVE NEURONS IN RAT SPINAL CORD SLICES

U. Pehl, H.A. Schmid and E. Simon

Max-Planck-Institut für physiologische und klinische Forschung, W.G. Kerckhoff-Institut, Parkstr. 1, 61231 Bad Nauheim, Germany; Supported by DFG Si 230/8-1.

Introduction

The local temperature sensitivity of the spinal cord (SC) of various homeothermic species has clearly been established in *in vivo* experiments. Local temperature changes of the spinal cord elicit adequate heat loss and heat gain mechanisms, like vasoconstriction and shivering during cooling and vasodilation or panting during warming (1,2,3). The existence of temperature sensitive neurons within the spinal cord (3) has been demonstrated in *in vivo* and recent *in vitro* (4) experiments. In recordings from spinal cord slices, a comparable number of warm-sensitive and temperature insensitive neurons has been observed in neurons recorded in the superficial laminae I and II and around the central canal (lamina X). Cold-sensitive neurons have only been found in the latter region.

Although a significant role for the diffusable gas nitric oxide (NO) in the central regulation of body temperature has not been described so far, an effect on temperature receptors originating in the spinal cord itself or deriving from cutaneous temperature receptors might be suggested by anatomical reasons. The existence of the enzyme NO-synthase, has been clearly demonstrated in spinal neurons and fibres, with the highest level of expression in laminae I, II, X and in areas with preganglionic sympathetic output neurons (5,6).

The virtual identity of the recording sites of temperature sensitive neurons in the spinal cord with the NO production sites suggests the possibility that temperature signals originating from spinal and from peripheral temperature receptors are modified by NO. This question was addressed by recording the reactivity of temperature sensitive and temperature insensitive neurons to sodium nitroprusside (SNP), which is known as a NO-donor, in an *in vitro* slice preparation.

Material and Methods

Adult male Wistar rats (150 - 320 g) were decapitated and a section of the upper lumbar spinal cord region (segments L2 to L4) was quickly removed and superfused with ice-cold artificial cerebrospinal fluid (aCSF) of the following composition (in mM): NaCl, 124; KCl, 5; NaH_2PO_4, 1.2; $MgSO_4$, 1.3; $CaCl_2$, 1.2; $NaHCO_3$, 26; glucose, 10; pH: 7.4, equilibrated with 95% O_2 and 5% CO_2, 290 mOsmol/kg). Several coronal slices (500 µm thick) were cut with a custom-made tissue slicer and preincubated at 35°C for at least 1 hour, before the first slice was transferred to the temperature controlled recording chamber. The chamber was continuously perfused with prewarmed (37.0°C) aCSF at a rate of 1.6 ml/min. Extracellular recordings were made with glass-coated platinum/iridium electrodes from various layers of the spinal cord. Spontaneously active neurons from laminae I,II and X were investigated for their local temperature sensitivity. The temperature was changed within 7 min either sinusoidally between 33 and 41°C, or slow (0.02°C/s) temperature ramps were applied to record the static temperature dependence. Fast step-like temperature changes (0.5°C/s) were also performed in order to detect possible dynamic temperature responses. A statistical program, which either fits one or two regression lines to a set of data calculated the set-point of a temperature response without requiring any further preconditions. Cells were classified as being warm- or cold-sensitive, when either the one (linear regression) or the steeper of the two regression lines (piecewise regression) had a slope (i.e. Temperature Coefficient, TC) exceeding +0.6 or -0.6 imp/s/°C, respectively. The spontaneous activity of each neuron was evaluated by averaging its activity for 60 seconds prior to the first temperature stimulus. After the temperature sensitivity of a neuron has been determined, sodium nitroprusside (SNP, schwarz pharma) was superfused with aCSF for approximately 7 min in concentrations between 10^{-7}M to 10^{-3}M. A neuron was considered as being sensitive to SNP, when the change in firing rate exceeded 20% of the baseline value at a given concentration of maximally 10^{-3}M. SNP solutions were made fresh daily from preweighed vials and stored as a stock solution at 0.1 M on ice and protected from light. Mean values given in the text are presented ±SEM. Statistical significance was calculated using students t-test or chi-square test where indicated.

Results

Recordings were made from 32 spontaneously active neurons in the laminae I,II and X from transversal spinal cord slices. The temperature sensitivity of these neurons was determined by at least two temperature stimuli, before they were tested for their responsiveness to the NO donor SNP. A comparable number of recordings was obtained from neurons in the laminae I and II and in lamina X (table 1). 13 neurons in lamina I and II were warm-sensitive and none was cold-sensitive. Like in the preceding study (4), cold-sensitive neurons were rare (n=2) and were observed only in lamina X, 9 neurons were warm-sensitive in this area.

All warm-sensitive neurons in lamina X increased their activities when superfused with SNP (10^{-5}-10^{-4}M). Similarly, all but one temperature insensitive neurons from this area were excited by SNP. Of the 2 cold-sensitive neurons recorded from lamina X, one was inhibited, the other was not affected. This result contrasts with the effect of SNP on warm-sensitive and temperature insensitive neurons in the superficial laminae, where excitatory effects were observed only rarely and 44% of the recorded neurons were inhibited (table 1). Despite this striking ($p<0.02$ for the effect of SNP on warm-sensitive neurons in laminae I,II and X) difference, the occurrence of warm- and temperature insensitive neurons were similar in both regions. Furthermore the average TC of all warm-sensitive neurons was not significantly different in the two spinal cord areas investigated (laminae I, II: 1.6±0.2 imp/s/°C; lamina X: 2.3±0.5 imp/s/°C). As in the preceeding study (4) the average spontaneous firing rate (FR) of warm-sensitive neurons from laminae I+II was significantly ($p<0.05$) lower (6.4±1.3 imp/s) as the FR of warm-sensitive neurons from lamina X (12.8±3.3 imp/s). Neurons displaying grouped discharges (bursts) were only found in the superficial dorsal horn (n=6). All but one of these bursting neurons were inhibited by SNP. Phasic responses to rapid temperature changes were observed in 50% of the tested neurons in laminae I and II (n=14), whereas in lamina X only one out of 11 neurons tested displayed a phasic response.

Fig. 1 shows an example of a warm-sensitive neuron recorded from lamina X of a transversal spinal cord slice. Superfusion with SNP for approximately 8 min caused a strong excitation which almost doubled the original firing rate of this neuron.

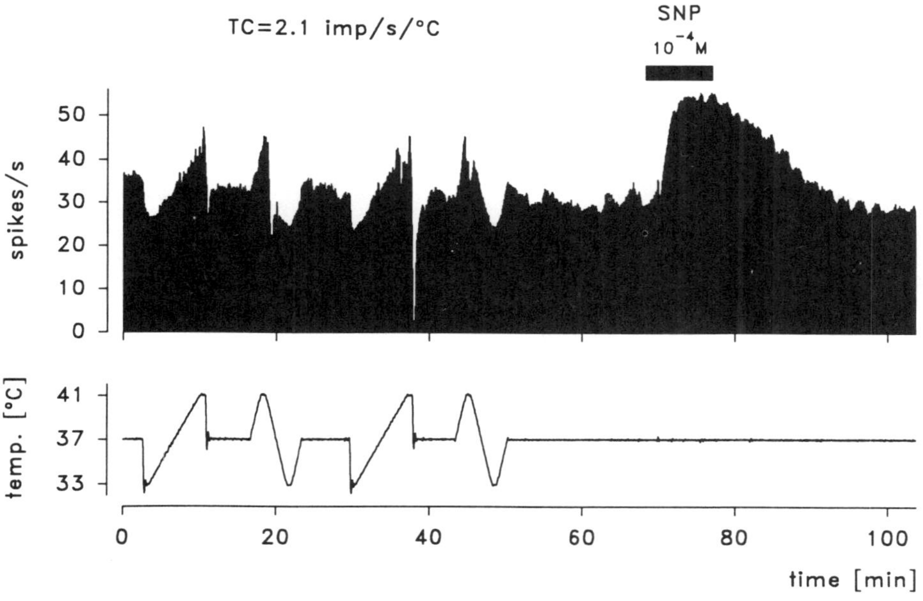

Fig. 1 Continuous rate meter recording of a warm-sensitive neuron (TC = temperature coefficient) recorded in lamina X from a transversal spinal cord slice. Superfusion with sodium nitroprusside (SNP) at the indicated time and concentration caused a reversible excitation of this neuron.

The FR returned to baseline within 16 min after the end of the superfusion. This excitatory response was typical for a SNP induced excitation of neurons in lamina X (n=13). On average, the excitatory effects of the neurons responsive to SNP at a concentration of 10^{-5}M (n=8) in lamina X started with a latency of 1.5±0.2 min and the response duration was on average 10.5±0.9min. Both, the latency and the response duration were concentration dependent. The excitatory effect to SNP (10^{-5}M) was on average 84±52% in this area (n=8). A 10-fold higher concentration caused an excitation of 141±61% (n=6). Such a strong excitatory effect was never observed in neurons recorded from lamina I and II.

The excitatory effect of SNP was dose dependent between 10^{-6} and 10^{-4}M. Concentrations >10^{-4}M caused strong and reversible effects (n=6) or sometimes irreversible responses, possibly due to unspecific actions of the drug and were not used in subsequent experiments.

Fig. 2 displays an example of a warm-sensitive neuron recorded from the lumbar superficial dorsal horn. This neuron is characterized by a relatively low spontaneous firing rate under control conditions, as it was typical for the majority of warm-sensitive neurons in the region of lamina II.

The temperature coefficient of this cell was determined from the slope of the regression line of the first ramp-like temperature change. A clear phasic response to rapid temperature changes could be observed in this neuron.

The inhibitory effects of 10^{-5}M SNP (n=6) reduced the FR on average by 74±8% Application of a 10-fold lower concentration caused a reduction of only 29±3%. Concentrations above 10^{-5}M were not tested routinely, because 10^{-5}M caused in all but one of the neurons tested a reduction in FR to 0 imp/s.

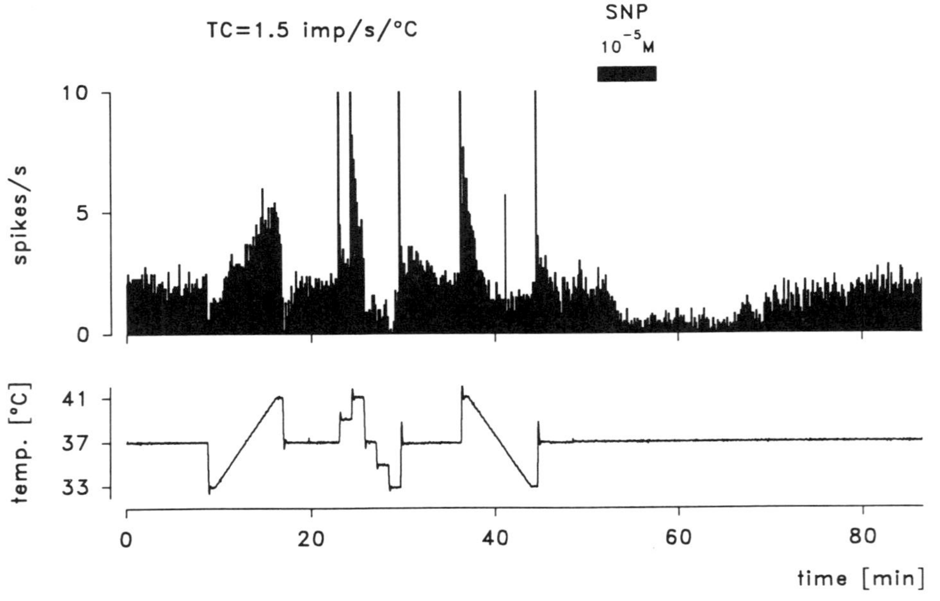

Fig. 2. Continuous rate meter recording of a warm-sensitive neuron recorded in lamina II of the dorsal horn from a spinal cord slice. Superfusion with sodium nitroprusside (SNP) at the indicated time and concentration caused a reversible inhibition of this neuron.

Table 1
Number of warm- and cold-sensitive and temperature-insensitive neurons in different laminae of rat spinal cord slices and their responsiveness (+=excitatory, -=inhibitory, 0=no effect on the FR) to superfusion with sodium nitroprusside (SNP, 10^{-5} - 10^{-4}M).

	warm-sens.			cold-sens.			temp. ins.			total
effect of SNP	+	-	0	+	-	0	+	-	0	n=
laminae I+II	3	6	4					1	2	16
near central canal	9				1	1	4		1	16
total n=		22			2			8		32

In 7 warm-sensitive neurons the effect of SNP on the temperature sensitivity was investigated by determining the temperature coefficient before, during and at least 10 min after the superfusion with SNP. In none of these neurons a clear tendency towards a shift in TC to higher or lower values could be observed. The average TC's of these warm-sensitive neurons calculated before, during and after superfusion with SNP were 1.8±0.4, 1.8±0.3 and 1.7±0.3 imp/s/°C, respectively.

Discussion

Despite the fact that NO has been shown to influence a huge variety of central and peripheral processes its implication in centrally mediated thermoregulatory responses has not been shown so far, neither for the preoptic/anterior hypothalamic area (PO/AH), nor for the spinal cord.
Warm-sensitive and temperature insensitive neurons were found in comparable quantities in both regions, but application of the NO donor SNP caused predominantly excitations of neurons in lamina X, while neurons in lamina I and II were predominantly inhibited. The rare (n=2) cold-sensitive neurons were only found in lamina X and were not excited by SNP. Based on histological studies (5,6,7,9) and based on the presented data, NO could in principle influence

a) the activity of preganglionic (visceral) sympathetic neurons
b) the activity of local temperature sensitive neurons in the spinal cord
c) the transmission of peripheral afferent temperature signals.

a) While neurons recorded from the superficial laminae I and II are generally believed to belong to the afferent sensory system, neurons in lamina X might belong either to the afferent sensory or efferent sympathetic system (5,6,7,9). With regard to the latter possibility it has been shown that preganglionic, sympathetic output neurons are abundant in the region dorsal and lateral to the central canal (also called central autonomic nucleus or dorsal commissural nucleus), which are thought to project via the splanchnic nerves and the inferior mesenteric ganglion to visceral structures (9,10). Functionally, it is tempting to speculate, that the activation of neurons in this area might be responsible for the increase in mass-activity seen in the splanchnic nerve during spinal cord warming of the rabbit (11,12). Other preganglionic sympathetic neurons which innervate primarily cutaneous blood vessels, e.g. neurons in the intermediolateral nucleus (IML) should consequently have a different temperature sensitivity as the neurons in lamina X, a hypothesis which is currently being tested. The fact that NO has almost exclusively excitatory effects on neurons in the lumbar lamina X suggests, that the visceral sympathetic output is stimulated by the release of NO in this spinal area.

b) The existence of central temperature receptors within the spinal cord has been clearly demonstrated by recording from temperature sensitive afferent spinothalamic fibres in the cat (3). The location of the somata of these temperature sensitive neurons are likely to be found in superficial laminae of the dorsal horn and around the central canal. Interestingly, about 10% of the NO-synthase containing neurons, which have their origin exclusively in lamina X, have direct spinothalamic projections (7). Other spinothalamic neurons in more superficial dorsal horn laminae, were shown to be in close proximity to NO-synthase containing processes (7). These NO-synthase containing terminals, especially the ones in lamina II, were shown to have an intraspinal origin, they contain GABA as transmitter and might therefore be responsible for inhibitory synaptic transmission (8). Depending on the location of the putative spinothalamic neurons in the spinal cord, local release of NO should have primarily excitatory (on neurons in lamina X) or inhibitory (on neurons in laminae I+II) effects.

c) Several recent studies (7,8,13) are concerned with involvement of NO in the transmission of pain signals to the spinal cord. Because peripheral pain as well as temperature signals are conveyed to the spinal cord via slowly conducting A∂ or C-fibers, it is likely, that peripheral temperature signals are also influenced by NO producing neurons in the spinal cord. As described for local temperature sensitive neurons in the spinal cord, NO could have primarily excitatory or inhibitory effects on the transmission of peripheral temperature signals depending on the location in the spinal cord where the peripheral temperature receptors synapse.

In summary, all of the above mentioned effects of NO on spinal cord neurons are possibilities and their relevance for thermoregulatory sensor- or effector- mechanisms, however, has to await functional *in vivo* studies.

References

1. Simon, E (1974): Temperature regulation: The spinal cord as a site of extrahypothalamic thermoregulatory functions. Rev Physiol Biochem Pharmacol 71, 1-76.

2. Simon, E, Pierau, Fr-K, Taylor, DCM (1986): Central and peripheral thermal control of effectors in homeothermic temperature regulation. Physiol Rev 66, 235-300.

3. Simon, E, Iriki, M (1971): Sensory transmission of spinal heat and cold sensitivity in ascending spinal neurons. Pflügers Arch 328, 103-120.

4. Schmid, HA, Pehl, U, Simon, E (1994): Temperature sensitivity of rat spinal cord neurons recorded *in vitro*. In: Temperature Regulation: Recent Physiological and Pharmacological Advances. Milton, AS ed. Basel: Birkhäuser, 109-114.

5. Anderson, CR (1992): NADPH diaphorase-positive neurons in the rat spinal cord include a subpopulation of autonomic preganglionic neurons. Neurosci Lett 139, 280-284.

6. Valtschanoff, JG, Weinberg, RJ, Rustioni, A (1992a): NADPH diaphorase in the spinal cord of rats. J Comp Neurol 321, 209-222.

7. Lee, J-H, Price, RH, Williams, FG, Mayer, B, Beitz, AJ (1993): Nitric oxide synthase is found in some spinothalamic neurons and in neuronal processes that appose spinal neurons that express Fos induced by noxious stimulation. Brain Res 608, 324-333.

8. Valtschanoff, JG, Weinberg, RJ, Rustioni, A, Schmidt, HHHW (1992b): Nitric oxide synthase and GABA colocalize in lamina II of rat spinal cord. Neurosci Lett 148, 6-10.

9. Jänig, W, McLachlan, EM (1986): Identification of distinct topographical distributions of lumbar sympathetic and sensory neurons projecting to end organs with different functions in the Cat. J Comp Neurol 246, 104-112.

10. Hancock, MB, Peveto, CA (1979): A preganglionic autonomic nucleus in the dorsal gray comissure of the lumbar spinal cord of the rat. J Comp Neurol 183, 65-72.

11. Simon, E (1971): Regional differentiation of vasomotor activity underlying thermoregulatory adjustments of blood flow. Int J Biometeor 15, 219-224.

12. Gregor, M, Jänig, W, Riedel, W (1976): Response pattern of cutaneous postganglionic neurones to the hindlimb on spinal cord heating and cooling in the cat. Pflügers Arch 363, 135-140.

13. Malmberg, AB, Yaksh, TL (1993): Spinal nitric oxide synthesis inhibition blocks NMDA-induced thermal hyperalgesia and produces antinociception in the formalin test in rats. Pain 54, 291-300.

Thermal Balance in Health and Disease
Advances in Pharmacological Sciences

EXCITATORY AMINO-ACID EFFECTS ON RAT HYPOTHALAMIC NEURONES WHICH RESPOND TO CHANGES IN SCROTAL SKIN TEMPERATURE

D.C.M. Taylor, R.J. Gayton, N.P. Cross, H.E. Miley and M.J. Parton

Physiological Laboratory, University of Liverpool, P.O.Box 147, Liverpool, L69 3BX, U.K.

There are populations of neurones in the rat hypothalamus which respond to changes in scrotal skin temperature. These neurones are thought to be involved not only in temperature regulation,but also in the autonomic arousal system (1). Their response is characteristically discontinuous.Raising the scrotal skin temperature above a threshold value of between 33 and 37°C causes an abrupt increase (or, occasionally, decrease) in firing rate to a new steady level (1,2). Furthermore the pattern of firing alters from bursting to more regular (3). Preliminary results indicate that the activity of the neurones is modulated by 5-hydroxytryptamine, but that this is probably not the primary transmitter (4). The experiments described here were performed to determine whether an excitatory amino-acid transmitter could also be involved, acting through the N-methyl-D-aspartate (NMDA) or non-NMDA receptors (5), as may be the case in the thalamus (6,7).

Methods

Male Sprague-Dawley rats, anaesthetized with urethane (1.25-1.5 g/kg) were used. End tidal CO2 and rectal temperature were monitored and maintained within normal limits (E.T.CO2 4-4.5%, Tre 37-38°C). Recordings of single anterior hypothalamic neurones were made using the centre barrel of a 7-barrelled glass micropipette (3M NaCl, impedance 4-6 MW). The electrode was positioned stereotaxically in the hypothalamus (6.4 mm AP, 0.5 mm ML and -2 to -3 mm deep (8)). Once a hypothalamic neurone was identified as responding to scrotal skin temperature, the following agents were applied ionophoretically; D-2-amino-5- phosphonopentanoic acid (D-AP5, 20mM in 160mM NaCl, pH8, 20nA, 40s-120s), 6-cyano-7-nitroquinoxaline-2,3-dione (CNQX, 1mM in 50 mM NaCl, pH8.8, 20nA, 40s), 2,3-Dihydroxy-6-nitro-7-sulfamoyl-benzo(F)quinoxaline (NBQX, 1mM in 50 mM NaCl, pH8, 20nA, 120s), L-AP5 (20mM in 160mM NaCl, pH8, 20nA, 40s), a-amino-3- hydroxy-5-methyl-4-isoxazole proprionic acid (AMPA, 10mM in 75mM NaCl pH8, 5-15nA, 5-20s), N-methyl-D-aspartate (NMDA, 100 mM, pH8, 5-15nA, 5-20s) and L-Na-glutamate (200 mM, pH8, 25nA, 20s), all ejected as anions. The barrel used for current balance contained pontamine sky blue (2.5% in 0.5M sodium acetate) to allow the recording site to be marked for subsequent histological verification. Scrotal skin temperature was controlled to within 0.1°C by a Peltier-effect thermode in close contact with the shaved scrotal skin (3,9).

Results

Thirty-six neurones responded to warming the scrotal skin to 39°C with an abrupt increase in firing rate. In twenty-nine out of thirty-one cases tested D-AP5 had no significant (P> 0.95, Student's t-test) effect on the spontaneous activity of the neurone at sub-threshold skin temperatures. In all cases, the thermally evoked increase in firing (figure 1) could be completely abolished by ionophoretic application of D-AP5, CNQX or NBQX at current levels and ejection times which were effective against the exogenous agonists.

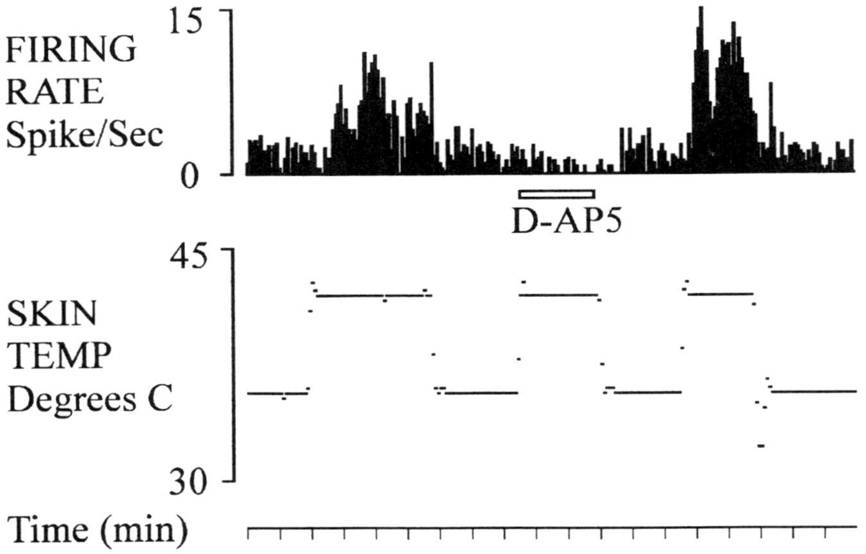

Fig. 1.: The firing rate (upper trace) of an anterior hypothalamic neurone averaged over successive 5s periods, showing the response to different steady scrotal skin temperatures (lower trace). The increase in activity which occured upon warming the skin was prevented by concurrent application of D-AP5 (open bar).

Ionophoretic application of glutamate excited all of the cells from which we were able to record and the effect was reduced or abolished by concurrent application of D-AP5 in all cases. Both AMPA and NMDA were also effective in exciting the neurones in doses which could be reversibly antagonized by the appropriate receptor antagonist (figure 2). The AMPA excitation could be prevented by concurrent application of CNQX or NBQX (8 out of 8 cases tested), but not by D-AP5. Conversely NMDA-induced excitation could be inhibited by D-AP5 but not by the inactive enantiomer L-AP5 or by NBQX (11 out of 11 cases tested). The stochastic activity of the thermoreactive neurones in response to warming the scrotal skin reverted, either wholly or in part, to bursting activity during either D-AP5 or NBQX application.

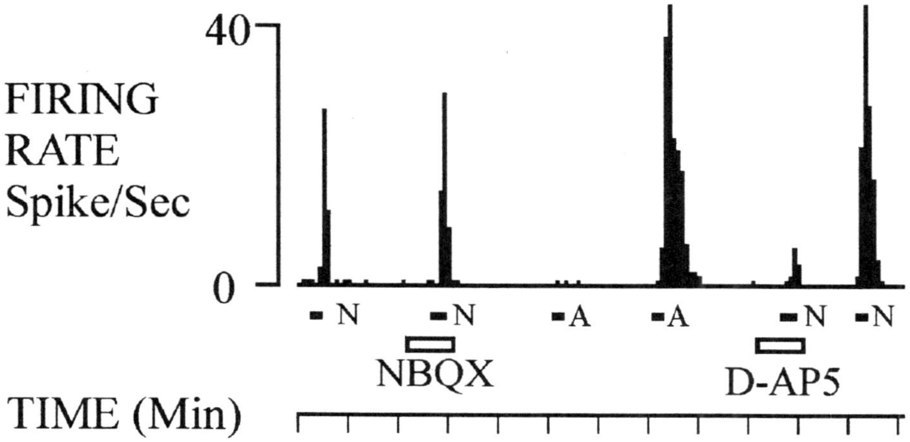

Fig. 2.: The response of an anterior hypothalamic neurone to NMDA (N) and AMPA (A) (closed bars). NBQX (first open bar) had no effect on the NMDA mediated excitation but had an effect on the AMPA receptor which lasted 3 to 4 minutes. Concurrent application of D-AP5 greatly reduced the NMDA mediated excitation.

Discussion

Recent work has shown that the responses of neurones in the ventrobasal thalamus to non-noxious somatosensory stimulation can be blocked by the application of excitatory amino-acid receptor antagonists (7). There are three excitatory amino acid receptor types directly coupled to ion channels which can be characterised by the specific agonists N-methyl-D-aspartate (NMDA), kainate and AMPA (RS-a-amino-3-hydroxy-5- methyl-4-isoxazole proprionic acid). The NMDA receptor can be blocked by D-2-amino- 5-phosphonopentanoic acid (D-AP5), whereas the non-NMDA receptor responses can be antagonised by 6-cyano-7-nitroquinoxaline-2,3-dione (CNQX) or 2,3-Dihydroxy-6-nitro- 7-sulfamoyl-benzo(F)quinoxaline (NBQX) which is more potent and selective than CNQX (10). The responses of thalamic neurones to non-noxious stimuli can be blocked by concurrent application of either D-AP5 or CNQX, whereas the responses of thalamic neurones to noxious somatic stimuli are blocked only by antagonists of the NMDA receptor (6).

The results from the present study demonstrate that the hypothalamic neurones which respond to scrotal skin temperature can be excited by excitatory amino acids acting at both AMPA and NMDA receptors. This makes an interesting parallel with the work by Salt and his colleagues who found similar chemical response profiles in thalamic neurones which reacted to non-noxious cutaneous stimulation. The location of the non-nociceptive neurones was the lateral edge of the ventrobasal thalamus (6,7), the same area of the thalamus to which the scrotal thermoafferent system projects (2). In previous studies on the rat scrotal thermoafferent system we have postulated that a proportion of the hypothalamic and thalamic neurones receive their input in parallel, diverging from the raphe nuclei (1,2). The similarity in chemical response profile between the hypothalamic thermoreactive neurones studied here, and the innocuous sensitive neurones studied by Salt (6,7) leads us to suggest that the thermoreactive neurones of the thalamus warrant ionophoretic study.

Both NMDA and non-NMDA receptor antagonists caused the regular activity of warm-activated hypothalamic neurones to revert to the bursting pattern they exhibited at lower scrotal skin temperatures. This was unexpected, since NMDA receptors are more usually associated with bursting responses. Further studies on the relationship between inter-burst and inter-spike intervals could well help to resolve this paradox.

NMDA receptors have been implicated in the nociceptive afferent system. The apparent importance of NMDA receptors in the rat scrotal thermoafferent system is interesting since there are several lines of evidence which suggest that the scrotal temperature-reactive neurones are not part of a specific thermosensory pathway, but behave functionally as an autonomic arousal pathway. Briefly, these are:

1. The number and distribution of the central neurones which respond to scrotal skin temperature (1).
2. The high degree of modality and spatial convergence they exhibit (11).
3. The discontinuous, switching nature of the response (1).
4. The desynchronization of the E.E.G. at supra-threshold temperatures, even though the skin temperatures are not noxious in this range (3,9).
5. The vasomotor effects of scrotal skin warming (12).

Work now in progress may help to elucidate the relative significance of the various excitatory amino acid receptor sub-types in modulating the activity of thermoreactive hypothalamic neurones.

References

1. Simon, E, Pierau, Fr-K, Taylor, DCM, (1986): Central and peripheral thermal control of effectors in homeothermic temperature regulation. Physiological Reviews 66, 235-300.

2. Hellon, RF, & Taylor, DCM, (1982): An analysis of a thermal afferent pathway in the rat. Journal of Physiology (London) 326, 319-328.

3. Taylor, DCM, Gayton, RJ, (1986): Alteration of the discharge pattern of rat diencephalic neurones with scrotal skin temperature. Neuroscience Letters 72, 59-63.

4. Taylor, DCM, Gayton, RJ, (1988): The effect of serotonin and cyproheptadine on thermoresponsive neurones in the rat. Pflügers Archiv 411,suppl.1, R114.

5. Parton, MJ, Gayton, RJ, Taylor, DCM, (1991): The effect of glutamate receptor antagonists on rat thermoreactive hypothalamic neurones. Journal of Physiology (London) 438, 78P

6. Eaton, SA, Salt, TE, (1990): Thalamic NMDA receptors and nociceptive sensory synaptic transmission. Neuroscience Letters 110, 297-302.

7. Salt, TE, (1987): Excitatory amino acid receptors and synaptic transmission in the rat ventrobasal thalamus. Journal of Physiology (London) 391, 499-510.

8. König, JFR, Klippel, RA, (1963): The Rat Brain. A Stereotaxic atlas of the forebrain and lower parts of the brain stem. The Williams and Wilkins Company, Baltimore

9. Taylor, DCM, Steele, J, Gayton, RJ, (1987): An analysis of the responses of rat striatal neurones to scrotal skin temperature. Brain Research 419, 352-356.

10. Randle, JCR, Guet, T, Cordi, A, Lepagnol, JM, (1992): Competitive inhibition by NBQX of kainate/AMPA receptor currents and excitatory synaptic potentials: importance of 6-nitro substitution. European Journal of Pharmacology 215, 237-244

11. Schingnitz, G, Werner, J, (1986): Significance of scrotal thermal afferents within the general thermoafferent system. Journal of Thermal Biology 11, 181-189.

12. Neya, T, Pierau, Fr-K, (1976): Vasomotor response to thermal stimulation of the scrotal skin in rats. Pflügers Archiv 363, 19-31.

Thermal Balance in Health and Disease
Advances in Pharmacological Sciences

HIBERNATION-RELATED CHANGES IN THE EFFECTS OF NEURO-MODULATORS: AN *IN VITRO* STUDY ON SYNAPTIC TRANSMISSION IN HIPPOCAMPAL SLICES

P. Igelmund, H. Spangenberger and F. Gh. Nikmanesh

Zentrum Physiologie und Pathophysiologie der Universität zu Köln, Institut für Neurophysiologie, Robert-Koch-Str. 39, D-50931 Köln, Germany

Introduction

The hippocampus is an important link in current models of neural control of entrance into and arousal from hibernation (1,4,8). Heller (4) suggests that the hippocampus is part of a feedback loop inhibiting the activity of the brain stem reticular formation during entrance into hibernation and during deep hibernation. Beckman and Stanton (1) assume a reciprocal inhibitory interaction between the hippocampus and the midbrain reticular formation as well as between the hippocampus and the hypothalamus. On the other hand, Pakhotin et al. (8) regard a tonic inhibitory influence of the hippocampus on the reticular formation during hibernation as unlikely.Instead, they suppose a modulatory action of the septohippocampal arousing system on hypothalamic thermoregulatory mechanisms during -activity. Excitability changes in neurons in the medial septum and the hippocampus are possibly involved in triggering the exit from a bout of hibernation (8). The transmission properties of hippocampal neurons are modified in connection with hibernation. The temperature below which synaptic transmission is blocked (the threshold temperature) is lowered (6,9) and paired-pulse depression at low temperatures is reduced (6).

In order to investigate further hibernation-related changes in the intrinsic transmission properties of the hippocampus, we studied the effects of the neuromodulators serotonin, histamine, adenosine, dopamine, and norepinephrine, and of the cholinergic agonist carbachol on stimulus-induced field potentials in hippocampal slices. As the role of the hippocampus during arousal from hibernation necessarily implies functioning at low temperatures, we investigated the effects of the neuromodulators and of carbachol at two temperatures, 37°C and 22°C. The experiments were carried out on hippocampal slices prepared from hibernating Turkish hamsters (HH), from warm-acclimated hamsters (WH) and, for comparison, from Wistar rats as non-hibernators. The present paper is a short synopsis of the main results of these studies; detailed reports will appear elsewhere (Nikmanesh et al., in prep., Spangenberger et al., in prep.).

Material and Methods

Transverse hippocampal slices were prepared from Turkish hamsters (*Mesocricetus brandtt)* iand Wistar rats taken from our institute's breeds. Our Turkish hamster breed is based on some animals provided by Dr. H. Pohl, Max-Planck-Institut für Verhaltensphysiologie, Seewiesen, Germany. HH slices were prepared from animals killed in deep hibernation - their brain temperatures were 6 to 10°C. WH and rat slices were prepared from euthermic animals killed under ether anesthesia. The slices were studied in an interface type chamber continuously perfused with artificial cerebrospinal fluid oxygenated with 95% O_2/5% CO_2 and containing (in mM): NaCl 124, NaH_2PO_4 1.25, $NaHCO_3$ 26, KCl 3, $CaCl_2$ 1.6, $MgSO_4$ 1.8, glucose 10. Neuromodulators were bath-applied for 15 min (carbachol for 3 min). Each slice was tested at 37°C and 22°C. Field potentials evoked by stimulation of Schaffer collaterals/commissural fibers with 100-µs pulses at 30-s intervals were extracellularly recorded in stratum pyramidale of area CA1. The stimulus intensity was adjusted to evoke population action potentials of either about 90% (*submaximal)* or, in some cases, 50% *(half-maximal)* of the maximal amplitude. Data were conventionally amplified and recorded on a chart writer for on-line control. Simultaneously they were digitized with a CED 1401 interface (Cambridge Electronics) and stored on a computer disk for further analysis. Stastistical significance was calculated with Student's t-test. Averaged values are indicated as mean ± standard error of the mean (SEM), with n = 6-14 experiments for each value. To allow recovery from the preparation, slices were incubated for at least 1 hour prior to a recording session. In each animal group, some of the slices were incubated at 37°C and others at 22°C. Similarly, some were first tested at 37°C and others first at 22°C. Data were pooled as there was no influence of the incubation temperature and the order of temperature changes detectable.

Results and Discussion

In transverse hippocampal slices, stimulation of the Schaffer collateral/commissural fiber bundle synaptically activates CA1 pyramidal cells. With extracellular electrodes positioned in the stratum pyramidale, this activity can be recorded as a positive going field EPSP superimposed by a negative going population action potential (*population spike, PS,* see Fig 1A). The PS amplitude was determined as the difference between the negative potential peak and the average of the two positive peaks preceding and following the negative peak. As a measure of the field EPSP, the initial slope of the positive peak preceding the population spike was determined. In our study, all the substances investigated were able to affect the PS amplitude, at least in a part of the experimental situations. Some substances also modified the field EPSP.

Carbachol

Carbachol (*CCh*), a cholinergic agonist, was applied in place of acetylcholine, which is not suitable for bath-application due to the high activity of cholinesterase. In order to minimize desensitization and to receive a full recovery from the effects, CCh - in contrast to the other substances tested in this study - was applied bolus-like for 3 minutes at a concentration of 10 µM.

At 37°C, application of CCh produced inconsistent results. Similarly in all three animal groups, about 60% of the slices showed a decrease in the PS amplitude in response to CCh, about 30% showed an increase (Fig. 1A,B, upper panel), and in about 10% the PS amplitude was not significantly affected. Independent of the direction of the effect on the PS amplitude, the EPSP was usually reduced by CCh. The reactions started within 2 min after application onset, reached a maximum after 3-4 min and then recovered within some minutes (Fig. 1B).

At 22°C, CCh consistently reduced the PS amplitude even in those slices which showed an increase of the PS amplitude at 37°C (Fig. 1A,B, lower panel). The reduction was 60±10%, 38±9%, and 54±5% for rats, warm-acclimated hamsters, and hibernating hamsters, respectively. There was no significant difference between the three animal groups.

These data suggest that the mechanisms, by which CCh enhances the excitability of hippocampal pyramidal neurons, are highly temperature dependent whereas the action of CCh on the field EPSP remains strong at low temperatures. Thus, the depressing action of CCh becomes consistently dominant at low temperatures. According to our results, there seems to be no species difference between hamsters and rats in the response of hippocampal neurons to cholinergic influences, nor does a change seem to occur in the intrinsic properties of cholinergic modulation in the hippocampus with hibernation. Nevertheless, the strong temperature dependence of the CCh action suggests a differential function of cholinergic modulation in connection with hibernation, especially during the cooling and rewarming phases.

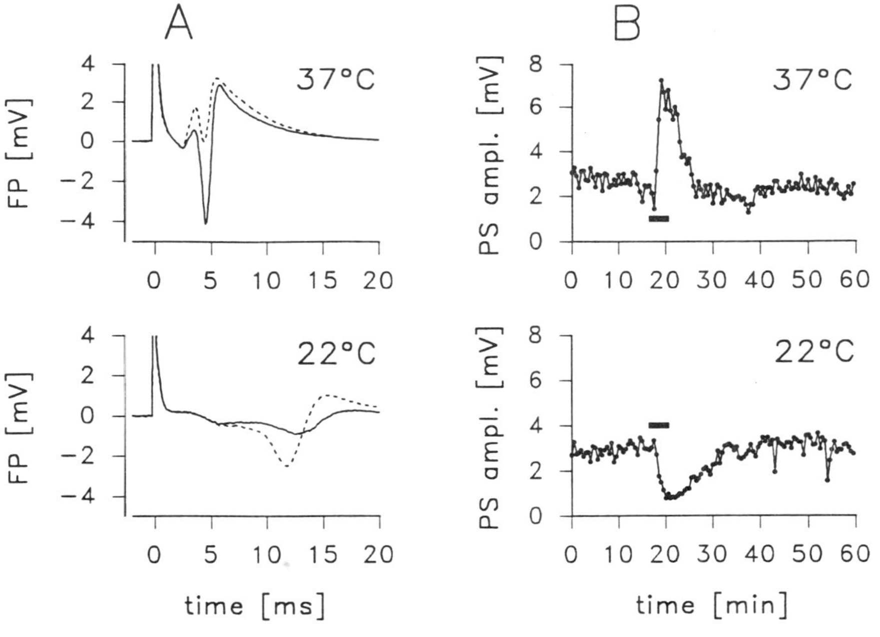

Fig. 1A,B: Effect of carbachol (CCh) on population spikes in a rat hippocampal slice at 37°C and 22°C. **A:** Example recordings showing the population spike in the control phase before application of CCh (dashed line) superimposed by the population spike during application of CCh (solid line) at 37°C (upper panel) and 22°C (lower panel). Each spike is the average of 7 consecutive responses. FP = field potential. **B:** Time course of the modulation of the population spike amplitude by CCh at 37°C (upper panel) and 22°C (lower panel). The application time is indicated by the horizontal bars. The data are from a single slice. Like this slice, about 30% of the slices of all 3 animal groups showed opposite reactions to CCh at 37°C and 22°C.

Adenosine

Applied in a concentration of 50 μM at 37°C, adenosine (*Ad*) reduced the amplitude of submaximal population spikes in slices from rats and from WH by an average of 25±13% and 44±14%, respectively. In contrast, Ad was ineffective at this concentration in slices from HH. With a concentration of 100 μM, however, comparable amplitude decreases were obtained in HH slices (64±12%). Reductions of the PS amplitude by Ad were associated with reductions in the field EPSP. The effects started immediately with substance application and reached their maximum within 4-5 minutes. Recovery began immediately at the end of substance application and was completed within 3-4 minutes.

At 22°C, the effects of Ad application were similar to those of 37°C, except for the slower time course. In a concentration of 50 μM the PS amplitude in rat slices and WH slices was reduced by 69±8% and 65±17%, respectively. Again, there was no effect in HH slices. With 100 μM, Ad reduced the amplitude in HH slices by 56±14%.

Experiments with half-maximal instead of submaximal stimulation revealed a higher sensitivity to Ad in all experimental groups. Nevertheless, these experiments confirmed the lower sensitivity to Ad of HH slices compared to WH and rat slices. The PS amplitude in HH slices was reduced by 50 μM but not by 10 μM Ad, whereas in WH and rat slices 10 μM Ad were clearly effective.

Endogenous Ad is thought to be one of the agents which produce the overall depression of the CNS during hibernation (7). The loss of sensitivity for Ad with hibernation, as shown in our data, may reduce this depression and thus help the hippocampus to build up its activity in the early arousal process.

Dopamine

Dopamine (*DA*) was applied in concentrations of 100 μM and 1 mM. At 37°C, the amplitude of submaximal population spikes was affected by DA application in less than half of the experiments, irrespective of the DA concentration (2 of 7 slices from rats, 3 of 8 from WH, 5 of 8 from HH). In these slices, DA consistently augmented the PS amplitude. The augmentation was mostly in the range of 10-40%. In some but not all experiments, the augmentation of the population spikes was associated with an increase in the slope of the field EPSP.

At 22°C, application of DA revealed remarkable species differences. In rat slices, DA was not effective in any of 6 slices, whereas 90% (9 of 10) and 79% (11 of 14) of the WH and HH slices, respectively, responded to DA with an increase in spike amplitude. Application of 0.1 mM DA in hamster slices increased submaximal population spikes by up to 82%. Half-maximal spikes (in some experiments with WH slices) were augmented by up to 295%.

From these results it is conceivable that the excitatory action of DA in the hamster hippocampus at low temperatures may contribute to the activation of hippocampal function during arousal from hibernation.

Norepinephrine

Norepinephrine (*NE*) was tested in concentrations of 20 µM and 100 µM. At 37°C, application of NE resulted in augmentation of submaximal population spikes by 10-94% without significant differences between HH, WH, and rats. The effect started 1-2 min after application onset. With submaximal stimulation, the augmentation usually lasted until some minutes after the end of application. With half-maximal stimulation, however, the effect often was biphasic: a short increase in amplitude was followed by a longer lasting decrease. Again, there was no significant difference between the three animal groups.

At 22°C, the effects of NE, like those of DA, were species dependent. In rat slices, NE application was ineffective in all of 6 cases with submaximal stimulation. In 3 slices investigated with half-maximal stimulation, NE decreased the PS amplitude by 25-67%. In no case was there an increase in amplitude. In experiments with HH slices, in contrast, 6 of 7 slices responded to NE application with an increase in amplitude of submaximal population spikes (15-66%). In no case the amplitude was decreased. In WH slices, half-maximal population spikes (n=4) were increased by 93-117%, whereas submaximal population spikes (n=3) were not affected.

The amplitude increase induced by application of NE at 22°C persisted for a long time (>30 min) in HH slices while in WH slices recovery started about 10 min after the end of NE application. Experiments with the ß-adrenergic agonist isoproterenol produced comparable long-term effects and thus suggest that the long-term enhancement is mediated by ß-receptors.

Like DA, the facilitating effect of NE in the hamster hippocampus makes synaptic transmission at low temperatures more viable. The long-lasting action of NE, as seen in HH slices, may further support this effect. The noradrenergic system in the CNS seems to have activating effects throughout the hibernation cycle. General depletion of NE in the CNS is reported to promote entrance into hibernation and to extend the duration of the hibernation bouts and the hibernation period (2).

Serotonin

Application of serotonin (*5-HT*) at 37°C revealed differences between hibernating and warm-acclimated hamsters as well as species differences between hamsters and rats. A summary of these results is shown in Fig. 2. In a concentration of 20 µM, 5-HT increased the PS amplitude in WH slices by 53±21% whereas in HH slices no reaction occurred (-2±4%). In rat slices, the reaction was opposite: the spike amplitude was reduced by 26±7%. With elevation of the 5-HT concentration to 100 µM, the augmenting effect was also present in HH slices (39±25%). In rat slices, however, as seen in two experiments, the PS amplitude was still reduced with elevated 5-HT concentration.

With lowering of temperature to 22°C, the enhancing effect of 5-HT in hamster slices became smaller and was preceded by a short (3-15 min) decrease in spike amplitude. As at 37°C, rat slices showed only depressing reactions.

High activity of the serotonergic system is thought to be essential for entrance into hibernation (3). The reduction of sensitivity measured in HH slices in our experiments could be based on a down-regulation of the 5-HT receptors resulting from high serotonergic acitvity during entrance into hibernation. The excitatory action of 5-HT in Turkish hamster slices, in contrast to the depressing effects in rat slices, points to a function of 5-HT in hippocampal activity during arousal. It should be noted that our results on Turkish hamster slices are contrary to results of Horrigan and Horowitz (5) on hippocampal slices from golden hamsters. In that study, 5-HT produced only depressant effects at all temperatures.

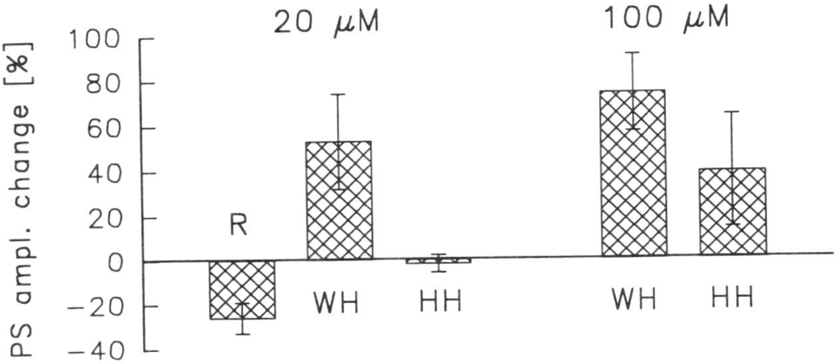

Fig. 2: Effect of 5-HT (20 and 100 μM) on the PS amplitude in slices from rats (R), warm-acclimated hamsters (WH), and hibernating hamsters (HH) at 37°C. The bars represent the changes (mean values ± SEM) relative to control.

Histamine

Histamine (*HA*), applied in a concentration of 50 μM, increased the population spikes in WH and HH slices. At 37°C, the amplitude increase in HH slices (145±24%) was significantly higher than in WH slices (67±26%). At 22°C, as compared to 37°C, the effect of HA was significantly reduced in HH as well as WH slices (augmentation by 26±7% and 17±6%, respectively). In rat slices, the spike amplitude was not affected by 50 μM HA at either temperature.

Like DA, NE, and 5-HT, the facilitating effect of HA on hippocampal activity points to a role in the arousal process. The higher sensitivity to HA during hibernation compared to the warm-acclimated state further supports the idea of a specific hibernation-related role of HA.

Concluding remarks

A comparison of the results reveals that DA and HA, if effective, always produced an augmenting effect. Ad, in contrast, decreased the PS amplitude. Finally, NE, 5-HT, and CCh showed augmenting as well as depressing effects, depending on the experimental situation.

Species differences between hamsters and rats were obvious for NE, DA, HA, and 5-HT. While at 22°C DA and NE had no excitatory action in rat slices, they enhanced the PS amplitude in hamster slices. HA, ineffective in rat slices at 37°C as well as 22°C, strongly enhanced the PS amplitude in hamster slices. 5-HT depressed the PS amplitude in rat slices at 37°C as well as 22°C. In hamster slices, in contrast, 5-HT augmented the PS amplitude at 37°C and had a biphasic action at 22°C. It is interesting to note that the species differences are all in the way of stronger facilitation in hamster slices compared to rat slices.

Like 5-HT in hamster slices, also CCh, NE, DA and HA showed a clear **temperature dependence** in their actions. While the effect of CCh at 22°C was always depressing, only 60% of the slices were depressed at 37°C and about 30% reacted to CCh with an increase in PS amplitude. NE and DA augmented the population spikes at 37°C. In contrast, no augmentation was induced by NE and DA in rat slices at 22°C. In hamster slices, however, the facilitating effect was equally present at 22°C. HA in hamster slices increased the PS amplitude at both temperatures but the effect was significantly stronger at 37°C than at 22°C.

Hibernation-related changes of the neuromodulatory action were found for 5-HT, HA, Ad, and NE. For 5-HT and Ad, higher concentrations were needed in HH slices compared to WH slices to produce similar effects. In contrast, HA was more effective in HH than in WH slices. A higher sensitivity in HH slices was also indicated for NE. Aditionnally, NE produced a long-term facilitation in HH slices which was not present in WH slices.

In summary, our results show stronger excitatory actions and weaker inhibitory actions of the investigated neuromodulators in the hippocampus of HH compared to rats and/or WH. This finding is in good agreement with the role of the hippocampus as a central structure important in the regulation of arousal from hibernation.

Acknowledgements

We are grateful to Mrs. E. Walde for technical assistance and to Dr. U. Heinemann and Dr. F.W. Klußmann for valuable discussions and reading the manuscript. This work was supported by the Deutsche Forschungsgemeinschaft (Ig10/1-1, Ig10/1-2).

References

1. Beckman, AL, Stanton, TL (1982): Properties of the CNS during the state of hibernation. In: The neural basis of behaviour. Beckman, AL ed. New York: MPTP Press, 19-45.

2. Cai, Y-P, Zhao, H-Q, Huang, Q-H (1989): Reevaluation of involvement of norepinephrine and serotonin in initiation of hibernation. In: Living in the Cold II. Malan, A, Canguilhem, B eds. Montrouge, London: John Libbey, 477-484.

3. Canguilhem, B, Miro, J-L, Kempf, E, Schmitt, P (1986): Does serotonin play a role in entrance into hibernation? Am J Physiol 251, R755-761.

4. Heller, HC (1979): Hibernation: Neural aspects. Ann Rev Physiol 41, 305-321.

5. Horrigan, DJ, Horowitz, JM (1990): Thermal dependence of serotonergic modulation of neural activity in the hamster. J Comp Physiol A 167, 79-88.

6. Igelmund, P, Heinemann, U, Klußmann, FW (1993): Hibernation-related modification of activity-dependent properties of synaptic transmission in hamster hippocampus. Neurosci Res Comm 13:167-173.

7. Lee, TF, Nürnberger, F, Wang, LCH (1993): Possible involvement of endogenous adenosine in hibernation. In: Live in the Cold: Ecological, Physiological, and Molecular Mechanisms. Carey, C, Florant, GL, Wunder, BA, Horwitz, B, eds. Boulder, Oxford: Westview Press, 315-322.

8. Pakhotin, P, Pakhotina, ID, Belousov, AB (1993): The study of brain slices from hibernating mammals *in vitro* and some approaches to the analysis of hibernation problems *in vivo*. Progr Neurobiol 40, 123-161.

9. Thomas, MP, Martin, SM, Horowitz, JM (1986): Temperature effects on evoked potentials of hippocampal slices from noncold-acclimated, cold-acclimated and hibernating hamsters. J Therm Biol 11, 213-218.

Thermal Balance in Health and Disease
Advances in Pharmacological Sciences

EFFECTS OF COLD ADAPTATION ON THE MODULATORY ACTION OF TRH UPON TEMPERATURE SENSITIVE AND INSENSITIVE HYPOTHALAMIC NEURONS OF RATS

B. Tzschentke, J. Schenda and Fr.-K. Pierau

Humboldt-Universität zu Berlin, FB Biologie, Invalidenstr. 43, D-10115 Berlin, FRG

Introduction

Adaptation to a cold environment changes the capacity of the autonomic and behavioural effector system and/or of the regulatory characteristics of the thermoregulatory system (1). The preoptic region of the anterior hypothalamus (PO/AH), known to be important in temperature regulation, contains a complex population of neurons that are sensitive to changes in local temperature (2). Besides it has been demonstrated that neuropeptides, e.g. bombesin and thyrotropin releasing hormone (TRH) change the temperature sensitivity of PO/AH neurons, resulting in an increase of the temperature coefficient (TC = imp $\cdot$ s^{-1} $\cdot$ $°C^{-1}$) of warmsensitive (WS) and temperature insensitive (TI) neurons and a transformation of TI into WS neurons (3, 4).

Earlier results about the influence of TRH on hypothalamic temperature sensitivity and thermoregulatory responses are controversial (5, 6, 7). Moreover, TRH administered into the brain may influence thermoregulatory response differently at different ambient temperatures. For example, intraventricular application of TRH in rats on the one hand produced hypothermia at low and medium ambient temperature (Ta) and on the other hand hyperthermia at high Ta (8).

The aim of the present study was to investigate whether adaptation to different Ta's affects the modulatory action of TRH on PO/AH neurons of rats' brain slices.

Method

Two groups of 9- to 12-month-old male Wistar rats (body weight 300-500 g) were used. One group of rats was exposed to Ta of 5°C (cold adapted) in a climatic chamber for 3-6 weeks. The other group was kept at 21°C (control rats).

On the day of the experiment, the rats were decapitated and the brain carefully removed. Brain slices including PO/AH (400 um thick) were prepared and incubated at 35°C in a chamber with oxygenated artificial cerebrospinal fluid (ACSF) for 2 hours. Then the brain slices were transferred into the recording chamber (detailed description 9) and continously perfused with ACSF using a peristaltic pump with a rate of 2 ml · min^{-1}. The bath temperature in the recording chamber (volume 0.5 ml) was at first maintained at a constant level of 38°C by means of a Peltier device and measured continously by a small thermocouple. Then the bath temperature was sinusoidally changed between 35 and 41°C at a velocity of about 0.02°C · s^{-1}. TRH was superfused (10ng/ml ACSF) over a period of 30-40 min. Sinusoidal temperature changes were applied 2-3 times before (control) and during TRH-superfusion.

Extracellular recordings were made with glass covered platinum-iridium electrodes. Firing rate (FR) and temperature were recorded and stored on an oszillograph and a digital recorder and also fed into a computer for off- and on-line data evaluation. The temperature response curve of each neuron was evaluated by a computer program relating FR to slice temperature and fitted to piecewise and/or rectilinear regression functions. Thermosensitivity of a neuron was defined by a TC of an absolute value >/= 0.6 Imp · s^{-1} · $°C^{-1}$, with a positive sign for WS and a negative sign for cold sensitivity (CS). All other neurons are temperature insensitive (TI) by this definition.

Results

The effect of TRH on FR and TC of 19 PO/AH neurons in cold adapted rats and 16 PO/AH neurons in control animals was investigated. Table 1 shows the number of WS, CS and TI neurons, found in the preoptic area of the investigated two groups of rats. Only one unit was cold sensitive and was not included in this study.

Tab. 1: Number of investigated warmsensitive (WS), coldsensitive (CS) and temperature insensitive (TI) PO/AH neurons of rats' brain slices

investigated PO/AH neurons (n)	total	WS	CS	TI
control rats	16	7	-	9
cold adapted rats	19	6	1	12

Fig. 1 shows an example of the influence of TRH on FR and TC of a TI neuron in an animal kept at 21°C. In animals kept at 21°C TRH superfusion increases the TC in most TI and WS neurons, resulting in a transformation of 5 of the 9 TI neurons into WS neurons (Fig. 2, left part).

In cold adapted rats the TC was only slightly increased in a few neurons, resulting in a transformation of 1 of the 12 TI neurons into a WS neuron (Fig. 2, right part).

Immediately after TRH superfusion the FR was elevated in 14 WS and TI neurons of the 16 investigated neurons of animals kept at 21°C and in 12 WS and TI neurons of the 19 investigated neurons of cold adapted rats as shown in Fig. 1. However, a relation between the elevation of FR after TRH superfusion and changes of temperature sensitivity was not ascertainable.

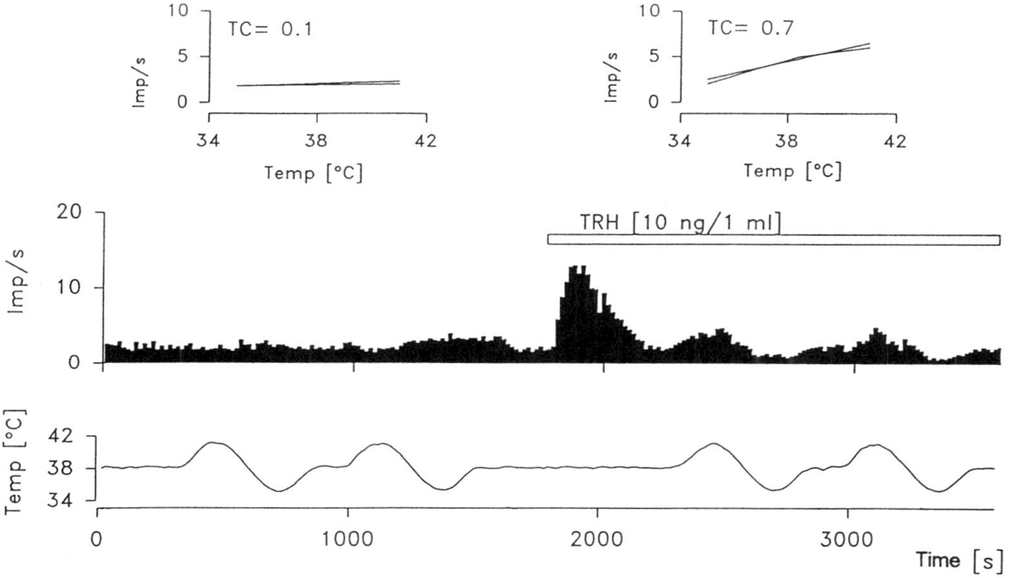

Fig. 1: Transformation of a temperature insensitive hypothalamic neuron into a warm sensitive neuron after TRH superfusion in a rat kept at 21°C. The lower part shows the temperature protocol and the temperature dependent firing rate before and after TRH superfusion. The upper part shows the relation between ambient temperature and firing rate. (TC=temperature coefficient)

Discussion

The effects of TRH in rats kept at 21°C are similar to findings about the influence of bombesin on WS and TI neurons in rats` hypothalamic slices (3, 4). Similar to bombesin TRH increases the TC of the majority of both WS and TI neurons, resulting in a transformation of 5 out of 9 TI neurons into WS neurons. Both endogenous substances produce a hypothermia under certain circumstances. Bombesin injected into the PO/AH produces a fall in body temperature in homoiothermic animals at room temperature and low Ta (3). A TRH induced hypothermia was decribed in cats (10) and ground squirrels (8). Using a behavioural paradigm in ground squirrels the TRH induced hypothermia appeared to be achieved by lowering the thermoregulatory set point (7). The influence of TRH on hypothalamic neurons depends on dose (10), loci (nuclei) of the PO/AH (11) and species (7, 8, 10). Intraventricular application of TRH produced hypothermia due to decreased metabolic heat production and cutaneous vasodilatation in rats at low and medium Ta`s (8 and 22°C) and hyperthermia due to increased heat production by behavioural excitation at 30°C (8). This suggests that the thermoregulatory reaction to TRH application into the brain is different at different Ta.

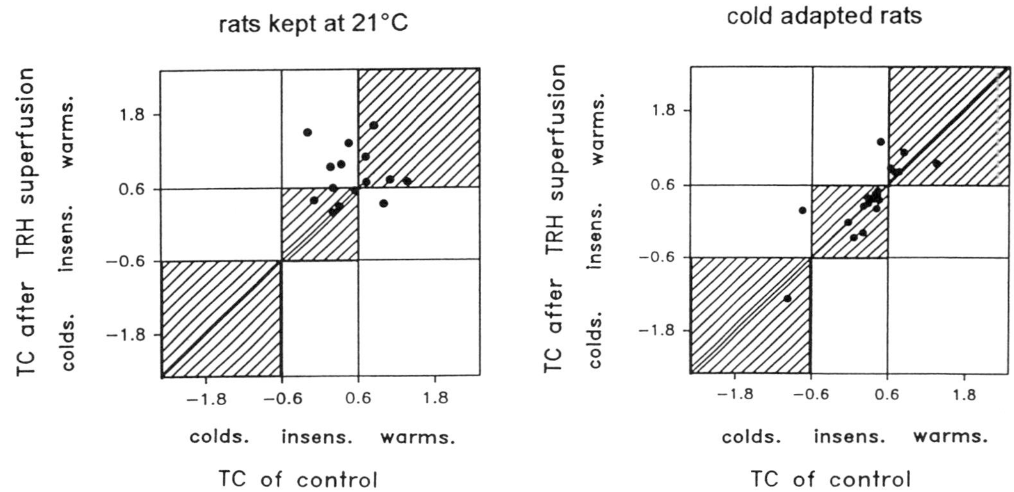

Fig. 2: Different effects of TRH superfusion on the thermosensitivity of PO/AH neurons in rats kept at 5°C (cold adapted) and rats kept at 21°C. The temperature coefficient (TC) after TRH superfusion (ordinate) is plotted against the TC during the control period (abscissa). The distance of single points from the line of identity correlates to the degree of the TC deviation after TRH superfusion. All points in the clear areas indicate neurons transformed into another category. The increase of the TC after TRH is more pronounced in control rats resulting in the transformation of the majority of TI into WS neurons.

Adaptation to different Ta appears to change the modulatory effect of certain peptides on hypothalamic neurons. For example, the proportion of TI neurons in which the TC was not changed after bombesin was higher in cold adapted rats (46%) as in animals kept at 21°C (30%). This could indicate a lower sensitivity of TI neurons to bombesin in cold adapted rats (12).

In the present experiments, the TC was only slightly increased by TRH in a few neurons of cold adapted animals. After TRH only 1 out of the 12 TI neurons was transformed into a WS neuron. The increase of TC and the transformation of a number of TI neurons into WS neurons was the main effect of TRH and bombesin observed in rats kept at 21°C suggesting an increase of the effectiveness of the regulatory "warm pathway" (3). Cold adaptation appears to decrease the sensitivity of TI neurons to TRH, similar to the change of responsiveness to bombesin observed after cold adaptation. A decrease of the effectiveness of the regulatory "warm pathway" after cold adaptation might be a possible adaptive change of the hypothalamic neuronal network involved in the control of body temperature. It appears that the reaction of the thermoregulatory neural mechanisms of the PO/AH depends more on temperature sensitivity of an individual neuron than on its change in tonic activity since the number of neurons in which the FR was increased was similar in cold adapted rats and control rats. Previous experiments have also suggested that the thermoregulatory effects of endogenous modulators is primarily due to their ability to change the temperature sensitivity of hypothalamic neurons (3, 12).

The results suggest that cold adaptation changes the sensitivity of PO/AH neurons to temperature as well as to the modulatory action of peptides, proposed to participate in the control of body temperature.

References

1. Brück, K (1986): Basic mechanisms in thermal long-term and short-term adaptation. J therm Biol 11, 73-77.

2. Boulant, JA, Dean, JB (1986): Temperature receptors in the central nervous system. Ann Rev Physiol 48, 639-654.

3. Pierau, Fr-K, Schmid, H (1990): Peripheral and central thermosensitivity. J Basic Clin Physiol Pharmacol 1, 323-335.

4. Schenda, J (1993): Einfluß von Temperatur, Neuropeptiden und Capsaicin auf die Aktivität und Temperaturempfindlichkeit von hypothalamischen Neuronen der Ratte - eine elektrophysiologische Studie an Hypothalamusschnitten. Inauguraldissertation, Giessen.

5. Salzman, SK, Beckmann, AL (1981): Effects of thyrotropin releasing hormone on hypothalamic thermosensitive neurons of the rat. Brain Res 7, 325-332.

6. Hori, T. (1991): An update on thermosensitive neurons in the brain: From cellular biology to thermal and non-thermal functions. Japan J Physiol 41, 1-22.

7. Hendriksen, JH, Bachelor, PA, Newman, RJ, Stanton, TL (1992): Thyrotropin-releasing hormone action in the preoptic/anterior hypothalamus decreases thermoregulatory set point in ground squirrels. Regulatory Peptides 38, 55-69.

8. Lin, MT, Chandra, A, Chern, YF, Tsay, BL (1980): Effects of thyrotrophic-releasing hormone (TRH) on thermoregulation in the rat. Experientia 36, 1077-1078.

9. Schmid, H, Jansky, L, Pierau, Fr-K (1993): Temperature sensitivity of neurons in sclices of the rat PO/AH hypothalamic area: effect of bombesin and substance P. Am J Physiol (Regulatory integrative Comp Physiol 33) 264, R449-R455.

10. Metcalf, G (1974): TRH: a possible mediator of thermoregulation. Nature 252, 310-311.

11. Boschi, G, Rips, R (1981): Effects of thyrotropin releasing hormone injections into different loci of rat brain on core temperature. Neuroscience Letters 23, 93-98.

12. Pierau, Fr-K, Schenda, J, Tzschentke, B, Konrad, M, Sann, H (1993): Adaptive changes of hypothalamic thermosensitive neurons after exposure to cold environmental temperature. In: Life in the cold. Carey, C, Florant, GL, Wunder, BA, Horwitz, B eds. Boulder, San Francisco, Oxford: Westview Press, 323-329.

Thermal Balance in Health and Disease
Advances in Pharmacological Sciences

THERMOREGULATORY CHANGES AFTER NEUROCHEMICAL LESIONS OF CATECHOLAMINERGIC AND SEROTONERGIC NEURONS IN THE LOWER BRAIN STEM OF THE GUINEA PIG

J. Jöckel and E. Zeisberger

Physiologisches Institut, Klinikum der JLU Giessen, Aulweg 129, 35392 Giessen, Germany

Much experimental evidence indicating an antagonistic modulatory action of the monoamines 5-hydroxytryptamine (5-HT) and noradrenaline (NA) on hypothalamic thermosensitive and thermointegrative neurons, as well as their roles in thermoregulation and thermal adaptation has been reviewed previously (1). Although our knowledge concerning the participating neurochemical transmitter systems has significantly increased, our understanding of their functional role in different species is yet very limited. In different species the thermoregulatory and adaptive strategies may be different and seem to depend on animals´ individual or the inherited experience (2). In the guinea pig, we described the changes in regulatory characteristics, defined as deviations in threshold and gain of thermoregulatory responses to thermal changes during the ontogenic development and adaptation to warm and cold climates (3). In previous experiments we tried to influence these responses by monoamines microinjected or microinfused into the thermointegrative hypothalamic area (cf. refs. 1 and 2 for reviews). Opposite changes could be evoked by microinjections of substances blocking the aminergic receptors in this area or by neurotoxins specifically destroying the synaptic structures (4, 5). From these experiments it was concluded that in the guinea pig the hypothalamic release of NA due to activation of brain stem afferents may increase body temperature, whereas the blockade of hypothalamic adrenergic receptors or a neurochemical lesion induced by microinjected 6-hydroxydopamine (6-OHDA) reduced body temperature. As a functional antagonist of NA, 5-HT also decreased body temperature. Since intrahypothalamically administered neurotoxins might have destroyed not only the terminals of specific aminergic afferents, but also modified the function of thermointegrative neurons in this area, the conclusions were verified by the present experiments. In these experiments the neurotoxins were injected into the areas in the lower brain stem from which the ascending catecholaminergic or serotonergic pathways originate, far away from the thermointegrative hypothalamic neurons.

Methods

Guinea pigs reared in our animal house at 23° C ambient temperature were used for the experiments. The adult animals (280-400 g) were anesthetized and submitted to a bilateral stereotactical implantation of guide cannulae into the catecholaminergic nucleus subcoeruleus, or into the serotonergic raphe nuclei of the lower brain stem. A week after this operation, the thermoregulatory experiments were started. During the experiments the animals were placed in a respiratory chamber controlled at 30° C. Metabolic rate (MR) calculated from oxygen uptake, integrated electrical activity (EMG) from the thigh muscle, heart (HR) and respiratory (RR) rates, chamber, various core and subcutaneous temperatures, as well as respiratory evaporative heat loss (REHL) calculated from the water content in the air inflow and outflow, were measured and recorded continuously using a computer system. Responses to external cooling (from 30° to 5° C) and heating (from 30° to 40° C) were compared before and 2 h after the neurochemical lesion (acute stage of lesion). The responses to similar cooling or heating were repeatedly tested 3, 8 and 14 days thereafter (chronic stage of lesion). The lesions of catecholaminergic structures were induced by bilateral injections (1 µl) of 6-OHDA (8 µg/µl) repeated three times in intervals of 30 minutes. In previous investigations (5) the effectiveness of the 6-OHDA dose (8 µg/µl) microinjected once in a volume of 1µl into the hypothalamus was demonstrated by a total loss of catecholaminergic fluorescence in this area and by the evidence of degenerative changes of the pre- and postsynaptic structures and the reduced stores of transmitter by means of electron microscopy. The lesions of serotonergic structures were induced by bilateral injections (1 µl) of 5, 6-dihydroxytryptamine (5, 6 DHT, 1 µg/µl) repeated twice in an interval of 60 minutes. Threshold temperatures for onset of shivering (TbSh) were determined by relating the cold-induced start of the increase in MR and EMG to mean body temperature, calculated from the arithmetical mean of subcutaneous and colonic temperatures. Threshold temperatures for onset of heat loss were estimated by relating the start of the increase in REHL during external heating to mean body temperature. The slopes (m) of thermoregulatory responses to the decrease resp. increase of mean body temperature during external cooling resp. heating were also estimated and compared. After the experiments the animals were killed, and the positions of injecting cannulae in the lower brain stem were verified.

Results

Effects of 6-OHDA lesions

The acute effects of 6-OHDA lesion on heart, respiratory and metabolic rates, EMG, core and subcutaneous temperatures in guinea pigs kept at 30° C are shown in Fig. 1 (left).

In 28 animals investigated in this series the Tb decreased during the first 2 h after the start of 6-OHDA administration from 38.45° C to 37.44° C due to strongly reduced activity in peripheral sympathetic nerves as evidenced by the decrease in heart rate from 275 to 113 and by the reduced difference between core and shell temperatures from 0.3 to 0.08° C, leading to an about 40% higher heat transfer to the skin. All these changes developed in spite of the constant ambient temperature of 30° C. Thereafter the responses of animals to cooling and heating were tested and compared with the control responses tested before the administration of the neurotoxin. The threshold temperatures for the cold-induced increase in MR and heat-induced increase in REHL as well as the slopes of the thermoregulatory responses to cooling and heating before and 2 h after the administration of 6-OHDA are compared in Fig. 2 (top).

In this acute stage of lesion the animals started to shiver and to increase their metabolic rate at a 1.2° C lower TbSh than during a control test before the lesion. They responded to cooling less sensitively, i.e. with t he decreased slope. On the other hand they started to pant and increase the

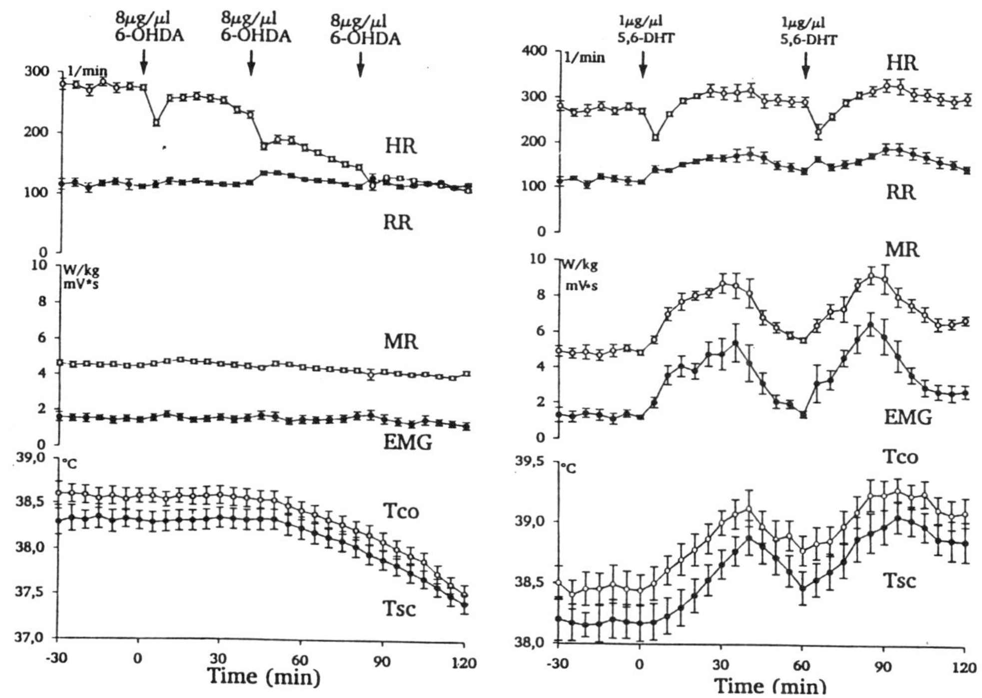

Fig. 1: The acute effects of 6-OHDA (left) and 5,6-DHT (right) microinjected into two different areas of the lower brain stem on heart, respiratory and metabolic rates, EMG, colon and subcutaneous temperatures in guinea pigs kept at 30° C.

REHL at a 0.6° C higher Tb. Thus 2 h after the 6-OHDA lesion Tb decreased by 1° C, but the interthreshold zone between TbSh and TbREHL has been widened from 1.16 to 3.01° C. These changes were only temporary, and 3, 8 and 14 days after the lesion the thresholds and slopes of thermoregulatory responses resembled those found before the lesion. A comparison of thresholds and slopes before and in different intervals after the 6-OHDA lesion is shown in Fig. 3 (top).

Effects of 5,6-DHT lesions

The acute effects of 5,6-DHT lesion on heart, respiratory and metabolic rates, EMG, core and subcutaneous temperatures in guinea pigs kept at 30° C are shown in Fig. 1 (right). In 10 animals investigated in this series the Tb increased during the first 2 h after the start of 5,6-DHT administration by 0.6° C due to increased heat production evidenced by increased electrical muscle activity as well as metabolic, heart and respiratory rates. Two hours after the lesion the animals responded more sensitively (higher slopes) to thermal loads and at elevated TbSh (by 1.4° C) and TbREHL (by 1.7° C). These acute changes are shown in Fig. 2 (below). They persisted also in the chronic stage of the lesion as shown in Fig. 3 (below).

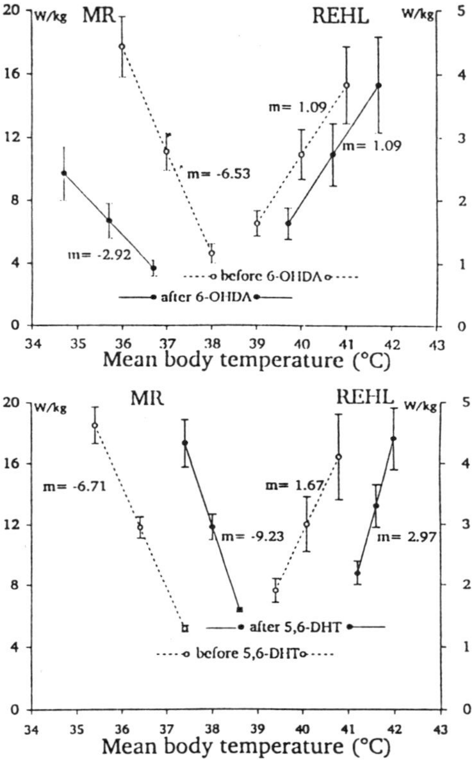

Fig. 2: Mean threshold temperatures (lowest values) for the cold-induced increase in MR and heat-induced increase in REHL as well as the slopes (m) of the thermoregulatory responses to cooling and heating before (open symbols,dotted lines) and 2 h after 6-OHDA (top), resp. 5,6-DHT (below) lesions (filled symbols, solid lines).

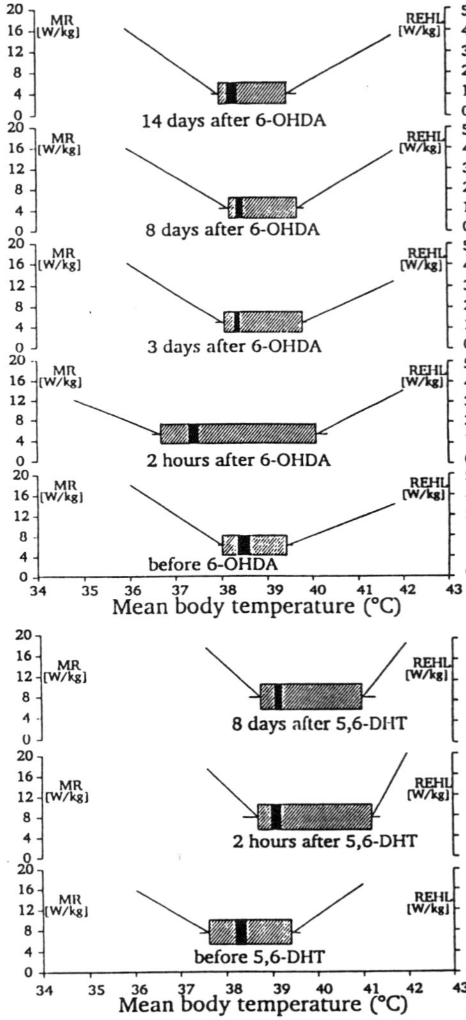

Fig. 3: A comparison of changes in thermoregulatory threshold temperatures for heat production (left borders of the shaded bars) and for heat loss (right borders of the shaded bars) before and in different time intervals after the neurotoxin lesions. The dark areas within the shaded interthreshold zones indicate the initiate mean body temperature of the guinea pigs before the start of cooling or heating tests. The lines indicate the slopes of the responses to cooling or heating. The effects of 6-OHDA lesion are shown on the top, those of 5,6-DHT lesion below.

Discussion

The results support previous experiments summarized in recent reviews (1,2). Electrolytic lesions in the serotonergic nucleus raphe magnus evoked similar increases of TbSh as specific neurochemical lesions by means of 5,6-DHT in our experiments, but failed to increase the TbREHL (6). This difference may be caused by damage of other than just 5-HT afferents by electrolytic lesion. Moreover, the electrolytic lesion may have been better focussed, and therefore some serotonergic pathways from other nuclei may have remained active thus keeping the TbREHL low. Concerning the functional roles of the brain stem aminergic afferents to the hypothalamus, the results support the view of a general neuromodulatory function of these systems (2). The neurochemical lesions of catecholaminergic afferents had only short-term effects and no permanent functional impact on the thermoregulatory system. This confirms the previous reports (4). The rapid compensation of the loss of one modulatory system can be explained by an adaptive change in the activity of the other antagonistic system, as it was already proposed for hypothalamic lesions (5). New is the finding that lesions of the serotonergic afferents in the lower brain stem had permanent functional impact on TbREHL which remained elevated also a week after the lesion. This may indicate that the neurons responsible for the activation of heat loss do not get antagonistic modulatory inputs, or that these inputs are not of the same importance. It is not clear whether this is a general organizational principle or only specific for the guinea pig which has a rather poor capacity for heat loss. The effects of lesions can be revealed only in functional tests (by determining threshold temperatures for initiation of heat production or dissipation), because the effects on resting body temperature are small (cf. Fig. 3). Besides in thermoregulation these modulatory systems seem to participate in a number of neurovegetative reflexes (7), stress-induced release of hormones (8) and in the control of antipyretic neuronal systems (9).

Conclusion

These results support the view that in the guinea pig catecholaminergic and serotonergic afferents to the hypothalamus have antagonistic, modulatory roles in the control of heat production. Activation of catecholaminergic afferents increases the TbSh and elevates body temperature. The destruction of this system leads therefore to a decrease of TbSh and body temperature. This decrease is transient, because it can be compensated by reduced activity in antagonistic, serotonergic afferents. The onset of heat loss is turned down merely by the activity of serotonergic afferents. Their destruction thus increases the TbREHL, a change which cannot be compensated.

Acknowledgement

This study was supported by the Deutsche Forschungsgemeinschaft (Project Ze 183).

References

1. Brück, K, Zeisberger, E (1990): Adaptive changes in thermoregulation and their neuropharmacological basis. In:Thermoregulation: Physiology and Biochemistry. Schönbaum, E, Lomax, P eds. New York: Pergamon Press, 255-307.

2. Zeisberger, E, Roth, J (1994): Central regulation of adaptive responses to heat and cold. In: APS Handbook of Physiology, Section 4, "Adaptation to the Environment". Fregly, MJ, Blatteis, CM eds. Bethesda: Oxford University Press, - in press-.

3. Zeisberger, E, Ewen, K (1983): Ontogenetic, thermoadaptive and pharmacological changes in threshold and slope for thermoregulatory reactions in the guinea-pig (Cavia aperea porcellus). J therm Biol 8, 55-57.

4. Behr, R, Zeisberger, E, Merker, G (1983): Response of the guinea-pig (Cavia aperea porcellus) to external cooling after aminergic denervation of the anterior hypothalamus. J therm Biol 8, 125-128.

5. Zeisberger, E, Behr, R, Ewen, K, Merker, G (1983): Threshold changes for heat production and dissipation after intrahypothalamic administration of 6-hydroxydopamine. In: Environment, Drugs and Thermoregulation. Lomax, P, Schönbaum, E eds. Basel: Karger, 58-61.

6. Szelényi, Z, Hinckel, P (1987): Changes in cold- and heat-defence following electrolytic lesions of raphe nuclei in the guinea pig. Pflügers Arch 409, 175-181.

7. Cunningham, JT, Nissen, R, Renaud LP (1992): Catecholamine depletion of the diagonal band reduces baroreflex inhibition of supraoptic neurons. Am J Physiol 263, R363-R367.

8. Richardson Morton, KD, Van de Kar, LD, Brownfield, MS, Lorens, SA, Napier, TC, Urban, JH (1990): Stress-induced renin and corticosterone secretion is mediated by catecholaminergic nerve terminals in the hypothalamic paraventricular nucleus. Neuroendocrinology 51, 320-327.

9. Unger, M, Merker, G, Roth, J, Zeisberger, E (1991): Influence of noradrenergic input into the hypothalamic paraventricular nucleus on fever in the guinea-pig. Pflügers Arch 419, 394-400.

Thermal Balance in Health and Disease
Advances in Pharmacological Sciences

FUNCTIONAL NEUROANATOMICAL CONSIDERATIONS ON AFFERENT AND EFFERENT PROJECTIONS OF THE NUCLEUS RAPHE MAGNUS

D. M. Hermann, P. Hinckel, P. H. Luppi* and M. Jouvet*

Physiologisches Institut der Justus-Liebig-Universität Giessen, Aulweg 129, 35392 Giessen, Germany, *CNRS URA 1195, 8 av. Rockefeller, 69373 Lyon, France

Introduction

The nucleus raphe magnus (NRM) is part of the thermoafferent system of the brainstem (1, 2). Thermoregulatory effects have been shown medially inside the NRM, not at all however in lateral parts of the NRM and in the gigantocellular medullary reticular formation (2). Adaptive effects to cold temperatures have been demonstrated in the NRM (3, 4). These effects are still partly maintained after complete microtranssection of the descending afferents to the NRM, suggesting that they are partly generated by descending influences from upper structures inside the brain, partly by structures of the lower brainstem. An aim of this study has been to find the structures involved in thermoadaptive effects inside the NRM.

Besides in thermoregulation, the NRM plays a role in the regulation of the spinal transmission of the pain response. By electrophysiological recording inside the NRM, we have obtained the impression that those neuronal units responsive to mechanical pain stimuli are mostly localized inside the serotonergic lateral caudal NRM, not at all however inside the dorsal caudal NRM, which contains very few serotonergic neurons. Neuronal units have been defined as pain stimulated, if their spike rate at least doubles when pinching the legs and ears of the animal as a pain stimulus, and as pain-inhibited, if the spike rate at least decreases by factor two.

Five of eight neuronal units for the lateral caudal NRM were pain-responsive (Fig. 1). Three of these units were inhibited by pain, two units were stimulated by the pain stimulus. The pain-inhibited units showed small-sized negative electrical spikes with a firing rate of about 10/s, which decreased to 2-5/s during application of the pain stimulus. One stimulated unit showed phasic activity during the pain stimulus and completely ceased firing after stimulation.

For the dorsal caudal part of NRM, none of 16 neuronal units were pain-responsive. Three pain-responsive units were found at the most ventral serotonergic limit of the NRM only; two of these were inhibited, one stimulated by a pain stimulus. The hypothesis, that mechanically pain-responsive neuronal units are equally distributed throughout caudal NRM, could not be maintained (chi-square=12.63; p less 0.1%).

In the serotonergic rostral part of the NRM, three of 15 neuronal units were pain-responsive; two units were pain-stimulated, one was pain-inhibited. One aim of this tracing study was to find out if there is a neuroanatomical correlate to these functional observations.

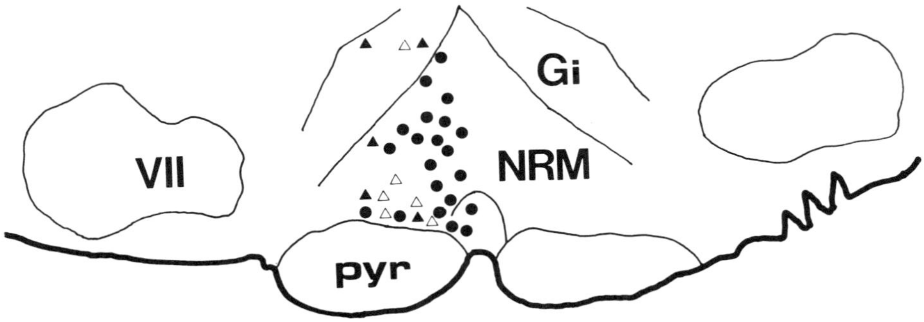

Fig. 1: Localization of pain-stimulated (filled triangle), pain-inhibited (empty triangle) and unresponsive (filled circle) neuronal units inside the caudal NRM of the rat

Unconjugated cholera toxin (subunit b) by iontophoretic application offers important advantages as a retrograde neuronal tracer (5). Unconjugated cholera toxin b is considered not to be taken up by passing fibre bundles without synaptic terminals, as long as these fibre bundles have not been damaged during the experiment. Iontophoretic application of unconjugated cholera toxin b largely guarantees injections without histological damage of the brain tissue. For application of the tracer very tiny micropipettes were used in combination with a very low current of injection in order not to harm the brain tissue.

For application to the central nervous system and immunocytochemical detection, cholera toxin b has a sensitivity which is substantially higher even compared with wheat germ agglutinin, another very sensitive retrograde tracer. Unconjugated cholera toxin by iontophoretic application allows to obtain very small and sharply limited injection sites; in our experiments, injection sites with a diameter of only 250 µm still offered excellent tracing results. Until now, it has neither been possible to distinguish the caudal NRM from the rostral nucleus raphe pallidus (NRP), nor to distinguish the rostral NRM from the caudal NRM by means of a tracing method.

Materials and Methods

Twenty-one adult male Sprague-Dawley rats were examined during pentobarbitone-anaesthesia. Unconjugated cholera toxin (subunit b, LIST Biological Laboratories) was applied iontophoretically by a positive current of 0.5-1 µA over 10-15 minutes in alternating function, using a glass-micropipette (external diameter of the tip 4-5 μm) containing a 1 % solution the tracer in 0.1 M phosphate buffer (PB). The cholera toxin b had been processed according to the method of Luppi before application at pH 6.0. The localization of the pipette inside the brain was verified by electrophysiological recording. Pinching the four legs and the ears was used as pain stimulus. After injection of the tracer, the micropipette remained in the brain for 10 minutes before removal in order not to trace the path of the pipette.

After surgery animals were allowed to survive for 6 to 19 days for retrograde transport, and 12 to 19 days for anterograde transport studies. The animals were sacrified in deep pentobarbitone anaesthesia by transcardial perfusion by Hartmann's solution, followed by 500 ml of fixative at 4 °C containing 4 % paraformaldehyde, 0.1 % glutaraldehyde and 0.2 % picric acid in 0.1 M PB. The brain was removed, postfixed overnight at 4 °C in 0.1 M PB with 2 % paraformaldehyde and 0.2 % picric acid, and rinsed for 48-72 hours in 0.1 M PB containing 30 % sucrose. Coronal 20 μm sections were cut on a freezing microtome, rinsed twice in phosphate buffered saline containing 0.1 % Triton X-100 (PBS, PBS-T) and 0.1 % sodium azide and stored for at least three weeks in PBS-T until immunohistochemical staining for the study of anterograde transport.

Immunohistochemical reactions of free-floating sections were carried out by sequential incubation in goat antiserum to cholera toxin (subunit b, LIST Biological Laboratories) over 60 hours at 4 °C, diluted 1:40000 in PBS-T containing 0.1 % sodium-acide, in biotinylated swine antigoat immunoglobulin (DAKOPATTS) overnight at 4 °C, diluted 1:2000 in PBS-T, and in streptavidin conjugated with horseradish peroxidase over 90 minutes at room temperature, diluted 1:40000 in PBS-T. Following each step sections were rinsed twice in PBS-T for 30 minutes each. The sections were immersed in 0.02 % 3,3'-diaminobenzidine .4HCl (DAB, Sigma), 0.3 % nickel ammonium sulfate and 0.003 % H_2O_2 in 0.05 M Tris-HCl buffer pH 7.6, processed under microscopical control, and rinsed twice in PBS-T with 0.1 % sodium azide to stop the staining reaction. The sections were mounted on gelatine-coated glass-slides, dried, dehydrated using graded alcohols, and cover-slipped.

Tracing Results

After injection of unconjugated cholera toxin b into the rostral NRM, intense terminal efferent fibres could be observed inside the nucleus accumbens, the preoptic area, the bed nucleus of the stria terminalis, the substantia innominata, the magnocellular preoptic nucleus, the paraventricular hypothalamic nuclei and the lateral hypothalamic area, particularly in its perifornical part (6). No terminal fibres at all were visible in these structures after injection of cholera toxin b into the caudal NRM and the rostral NRP. Experiments of control with injection of cholera toxin b into the preoptic area and the lateral hypothalamic area clearly showed that efferents of the NRM originated from the rostral NRM up to the level of the genu of the facialis nerve, not at all however in the caudal NRM and in the NRP.

It is noteworthy that terminal efferent fibres were imaged by thin thread-shaped structures, on which bead-like varicose formations, which obviously seemed to represent synaptic formations including both boutons en passant and terminal synapses, were clearly visible. Terminal efferents could often be traced for some hundred micrometers inside a single slice of brain tissue. After injections into the caudal NRM and the rostral NRP, almost no efferent fibres were visible throughout the forebrain.

For the dorsal caudal part of the NRM, substantial afferents could be found inside the preoptic area, the bed nucleus of the stria terminalis and the lateral hypothalamic area; for the rostral NRM additional afferents could be observed in the nucleus accumbens, the septal nuclei, the horizontal limb of the diagonal band, the substantia innominata, in prefrontal cortical structures and the insular as well as the perirhinal cortex (Fig. 2). Very few afferents were seen throughout the forebrain after injection of cholera toxin b into the lateral caudal part of the NRM.

For the rostral NRM and the dorsal caudal NRM, substantial afferents were found in the subcoeruleus area, the medial parabrachial area and the lateral parabrachial area (Fig. 3). These afferents were much less intense for the lateral caudal NRM. The subcoeruleus area is involved in the thermoafferent system of the lower brainstem (2).

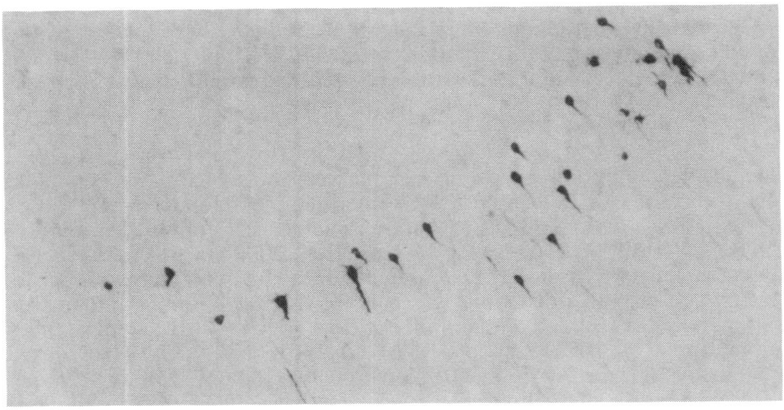

Fig. 2: Photograph illustrating blackly-stained afferent cells inside the insular cortex after injection of cholera toxin b into the rostral NRM

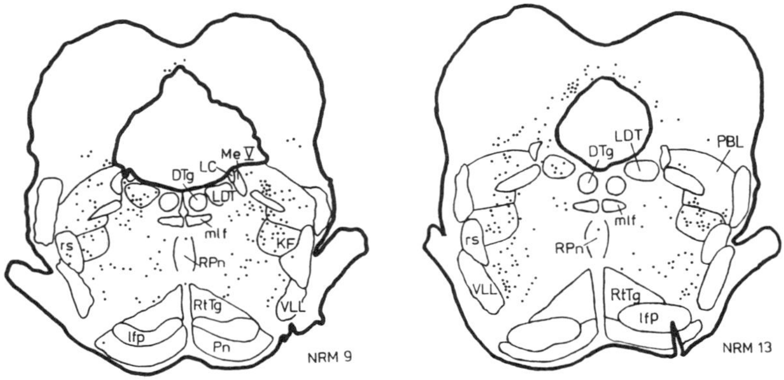

Fig. 3: Afferents to the dorsal caudal NRM (NRM 9) and the lateral caudal NRM (NRM 13) at the level of the subcoeruleus area

For the lateral caudal NRM, strong afferents with ipsilateral dominance were observed in the midbrain periaqueductal gray and the adjacent deep mesencephalic reticular formation, projections which play an important role in descending antinociception, further in the oral pontine reticular formation, in the Kölliker-Fuse nucleus, in the caudal pontine reticular formation and in the gigantocellular medullary reticular formation.

Discussion

Our study shows that unconjugated cholera toxin b by iontophoretic application is an excellent, highly sensitive retrograde neuronal tracer. Performing restricted injection sites into the NRM we have been able to show substantially different afferent projection patterns for the rostral NRM, the dorsal caudal NRM, the lateral caudal NRM and the rostral NRP. Our study further shows that unconjugated cholera toxin b by iontophoretic application is very useful besides retrograde for anterograde neuronal tracing to study efferent connections inside the brain. Combined retrograde and anterograde use might have an important application in the study of reciprocal neuronal connections, it allows to superpose afferent and efferent projection patterns, giving an accurate idea of bidirectional connections in the brain.

The fact that both specific efferent and afferent connections were described for the rostral, but not the caudal NRM, suggests closed loops of these efferent and afferent connections of the rostral NRM. This raises the question if these specific projections might also be involved in thermoadaption. At forebrain levels, thermoacclimatic effects might be generated by parts of the preoptic area or the basal forebrain, as these structures are key structures in central nervous thermoregulation.

Our tracing results give support to the idea that the rostral and the dorsal caudal NRM, but not the lateral caudal NRM, are involved in thermoregulation, since both the rostral and the dorsal caudal NRM in contrast to the lateral caudal NRM receive substantial afferents from the preoptic area, the lateral hypothalamic area and the subcoeruleus area, which play a central role in thermoregulation.

It is quite remarkable that efferents of the NRM to preoptic and lateral hypothalamic structures originate from the rostral, not at all however from the caudal NRM. This suggests that the rostral NRM rather than the caudal NRM is part of the thermoafferent system.

Our finding that the dorsal caudal part of the NRM might not at all be involved in the response to mechanical pain stimuli is in contrast to the traditional view, regarding the NRM as a homogeneous structure with a major role in descending antinociception.

Thermoadaptive effects inside the NRM are partly generated by structures of the lower brainstem. At this level, the subcoeruleus area, the parvocellular medullary reticular formation and the ventrolateral caudal medullary reticular formation must be considered to be involved in thermoadaption. Further electrophysiological experiments will be necessary to define those structures of the forebrain and the brainstem which are responsible for thermoadaptive effects in the NRM.

Summary

Afferent and efferent structures of the NRM have been examined in the rat by iontophoretic application of unconjugated cholera toxin b in combination with electrophysiological recording. Substantial differences of the afferents and efferents are described for the rostral NRM, the dorsal caudal NRM, the lateral caudal NRM and the rostral NRP. Structures are discussed which might be involved in thermoadaption. In contrast to the caudal NRM, the rostral NRM sends off intense efferents to forebrain structures, suggesting that rostral NRM rather than caudal NRM is part in the thermoafferent system of the brainstem. Neuroanatomical results give support to the idea that the caudal NRM itself consists of two functionnally different substructures, the dorsal caudal and the lateral caudal NRM.

References

1. Dickenson, A. H. (1977) Specific responses of rat raphe neurones to skin temperature. J. Physiol. (Lond.) 273:277-293

2. Brück, K., and P. Hinckel (1980) Thermoregulatory noradrenergic and serotonergic pathways to hypothalamic units. J. Physiol. 304:193-202

3. Hinckel, P., and K. Schröder-Rosenstock (1982) Central thermal adaption of lower brain stem units in the guinea-pig. Pflügers Arch. 395:344-346

4. Hinckel, P., and W.T. Perschel (1986) Influence of cold and warm acclimatisation on neuronal responses in the lower brain stem. Can. J. Physiol. Pharmacol. 65:1281-1289

5. Luppi, P.-H., P. Fort, and M. Jouvet (1990) Iontophoretic application of unconju-gated cholera toxin b subunit (CTb) combined with immunohistochemistry of neurochemical substances: a method for transmitter identification of retrogradely labeled neurons. Brain Res. 534:209-224

6. Hermann, D.M., P. Hinckel, P.H. Luppi, and M. Jouvet (in preparation) Efferents of the rostral nucleus raphe magnus in the rat by iontophoretic application of unconjugated cholera-toxin (subunit b)

Section II:

HEAT PRODUCTION

Thermal Balance in Health and Disease
Advances in Pharmacological Sciences

THE ROLE OF ADRENERGIC STIMULATION IN REGULATION OF HEAT PRODUCTION AND RECRUITMENT OF BROWN ADIPOSE TISSUE

B. Cannon, T. Bengtsson, A. Dicker, A. Jacobsson, P. Kuusela, H. Thonberg, P. Tvrdik, J. Zhao and J. Nedergaard

The Wenner-Gren Institute, The Arrhenius Laboratories F3, Stockholm University, S-106 91 Stockholm, Sweden

Despite the fact that it is only 30 years since thermogenesis was suggested to be the function of brown adipose tissue, the role of brown adipose tissue as a unique thermogenic organ in mammals is now well established. Kurt Brück (1) early realised its significance for thermoregulation of the newborn of all mammalian species, including the human infant, and it is therefore very appropriate to present an overview of present knowledge on brown adipose tissue in this symposium, dedicated to his memory. The role of the sympathetic nervous system as the means of activation of the acute heat production in brown adipose tissue has long been realised, but as the result of continuous study of the tissue, it has been understood that the sympathetic nervous system, through norepinephrine release, also plays a dominating role for the development of the tissue in response to e.g. prolonged cold exposure – i.e. for the so-called recruitment process. Certain circulating hormones, and some growth factors, may also influence tissue metabolism and development, but the role of these agents may mainly be seen as permissive, i.e. their presence may be necessary for the recruitment process but they do not in themselves constitute a sufficient signal for tissue recruitment; rather, the chronic adrenergic stimulation seems to have this role. It is therefore clear that an understanding of the responses of the tissue to adrenergic stimulation is the predominant prerequisite for an understanding both of the acute functioning of the tissue and of its recruitment in response to physiological stimuli. We will therefore in the following concentrate on the adrenergic receptors in the tissue and on their effects, both on the acute thermogenic response and on cell proliferation and differentiation during the recruitment process. As the number of permitted references is limited, we will generally refer the reader to more detailed reviews on each aspect. A comprehensive overview with detailed references is found in (2).

Adrenergic receptors

The plasma membrane receptors responsive to adrenergic stimulation belong to the family of G-protein-coupled receptors, consisting of a single polypeptide chain with seven putative membrane-spanning domains. The adrenergic receptors can be divided into three main categories – α_1, α_2 and β – based both on pharmacological criteria and on the coupling of these receptors via different G-proteins to different intracellular signal transduction pathways (3). There are three to four subtypes of each of these receptor categories, and most of the receptors have now been cloned and expressed in artificial systems allowing for a detailed analysis of the characteristics of each receptor. Of the members of this large family of adrenergic receptors, many have been reported to exist in brown adipose tissue, of at least one species. We will here summarise what is known today about the presence of these receptors in brown adipose tissue.

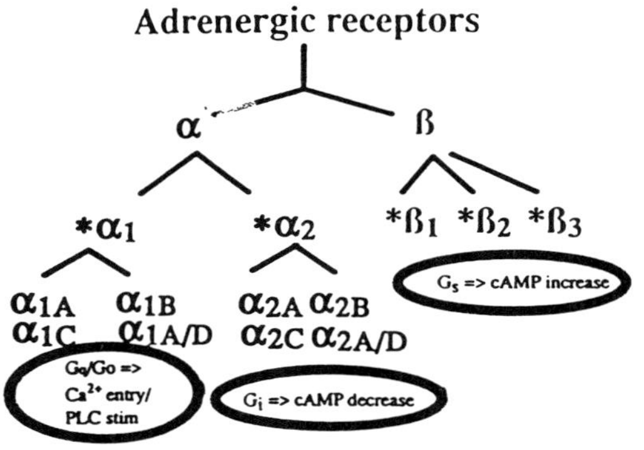

Fig. 1. The adrenergic receptor family ≈ 1993. Those members of the family which have been identified in the brown adipose tissue of at least one species are indicated with *.

α_1-Adrenergic receptors

With the help of the ^{3}H-labelled α_1-antagonist prazosin, α_1-receptors have been identified in hamster and rat brown adipose tissue. Their density is apparently positively correlated with the degree of recruitment of the tissue, such that the density is increased by cold acclimation and decreased in obesity (4). These observations can be interpreted to indicate an important role for α_1-receptors in tissue function. The α_1-receptors have not been studied sufficiently pharmacologically or by molecular biological techniques to enable identification of which α_1-receptor subtype is present in the tissue. Functionally, the α_1-receptors have been shown to be coupled via a G-protein, presumably G_q, to breakdown of phosphatidyl-inositol-bisphosphate, leading to inositol-trisphosphate production and through this to an elevation of intracellular Ca^{2-} concentration and to activation of Ca^{2+}-dependent K^+-channels. Simultaneously, protein kinase C is activated. Also plasma membrane Cl^- channel activity seems to be activated due to α_1-adrenergic stimulation.

The ultimate role of the α_1-receptors is not understood. It is evident that they may participate synergistically in activation of the expression of certain genes (see below), but a unique role in this context is still lacking, as is a physiological role for the stimulated ion fluxes.

α_2-Adrenergic receptors

The presence of α_2-receptors in brown adipocytes from rats and mice is indicated by functional observations, such as inhibition of cyclic-AMP production and inhibition of lipolysis through activation of these receptors. No studies have analysed which of the receptor subtypes is active. Verification of the presence of α_2-receptors on the cells has not been performed with radioligand binding. Brown adipose tissue from hamster does not appear to possess these receptors, at least not in animals living in control conditions, so they are not essential for tissue function. Whether their density changes as a result of physiological conditions is presently unknown, and thus the role of α_2-receptors in the control of the metabolism of brown adipose tissue is currently unknown.

β-Receptors

The three subtypes of β-receptors all couple positively to cyclic-AMP production through G_S-proteins, and all three subtypes have been identified in brown adipose tissue, at least of some species. Initially, by radioligand binding studies, β_1- and β_2-receptors were characterised in the tissue. Their presence has subsequently been confirmed by demonstration of the presence of mRNA for these subtypes in total tissue extracts. The presence of β_3-receptors can only be inferred from radioligand binding, but the presence of these receptors is unequivocal in that the mRNA is readily observed in the tissue (5), and the cells show β_3-stimulated responses.

The relevance of the tissue having all 3 β-receptor subtypes which all couple to cyclic-AMP production is not clear. While there are results indicating that cells in different stages of maturation may possess different relative concentrations of the receptors (see below), this does not in itself provide a functional explanation for their presence. Although the different subtypes also seem to desensitize by different mechanisms, it is not clear under what circumstances the tissue would use this distinction.

THE ACUTE THERMOGENIC RESPONSE

Newborn mammals respond to the experienced cold stress of the postnatal period by an elevation of oxygen consumption (= heat production). Hull and coworkers and Brück and coworkers showed convincingly in the early sixties that this response was mediated by norepinephrine and emanated mainly from the brown adipose tissue. Adult animals acclimated to a cold environment also defend body temperature by a norepinephrine-mediated increase in oxygen consumption, but it was not until 1977, through the work of Foster and Frydman, that it became generally accepted that also in adult animals, this sympathetically mediated heat production was localised to brown adipose tissue. A major role in the mediation is ascribed to β-adrenergic receptors, although α-receptors appears to have an accessory role, which has not been entirely clarified.

Thermogenic effects of stimulation of α_1-adrenergic receptors

Early experiments with isolated, mature brown adipocytes demonstrated that addition of norepinephrine to suspensions of such cells could mimic stimulation by the sympathetic nervous system, causing increased oxygen consumption. This was initially believed to be a fully β-receptor mediated effect, in that it was inhibited by the β-blocker propranolol. Later detailed analysis in hamster brown adipocytes elucidated a role also for α_1-receptors. In these cells, stimulation of α_1-receptors contributes some 15-20 % to the total norepinephrine-stimulated oxygen consumption. This stimulated oxygen consumption is to a large extent secondary to the energetic costs of the stimulated ion fluxes referred to above, i.e. due to an indirect activation of the Na^+/K^+-ATPase leading to increased ATP consumption and activation of oxidative phosphorylation in the mitochondria (6). It is doubtful that this process has thermogenesis at its primary goal.

Thermogenic effects of stimulation of β–adrenergic receptors

As understood above, although α_1-adrenergic receptors are present in rat brown adipocytes, their activation by norepinephrine does not contribute significantly to the stimulated oxygen consumption. Consequently, the thermogenic role of α_1-receptors would appear to be minor, and an understanding of the acute thermogenic mechanism can indeed be found primarily in the β-receptor mediated actions on the adipocytes.

Until recently, the β-adrenergic thermogenic response was classifed as β_1. This was initially based on analysis of dose-response curves for norepinephrine, epinephrine and isoprenaline. This classification was further verified by radioligand binding studies with the β-antagonist dihydroalprenolol and studies of competition for the dihydroalprenolol binding site by β-adrenergic agonists and antagonists. Extensions of this work showed that a fraction of the receptors in rat could be classified as β_2, while the majority were β_1. These binding studies were performed generally in membrane fractions of total tissue, but also on isolated, mature adipocytes.

Although the pharmacological characterisation of the dihydroalprenolol binding site was unequivocal, it did not agree in detail with results obtained for oxygen consumption studies and measurements of cyclic-AMP production in isolated cells. One serious point was that inhibition of the physiological responses required much higher doses of β-antagonists than did inhibition of radioligand binding. This discrepancy received its explanation in 1984 when Arch et al. demonstrated that the β-receptor which stimulates lipolysis in brown adipose tissue is atypical.

Many further studies have subsequently led to the cloning and expression of what is now termed the β_3-adrenergic receptor. While being some 50 % homologous to the β_1- and β_2-receptors, and having not dissimilar affinity to the β_1-receptor for classical agonists, it has several distinctive features (7). It demonstrates high affinity for some selectively developed agonists (e.g. the so-called BRL-compounds, such as BRL-37344); it has low affinity for nonselective β-antagonists, such as propranolol; and it shows agonist responsiveness with certain drugs earlier considered to be β_1- and β_2-antagonists (e.g. CGP-12177) (8). The β_3-receptor lacks many of the amino acid residues found in the third cytoplasmic loop and C-terminal region of the β_1- and β_2-receptors, which serve as substrates for the β-adrenergic receptor kinase and protein kinase A and whose phosphorylation leads to desensitisation of the receptor response. The β_3-receptor is therefore presumably less sensitive to this means of desensitisation. The number of β_3-receptors can, though, be regulated at the level of gene expression.

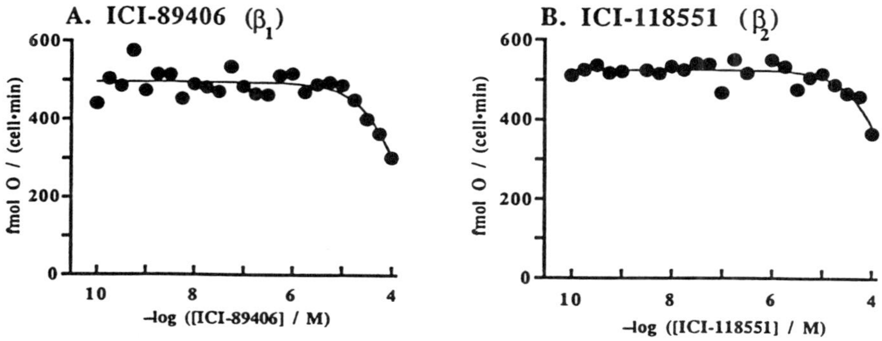

Fig. 2: The inability of selective β_1 and β_2 antagonists to inhibit norepinephrine-induced oxygen consumption in isolated hamster brown-fat cells. From (9).

The point of the existence of all these three β-receptors in the tissue is still not resolved. Our recent results do indicate that oxygen consumption in isolated brown adipocytes from hamster is exclusively stimulated by the β_3-receptors. We have been unable to find evidence for a contribution from either β_1- or β_2-receptors for the thermogenic response but these receptors may be coupled to other processes in the tissue. Thus, although β_1-receptors can be identified on the isolated adipocytes by radioligand binding, we have been unable to find evidence for any role in stimulation of oxygen consumption. Since the only well established second messenger for β-receptors is cyclic-AMP, and elevation of cyclic-AMP levels indeed correlates well with increases in oxygen consumption, it would appear that the β_1-receptors are not coupled via G_S-proteins to adenylyl cyclase in the isolated, mature cells. However, in membrane preparations from the entire tissue, β_1-receptor stimulation does lead to cyclic-AMP production (10). This would indicate that there are at least two populations of β_1-receptors in the tissue, one coupled to cyclic-AMP production (probably on young cells which still have a proliferative potential, see below), and one no longer coupled to cyclic-AMP production (on mature cells).

Intracellular signal transduction

While the localisation and functional coupling of β_1/β_2-receptors in the tissue is still unclear, β_3-receptors appear to exist only on mature adipocytes. The intracellular signal transduction of the acute thermogenic response in the mature adipocytes, via the β_3-receptor, is through the G_S-protein to adenylyl cyclase. This is confirmed by experiments with both cholera toxin and forskolin, and determination of cyclic-AMP production. Both long and short splicing variants of the $G_S\alpha$-protein are found in the tissue, and they may play a differential role in signal transduction, perhaps being active in different cells or at different times.

Lipolysis

Subsequent activation of protein kinase A by cyclic-AMP leads – among other things – to activation of hormone-sensitive lipase. This lipase appears to be identical to the one in white adipose tissue, and its magnitude is not massively increased by cold acclimation, presumably because its activity is well regulated by protein kinase A (11). The ensuing lipolysis provides free fatty acids, some of which seem to be exported from the tissue when lipolysis is maximally stimulated. Those remaining within the brown adipocyte are substrates for mitochondrial combustion and may play a direct or indirect role in activation of the mitochondrial uncoupling protein thermogenin.

The mitochondrial uncoupling protein thermogenin

It is the activation of the uncoupling protein thermogenin which leads to heat production. Despite intensive research over several decades, first on isolated brown-fat mitochondria, and later on isolated and reconstituted thermogenin, a molecular understanding of the function of this protein has not been reached (12).

For the uncoupling process to take place, it is clear that the activity must correspond functionally to a transfer of H^+ over the mitochondrial membrane. Whether this is in the form of H^+ itself or in the form of OH^- is not known; as thermogenin is able to transport Cl^- and even large anions, a simplistic view would favour OH^- transport. From studies of isolated thermogenin or heterologously expressed thermogenin it would seem that the uncoupling activity is only observable in the presence of free fatty acids, and the regulation would therefore be fairly simple to understand, had it not been that thermogenin also possesses an inhibitory purine-nucleotide binding site (often referred to as the GDP-binding site). Even in the presence of fatty acids, GDP can still inhibit the uncoupling action. Thus, a reasonable formulation would be that the presence of free fatty acids is necessary for uncoupling to occur, but the regulation occurs via the GDP-binding site. If this is so, a physiological activator of thermogenesis, different from free fatty acids, must be postulated; the nature of such an activator is not presently known.

There is now good agreement as to the supposed structure of thermogenin but this gives few clues to its function. In certain respects, the suggested structures are even difficult to reconcile with known features of thermogenin; it would e.g. seem that the GDP-binding site is found in a loop of the protein which is depicted as being close to the matrix site of the mitochondrial membrane, but it is functionally clear that GDP acts from the cytosolic side.

Lipoprotein lipase

As understood from above, activation of the β_3-adrenergic receptors leads to activation of hormone-sensitive lipase and subsequent lipolysis and this provides the substrate for thermogenesis. However, the triglyceride stores in the tissue are limited, and any prolonged thermogenic response necessitates supplementation of these stores from external sources. The most important external source is probably circulating chylomicrons and very-low density lipoproteins, which are hydrolysed in the capillary lumen by lipoprotein lipase. The released fatty acids are transported into the brown adipocytes for combustion.

Lipoprotein lipase is synthesised in the adipocytes, glycosylated and then secreted through the endothelial cells to bind finally as an active homodimer to heparan stalks on the capillary wall, where it then gains access to its circulating substrates. It requires core glycosylation for activation and transport from the reticulum, but the further processing to the complex oligosaccharide chains which normally occurs is neither required for activity, nor for secretion (13). Its secretion from the adipocyte wall and into the endothelial cell is facilitated by products of triglyceride degradation (14).

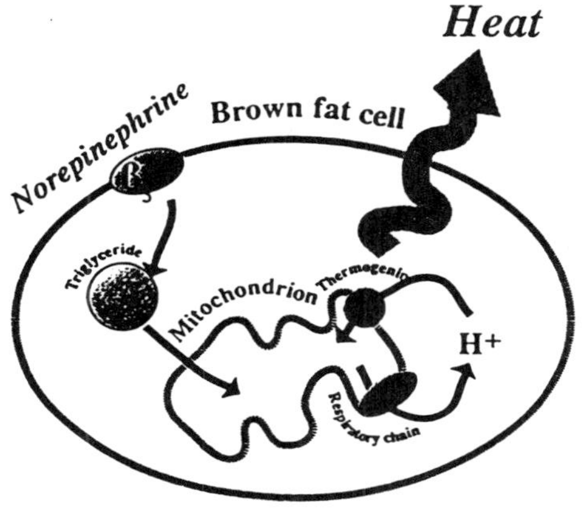

Fig. 3A: The proton circuit underlying heat production in the mitochondria of the brown fat cell.

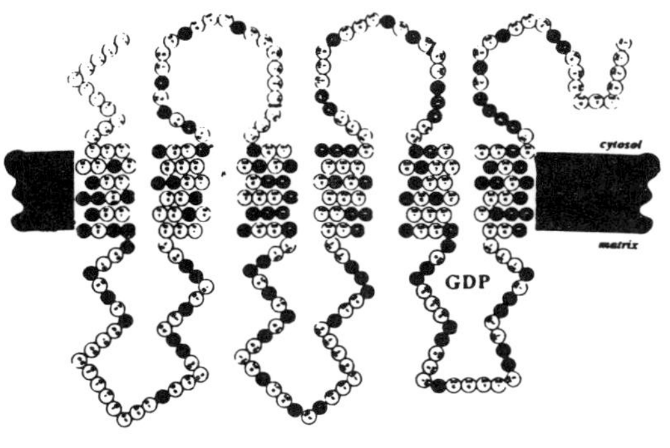

Fig. 3B: Suggested structure of the uncoupling protein thermogenin. The part of the molecule which has been demonstrated to be the binding site for purine nucleotides (GDP) is indicated. The site of interaction of fatty acids and the localization of the proton transportation have not been identified. Dark residues are hydrophobic. Adapted from (12).

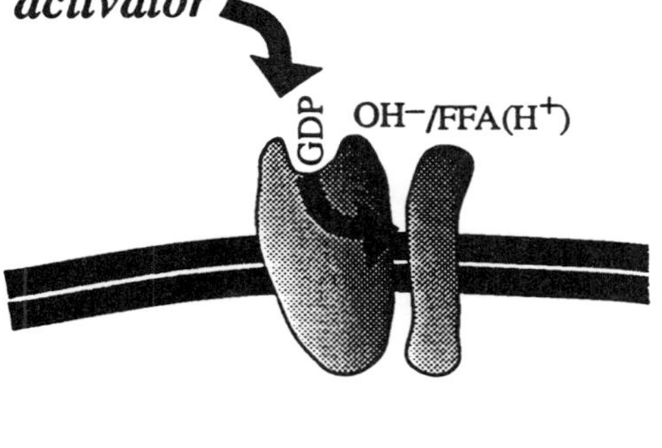

Fig. 3C: The interaction of fatty acids and the unidentifed activator with thermogenin. It is possible that the fatty acids function in the transport process as H+ carriers whereas the activity of the protein is regulated by the GDP-binding site.

Adrenergic stimulation of animals (15) or of cultured brown adipocytes increases the activity of lipoprotein lipase and this response is primarily β-adrenergic. The β-receptor involved appears to be the β_3, since β_3-selective agonists are also able to stimulate gene expression acutely in cultured brown-fat cells. Published results from intact animal studies on a role for α_1-adrenergic receptors in stimulation of lipoprotein lipase are not concordant, and the question will have to await elucidation in cultured cells. Also insulin can stimulate lipoprotein lipase gene expression and increase enzyme activity, as is the case in white adipose tissue (16).

The half-life of the enzyme activity is very short and maintenance of an elevated level requires therefore continuous adrenergic stimulation. There is an ongoing discussion as to whether the elevated level is achieved only by activation of gene expression or whether also translational and/or posttranslational mechanisms are involved. The discussion is complicated by the different half-lives of the enzyme and the mRNA, and by measurements made under nonequilibrium conditions. Stimulation of gene expression is certainly an important step and probably the predominant step in the elevation of enzyme activity.

DESENSITIZATION

When a tissue is exposed to prolonged stimulation by an agonist, the response of the tissue to subsequent agonist stimulation is often diminished, a phenomenon known as desensitization. The function of desensitization is often described as being to prevent overstimulation of the tissue, an 'explanation' which hardly seems satisfactory in any physiological context.

Cold-induced desensitization of the thermogenic response

Brown adipocytes isolated from cold-acclimated hamsters or rats are less responsive to norepinephrine stimulation than those from control animals (17). This is true irrespective of which component of the stimulated pathway is determined: cyclic-AMP production, lipolysis, oxygen

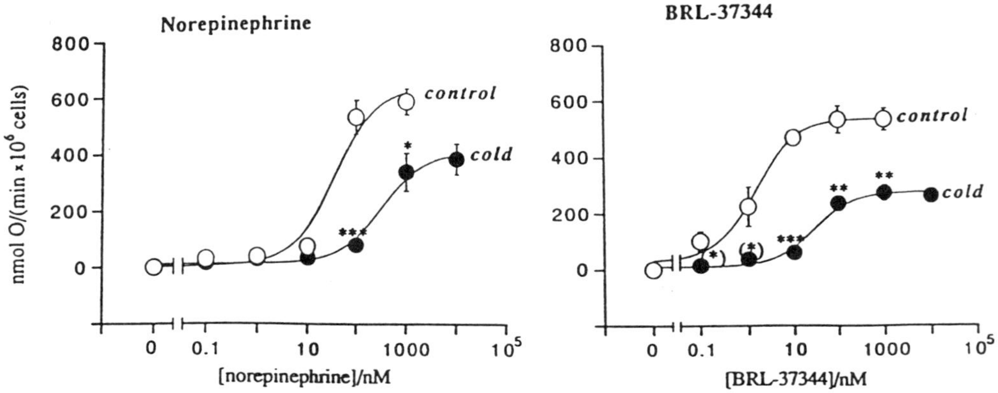

Fig. 4: Functional desensitization of the response of isolated hamster brown-fat cells due to cold acclimation of the animal. Left: The response to norepinephrine; right: the response to the β_3-agonist BRL-37344. Adapted from (18)

consumption and heat production are all reduced. As explained above, stimulation of these components is mediated in the mature adipocytes by β_3-receptors which in themselves are less prone to desensitization than β_1 or β_2 receptors. It is therefore of interest to know which intracellular steps in the transduction pathway are responsible for the functional desensitization observed.

At the receptor level

In other tissues with β-adrenergic receptors, desensitization can be observed at the receptor level and seen to occur within minutes. In general terms, two types of desensitization are discussed: homologous and heterologous desensitization. The former describes the decreased responsiveness of a tissue to further stimulation by the same agonist as that to which it was first exposed, while the latter describes the decreased responsiveness to any agonist which is coupled to the same effector system as the original agonist.

Homologous desensitization occurs through the action of the β-adrenergic receptor kinase which, at high agonist concentration, phosphorylates receptors occupied by agonist. This induces binding of the protein β-arrestin, such that receptor/G_S protein interaction is uncoupled. Heterologous desensitization occurs through the action of protein kinase A, already at low agonist concentration, on occupied or nonoccupied receptors, and the phosphorylation by the kinase causes receptor/G_S-protein uncoupling. The phosphorylated receptors may be subsequently internalised and reactivated. In several experimental systems, such mechanisms of desensitization have been elucidated for β_2, and also for β_1-receptors, but, as noted above, most of the relevant amino acid residues required for these types of desensitization are lacking in the β_3-receptor. Consequently, these processes cannot reasonably be major mechanisms of desensitization in brown adipose tissue.

Even though decoupling between β_3 receptor and G_S protein probably does not occur, functional desensitization may still take place at the receptor level. This could be due to a reduction in the receptor number. Since there are no highly selective β_3-antagonists available which would be suitable for radioligand binding, β_3-receptor number can only be estimated with difficulty, and such studies have not been performed for a comparison between cells from cold-acclimated and control animals. It is thus not known whether receptor number is changed between these two states. The down-regulation of receptor number could be achieved at the mRNA level. Short-term treatment with a β_3-agonist (19) or short-term cold exposure (20) decrease mRNA levels for the β_3-receptor. However, there are presently no reports of β_3-receptor mRNA levels in cold-acclimated animals, so also here it is unclear whether there is a desensitization at this level.

Post-receptor desensitization

Functional desensitization can occur at a post-receptor level, and could thus include changes in the activity or content of the transducing G-proteins, in adenylyl cyclase activity, in cyclic-AMP phosphodiesterase activity etc.

The activity of $G_S\alpha$ (as estimated through reconstitution experiments with cyc^- membranes) in membranes of brown adipose tissue from cold-acclimated animals is lower than that from controls; also the total $G_s\alpha$ content is reduced in parallel with this (21). This result in cold-acclimated animals is in accordance with decreases seen in $G_s\alpha$ mRNA levels after short-term cold exposure. The decrease in $G_s\alpha$ content could thus contribute to the observed desensitization.

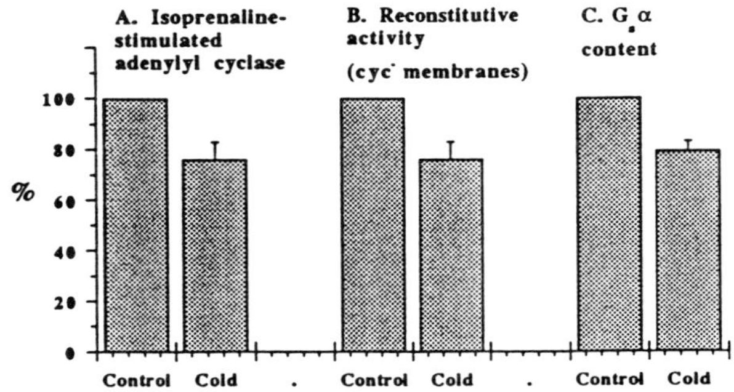

Fig. 5: Parallel decrease due to cold acclimation in ability of isoprenaline to stimulate adenylyl cyclase in isolated plasma membranes from brown adipose tissue (A), in ability of a cholate extract of brown adipose tissue plasma membranes to mediate isoprenaline stimulation in cyc^- membranes (which lack $G_s\alpha$) (B), and in amount of $G_s\alpha$ in brown-adipose-tissue plasma membrane, as assayed immunologically (C). Adapted from (21).

Although changes in adenylyl cyclase activity could also bring about desensitization, there are no reports which indicate that this should occur, and adenylyl cyclase activity seems to be unchanged before and after cold acclimation. However, the oxygen consumption which can be evoked by stimulation with the adenylyl cyclase activator forskolin is markedly desensitized (18). This implicates a site of desensitization at the post-cyclic AMP level. In accordance with this, the cyclic nucleotide phosphodiesterase activity is significantly elevated in cells from cold-acclimated animals. This increased activity would thus also contribute to the desensitization phenomenon (18).

Although most further distal steps from cyclic-AMP have not been experimentally excluded, it is clear that the relationship between the cellular cyclic-AMP concentration and the level of oxygen consumption is similar in cells from control and cold-acclimated animals (18), indicating that further desensitization steps are unlikely.

Inhibitory receptors

It has been hypothesised that the desensitization event could involve an increase in the content of $G_i\alpha$ or an increased activity of inhibitory receptors coupled to the G_i-protein. Changes in the content of $G_i\alpha$ in brown adipose tissue have not been studied after cold acclimation, but this could be of interest. One G_i-protein-coupled receptor, the adenosine A_1-receptor, has been studied in detail (22). However, no change in A_1-receptor number could be detected after cold acclimation, and the responses to A_1 stimulation proved, in fact, to be desensitized. Thus, clearly, the functional β-adrenergic desensitization is not due to an increase in the activity of the inhibitory A_1-receptors. In some species, α_2-adrenergic receptors, which also couple to G_i-proteins, appear to be present. There is currently no evidence that activity of these receptors increases after cold acclimation, although this has not been excluded. There may indeed be other inhibitory receptors involved. This can e.g. be implied from the fact that pertussis toxin treatment indicates the presence of an inhibitory G_i-protein even in hamster brown fat, where α_2-receptors have not been demonstrated.

In conclusion, the desensitization event occurring in brown adipose tissue as a result of physiological stimulation appears to involve changes at several levels in the signal transduction pathway, and the physiological relevance of the phenomenon is still unclear.

RECRUITMENT OF BROWN ADIPOSE TISSUE

Cell proliferation

When animals acclimate to cold, they show a marked increase in their capacity for nonshivering thermogenesis and in their ability to respond at thermoneutral temperatures to norepinephrine by an increase in oxygen consumption. This increase in thermogenic capacity of the entire tissue thus occurs in spite of the fact that desensitization to norepinephrine develops in the single brown adipocytes following acclimation to cold. This implies that as a consequence of the recruitment process, the tissue obtains more active cells.

Already in the early sixties, Cameron and Smith demonstrated that the mitotic index of the tissue was dramatically increased by cold exposure, and this work has been confirmed and extended by several later studies. The mitotic activity is found in endothelial cells, interstitial mesenchymal cells and adipocyte precursors but not in the mature adipocytes.

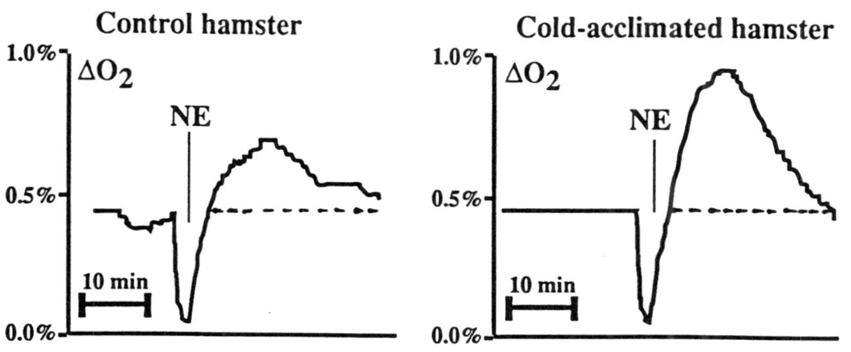

Fig. 6: The thermogenic response to norepinephrine injection in control and cold-acclimated hamsters. Despite the desensitization observed at the level of the isolated brown-fat cell, the response of the intact animal is increased due to cold acclimation, even in the hamster. A. Dicker et al., unpubl. obs.

It has also been demonstrated in intact animals that the physiological initiator of the increased cell proliferation is again norepinephrine. The rather unusual aspect of the norepinephrine action is that the receptor involved is a β-receptor, with cyclic-AMP as the intracellular mediator. In most other cell systems, increased cyclic-AMP is rather a negative modulator of cell division, and there are only a few other cases of cyclic-AMP action being positive.

Studies have been performed with cultured brown adipocytes to further characterise the nature of the β-adrenergic stimulation of DNA synthesis. In these studies, norepinephrine-stimulated DNA synthesis could be inhibited by CGP-12177 which, as commented above, is a β_1-antagonist but a β_3-agonist. Thus, DNA synthesis is mediated by β_1-receptors (23). In the young, proliferative cells β_1-mRNA is present, and β_3-mRNA does not appear until the cells commence differentiation.

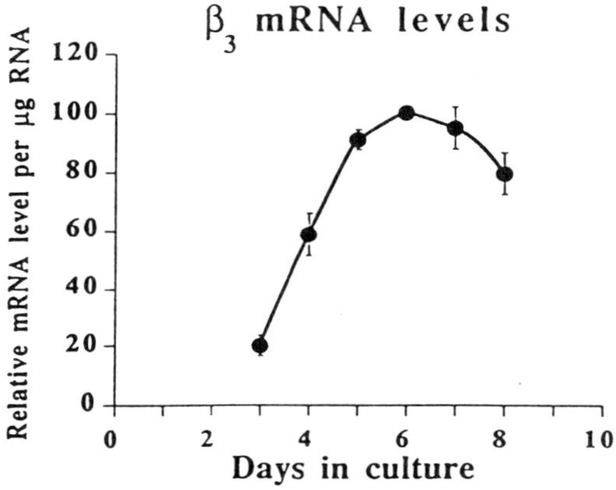

Fig. 7: Appearance of β_3 adrenoceptor mRNA during differentiation of brown-fat cells in cell culture. T. Bengtsson, unpubl. obs.

It can therefore be concluded that, in the proliferative cells, β_1-receptors are coupled to cyclic-AMP production, and cyclic-AMP is thus the intracellular mediator for cell proliferation.

Cell proliferation is often associated with increase in expression of the proto-oncogene *c-myc*. In brown adipose tissue, *c-myc* has been shown to be expressed during fetal development of the tissue, but there does not seem to be a requirement for a dramatic increase in *c-myc* expression for the cold-induced increase in proliferation to commence. A mediatory role for *c-myc* when proliferation is stimulated via a cyclic-AMP-dependent pathway does not therefore seem to exist. The proto-oncogene *c-fos* is, however, markedly increased due to cold exposure, peaking after only about 30 min in the cold, and returning to basal levels already after 90 min. Although enhanced DNA synthesis seems only to require β_1-mediated signalling, *c-fos* gene expression is increased both by α_1-stimulation and β-stimulation, and these actions are at least additive if not synergistic. Both pathways will, of course, be stimulated upon physiological activation. The exact role of the elevated *c-fos* expression has so far not been clarified.

Also the involvement of the transcription factors C/EBPα and C/EBPβ in proliferation of brown adipocytes has been investigated (24). C/EBPβ apparently responds in a manner similar to *c-fos*, in that norepinephrine stimulates gene expression at all stages of cell development. The response is primarily mediated via β-receptors. In contrast, C/EBPa mRNA is decreased after norepinephrine stimulation in proliferatively competent cells, although it is increased by norepinephrine stimulation later. This decrease is also seen in brown adipose tissue immediately upon cold exposure of animals and is perhaps a prerequisite for the initiation of proliferation, such that high C/EBPα levels serve as inhibitors of mitogenesis. Both α_1- and β-pathways seem to be required to bring about the decrease. There are doubtless numerous other transcription factors whose expression is influenced by norepinephrine and, with time, their roles will hopefully be clarified.

In addition to activation of these pathways, norepinephrine also activates numerous ion fluxes. Through activation of the α_1-pathway, intracellular Ca^{2+} levels are increased and a Ca^{2+}-activated K^+ channel is opened. The membrane potential depolarises, probably initially due to enhanced Cl^- movement. Norepinephrine also stimulates Na^+ influx in brown adipocytes and this is a β-mediated event and cyclic-AMP dependent. It has been suggested that ionic events may play a permissive and/or a triggering role in cell proliferation. The ion fluxes described in brown adipocytes have been determined in mature, post-mitotic cells but it may be extrapolated that these events also occur upon norepinephrine stimulation of immature cells. Their role in cell proliferation is, though, unexplained, but there are observations that inhibition of these ionic fluxes may also inhibit norepinephrine-induced cell proliferation (25).

A requirement for certain growth factors for successful proliferation can also be anticipated. These include IGF-1, IL-1α and bFGF. These compounds would certainly be present in the serum generally used for culture of brown adipocytes, and 'spontaneous' cell proliferation occurs under these conditions. Under serum-free conditions, IGF-1 can enhance DNA synthesis in the absence of norepinephrine (26). However, both in the presence of serum, when cell proliferation is anyway occurring, and in its absence, when cell proliferation does not occur spontaneously, norepinephrine can still stimulate DNA synthesis. Thus, norepinephrine appears to be a sufficient signal to stimulate proliferation.

CELL DIFFERENTIATION

The progression of cell differentiation is generally inhibitory of cell proliferation. This is probably also true for the single brown-fat cell, but in the mixed cell population found in brown adipose tissue, these processes occur simultaneously during the recruitment process, in order to achieve an enhanced number of mature, heat-producing brown adipocytes.

In cells differentiating in culture, adipose conversion appears to be initiated concomitantly with the expression of the β_3-adrenergic receptors. It is currently difficult to comprehend this transition from adenylyl cyclase-coupled β_1-receptors to β_3 and whether this is involved in initiating the differentiation programme or merely occurs in parallel.

Early in the differentiation process, brown adipocyte-specific cell surface antigens appear. Further, lipoprotein lipase, and also glycerol-3-phosphate dehydrogenase activities are increased, indicating that adipose conversion has occurred and lipid accretion begins. These activities are high in cultured cells around confluence and not dramatically enhanced by norepinephrine. They may be considered as early markers of adipose conversion. In the intact animal, however, these enzymes are markedly stimulated by norepinephrine or cold.

There are relatively few mitochondria in adipocyte precursor cells, as seen both in intact animals and in cultured cells. Active mitochondriogenesis occurs in parallel with differentiation, as evidenced by increases in cytochrome-*c*-oxidase and ATP synthase proteins. Electron microscopic studies also demonstrate the remarkable mitochondriogenesis occurring, both with and without norepinephrine.

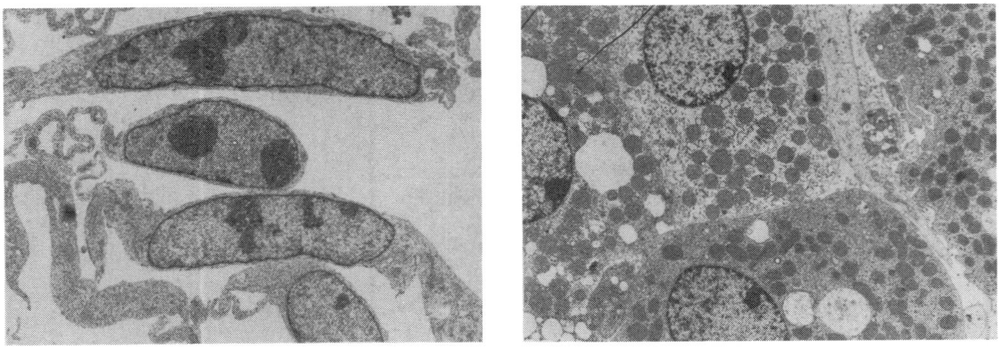

Fig. 8: Differentiation of brown-fat cells in culture. Left: cells cultured for 2 days in the standard medium; right: cells cultured for 10 days in the standard medium (27), in the presence of norepinephrine from day 4. Same magnification in both micrographs. Note the large increase in the mitochondrial complement. Our unpubl. obs.

The ultimate differentiation step in the achievement of an active brown adipocyte is the expression of the gene *ucp* coding for the mitochondrial uncoupling protein thermogenin (UCP) and incorporation of thermogenin into the mitochondrial inner membrane (28). Under most culture conditions, thermogenin gene expression is absent or minimal. In some cases, e.g. a brown adipocyte cell line HIB-1B, or ovine brown adipocytes in culture, thermogenin gene expression is enhanced by the addition of glucocorticoids. However, in most culture systems, as in the intact animal, thermogenin gene expression is induced by norepinephrine addition. Even for this process, norepinephrine acts primarily on β_3-receptors and is most effective in the cell cultures around the period of confluence. In the intact animal, the norepinephrine effect is also large and rapidly observed and mainly of β_3-origin. Although α_1-receptors only seem to contribute marginally to *ucp* gene expression in themselves, their action appears to facilitate and enhance the β_3-stimulated expression.

In parallel with the norepinephrine-induced recruitment in the tissue, there is also a need for a massive angiogenesis, the control of which has virtually not been studied. It is possible that it is, at least in part, regulated in a paracrine fashion by growth factors secreted from the brown fat cells as a result of their enhanced development. In this way, a functionally more active tissue is generated.

It may thus be concluded that upon norepinephrine stimulation of both precursor cells and mature adipocytes in brown adipose tissue, a multitude of signal cascades are activated, giving rise simultaneously in the tissue to increased DNA synthesis, to increased mitochondriogenesis and the expression of certain genes, and also to increased heat production. This large array of events induced by norepinephrine all function in concert to produce a tissue better able to fulfil the requirements of the animal with respect to thermogenesis and energy balance.

REFERENCES

1. Brück, K, Wünnenberg, B (1965): Untersuchungen über die Bedeutung des multiloculären Fettgewebes für de Thermogenese des neugeborenen Meerschweinchens. Pflügers Arch 283, 1-16.

2. Himms-Hagen, J (1990): Brown adipose tissue thermogenesis: role in thermoregulation, energy regulation and obesity. In: Thermoregulation: Physiology and Biochemistry. Schönbaum, E, Lomax, P eds. New York: Pergamon Press, Inc., 327-414.

3 Kobilka, B (1992): Adrenergic receptors as models for G protein-coupled receptors. Annu Rev Neurosci 15, 87-114.

4. Raasmaja, A (1990): α_1- and β-adrenergic receptors in brown adipose tissue and the adrenergic regulation of thyroxine 5'-deiodinase. Acta Physiol Scand (Suppl 590) 139, 1-64.

5. Strosberg, AD (1992): Biotechnology of β-adrenergic receptors. Mol Neurobiol 4, 211-250.

6. Mohell, N, Connolly, E, Nedergaard, J (1987): Distinction between mechanisms underlying α- and β-adrenergic respiratory stimulation in brown fat cells. Am J Physiol 253, C301-C308.

7. Arch, J (1989): The brown adipocyte β-adrenoceptor. Proc Nutr Soc 48, 215-223.

8. Mohell, N, Dicker, A (1989): The β-adrenergic radioligand (^{3}H)CGP-12177, generally classified as an antagonist, is a thermogenic agonist in brown adipose tissue. Biochem J 261, 401-405.

9. Zhao, J, Unelius, L, Cannon, B, Nedergaard, J (1993): Coexisting β-adrenoceptor subtypes: significance for the thermogenic process in brown-fat cells. Subm for publ.

10. Granneman, JG (1990): Norepinephrine and BRL 37344 stimulate adenylate cyclase by different receptors in rat brown adipose tissue. J Pharmacol Exp Ther 254, 508-513.

11. Holm, C, Fredrikson, G, Cannon, B, Belfrage, P (1987): Hormone-sensitive lipase in brown adipose tissue: Identification and effect of cold exposure. Biosci Rep 7, 897-904.

12. Nedergaard, J, Cannon, B (1992): The uncoupling protein thermogenin and mitochondrial thermogenesis. In: New Comprehensive Biochemistry vol. 23: Molecular Mechanisms in Bioenergetics. Ernster, L eds. Amsterdam: Elsevier, 385-420.

13. Masuno, H, Schultz, CJ, Park, JW, Blanchette-Mackie, EJ, Mateo, C, Scow, RO (1991): Glycosylation, activity and secretion of lipoprotein lipase in cultured brown adipocytes of newborn mice. Effect of tunicamycin, monensin, 1-deoxymannojirimycin and swainsonine. Biochem J 277, 801-809.

14. Sasaki, A, Goldberg, IJ (1992): Lipoprotein lipase release from BFC-1β adipocytes. Effects of triglyceride-rich lipoproteins and lipolysis products. J Biol Chem 267, 15198-15204.

15. Carneheim, C, Nedergaard, J, Cannon, B (1988): Cold-induced β-adrenergic recruitment of lipoprotein lipase in brown adipose tissue is due to increased transcription. Am J Physiol 254, E155-E161.

16. Carneheim, C, Alexson, S (1989): Refeeding and insulin increase lipoprotein lipase activity in rat brown adipose tissue. Am J Physiol 256, E645-E650.

17. Nedergaard, J (1982): Catecholamine sensitivity in brown fat cells from cold-adapted hamsters and rats. Am J Physiol 242, C250-C257.

18. Unelius, L, Bronnikov, G, Mohell, N, Nedergaard, J (1993): Physiological desensitization of β3 adrenergic responsiveness in brown fat cells. Involvement of a post-receptor mechanism. Am J Physiol 265, C1340-C1348.

19. Revelli, J-P, Muzzin, P, Giacobino, J-P (1992): Modulation *in vivo* of β-adrenergic-receptor subtypes in rat brown adipose tissue by the thermogenic agonist Ro 16-8714. Biochem J 286, 743-746.

20. Granneman, JG, Lahners, KN (1992): Differential adrenergic regulation of β_1- and β_3-adrenoceptor messenger ribonucleic acids in adipose tissues. Endocrinology 130, 109-114.

21. Svoboda, P, Unelius, L, Cannon, B, Nedergaard, J (1993): Attenuation of G_s-protein coupling efficiency in brown adipose tissue plasma membranes from cold-acclimated hamsters. Biochem J 295, 655-661.

22. Unelius, L, Mohell, N, Nedergaard, J (1990): Cold acclimation induces desensitization to adenosine in brown fat cells without changing receptor binding. Am J Physiol 258, C818-C826.

23. Bronnikov, G, Houstek, J, Nedergaard, J (1992): β-Adrenergic, cAMP-mediated stimulation of proliferation of brown fat cells in primary culture. Mediation via β_1 but not via β_3 receptors. J Biol Chem 267, 2006-2013.

24. Rehnmark, S, Antonson, P, Xanthopoulos, KG, Jacobsson, A (1993): Differential adrenergic regulation of C/EBPa and C/EBPb in brown adipose tissue. FEBS Lett 318, 235-241.

25. Pappone, PA, Ortiz-Miranda, SI (1993): Blockers of voltage-gated K channels inhibit proliferation of cultured brown fat cells. Am J Physiol 264, C1014-C1019.

26. Lorenzo, M, Valverda, AM, Teruel, T, Benito, M (1993): IGF-1 is a mitogen involved in differentiation-related gene expression in fetal rat brown adipocytes. J Cell Biol 123, 1567-1575.

27. Rehnmark, S, Néchad, M, Herron, D, Cannon, B, Nedergaard, J (1990): α- and β-adrenergic induction of the expression of the uncoupling protein thermogenin in brown adipocytes differentiated in culture. J Biol Chem 265, 16464-16471.

28. Klaus, S, Casteilla, L, Bouillaud, F, Ricquier, D (1991): The uncoupling protein UCP: A membraneous mitochondrial ion carrier exclusively expressed in brown adipose tissue. Int J Biochem 23, 791-801.

Thermal Balance in Health and Disease
Advances in Pharmacological Sciences

PHARMACOLOGY AND REGULATION OF THE ION CHANNELS OF THE BROWN ADIPOCYTE PLASMA MEMBRANE

A. Koivisto*, T. Ringer, U. Ruß, J. Nedergaard* and D. Siemen

Institut für Zoologie, Universität Regensburg, D-93040 Regensburg, Germany. * The Wenner-Gren-Institute, Stockholm University, S-10691 Stockholm, Sweden.

Brown adipose tissue is responsible for regulatory nonshivering thermogenesis in most mammals of small body size. Sympathetic nerve endings throughout the tissue allow neuronal control. After stimulation by their transmitter noradrenaline, the adipocyte shows a biphasic response of its membrane potential. An early brief depolarisation and a late long-lasting depolarisation are separated by a hyperpolarisation about 1 min after the stimulus (1, 2). The underlying ion channels have at least partly been identified by single channel analysis. A nonselective channel for monovalent cations probably accounts for the tetrodotoxin(TTX)-insensitive sodium influx (3, 4, for review see 5). Part of the potassium efflux is mediated by a voltage-dependent K-channel (6, 7). Another major contribution to the potassium efflux derives from a Ca^{2+}-activated K-channel first demonstrated by flux measurements (8) and later shown by whole-cell patch clamp experiments (9). Recently a further K-channel was shown that is voltage-insensitive (7). There are indications that further ion channels contribute to the response. In this paper we will focus on pharmacological mechanisms which can modify the nonselective cation channel as well as some of the K-channels.

Methods

Mature brown adipocytes were prepared from the interscapular tissue of 3-5 week old male Sprague-Dawley rats. Cells were isolated by collagenase treatment (5mg/ml) and grown in Dulbecco's Modified Eagle's Medium, supplemented by 10% Newborn Calf serum, 4 nM insulin, 50 IU/ml penicillin and 50 µg/ml streptomycin. Culture dishes were kept in an incubator at 37 °C, 8% CO_2 in air and 100% humidity. We used the patch-clamp technique to study single channel currents in excised inside-out and outside-out patches. Details of the methods in general are given by Hamill et al. (10). Pipettes were made from borosilicate glass (Clark, Pangbourne, England). Currents were recorded by a L/M-EPC7 patch-clamp amplifier (List, Darmstadt, Germany) and stored on video tape or on computer. For data acquisition and analysis we used the P-CLAMP software (Axon instruments, Foster City, CA, USA).

For inside-out patches we used a modified Ringer's containing 134 NaCl, 6 KCl, 1.2 $CaCl_2$, 1.2 $MgCl_2$, and 10 HEPES (pH 7.4 with NaOH, all concentrations in mM) in pipette and bath. Sulfhydryl-group modifying reagents were dissolved into a solution containing 130 KCl, 10 NaCl, 1.2 $CaCl_2$, 1.2 $MgCl_2$, and 10 HEPES (pH 7.2 with KOH) and used for superfusing the cytoplasmic face of the excised patches. K-channels were studied in outside-out patches with a pipette solution containing 139 KF, 4 KCl, 10 HEPES, 5 EGTA (pH 7.4 with KOH). Experiments were done at room temperature. More details of our work can be found in (11).

Results

Nonselective Channel for Monovalent Cations

Sodium influx into brown adipocytes cannot be blocked by TTX. Therefore the influx is not due to a voltage-dependent Na-channel such as that in nerve or in muscle cells. Quantitatively a 25-30 pS (25 °C) nonselective cation channel (NSC-channel), activated by a rise in intracellular Ca^{2+}, could account for the Na-influx (12). In inside-out patches with 1.2 mM Ca^{2+} at the inside, this channel was predominant. There may be about 3000 channels within the plasma membrane of a 30 µm-cell. The activation of the NSC-channel could be modulated by the membrane voltage, increasing with depolarisation. The steepest increase of the open probability occured in the physiological range at about -45 mV.

Understanding of the activation mechanism of the NSC-channel is complicated by the observation that effects leading to increased cAMP levels may also raise the Na-influx (13). We thus tested the effect of 100 µM cAMP, applied at the internal side of the membrane and found a complete and reversible block. 100 µM cGMP gave a similar result. The metabolically important purine nucleotides ATP and ADP did block the NSC-channel as well. Their effect was a reduction of the mean open time with a complete block at 1 mM. Also the nonhydrolyzable ATP-analogue AMP-PCP showed a block. Although 100 µM GTP-c-S inactivated the NSC-channel in brown adipocytes, control experiments did not prove an involvement of G-proteins.

For several years no blockers were known with which NSC-channels could be identified. Gögelein et al. (14) had found a blocking effect of some antiphlogistics. We were able to demonstrate that also in brown adipocytes 100 µM flufenamic and 100 µM mefenamic acid applied to the inner side caused a complete and reversible block.

Only recently mercury and thimerosal, two sulfhydryl-group modifying reagents, were shown to influence NSC-channels at brown adipocytes quite potently (11). The IC_{50}-values for their internal blocking action were 21 nM, and 1.5 µM, respectively. The block was irreversible upon washout with Ringer's but could be relieved significantly by dithiotreitol.

Voltage-Dependent K-Channel

In about every second outside-out patch with KF-solution at the inside of the membrane we found single channel events of a voltage-dependent K-channel of the delayed-rectifier type, as was first shown by Lucero and Pappone (6). With a physiological distribution of the Na^+- and the K^+-ions we observed a single channel conductance of c = 8 pS. Extrapolation of the current voltage relationship to negative potentials yielded a reversal potential of -63 mV. c increased to 20.5 pS for inward currents in symmetrical K^+-concentrations on either side of the membrane. Outward currents showed a c of 17 pS under these conditions.

These single channel events could be reduced in amplitude by increasing concentrations of the K-channel blocker tetraethylammonium (TEA) at the extracellular side. The dose response curve yielded an IC_{50} of 1.8 mM and a Hill coefficient of 1.0. Neither 0.5 µM apamin, known to block Ca^{2+}-activated K-channels of the SK-type, nor 1 mM Ba^{2+} (blocker of several K-channels) or 10 µM dendrotoxin (blocker of delayed rectifiers of the I-type) were able to cause any change in channel activity. Exchanging part of the Na^+-ions at the outside against Cs^+- or Ca^{2+}-ions led to a upward bent i-E curve for the inward currents. Such curves were interpreted as signs of a voltage-dependent block by these ions (7).

Voltage-Independent K-Channel

Another as yet unknown K-channel was observed in 13 outside-out patches, in 7 cases together with the voltage-dependent K-channel. In solutions with symmetrical K^+-concentrations this channel showed a c of 12 pS for outward currents and a c of 25 pS for inward currents. Kinetics were much faster than those of the delayed rectifiers giving the events a more flickery appearance. This channel was insensitive to apamine (0.5 µM) as well as to TEA (10 mM).

Discussion

A biphasic depolarization interrupted by a small hyperpolarization of the plasma membrane of nor-adrenaline-activated brown adipocytes was described several years ago (1, 2). This paper gives a brief summary of pharmacological or toxicological tools by which the underlying ion channels can be identified and modified in their behaviour. Part of the results is summarized in Table 1.

Until now, four ion channels were described, three of them in more detail:

The **NSC-channel** is the most frequently observed. It is likely that it is this channel which, when activated via stimulation of β_3-receptors, conducts a net Na^+-influx at physiological ranges of the membrane potential. This channel is activated by elevated Ca^{2+}-levels of more than 10 µM and modulated by voltage. (In mouse pancreatic acinar cells the Ca^{2+}-sensitivity is higher if the inside of the channel-containing patches is continuously exposed to ATP (15)). Activity seems to be linked to cell metabolism by its sensitivity to purine nucleotides. ATP, ADP, GTP, all show a blocking action. Also cAMP and cGMP do block, although this is difficult to understand as Connolly et al. (13) made it likely that the ß-mediated increase in intracellular Na^+ is caused by increased cAMP. Block of NSC-channels by various purine nucleotides has been shown in other preparations before (e.g. endocrine, exocrine and kidney cells). Also its sensitivity to flufenamic acid and mefenamic acid compares to that of other NSC-channels (14). A new exciting finding is the high sensitivity of the NSC-channel to the sulfhydryl group modifying reagents mercury and thimerosal. It could indicate that SH-groups play an important role in gating of these channels.

Table 1 Ion channels and modifying agents in brown adipocytes

channel	modifying agent (block if nothing mentioned)	concentration	Ref.
nonselective	Ca^{2+}_i (agonist)	>10 µM	5
	purine nucleotides		5
	antiphlogistics		5
	(mefen.-, flufen. acid)	100 µM	5
	Hg^{2+}	IC_{50}=21 nM	11
	thimerosal	IC_{50}=1.5 µM	11
delayed rectifier	TEA	IC_{50}=1.8 mM	6, 7
	4 AP	1-2 mM	6
	Ca^{2+}		7
	Cs^{+}		7
	(DTX does not block)		7
Ca^{2+}-activated	apamin	500 nM	9
	charybdotoxin	200 nM	9
	TEA	<20 mM	9
	4 AP	<5 mM	9
voltage insensitive	(TEA does not block)		7

In brown fat cells K-channels were first studied by Lucero and Pappone with the whole cell method as macroscopic currents (6). Due to their voltage dependence one type of them clearly belongs to the **delayed rectifier** family. Recently, further single-channel measurements showed additional results (7). Like the typical delayed rectifier it can be blocked by TEA with an IC_{50} close to the value of 1 mM found for F-type channels in the node of Ranvier (16). TEA block has a Hill coefficient of 1.0 and is fast as it is usually in K-channels. Kinetics were not influenced by the TEA block. The IC_{50} for the 4-aminopyridine (4-AP) block is between 1 and 2 mM. Both substances were washed out almost completely within 2 min (6). Neither dendrotoxin (DTX) from the Eastern Green Mamba (*Dendroaspis angusticeps*) nor the crude venom of the scorpion *Pandinus imperator* have any effect on these channels (7, 6), different to nodal I-channels or to a pituitary tumor cell line.

The **Ca^{2+}-activated K-channel** seems to be unique as it is blocked both by the bee polypeptide apamin and by the scorpion-derived charybdotoxin. Also TEA and 4-AP are effective. The **voltage-insensitive channel**, however, is not TEA sensitive. Since long, attempts to identify such K-channels have been made in order to explain the missing TEA-effect on the resting potential. We suggest that this channel could be responsible for the resting potential of brown adipocytes.

Unfortunately, little is known about a role of these channels for the brown adipocyte function. There is e.g. almost no change in noradrenaline-induced respiratory activity of these cells in solutions with blocking concentrations of TEA (17) or with blocking concentrations of mefenamic acid (Koivisto, unpublished result). On the other hand, it could be possible that the above mentioned change in membrane potential is part of the signal cascade forwarding the mitogenic signal as in fibroblasts (18, see 5). As two of the involved channels are activated by intracellular Ca^{2+}, Ca^{2+}-signalling seems to play a central role for the changes in membrane potential after noradrenaline application.

Acknowledgements

We are indebted to E. Dotzler for reading the manuscript and to M. Dietl and U. Schmitt for expert technical assistance. Financial support by the the Swedish Natural Science Council, the Hasselblad Foundation, and the Deutsche Forschungsgemeinschaft is gratefully acknowledged.

References

1. Horwitz, BA, Hamilton, J (1984): Alpha-adrenergic-induced changes in Hamster (*Mesocricetus*) brown adipocyte respiration and membrane potential. Comp Biochem Physiol 78C, 99-104.

2. Schneider-Picard, G, Coles, JA, Girardier, L (1985): α- and ß-adrenergic mediation of changes in metabolism and Na/K exchange in rat brown fat. J Gen Physiol 86, 169-188.

3. Siemen, D, Reuhl, T (1987): Non-selective cationic channel in primary cultured cells of brown adipose tissue. Pflügers Arch 408, 534-536.

4. Weber, A, Siemen, D (1989): Permeability of the non-selective channel in brown adipocytes to small cations. Pflügers Arch 414, 564-570.

5. Koivisto, A, Dotzler, E, Russ, U, Nedergaard, J, Siemen, D (1993): Nonselective cation channels in brown and white fat cells. In: Nonselective Cation Channels. Siemen, D, Hescheler, J eds. Basel: Birkhäuser, 201-211.

6. Lucero, MT, Pappone, PA (1989): Voltage-gated potassium channels in brown fat cells. J Gen Physiol 93, 451-472.

7. Ruß, U, Ringer, T, Siemen, D (1993): A voltage-dependent and a voltage-independent potassium channel in brown adipocytes of the rat. Biochim Biophys Acta 1153, 249-256.

8. Nånberg, E, Connolly, E, Nedergaard, J (1985): Presence of a Ca^{2+}-dependent K^+-channel in brown adipocytes. Possible role in maintenance of α_1-adrenergic stimulation. Biochem Biophys Acta 844, 42-49.

9. Lucero, MT, Pappone, PA (1990): Membrane responses to norepinephrine in cultured brown fat cells. J Gen Physiol 95, 523-544.

10. Hamill, OP, Marty, A, Neher, E, Sakmann, B, Sigworth, FJ (1981): Improved patch-clamp techniques for high-resolution current recording from cells and cell-free membrane patches. Pflügers Arch 391, 85-100.

11. Koivisto, A, Siemen, D, Nedergaard, J (1993) Reversible blockade of the calcium-activated nonselective cation channel in brown fat cells by the sulfhydryl reagents mercury and thimerosal. Pflügers Arch 425, 549-551.

12. Siemen, D, Weber, A (1989): Voltage-dependent and cation-selective channel in brown adipocytes. In: Thermal physiology. Mercer J ed. Elsevier Science Publishers B.V., 241-246.

13. Connolly, E, Nånberg, E, Nedergaard, J (1986): Norepinephrine-induced Na^+ influx in brown adipocytes is cyclic AMP-mediated. J Biol Chem 261, 14377-14385.

14. Gögelein, H, Dahlem, D, Engelert, HC, Lang, HJ (1990): Flufenamic acid, mefenamic acid and niflumic acid inhibit single nonselective cation channels in the rat exocrine pancreas. FEBS Lett 268, 79-82.

15. Thorn, P, Petersen, OH (1992): Activation of nonselective cation channels by physiological cholecystokinin concentrations in mouse pancreatic acinar cells. J Gen Physiol 100, 11-25.

16. Safronov, BV, Kampe, K, Vogel, W (1993): Single voltage-dependent potassium channels in rat peripheral nerve membrane. J Physiol 460, 675-691.

17. Pappone, PA, Lucero, MT (1992): Potassium channel block does not affect metabolic responses of brown fat cells. Am J Physiol 262, C678-C681.

18. Pappone, PA, Ortiz-Miranda, SI (1993): Blockers of voltage-gated K channels inhibit proliferation of cultured brown fat cells. Am J Physiol 264, C1014-1019.

Thermal Balance in Health and Disease
Advances in Pharmacological Sciences

SYMPATHETIC OUTFLOW TO BROWN ADIPOSE TISSUE AND METABOLIC HEAT PRODUCTION IN ADULT, AGED, AND COLD-ACCLIMATED ADULT MICE.

M. I. Talan and S. A. Kirov

Laboratory of Behavioral Science, National Institute on Aging, National Institute of Health, Gerontology Research Center, 4940 Eastern Avenue, Baltimore, Maryland, USA, 21224.

Introduction

We have shown that in C57BL6J adult mice repeated cold stress results in cold acclimation (1). Cold stress was provided by placement of mice for 3 hours at 2-wk intervals for 3 occasions in individual restrainers that prevented gross motor activity in the 6°-10°C environment (2). Cold acclimation was evidenced by a greater cold tolerance (reduced rate of decline of colonic temperature) and increased heat production in the third test relative to the first. We also reported that aged mice have poorer cold tolerance (higher rate of decline of colonic temperature) than adults and demonstrate a reduced ability to increase heat production in response to cold stress. Moreover, repeated cold stress tests fail to improve cold tolerance and to increase heat production in aged mice (3).

Thermogenesis by brown adipose tissue (BAT) is a major source of heat production during cold exposure in rodents. Characteristic biochemical and anatomical changes in BAT appear at the onset of cold exposure and as a result of cold acclimation (4). The reduced ability of aged rodents to produce heat during cold exposure, and to withstand cold is also attributed to age-related changes in BAT (5). Since it is generally agreed that sympathetic outflow to BAT controls BAT heat production during cold exposure (6), it is reasonable to expect that greater thermogenic response during cold exposure in cold-acclimated mice is associated with increased nervous activity in efferent sympathetic nerves to BAT, while reduced heat production in aged mice is related to decreased nervous activity.

In this work we measured the sympathetic outflow to BAT and metabolic heat production during cold exposure in anesthetized, curarized, C57BL/6J mice. Sympathetic nervous activity of adult mice maintained at room temperature was compared with nervous activity measured in aged mice (7) and cold-acclimated adult mice (8). In separate experiments heat production was measured and compared between adult and aged mice.

Methods

Subjects

Three groups of C57BL/6J male mice were used for each of these experiments: naive adults **(NAIVE)** - 10-month old animals maintained at room temperature, aged **(AGED)** - 30-month old mice maintained at room temperature, and cold acclimated adults **(ACCLI)** - 10-month old mice subjected to three repeated cold stress tests with 2 weeks intervals immediately prior to the experiment. Only animals that demonstrated an improvement in cold tolerance during the third test were used in experiment. The number of mice which completed the experiment 1 was: NAIVE - 9, AGED - 8, and ACCLI - 9. In experiment 2 we used 19 NAIVE and 12 AGED mice.

Procedure

Anesthesia was induced by intraperitoneal administration of urethane (1.4 g/kg for adult and 1.1 g/kg for aged mice) and maintained by inhalation of isoflurane (0.5% for adult and 0.4% for aged mice). Mechanical ventilation was provided through a tracheal cannula with a rate of 120/min and tidal volume of 0.4-0.6 ml. Vecuronium bromide (1 mg/kg) was injected to block somato-motor activity. A small branch from one of the intercostal nerves entering BAT was traced as distally as possible and cut at the peripheral end. Efferent mass discharges were recorded from the proximal end with bipolar, platinum-iridium electrodes. The nerves signals were amplified, filtered, and rectified, nerve activity was counted every five sec.

The animal's colonic temperature (Tco) was constantly measured and maintained by a heating pad and an infrared lamp. At the end of one hour after surgery the baseline Tco and nervous activity (SNA) was recorded, then the heating devices were turned off, and the mouse was covered with a plastic bag containing a slurry of iced water. The cold was applied until Tco dropped to 7°C below baseline level; then cold was withdrawn and heating was resumed. Physiological recording was continued during the entire test until Tco returned to baseline level. The second experiment was similar to the first, but SNA was not recorded, instead O_2 and CO_2 concentrations were measured in expired air taken from the mechanical ventilator. Heat production (HP) was calculated on the basis of O_2 consumption and CO_2 production.

Statistical procedure

Statistical analyses were conducted only for data during cold stimulation. The linear regression of Tco over time of cold stimulation for each mouse was calculated to determine the slopes of Tco - our definition of cold tolerance. Slopes of Tco were compared between NAIVE and AGED and between NAIVE and ACCLIM groups using the t-test. Changes in SNA and HP from baseline level for each group were evaluated using ANOVA with repeated measures and mean differences between NAIVE and AGED and between NAIVE and ACCLI groups were compared using Waller-Duncan Bayes least significant difference test. ANCOVA also was conducted to analyze differences between groups with regard to SNA as a function of decrease of Tco and with regard to increase of SNA during the first-degree decrease in Tco. Appropriate baseline values were used as a covariate in the ANCOVA.

Results

NAIVE and ACCLI groups were equivalent with respect to body weight (36.3 ± 0.66 and 36.9 ± 0.96 respectively); however, AGED mice weighed significantly less than NAIVE (30.4 ± 0.72; t(15)=6.08, $p<0.00005$).

Baseline levels of SNA were significantly higher in AGED and ACCLI groups than in NAIVE. NAIVE: 11.8 ± 2.53 imp/sec, AGED: 44.0 ± 12.1 imp/sec (t(15)=2.77, $p<0.05$), ACCLI: 47.0 ± 11.11 imp/sec (t(16)=3.09, $p<0.01$).

Figure 1 shows the slopes of Tco over time. AGED mice lost Tco at a greater rate than NAIVE while ACCLI mice lost Tco at a lower rate than NAIVE animals. However, the differences were statistically significant only between NAIVE and ACCLI groups (t(16)=2.36, $p<0.05$), indicating that old mice have cold tolerance similar with adults, and that cold-acclimated mice have greater cold tolerance than non-cold-acclimated adults.

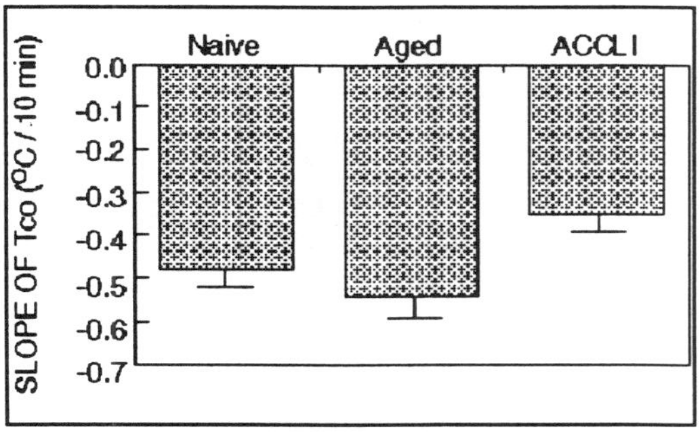

Fig. 1: The rate of decline of Tco during cold stress

Figure 2 shows the SNA response of as a function of change in Tco. SNA increased significantly during cold and returned to prestimulus levels in all three groups. ANCOVA revealed that there were significant differences between NAIVE and ACCLI groups and between NAIVE and AGED groups in the average levels ($F(1,15)=11.1$, $p<0.01$ and $F(1,14)=9.25$, $p<0.01$). There was also a significant linear increase in SNA over time for each group, but the slopes of SNA were similar between NAIVE and AGED ($F(1,15)=0.09$, $p>0.1$) and between NAIVE and ACCLI ($F(1,16)=1.98$, $p>0.1$). The increase in SNA from control level during the first-degree increase in Tco was greater in the ACCLI group than in NAIVE ($F(1,15)=6.23$, $p<0.05$) and it was greater in the AGED group compared to NAIVE ($F(1,14)=9.23$, $p<0.05$). These analyses indicate that both aged and cold-acclimated adult mice had greater average SNA before and during cold stimulation that non-cold-acclimated adults. Moreover, both aged and cold-acclimated adult mice had greater response to cold at the onset of cold stimulation. After the onset of cold stimulation the change of SNA over Tco was similar between groups.

Figure 3 shows the HP as a function of change in Tco. HP increased significantly during cold and returned to prestimulus level in both NAIVE and AGED groups. There was a significant main effect of age ($F(1,29)=8.31$, $p<0.01$), and a significant main effect of change of temperature ($F(13,377)=32.95$, $p<0.001$) indicating that a decrease in body temperature results in an increase in heat production, and that aged animals produce more heat in response to cold than adults. There was also an interaction between groups and change of temperature ($F(13,377)=6.97$, $p<0.001$) indicating that aged and adult mice increase heat production in response to cold differently. Nevertheless, simple main effects of change of body temperature at the NAIVE and the AGE groups both were significant ($F(13,377)=6.28$, $p<0.001$; $F(13,377=28.60$, $p<0.001$) indicating that both age groups significantly increased HP during cold.

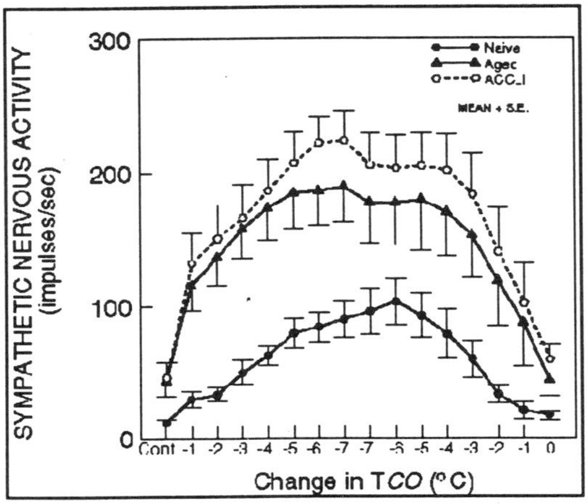

Fig. 2: Sympathetic nervous outflow to BAT during cold stress

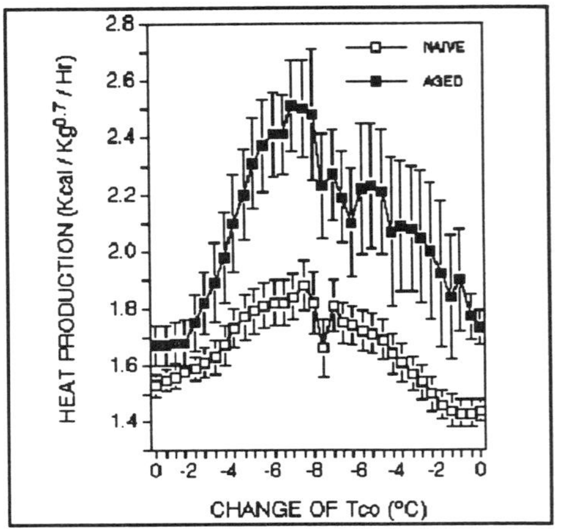

Fig. 3: Heat production during cold stress

Discussion.

The main findings of these studies are: (a) during cold stimulation SNA to BAT increased in all three groups of mice; (b) during cold stimulation the rate of decline of Tco was less in ACCLI than in NAIVE mice, while decline of Tco in AGED was not significantly different from NAIVE mice; (c) before cold stimulation SNA to BAT was significantly higher in both ACCLI and AGED mice compared to NAIVE; (d) SNA response to acute cold exposure was greater in both ACCLI and AGED mice compared to NAIVE; (e) HP response to acute cold was greater in AGED mice comparing to NAIVE.

The effect of cold stimulation on SNA to BAT was similar to that reported by Nijima et al. (9). Engel et al. (10), using a preparation similar to ours, showed that during cold stimulation SNA to BAT decreased, while SNA to skin increased. They concluded that decreased SNA in their animals was associated with recording from branches that ended on blood vessels of BAT and represented the central command to inhibit vasoconstriction and to provide increased blood flow to BAT. In the present experiments we intentionally recorded from different BAT sympathetic nerves that we believe ended on adipocytes. Our findings suggest that there are two functionally different types of sympathetic nerves innervating BAT: those that terminate on blood vessels and decrease their firing rate during cold stress and those that terminate on adipocytes and increase their firing rate.

We reported previously that repeated cold exposure of partially restrained mice resulted in cold acclimation (1). In the present studies the rate of fall of Tco also was attenuated in cold-acclimated animals. The fact that the manifestations of cold-acclimation were present under anesthesia and myorelaxation, i.e. in a condition when shivering thermogenesis was inhibited, indicates that cold acclimation resulted in increase of nonshivering thermogenesis and/or better heat conservation. Indeed, the sympathetic outflow to BAT in response to cold was greater in cold-acclimated mice which supports the idea that nonshivering thermogenesis in BAT increases. Whether heat conservation was also enhanced in these mice require additional study.

In aged animals SNA to BAT prior to cold stress and SNA response to cold were also greater than in adult, non-cold-acclimated mice. This was a surprising finding which is not easily reconciled with our previous observations on non-anesthetized mice. We have reported previously (3), as have others (11), that aged mice have lower heat production in response to cold than adult animals. Since a significant portion of heat production in rodents is attributed to non-shivering thermogenesis, it is a general consensus, that cold-induced non-shivering thermogenesis is reduced in old mice. However, based on some of our previous findings, we reported (12) that aged mice can sense the cold and can produce sufficient heat in response to mild cold. In fact, aged mice produced as much heat as adult animals in response to 18°C environmental temperature. In the study presented here, SNA to BAT in response to cold was increased. This confirmed that the central nervous system of aged animals respond to cold reliably. Increased SNA response to cold in conjunction with previously reported reduced heat production in aged mice could indicate an age-related deficit in peripheral mechanisms of heat production. However, in the experiments reported here, we showed increased heat production in aged mice in response to cold. The mice in this experiment were under anesthesia and myorelaxation, all sources of heat production except non-shivering thermogenesis were suppressed. Moreover, since the animals were on mechanical ventilation and expired air was analyzed directly, any possible age-related changes in breathing patterns could not contribute to results. These findings indicate that the non-shivering thermogenic response to cold in old mice is enhanced. Thus, we now believe that the reduced total cold-induced heat production in aged mice may reflect reduced thermogenic capacity of somatomotor system - i.e., reduced lean body mass. It is necessary to remember, that age-related reduction in lean body mass is well documented for mice, but rat's body weight is increased with age, therefore, our considerations are not necessarily applicable to research conducted on rats.

References.

1. Talan, MI, Engel, BT (1988): Habituation and dishabituation to repeated mild cold exposures in C57BL/6J mice. Physiol Behav 44, 753-757.

2. Talan, MI, Engel, BT (1984): An improved cold stress test for aging mice. Exp Gerontol 19, 79-86.

3. Tatelman, HM, Talan, MI (1990): Metabolic heat production during repeated cold stress in adult and aged male C57BL/6J mice. J Gerontol 45, B215-219.

4. Ricquier, D, Bouillaud, F (1986): The brown adipose tissue mitochondrial uncoupling protein. In: Brown Adipose Tissue. Trayhurn, P, Nicholls, DG eds. London: Edward Arnold Ltd, 86-104.

5. Scarpace, PJ, Mooradian, AD, Morley, JE (1988): Age-associated decrease in beta-adrenergic receptors and adenylate cyclase in rat brown adipose tissue. J Gerontol 43, B65-70.

6. Flaim, KE, Horowitz, JM, Horowitz, BA (1976): Functional and anatomical characteristics of the nerve-brown adipose interaction in the rat. Pfluger Arch 365, 9-14.

7. Kawate, R, Talan, MI, Engel, BT (1993): Aged C57BL/6J mice respond to cold with increased sympathetic nervous activity to interscapular brown adipose tissue. J Gerontol 48, B180-183.

8. Kawate, R, Talan, MI, Engel, BT (1993): Sympathetic nervous activity to brown adipose tissue increases in cold-tolerant mice. Physiol Behav 28, 579-584.

9. Nijima, A, Rohner-Jeanrenaud, F, Jeanrenaud, B (1984): Role of ventromedial hypothalamus on sympathetic efferents of brown adipose tissue. Am J Physiol 247, R650-654.

10. Engel, BT, Sato, A, Sato, Y (1992): Responses of sympathetic nerves innervating blood vessels in interscapular, brown adipose tissue and skin during cold stimulation in anesthetized C57BL/6J mice. Jpn J Physiol 42, 549-559.

11. McDonald, RB, Horowitz, BA, Hamilton, JS, Stern JS (1988): Cold- and norepinephrine-induced thermogenesis in younger and older Fisher 344 rats. Am J Physiol 254, R457-462.

12. Tatelman, HM, Talan, MI (1993): Metabolic heat production and cold tolerance during cold stress of different intensity of adult and aged male C57BL/6J mice. Exp Gerontol 28, 579-584.

EFFECTS OF FEMALE SEX STEROIDS IN THERMOGENIC TISSUES

M. Puerta, M. P. Nava, M. Abelenda and A. Fernández

Dept. Animal Biology II (Physiology), Faculty of Biological Sciences, Complutense University, 28040 Madrid, Spain

The great heat generating power of brown adipose tissue (BAT) has a profound influence in energy balance. Cold-acclimated animals become hyperphagic to sustain the high substrate combustion in brown fat depots without reducing body energy stores. On the other hand, the low activity of brown fat in obese animals contributes to the low energy expenditure and the development of obesity. Reproduction is one of the major energy challenges in the mammalian life cycle. On viewing the profound influence of BAT in energy balance, it seems not surprising that brown adipose tissue is suppressed during late pregnancy and lactation as a way to save substrates for growth of mother and litter (1,13). Female sex steroids, 17ß-oestradiol and progesterone seem to be involved in such disactivation. Brown adipocytes isolated from female rats with high plasma levels of sex steroids show a lower responsiveness to noradrenaline (NA) - as revealed by the lower increase in respiration rate after NA addition to the incubation medium- than that displayed by adipocytes isolated from untreated controls (10). The decreased responsiveness of brown adipocytes when plasma levels of oestradiol are high leads to a reduction of BAT thermogenic activity in cold-acclimated rats (11). This poses two additional questions, i.e., are oestradiol-treated females rats hypothermic? or are there alternative adjustments that compensate for the low BAT thermogenesis? On the contrary, progesterone impairment of responsiveness to NA does not decrease BAT thermogenesis (9). This raises another question, how progesterone treated animals maintain BAT thermogenesis despite the low responsiveness of brown adipocytes to NA? As in the case of oestradiol a compensatory adjustment was hypothesized. The results we have obtained in our laboratory are presented below.

Materials and Methods

Oestradiol studies. Female Wistar rats, 215±1 g body weight, were preacclimated for two weeks to cold (6°C). Once acclimated, one Silastic capsule (5 mm long and with 3.2 outer and 1.5 inner diameter) empty or filled with 17ß-oestradiol was subcutaneously implanted. After 15 days of hormonal treatment, rectal temperature and O_2 were recorded at 6°C. To avoid the obligatory postprandial thermogenesis food was removed 12-14 hours before oxygen consumption measurements that were carried out in a closed circuit system, soda lime serving as the CO_2 absorbent. Rectal temperature was measured twice with an electronic probe inserted 5 cm into the rectum, at the start of food withdrawal and in the food deprived state just before the start of the O_2 measurement. Samples of liver, heart, uterus, kidney and two skeletal muscles (*vastus lateralis*, *rectus abdominis*) were removed and analyzed for protein content and cytochrome c oxidase activity.

Table 1. Oxygen consumption, rectal temperature and oxidative capacity of several tissues in cold-acclimated rats with high plasma levels of oestradiol.

	CONTROL	TREATED
Rectal temperature (°C)		
Unfasted	38.4±0.2 (11)	38.2±0.1 (18)
Fasted	37.2±0.2 (11)	37.5±0.1 (18)
O_2 (ml O_2/min.Kg$^{0.75}$)	25.2±1.1 (8)	23.9±0.6 (8)
COX activity (μmol ox. min^{-1})		
Vastus lateralis		
per 100 mg tissue	0.5±0.1 (5)	0.3±0.1 (5)*
per mg protein	0.06±0.01 (5)	0.04±0.01 (4)
Rectus abdominis		
per 100 mg tissue	0.2±0.0 (6)	0.3±0.0 (6)*
per mg protein	0.04±0.01 (6)	0.05±0.00 (6)*
Kidney		
per 100 mg tissue	2.1±0.3 (6)	2.9±0.3 (6)
per mg protein	0.18±0.02 (6)	0.24±0.02 (6)*

COX: Cytochrome c oxidase activity; O_2 expressed in STPD conditions; * $P<0.05$ when compared with the same tissue in untreated animals (one-way ANOVA).

Progesterone studies. When progesterone is administered systemically it reaches both brown adipocytes -where it depresses NA responsiveness- and the central nervous system. Since BAT thermogenesis is not depressed in progesterone-treated rats (9) despite the reduced responsiveness of brown adipocytes to NA (Abelenda et al. unpublished results), we reasoned

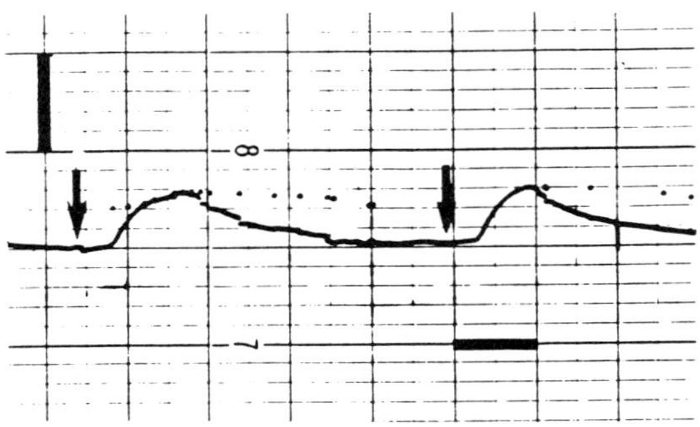

Fig. 1. Increases in IBAT temperature after i.c.v. administration of glutamate. Arrows indicate time of injection. Horizontal bar, 10 minutes. Vertical bar, 2.5°C.

that progesterone might act centrally facilitating or increasing the sympathetic discharge to the BAT after cold-exposure. If so, it may counterbalance the reduced NA responsiveness of brown adipocytes, so that BAT thermogenesis might remain unaltered, as it is. Accordingly, to differentiate possible central from systemical effects, we assessed BAT function in two kinds of animals, those which received the hormone systemically and those which only received it centrally.

Systemic administration was carried out with two Silastic capsules (4 cm) implanted subcutaneously whereas central administration was accomplished by i.c.v. administration of a progesterone solution in sesame oil. In both cases we used female rats (215 g b.w.) preacclimated to thermoneutrality (28°C). BAT function was assessed by measuring the changes in temperature and in lipolysis in the interscapular depot after a central stimulation by means of the excitatory neurotransmitter glutamate given stereotaxically in a lateral ventricle. BAT temperature was measured by placing a large-surface thermocouple below the interscapular depot (IBAT). Lipolysis was measured by comparing the arterio-venous differences in free fatty acids (FFA) content between the ventral aorta and the Sulzer's vein blood. Blood was collected when the increase in IBAT temperature was at its maximum after glutamate administration. FFA were determined spectrophotometrically in plasma samples.

Results and Discussion

Oestradiol studies. Oestradiol treatment did not induce hypothermia in the cold-acclimated rats. Further, the degrees of hypothermia induced by food deprivation in the cold-exposed rats acclimated to 6°C did not differ between oestradiol-treated rats and controls (Table 1). These results indicate that oestradiol does not alter the levels at which core temperature is regulated under various thermal conditions and metabolic states (5).

When considering oxygen consumption as the most important thermoregulatory effector in conditions of cold-exposure, the results of the present study suggest that its efficiency was not impaired by the oestradiol treatment as shown in Table 1. This is at variance with the observations of Laudenslager et al. (8) but agrees with those of Richard (12) and Fregly et al. (6). The conclusion that metabolic cold defence is equally efficient in cold-acclimated rats with and without oestradiol treatment does, however, not exclude the possibility that the modes of metabolic heat generation may be different. Our own preceding studies (10,11) showed that high oestradiol levels depress the oxidative capacity of BAT and reduce its responsiveness to NA. If this oestradiol effect is taken into account for the present study, it must be hypothesized that other tissues are capable of compensating for BAT as a source of regulatory heat production in the cold. Support for this hypothesis may be derived from the observation of increased cytochrome c oxidase activity per mg of tissue protein in the kidney and the *rectus abdominis* muscle (Table 1) among the analyzed tissues (5). In the *vastus lateralis* cytochrome c oxidase activity was found to be reduced, but it remains to be determined whether this was a consequence of the general decrease in protein content recorded in that muscle (data not shown). Since our experimental approach cannot exclude the contribution of shivering in the cold defence of oestradiol-treated animals, differences in metabolic capacity of a trunk muscle (*rectus abdominis*) as compared to a leg muscle (*vastus lateralis*) might be due to their different involvement in the shivering response. Nonetheless, skeletal and vascular smooth muscle have also been considered as sites of non-shivering heat production (4,7). Therefore, results presented here suggest that oestradiol changes the relative contribution of metabolically relevant tissues to the maintenance of body temperature in cold-acclimated rats. However, the processes -shivering and non-shivering thermogenesis- and tissues involved remain to be elucidated.

Table 2. Increases in IBAT temperature and arterio-venous differences in plasma free fatty acids after either central or systemic progesterone administration.

	CONTROL	TREATED
Changes in IBAT temperature (°C)		
Systemic administration	0.96±0.09 (16)	0.95±0.09 (16)
Central administration	0.62±0.04 (11)	0.87±0.08 (18)*
Arterio-venous differences in FFA (μmol/ml)		
Systemic administration	-0.08±0.06 (4)	-0.21±0.11 (4)
Central administration	-0.08±0.08 (8)	-0.14±0.03 (8)†

* $P<0.01$ when compared with control animal (Student's t test). † $P<0.01$ when compared with zero (two-tailed Student's t test).

Progesterone studies. Glutamate is an excitatory neurotransmitter which is known to increase BAT temperature when applied to the ventromedial and the paraventricular hypothalamic nuclei by increasing sympathetic discharge (2,3). In agreement with these previous observations, the administration of glutamate into a lateral ventricle led to a well-defined increase in BAT temperature in the four animal groups we used (Fig. 1). However, when it was given to animals which had received progesterone systemically, no IBAT temperature differences were recorded between control and treated animals (Table 2). Since our previous in vitro studies revealed that brown adipocytes of progesterone-treated rats are less responsive to NA, this lack of differences raised the possibility of an increase in sympathetic discharge as a possible compensatory adjustment. The results with central administration corroborate this view. In fact, rats receiving progesterone just at a central level showed a greater increase in IBAT temperature after glutamate administration than those receiving vehicle alone (Table 2). This suggests that progesterone acting centrally increases the sensitivity to a chemical stimulus which is known to increase sympathetic discharge to BAT and hence BAT temperature. This explanation fits well with the unaltered BAT thermogenesis of cold-acclimated rats with high plasma progesterone levels (9). Moreover, when we measured the plasma concentration of FFA and calculate the arterio-venous differences after central glutamate administration, these differences were not statistically different from zero either in the two groups treated systemically or in the control animals receiving vehicle just centrally (Table 1). But an arterio-venous difference significantly different from zero was obtained in those rats which received progesterone only centrally. This shows a net exchange between blood and IBAT in this latter case. In other words, lipolysis in response to central administration of glutamate is enhanced in rats receiving progesterone only at a central level, reinforcing the view of a facilitatory action of progesterone for stimuli that drive sympathetic discharge to BAT.

In summary, our studies have shown that female sex steroids reduce the responsiveness to NA of brown adipocytes. Oestradiol leads to a reduced BAT thermogenesis. However, an increased thermogenesis in other thermogenically relevant tissues of oestradiol-treated animals seems to maintain homeothermia. On the contrary, progesterone appears to have a central effect that could enhance sympathetic discharge to BAT so compensating the decreased NA responsiveness of brown adipocytes.

Acknowledgements

This work was supported by CICYT Grant SAL 91-0502. A. Fernández was the recipient of a fellowship from the Complutense University. Thanks are given to C. Castro, R. Martínez and C. Venero.

References

1 Abelenda, M, Puerta, M (1987): Inhibition of diet-induced thermogenesis during pregnancy in the rat. Pflügers Arch 409, 314-317

2 Amir, S (1990): Stimulation of the paraventricular nucleus with glutamate activates interscapular brown adipose tissue thermogenesis. Brain Res 508, 152-155

3 Amir, S (1990): Intraventromedial hypothalamic injection of glutamate stimulates BAT thermogenesis in the rat. Brain Res 511, 341-344

4 Colquhown, EQ, Clark, MG (1991): Open question: Has thermogenesis in muscle been overlooked and misinterpreted? NIPS 6, 256-259

5 Fernández, A, Abelenda, M, Nava, MP, Puerta, M: Oxygen consumption in oestradiol treated rats. Pflügers Arch (in press)

6 Fregly, MJ, Black, DJ, Kelleher, DL, MacArthur, SA (1977): Tolerance of oestrogen-treated rats to acute cold-exposure (Abstract). Federation Proc 36, 419

7 Himms-Hagen, J, Behrens, W, Hbous, A, Greenway, D (1976): Altered mitochondria in skeletal muscle of cold-acclimated rats and the adaptations for nonshivering thermogenesis. In: Regulation of depressed metabolism and thermogenesis. Jansky, L, Musacchia, XJ eds. Springfield: Thomas Ch Publisher, 243-260

8 Laudenslager, ML, Wilkinson, CW, Carlisle, HJ, Hammel, HT (1980): Energy balance in ovariectomized rats with and without estrogen replacement. Am J Physiol 238, R400-R405

9 Nava, MP, Abelenda, M, Puerta, M (1990): Cold-induced and diet-induced thermogenesis in progesterone-treated rats. Pflügers Arch 415, 747-750

10 Puerta, M, Abelenda, M, Nava, MP, Fernández, A (1993): Reduced noradrenaline responsiveness of brown adipocytes isolated from oestradiol-treated rats. Can J Physiol Pharmacol 71 (in press).

11 Puerta, M, Nava, MP, Abelenda, M, Fernández, A (1990): Inactivation of brown adipose tissue thermogenesis by oestradiol treatment in cold-acclimated rats. Pflügers Arch 416, 659-662.

12 Richard, D (1986): Effects of ovarian hormones on energy balance and brown adipose tissue thermogenesis. Am J Physiol 250, R245-R249.

13 Trayhurn, P, Douglas, JB, McGuckin, MM (1982): Brown adipose tissue thermogenesis is "suppressed" during lactation in mice. Nature 298, 59-60.

Thermal Balance in Health and Disease
Advances in Pharmacological Sciences

EFFECT OF TUMOR NECROSIS FACTOR ON THE LIPOPROTEIN LIPASE GENE EXPRESSION IN BROWN ADIPOCYTES DIFFERENTIATED IN CULTURE

Y. Uchida , K. Irie , F. Tsukahara , K. Ohba , T. Nomoto and T. Muraki

Department of Pharmacology, Tokyo Women's Medical College, 8-1 Kawada-cho, Shinjuku-ku, Tokyo 162, Japan.

Brown adipose tissue (BAT) has a very high thermogenic capacity in response to cold (1) and diet (2). It dissipates energy in the form of heat which is produced via the specific mitochondrial uncoupling protein thermogenin and controls energy balance (1). A reduction in thermogenic function of BAT is highly associated with the development of obesity (3), while an increase in thermogenic function of BAT may have a causal relationship to a catabolic state such as cachexia of cancer or infection. However, the role of BAT in the enhanced catabolism has not been defined clearly.

Tumor necrosis factor-α (TNF-α) /cachectin, which acts as mediator of cachexia, elicits wasting of adipose tissue triglyceride stores (cachexia) accompanied by a decrease in lipoprotein lipase (LPL) activity (4,5) and an increase in intracellular lipolysis (5,6). It is generally thought that TNF exerts the effects by acting on TNF receptors on the cell membrane and subsequently stimulates intracellular events such as activation of protein kinase C (7) and release of arachidonic acid (6).

In the present study, we have examined the effect of TNF on LPL enzyme activity and LPL mRNA amounts in primary culture of brown adipocytes, in an attempt to clarify whether the effect of TNF is involved in gene expression. Furthermore, the possibility of the involvement of cyclooxygenase metabolites as a second messenger in the TNF effect was examined by using a cyclooxygenase inhibitor, indomethacin.

Methods

Cell isolation and Culture Brown fat precursor cells were isolated from interscapular, cervical and axillary brown adipose tissue of 4-week-old male ICR mice, as described by Rehnmark et al. (8). Cells were cultured in 6-well multiplates (Corning, 35 mm) in Dulbecco's modified Eagle's medium supplemented with 10% newborn bovine serum, 4 nM insulin, 10 mM Hepes (pH 7.4), 50 IU/ml penicillin, 50 μg/ml streptomycin, and 25 μg/ml sodium ascorbate at

37°C in an atmosphere of 5% CO_2. The medium was changed on days 1, 3 and 6 after plating and the cells were analysed on day 9. The cells were exposed to 5 or 50 ng/ml of TNF-α (recombinant human TNF-α: 2.55 x 10^6 JRU/mg of protein, Dainippon Pharmaceutical Co., Ltd., Osaka) for 3, 6 or 24 h and to indomethacin for 24 h (18 h prior to the addition of TNF 5 ng/ml, 6h).

Cell harvest and measurement of LPL activity At the end of the incubation with TNF and/or indomethacin, the medium was changed with fresh medium without TNF or indomethacin and the cells were exposed to 10 IU/ml heparin for 1 h. Then LPL activity was measured in two fractions; LPL released by heparin into the culture medium (heparin-releasable: HR) and LPL activity remaining in cells (extractable: EXT). LPL was assayed by the method of Nilsson-Ehle and Schotz (9) with minor modifications using a [^{3}H]triolein (NEN, SpA 20.8 Ci/mmol)-containing substrate emulsified with lecithin and containing normal rat serum as a source of apo-

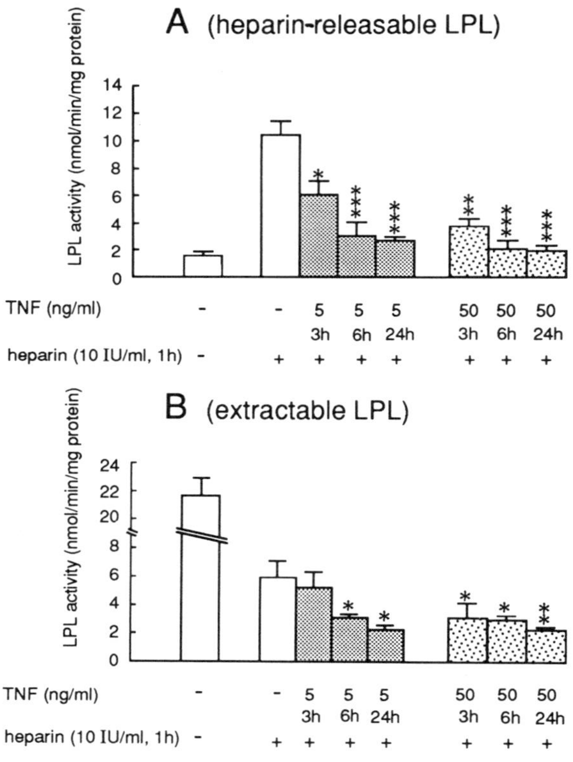

Fig. 1: The effect of TNF on LPL activity in cultured brown adipocytes. Primary culture cells on day 9 were treated with the indicated concentrations of TNF (5, 50 ng/ml) for the indicated times (3, 6, 24 h) followed by 1 h heparin exposure, before determination of LPL activity. The amount of LPL activity was determined in two fractions: A, LPL released by heparin (10 IU/ml, 1 h) into the medium (heparin-releasable: HR) and B, LPL activity remaining in cells (extractable: EXT). Values are means ± SE (bars) obtained from 4-7 experiments. * $P< 0.05$, ** $P< 0.01$, *** $P < 0.001$ vs. no additon (TNF -, heparin +). Student's unpaired t-test was used for statistics.

lipoprotein-CII. LPL activity was expressed as nanomoles of FFA released per mg of protein measured by the method of Lowry et al. (10).

RNA extraction and Northern analysis The cells were washed once in 2 ml of saline and dissolved in 1 ml of hot guanidine HCl extraction buffer. Total RNA was isolated by the guanidine hydrochloride method (11) and 5 μg of total RNA per lane was separated by electrophoresis on 1 % agarose gel containing 6 % formaldehyde, followed by blotting onto nylon filter (Hybond N, Amersham). The blots were hybridized with ^{32}P-labeled probes of mouse LPL cDNA (mL-5, courtesy of Dr. M.C.Schotz) and bovine β–actin cDNA (pBA1, generous gift of Dr.D.R.Morris). After hybridization with [^{32}P]LPL cDNA, the membranes were washed at room temperature in 2xSSC (sodium chloride/sodium citrate buffer) 0.2% SDS (sodium dodecyl sulfate), and at 50°C in 0.1xSSC, 0.2% SDS before autoradiography and quantitative densitometric scanning. Then, membranes were stripped of hybridized cDNA by boiling in 0.1 % SDS solution for 5 min to allow reprobing with a β-actin [^{32}P] cDNA to check for consistency of RNA amount.

Results

The addition of heparin (10 IU/ml) for 1 h produced a prompt release of LPL into the medium. LPL activity in medium after heparin exposure (heparin-releasable LPL) was several times higher than basal LPL activity secreted spontaneously into the medium (Fig 1A, open column), while LPL activity remained in cells (extractable LPL) was reduced by 30 % (Fig 1B, open column) showing the opposite response to heparin in both fractions (HR, EXT). After 3 h of incubation, TNF in doses of 5 and 50 ng/ml reduced HR activity by 42 % and 64 %, respectively. Time-course studies showed that a reduction of HR activity by TNF (5 ng/ml) was apparent as early as 3 h and that a submaximal effect was observed 6 h later. When the 10 times higher

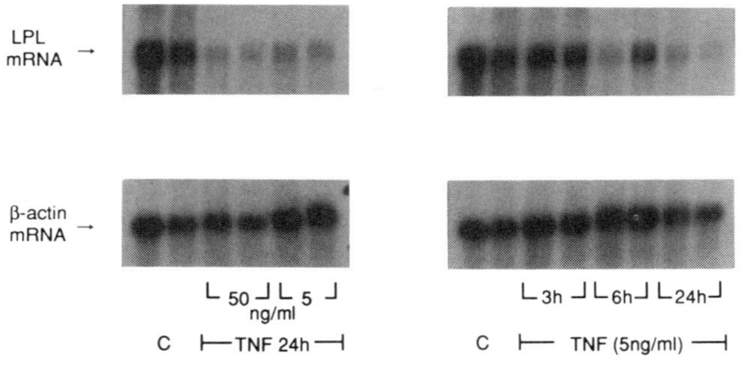

Fig. 2: Northern analysis of LPL mRNA levels in brown adipocytes differentiated in culture. Primary culture cells on day 9 were treated with TNF (5, 50 ng/mg) for 24 h (on the lefthand side) and 5 ng/ml TNF for 3, 6, and 24 h (on the righthand side). Five μg of total RNA per each lane was electrophoresed on a formaldehyde-agarose gel. mRNAs of LPL and β-actin were detected using P32-labeled mouse LPL cDNA or bovine β-action cDNA.

amount of TNF (50 ng/ml) was added, a greater decrease was observed (Fig. 1A). We saw almost the same tendency with the effect of TNF on LPL activity in the EXT fraction, except that addition of 5 ng/ml of TNF for 3 h was insufficient to cause a significant decrease in LPL activity (Fig. 1B). As shown in Fig. 2, TNF in doses of 5 and 50 ng/ml caused a marked decrease in LPL mRNA, whereas β-actin levels were not reduced in the presence of TNF.

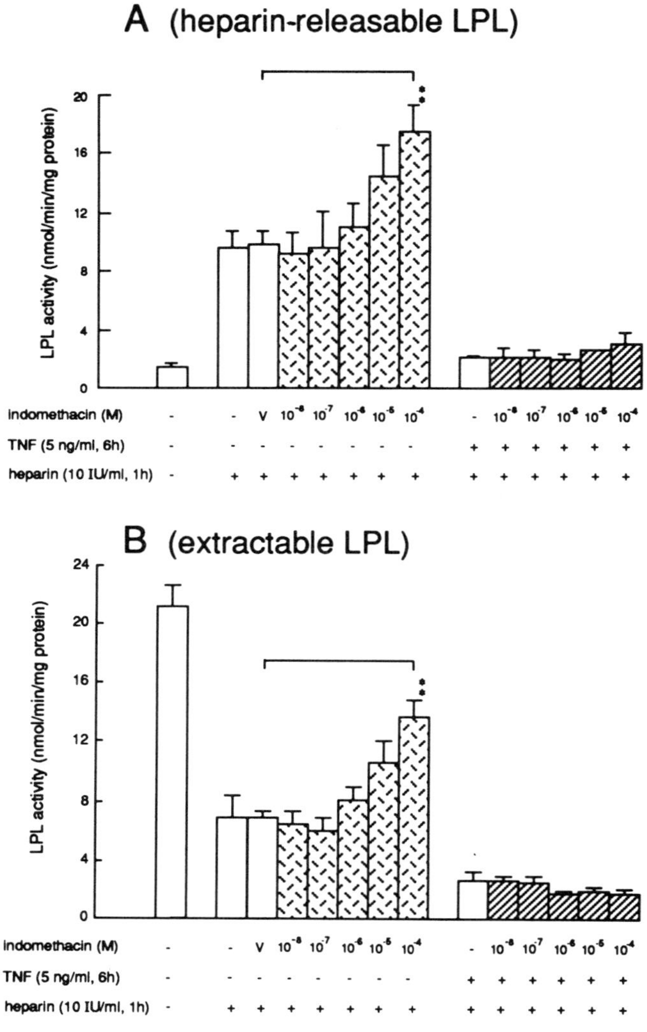

Fig. 3.: The effect of various doses (10^{-8}-10^{-4}, 18 h) of indomethacin on TNF (5 ng/ml, 6 h) induced suppression of LPL activity. V represents vehicle of indomethacin (4 % ethanol solution). Experimental conditons are the same as described in Fig. 1.

As it is reported that many in vitro effects of TNF are mediated by activation of the cyclooxygenase pathway (12, 13), we have investigated possible involvement of prostaglandins in the inhibitory effect of TNF on LPL activity in brown adipocytes. Indomethacin by itself caused a dose-dependent increase in LPL activity in both fractions (HR, EXT). Pretreatment of brown adipocytes with indomethacin (10^{-8}-10^{-4} M, 18 h) did not cause any significant changes in suppressive effects of TNF (5ng/ml, 6 h) on LPL activity in either fraction (Fig. 3).

Discussion

These results show that TNF in doses of 5 and 50 ng/ml down-regulates the LPL activity of brown adipocytes differentiated in vitro. The effect of TNF on LPL activity in mouse brown adipocytes involves a specific and parallel reduction in levels of mRNA. The inhibition of LPL mRNA levels is specific, as β-actin levels were not reduced in the presence of TNF. Moreover, the magnitude of the reduction in LPL activity was quite similar to the decrease in levels of LPL mRNA. Thus, the primary effect of TNF on LPL in mouse brown adipocytes is apparently at the level of LPL gene expression.

There are some studies about TNF effects on thermogenesis in BAT, although they are limited to in vivo experiments only. Most of the actions of TNF are believed to be mediated peripherally, but it seems likely that activation of thermogenesis by TNF involves central mechanisms to produce an increase in sympathetic outflow. Coombes et al. reported that an i.v. injection of TNF-α stimulates sympathetic outflow to BAT in rats and subsequently elevates BAT thermogenic activity (mitochondrial GDP binding) (14). Actually, sympathetic activation of BAT thermogenesis in cachexia was demonstrated in tumor-bearing mice (15). Moreover, when administered intracerebroventricularly, TNF-α elicits a direct central action to modulate the sympathetic efferent outflow to interscapular BAT (16, 17). Our in vitro data obtained from brown adipocytes showed that TNF reduced directly the LPL activity without being under control of the sympathetic nervous system. This confirms previous reports on a direct in vitro effect of TNF on LPL activity of white adipocytes cell lines such as 3T3-L1, TA1 and on primary culture. It is well known that LPL catalyzes the hydrolysis of circulating triglycerides to free fatty acids that can be taken up, esterified, and stored in adipose cells. Based on our data showing the suppressive effect of TNF on LPL activity, it was demonstrated that brown adipocytes act as one of the factors for development of the cachectic state.

In our in vitro culture cells, brown fat cellular responses to TNF seem to be mediated by the binding of TNF to specific cell surface receptors. Recently two distinct TNF receptor subtypes (TNF-R1 55kDa, TNF-R2 75kDa) were identified(18). However, it remains uncertain which receptor is closely associated with this TNF effect. Furthermore, little is known of the intracellular mediators of the TNF action. It is reported that TNF activates two second messenger pathways, including activation of protein kinase C and release of arachidonic acid. Here we demonstrated that treatment of brown adipocytes with a cyclooxygenase inhibitor, indomethacin, did not cause any effect on TNF-induced suppression of LPL activity. This result is consistent with the finding reported by Hardardottir et al.(19), who demonstrated that TNF enhances lipolysis and decreases LPL acitvity in 3T3 adipcytes by a PG independent mechanism. Some evidence obtained by Haliday et al. (20) supports another explanation, i.e. arachidonic acid released by TNF is required to be converted to a lipoxygenase metabolite but not to a cyclooxygenase metabolite for mediating the TNF action. In our experiments demonstrating that a cyclooxygenase inhibitor has no effect, the same explanation could be valid.

In conclusion, we have demonstrated that TNF-α potently suppresses LPL activity by inhibiting the LPL gene expression in brown adipocytes differentiated in culture, possibly via a cyclooxygenase pathway independent mechanism. This may play an important role in the pathogenesis of an enhanced catabolic state such as cachexia associated with infection or malignancy.

Acknowlegements

We are grateful to Drs. B.Cannon and J. Nedergaard (Wenner-Gren Institute, Stockholm) for giving Y.U. an opportunity to collaborate and be trained brown fat cell culture techniques.

References

1. Cannon, B, Nedergaard, J (1985): The biochemistry of an inefficient tissue: brown adipose tissue. Essays Biochem 20, 110-164.

2. Rothwell, NJ, Stock, MJ (1979): A role for brown adipose tissue in diet-induced thermogenesis. Nature 281, 31-35.

3. Trayhurn, P (1986): Brown adipose tissue and energy balance. In: Brown Adipose Tissue. Trayhurn, P, Nicolls, DG eds. London: Edward Arnold, 299-338.

4. Price, SR, Olivecrona, T, Pekala, PH (1986): Regulation of lipoprotein lipase synthesis by recombinant tumor necrosis factor - The primary regulatory role of the hormone in 3T3-L1 adipocytes. Arch Biochem Biophys 251 738-746.

5. Kawakami, M, Murase, T, Ogawa, H, Ishibashi, S, Mori, N, Takaku, F, Shibata, S (1987): Human rcombinant TNF suppresses lipoprotein lipase activity and stimulates lipolysis in 3T3-L1 cells. J Biochem 101, 331-338.

6. Feingold, KR, Doerrler, W, Dinarello, CA, Fiers, W, Grunfeld, C (1992): Stimulation of lipolysis in cultured fat cells by tumor necrosis factor, interleukin-1, and the interferons is blocked by inhibition of prostaglandin synthesis. Endocrinol 130, 10-16.

7. Galeotti, T, Boscoboinik, D, Azzi, A (1993): Regulation of the TNF-α receptor in human osteosarcoma cells: role of microtubules and of protein kinase C. Arch Biochem Biophys. 300, 287-292.

8. Rehnmark,S, Kopecky,J, Jacobsson, A, Nechad,M, Herron,D, Nelson, BD, Obregon, M-J, Nedergaard,J, Cannon, B (1989): Brown adipocytes differentiated *in vitro* can express the gene for the uncoupling protein thermogenin: effects of hypothyroidism and norepinephrine. Exp Cell Res 182, 75-83.

9. Nilsson-Ehle, P, Schotz, MC (1976): A stable, radioactive substrate emulsion for assay of lipoprotein lipase. J Lipid Res 17, 536-541.

10. Lowry, OH, Rosebrough, NJ, Farr, AL, Randall, RJ (1951): Protein measurement with the folin phenol reagent. J Biol Chem 193, 265-275.

11. MacDonald,RJ, Swift,GH, Przybyla,AE, Chirgwin,JM (1987): Isolation of RNA using guanidinium salts. Methods Enzymol 152, 219-227.

12. Dayer, J-M, Beutler, B, Cerami, A (1985): Cachectin/tumor necrosis factor stimulates collagenase and prostaglandin E_2 production by human synovial cells and dermal fibroblasts. J Exp Med 162, 2163-2168.

13. Suffys, P, Beyaert, R, Van Roy, F, Fiers, W (1987): Reduced tumour necrosis factor-induced cytotoxicity by inhibitors of the arachidonic acid metabolism. Biochem Biophys Res Commun 149, 735-743.

14. Coombes, RC, Rothwell, NJ, Shah, P, Stock, MJ (1987): Changes in thermogenesis and brown fat activity in response to tumour necrosis factor in the rat. Biosci Rep 7, 791-799.

15. Brooks, SL, Neville, AM, Rothwell, NJ, Stock, MJ, Wilson, S (1981): Sympathetic activation of brown-adipose-tissue thermogenesis in cachexia. Biosci Rep 1, 509-517.

16. Rothwell, NJ (1988): Central effects of TNFα on thermogenesis and fever in the rat. Biosci Rep 8, 345-352.

17. Holt, SJ, Grimble, RF, York, DA (1989): Tumour necrosis factor-α and lymphotoxin have opposite effects on sympathetic efferent nerves to brown adipose tissue by direct action in the central nervous system. Brain Res 497, 183-186.

18. Tartaglia, LA, Goeddel, DV (1992): Two TNF receptors. Immunol Today 13, 151-153.

19. Hardardottir, I, Doerrler, W, Feingold, KR, Grünfeld,C (1992): Cytokines stimulate lipolysis and decrease lipoprotein lipase activity in cultured fat cells by a prostaglandin independent mechanism. Biochem Biophys Res Commun 186, 237-243.

20. Haliday, EM, Ramesha, CS, Ringold, G (1991): TNF induces *c-fos* via a novel pathway requiring conversion of arachidonic acid to a lipoxygenase metabolite. The EMBO J 10, 109-115.

Thermal Balance in Health and Disease
Advances in Pharmacological Sciences

ETHANOL AND BROWN ADIPOSE TISSUE THERMOGENESIS IN RATS EXPOSED TO THE COLD: UNCOUPLING PROTEIN m-RNA

P. Huttunen[1], M. Sämpi[1] and R. Myllylä[2]

[1]Department of Forensic Medicine, University of Oulu, Kajaanintie 52 D, FIN-90220 Oulu, Finland and [2]Department of Biochemistry, University of Oulu.

Introduction

Many drugs such as ethanol act centrally to alter the processes of thermoregulation. The magnitude of the effect of ethanol depends on both dose and ambient temperature. At ambient temperatures below the thermoneutrality of the animal ethanol lowers the core temperature through increased heat loss and/or decreased heat production. Since brown adipose tissue is the only tissue whose main function is to produce heat, we examined whether the ethanol-induced drop in the core temperature is associated with the mitochondrial uncoupling protein (UCP), which is responsible for energy dissipation as heat in brown fat.

Material and methods

Male Sprague-Dawley rats at the age of 3 months were used. The dose of ethanol was 2 g/kg (i.p.), and the rats were exposed to the cold (+4°C) for 30 min. Controls received 0.9% NaCl solution. Colonic temperature was measured prior to the injection of drugs and after cold exposure. After cold exposure the rats were killed with CO_2 and the interscapular brown adipose tissue (IBAT) was dissected out. Uncoupling protein m-RNA in the brown fat was measured by the Northern blot and dot blot methods using a cDNA probe coding for rat UCP (a gift from Dr. D. Ricquier, Meudon-Bellevue, France). The validity of the UCP m-RNA variations was verified by hybridization of brown fat RNA to a rat 18 S ribosomal RNA oligonucleotide probe.

Results

The mean blood alcohol concentration of the rats after cold exposure was 2.4 ± 0.2 g/l, and their colonic temperature had dropped about 2° in half an hour, while that of the controls remained unchanged.

The UCP m-RNA level in the IBAT had decreased by 40% relative to the controls ($p < 0.01$, t-test).

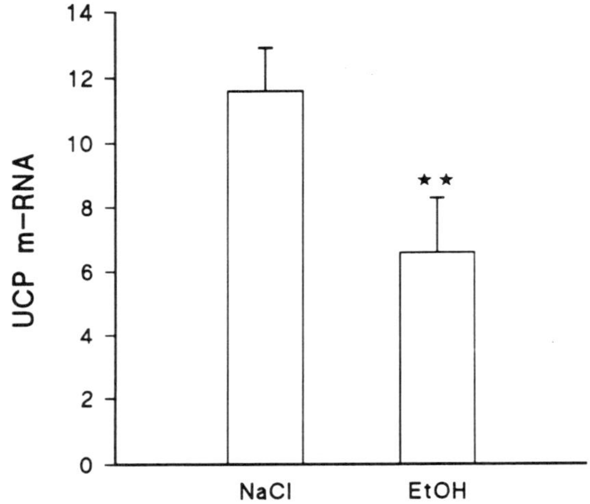

Fig. 1: Effect of ethanol (2g/kg i.p.) on UCP m-RNA level in the IBAT of rats exposed to the cold (4°) for 30 min. Controls received 0.9% NaCl solution. The amount of UCP m-RNA is expressed as the ratio of densitometric readings of UCP m-RNA to those of 18 S ribosomal RNA. Values are expressed as means ± SE. ** = $p < 0.01$.

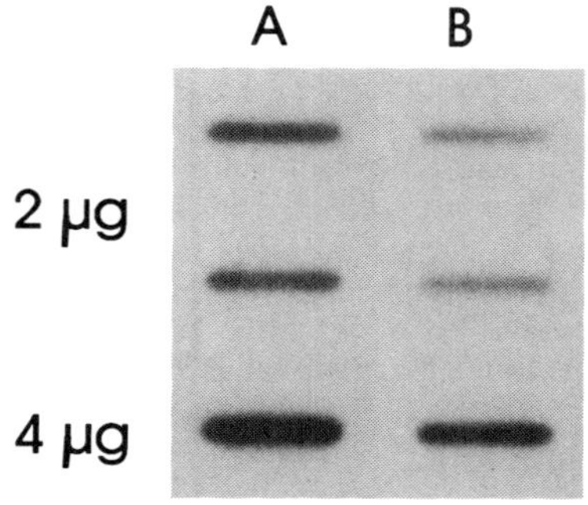

Fig. 2: Assay of UCP m-RNA level in the IBAT; dot-blot hybridization. Dots correspond to 2 or 4 µg of RNA. RNA was hybridized to the ^{32}P-labelled UCP cDNA. The rats were exposed to 4° for 30 min. (A) controls, (B) ethanol-treated rats.

Discussion

Ethanol may have inhibited cold-induced UCP gene expression, since the UCP m-RNA level in the ethanol-treated rats had significantly decreased. It has been shown that the UCP gene is transcriptionally regulated by norepinephrine and that triiodothyronine amplifies this effect on transcription. Induction UCP m-RNA is rapid following stimulation of the brown adipose tissue by exposure of the rats to the cold. Acute exposure to 4° for 15 min has been found to induce a 10-fold increase in the initiation on rate of UCP gene transcription (1), and a small but significant increase in the proportion of UCP protein in brown fat in the rat has been reported within the first hour of exposure to 5° (2). It is possible that ethanol reduces the brown fat response to catecholamines rather than suppressing cold-induced activity in the sympathetic nervous system (3). It is also possible that ethanol may directly inhibit UCP gene expression or that it may centrally diminish other signals that stimulate UCP gene expression.

In conclusion, the present results show that a moderate dose of ethanol suppresses brown adipose tissue thermogenesis in rats exposed briefly to mildly cold conditions by inhibiting uncoupling protein transcription.

References

1. Ricquier, D, Casteilla, L, Bouillaud, F (1991): Molecular studies of the uncoupling protein. FASEB J 5,2237-2242.

2. Ricquier, D, Mory, G, Bouillaud, F, Thibault, J, Weissenbach, J (1984): Rapid increase of mitochondrial uncoupling protein and its mRNA in stimulated brown adipose tissue. FEBS Letters 178, 240-244.

3. Kortelainen, M.-L, Lapinlampi, T, Huttunen, P (1986): Hypothalamic and serum catecholamines in ethanol and acetaldehyde treated guinea-pigs. Relation to moderate short-term cold exposure. Alcohol 3, 351-355.

AN ARTIFICIAL INTELLIGENCE APPROACH TO SELECTIVELY SEARCH, ANALYZE, AND DISPLAY THERMOREGULATORY DATA

T. J. Malkinson, K. E. Cooper, W. L. Veale and Q. J. Pittman

Faculty of Medicine, University of Calgary, Calgary, Canada. T2N 4N1

Artificial intelligence (AI) techniques using sophisticated computer software are increasingly being used in the analysis of biological systems and for the design of biological models. The term artificial intelligence refers to machine-based processes that are functionally equivalent to human cognitive tasks for analyzing data in the domain for which they were designed. They are considered intelligent because of human-like reasoning processes. The expert system and the artificial neural network are two examples of such techniques (1). We are in the process of adapting these new computer information processing techniques into a knowledge-based modelling system which will be used to selectively search, analyze, retrieve and display thermoregulatory data in response to operator requests. We have developed a prototype system in which a rule-based expert system integrates its knowledge base with our accumulated historical database of experimentally obtained information on thermoregulation to provide new retrieval and analysis capabilities which will be useful for both research and teaching. This enables you to address otherwise unmanageable problems. Heuristics are often used. These types of systems are able to work with uncertainty, reason about what is unknown, and handle multivalued information with multiple conclusions. They are very effective at handling the problems that they were designed to solve, and are excellent for situations in which many people require access to a body of knowledge. Expert systems or neural networks when used alone are not optimal. The deficits in each can be filled by designing an integrated hybrid system in which neural networks are embedded within the rule based expert system and so both synergistically fill the gaps in each others technology. These hybrid systems are excellent for data-rich problems such as modelling and prediction. This prototype will in the future be enhanced by the embedding of neural network topology within the expert system, and by incorporating the exciting new man\machine graphical interfaces and modeling techniques which are now available to display and effectively explain the results of AI investigations.

Expert systems is a type of AI computer program that attempts to emulate the behaviour and interaction that a user might have with a human expert to solve a defined problem within a clearly defined knowledge domain (2,3). Once an expert system is created, a user can interact with it by answering computer generated multiple choice questions and requests for data. Based on the users input and available data, the program will analyze and present a report using declarative programming that focuses on encoding the knowledge that the system is to represent and which can reason about words and symbols in a manner that is much different from conventional programming. The expert system programming method consists of three elements, a rule base, an inference method and a user interface. The rule base consists of a series of IF/THEN structures expressed as logic statements which represent the "knowledge" about how to solve a specific kind of problem. These independent and transparent rules may be easily added, changed or deleted with the user interface. The inference method applies these rules to obtain a solution by dynamically asking converging questions. This software chooses which rules to consider, what additional information is required, how best to get this information, and recognizes when a solution is reached. Rules are linked together into decision trees. The inference engine dynamically directs or controls the system when it searches the knowledge base and determines the search strategy. Inference in rule based systems is based on the logic principle of "modus pones" using the logic IF A + B THEN C. Should you find that in a particular situation that A and B are both TRUE then you are justified in inferring with a high degree of specified confidence that C is also TRUE, and the THEN condition will be added to the working memory as knowledge which has been proven to be TRUE. This ability to derive information allows the programmer to analyze a collection of knowledge to arrive at logical conclusions about complex problems. A well designed system can explain the reason for its conclusions. In our prototype system we are able to enter and permanently store data obtained from our past and current thermoregulation experiments in a random access data file which serves as a repository of experimentally derived "real" data which can be used in the inferencing process. When a new problem or request is presented to the system it will reason about the information, apply the previous knowledge and produce a report of the results.

The artificial neural network is an AI computer architecture implemented in software or hardware that attempts to simulate the neurological (information) processing abilities of biological systems (4,5). The basic processing element has characteristics similar to a neuron where it is able to sum inhibitory and excitatory inputs to produce a single output which in turn "synapses" with one or more similar processing elements in a network topology. The learning and processing capability is determined by the architecture of the neuronal interconnections and the training algorithm (input-output function) used. The signal modulation characteristics of the neuronal interconnections are established during the network training process in which confirmed data is presented to the inputs together with the corresponding output. An artificial neural network differs from other statistically derived algorithms in that the relationship between the input elements and the output is neither defined or quantified. They can recognize and identify patterns based on past experience without explicitly identifying the basis on which the task is performed. Neural networks are often used to create predictive models from data. They can recognize patterns in data and adapt to changing conditions allowing the system to be a dynamic model. An example for this application to this system could be the prediction of the febrile response to a specified dose of a thermoactive agent under specified conditions based upon data from previous responses present in the database. AI also allows for the capability to apply dynamic modeling (simulation) to expert-system problems. Recent advances in computer based modeling now make it possible to easily and interactively create realistic images in 3D representation and use these to make interactive dynamic presentations (6,7). These sophisticated new technologies are rapidly becoming the norm in science and industry because of the reduction in the cost\performance ratio and learning curve using high-level device-independent software. 3D design is now highly accurate and completely detailed (photo-physically- and materially-realistic). The operator can easily interact with the model automatically generating new views, applying motion dynamics (animation), determining the relationship between spaces and parts, and 3D presentations simulating time-varying geometry and object behaviour of real and simulated objects which allow for better scientific understanding. Solid modeling capabilities which define the entire volume that the object occupies allow you to create a materially realistic 3D representation of nearly any object and extract visual and mathematical information about the objects surface and interior (8). Analysis of material properties can extend to the creation of new biological material definition files which could include properties such as thermal conductivity, thermal gradients and specific heats to name but a few. The brain is a complex, structured 3-D organ consisting of a large number of anatomically distinguishable elements. Full scale three-dimensional computer reconstruction of complex subcortical structures will allow for innovative methods of organizing available data on neuroanatomy, neural interaction, neural interconnections, simulating the interaction of neurons, physiology and pharmacology to build topographical models which may lead to increased understanding and the design of new experiments. This could also be very useful in the planning and conducting of stereotaxic operations. Computer modeling of time-varying behaviour of real and simulated objects is increasingly being used to examine physiological models of system and organ function. The computer can be used to model a real biological system by establishing a clocked sequence of discrete thermoregulatory events. When used in conjunction with multimedia systems it allows for a particularly effective presentation of data that changes over time. We propose to incorporate these techniques in our AI thermoregulatory system.

Our prototype system uses the personal computer hardware environment. The expert system shell provides for control and has the facilities for building a knowledge base including the tools of editors, external program interfaces, and uses open architecture for flexibility in system design. This shell enables the system to link and interact directly with other software packages such as spreadsheets and databases as well as to user developed conventional program fragments which are used extensively in our system. We have chosen to utilize the multiple-module programming structure available in the QuickBASIC programming language for conventional external program fragments (9). QuickBASIC combines the simplicity of the original BASIC with powerful new features that force structured programming. It is menu driven with a full screen editor in an integrated programming environment that speeds up program development. It is possible to compile an executable (.EXE) version of the file which will run faster and can be called as external programs from the expert system shell. QuickBASIC provides for multiple-module programming, named common blocks, procedure libraries, and enhanced debugging tools. These external program fragments may be called at the start of the run, from the THEN part of the rule or associated with the determination of the value for a variable or qualifier. Data may be passed to and from these external programs and the expert system shell. This portion of the knowledge base consists of "real" experimentally derived data obtained from our thermoregulatory investigations and stored in the programs database. The QuickBASIC user defined composite data structure (record) allows the programmer to combine elements of different types and associate them with a common name. Unlike arrays, record elements can have different data types each of which has a fixed number of bytes and so the record is a perfect interface to random access files and is used to construct our database. New data may be easily added. A user friendly interface provides the means by which the operator can interact with the system. Once the processor is satisfied that enough input information is present to make a logical decision other modules will extract from the database information applicable to the question the operator wants answered. An appropriate report (graphical, textual or model) will then be output to the operator. The advantage of the AI expert system approach is that the system can analyze the operators requests, search the database and use existing data to predict the result of new investigations with an indicated confidence factor. Examples of the results of an operator's search for information from our thermoregulatory database follow. Data were compiled from control experiments conducted under the same experimental conditions.

This analysis shows that this febrile response fits a normal distribution. That this frequency distribution is Gaussian validates the usual statistical procedures used often in fever work. It would also be valuable to examine the distribution for fever index to see if it formed a similar distribution.

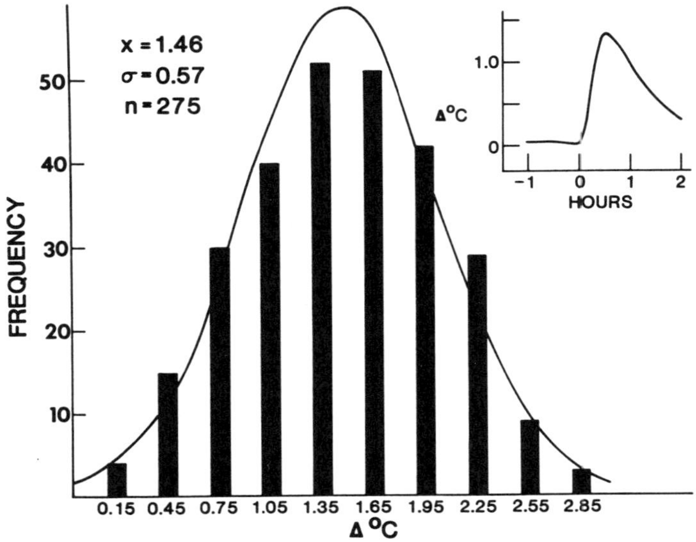

Fig. 1:The frequency distribution of the maximum change in body temperature to 400 ng Prostaglandin E_1 given intracerebroventricularly to 275 conscious adult male Wistar Rats. The inset illustrates the fever profile.

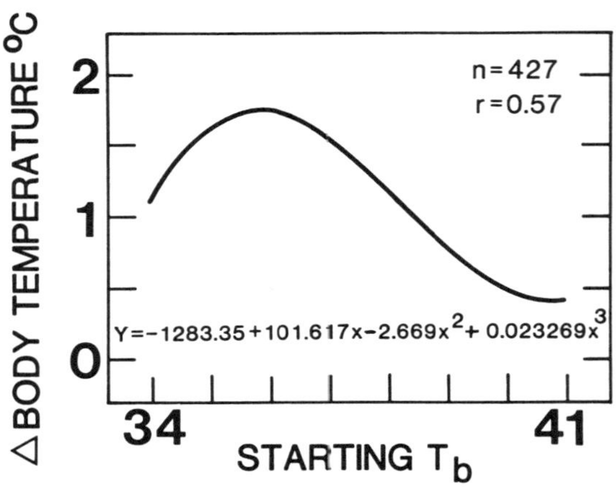

Fig. 2: The maximum change in body temperature as a function of starting body temperature from 427 adult male urethaneanesthetized (1.5 gm/kg) Wistar rats to 400 ng Prostaglandin E_1 given intracerebroventricularly at various starting body temperatures within the range 34-41°C.

In this analysis the polynomial regression illustrated above shows that the maximum fever reached varies according to the starting body temperature. The maximal febrile response occurred at a starting body temperature of 36°C. These responses above a starting temperature of 36°C might suggest a limiting "set point" in fever. They are also consistent with investigations which showed that PGE_2 raises body temperature to a level, largely independent of either the initial body temperature or the phase of the light-dark cycle (10).

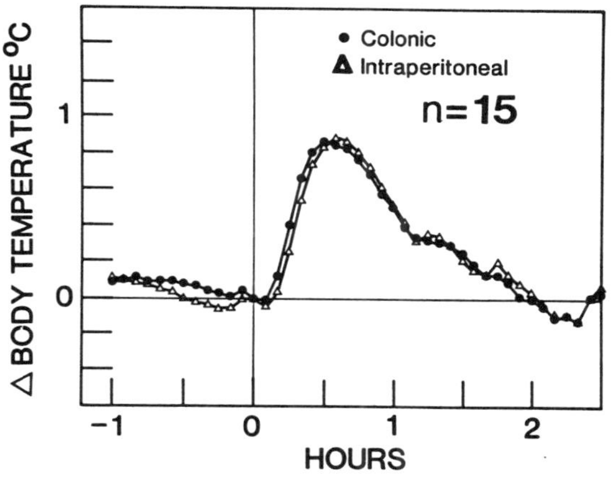

Fig. 3: Changes in intraperitoneal and colonic temperature to 200 ng Prostaglandin E_1 given intracerebroventricularly to two groups of 15 adult male Wistar rats. The SEM was < 0.15.

With the increasing utilization of telemetric body temperature measurement it is of interest to determine whether there was any difference between the simultaneous telemetric measurement of intraperitoneal temperature and thermistor measurement of colonic body temperature, during a Prostaglandin E_1 fever. As the above graph shows there was no significant difference in the change in body temperature as measured from the two sites. There was however a slight difference between the absolute colonic and intraperitoneal body temperatures with the colonic temperature being warmer (0.13°C). This difference is similar to that occurring in the human (11). The results validate the use of intraperitoneal temperature telemetric devices.

Artificial intelligence software and mathematical modelling can play an important role in developing an understanding about real biological systems. This technology aims at improving qualitative factors and adds new capabilities which do not now exist to assist the scientist in his understanding and problem solving capabilities. AI offers the opportunity or organize human expertise and experience into a form that the computer can manipulate. These new techniques however can aid the user only to the extent that the system has been designed and the results have validity only over the applicable range. AI systems are evaluated during development by testing with a large number of sample cases and by comparing performance to that of human experts and the results of experimentation. This communication is a first attempt at exploring how the power of AI and specifically expert reasoning can be applied to thermoregulation by taking advantage of the resource of a large database of existing experimental data.

Acknowledgements

Supported by the Medical Research Council of Canada. The authors would like to thank J.R. Catley, N.A. Rothfus and B. Clark of the Southern Alberta Institute of Technology, Calgary, Canada, for their instructional support. All animal experiments were conducted in accordance with the American Physiological Societies "Guiding Principles for Research Involving Animals and Human Beings."

References

1. Harmon, P. Hall, C. (1993): Intelligent Software Systems Development. New York: Wiley. 173-232.

2. Payne, EC. McArthur RC. (1990): Developing Expert Systems. New York: Wiley.

3. Pederson, K. (1989): Expert Systems Programming. New York: Wiley.

4. Hinton, GE. (1992): How neural networks learn from experience. Scientific American 267,145-151.

5. Hammerstrom, D. (1993): Working with neural networks. IEEE Spectrum. 30,46-53.

6. Foley, JD. Van Dam A. Feiner SK. Hughes JF. (1992): Computer Graphics: Principles and Applications, 2 ed. New York: Addison-Wesley. 1-23

7. Bousquet, M. Hester J. (1991): AutoCAD 3D Design and Presentation. Carmel: New Riders Publishing.

8. Leigh, RW. (1992): Solid Modelling with AutoCAD. Chapel Hill: Ventana Press.

9. Nameroff. S. (1989): QuickBASIC:The Complete Reference New York: McGraw-Hill.

10. Feng, J. Price, M. Cohen, J. Satinoff E. (1989): Prostaglandin fevers in rats: regulated change in body temperature or change in regulated body temperature? Am. J. Physiol. 257, R695-R699.

11. Taylor, PJ. Cooper KE. Malkinson TJ. (1976): Temperature gradients at various points within the human pelvis as measured during laparoscopy. Journal of Reprod. Med. 16,163-166.

Section III:

ADAPTATION

Thermal Balance in Health and Disease
Advances in Pharmacological Sciences

BENEFICIAL AND DETRIMENTAL EFFECTS OF THERMAL ADAPTATION

J. Werner

Institut für Physiologie, Ruhr-Universität, MA 4/59, D-44780 Bochum, Germany

Basic definitions

The heat balance of the human body depends on the one hand on the amount of energy produced by metabolism, mainly in organs and muscles, and on the other hand on the amount of heat transferred to or from the environment. In basic conditions (70...100 W metabolic heat production) and in a thermoneutral environment (around 22°C, low humidity and wind speed, light clothing) the heat balance of the body will yield a mean body temperature between 36 and 37°C. The body as a thermal system without activated thermoregulatory control, described by the physical laws of heat transfer, is generally called the "*passive* (thermal) system" (see Fig. 1), because it would *passively* accept thermal changes induced by the environment (external cold/heat) and by muscle activity (endogeneous heat). No optimal functioning of the internal processes around a core temperature of 37°C, not even survival, could be expected, because in a "passive system" the thermoregulatory effector mechanisms, metabolic heat production, sweat production and vasomotor activity, would not be adjusted to thermal needs of the body. Fortunately, body temperatures are sensed by spatially distributed thermal receptors and fed back to parts of the central nervous system which may be called the "*active* (thermoregulatory) system" (see Fig. 1), because it *activates* effector mechanisms reducing mean body temperature deviations: mainly vasoconstriction and shivering as cold defense mechanisms or vasodilatation and evaporative heat loss (in humans via sweat production) as heat defense. The neuronal outputs of the active system are transformed into heat gain or heat loss mechanisms as inputs of the passive system compensating, at least partially, the "disturbing" inputs from environment and muscle exercise. This is achieved by the basic control loop of thermoregulation.

Within the capacity of the system components, an external or internal "disturbance" evokes a dynamic response which, within minutes or hours, transfers the system into another steady state. The adjustments within this dynamic range, sometimes called "short-term" adaptation, are due to the dynamics of the components and their non-linear relationships. They will not be treated within the scope of this contribution. For a survey see e.g. (1), (2). A higher level of thermoregulatory system performance are "long-term" adaptations (see Fig. 1) which evidently, as a consequence

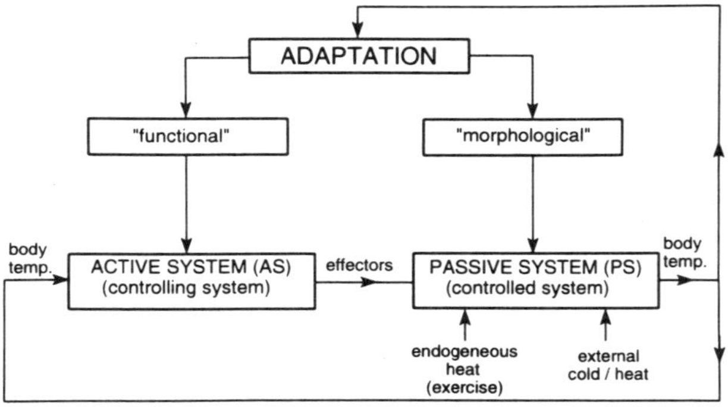

Fig.1: Basic thermoregulatory control loop and "long-term" adaptation.

of a higher or lower body temperature persisting or occurring repetitively during weeks or longer, alter either the properties of the passive system (often called "morphological" adaptation) or of the active system (often called "functional" adaptation), aiming either at a better efficiency or a better economy of the system. However, we shall see later on that beneficial adaptive adjustments induced by a given stressful environment may turn out to be detrimental if new stressors occur.

If thermal adaptation is acquired by living continuously in a natural environment, this thermal adaptation is referred to as "accclimatization", if induced intentionally by (maybe repetitive) exposure to a stressful climate (mostly artificially produced in climatic chambers), it is named "acclimation". The differentiation seems to be justified as the two processes may follow different strategies, the latter relying more or less exclusively on adjustments of physiological (autonomic) control, the first determined very strongly also by behavioural mechanisms. This report is confined to acclimation processes.

Morphological and functional adaptation

Typical morphological adjustments (those affecting the properties of the passive system) are processes which alter the heat transfer body core / body shell / environment, e.g. long-term adjustments of the peripheral circulatory system, development of the fat layer, and (in many animals) a change of fur insulation. Also adaptive changes of basal values of metabolic or evaporative rate and of heat capacity should be attributed to the "morphological" ones, because they are present even if the control mechanisms are not activated. A possible "functional" adjustment would be a change of the weights of the sensory inputs (e.g. shift of the impact of core versus peripheral signals), which, however, has not been clearly demonstrated up to now. A mean body temperature constituting a 9:1 (up to 8:2) weighting of the central and the peripheral sensors, that is more or less of "core temperature" and of "mean skin temperature", is generally considered to be the decisive summarized input signal to the thermal controller. A more evident functional adaptation is a change of the gain factor transforming the overall sensory input into an effector signal: a "change of gain", which in a temperature/effector diagram alters the slope of the functional relationship. The functional effect which has been demonstrated most commonly and seems to play the dominant role, is the change of effector threshold, meaning that the value of body temperature at which an effector mechanism is activated above its basic level is shifted down- or upwards. An analysis of such "gains" and "thresholds" takes place at rather a macroscopic system level, and many efforts have been devoted to the question of their neurophysiological and pharmacological correlatives.

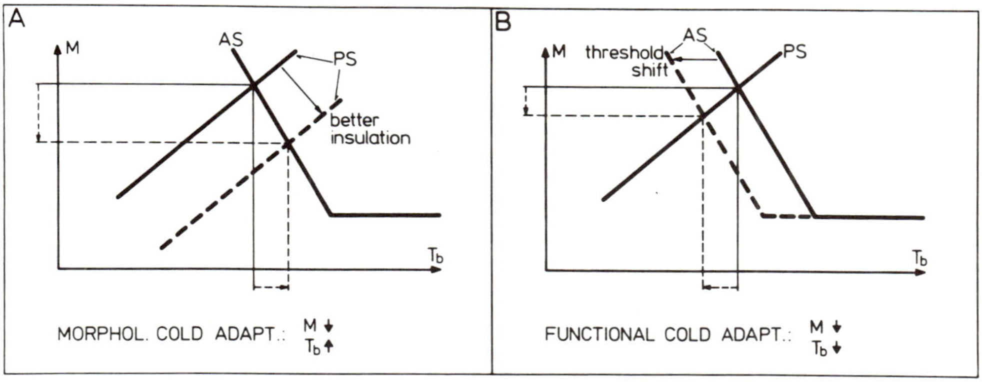

Fig. 2: A,B: Different consequences of morphological (A) and functional (B) cold adaptation. Decrease of metabolic heat production M is accompanied in case (A) by higher mean body temperature T_b, in case (B) by lower T_b. AS denotes functional relationship of the controlling system (active system), PS that of the controlled (passive) system. Broken lines are valid for the post-adaptation state, unbroken lines for pre-adaptation state.

These may be found on the sensory and / or the central (integrative) and / or the effector level. Some evidence has been gathered that such a change of "gain" and "threshold" may be traced back to ionic and transmitter processes of central neuronal networks. This contribution while staying mainly on the system level, will also try to give some hints on both the neurophysiological and to the pharmacological basis of adaptation. However, for a more elaborate survey on this topic, see (2), (3). Moreover this analysis has to be restricted to the effects demonstrated in adult humans or effects particularly relevant to human physiology, so the problems of non-shivering thermogenesis in brown adipose tissue which are of great importance for the neonate human and also for some adult animal species, cannot be dealt with here, see e.g. (4). The phenomenology of adaptive effects covers an extremely wide spectrum, and many findings reveal real or apparent discrepancies. One of the reasons is that the experimental procedure for adaptation is composed of so many factors, which may often be responsible for differing results. As thermal adaptation is a long-term process, there should hardly be a study reporting on post-adaptive *steady-state*. So all those studies demonstrating the status "after" adaptation may refer to very different phases of the adaptational process, which moreover depends strongly on the protocol: kind and combination of stressors (external temperature, humidity, wind speed, exercise, posture, clothing etc.), level of stressors, time course (continuous/intermittent, constant/variable) and duration of stressors, individual factors etc., not to mention the many problems induced by using different or inadequate measuring techniques.

However, a special kind of discrepancy reported may be eliminated very easily. For example, let us look at the two apparently incompatible findings: #1. cold adaptation evokes a lower metabolic rate and (consequently) a lower mean body temperature, and #2. cold adaptation evokes a lower metabolic rate, but (nevertheless) a higher mean body temperature. Both apparently contradictory statements are true, if #1 refers to functional adaptive effects and #2 to morphological effects. For an explanation see Fig. 2: In the open-loop passive system (PS) mean body temperature T_b would increase with *increasing* metabolic heat production M. This is documented in Fig. 2 by a (not necessarily) linear relationship, a line (PS) with positive slope. On the other hand, the controller (the active system AS) has to respond with an increase of metabolic heat production M to *falling* body temperature T_b, represented in the diagram by a line AS with negative slope. This is already the only necessary requirement for negative feedback control. The only steady-state which in closed-loop operation is possible under the given environmental conditions, is determined by the point of intersection of the two unbroken lines in Fig. 2A and B, because it is the only state which fullfils the requirements of both subsystems. Now, Fig. 2A deals with exclusively morphological adaptation in the form of a better insulation, altering the situation in a way that at equal metabolic rate, temperature would rise in the passive system. This is represented in Fig. 2A by a shift of the PS-characteristic (to the broken line). Hence, a steady-state can only be achieved, if the closed control loop in a dynamic process changes to the state given by the point of intersection between the unchanged AS-characteristic and the changed (broken) PS-line.

Then the new state (after morphological adaptation) is inevitably characterized by higher body temperature *in spite of* decreasing metabolic rate (Fig. 2A). On the other hand, if there were a purely functional adaptation in the form of a downward shift of the shivering threshold (in Fig. 2B shift of AS-characteristic to the broken line), a new state would necessarily mean a lower body temperature *because of* lower metabolic rate (Fig. 2B). Later on, when cold adaptation will be discussed explicitly, we shall see that, if both morphological and functional adaptation are present, there may be not any change of body temperature at all. So, a class of very different findings may be explained by the fact that they are due to a different proportion of morphological and functional adaptive adjustments.

Cold adaptation

The predominant effect of cold adaptation, reported both in studies on humans and on animals without significant brown adipose tissue, is the downward shift of the shivering threshold. For example Brück and co-workers (5) analysed cold modifications in humans, induced by repeated cold exposures (-5...+5 °C, 1 h, unclothed, resting position). It turned out that in 2/3 of the subjects metabolic reactions and shivering threshold were shifted to a lower mean body temperature (see Fig. 3) which also demonstrates that this modification is linked to a reduction in thermal discomfort. This tendency to hypothermic reaction (reduced mean body temperature) has already been found in naturally cold acclimatized humans and is clearly supported by further studies (see e.g. (6)). In most of these studies there is also evidence of a reduction of skin blood flow.

Hensel and Schäfer (7) tried to clarify whether at least part of such a threshold shift could be reduced to changes on the sensor level in experiments on cats exposed to cold for some weeks. Although this is a time period in which the effector threshold shift can be observed, it was not possible to demonstrate changes in the functional characteristics of cutaneous cold receptors. However, in former investigations on cats reared for about four years in the cold, these authors were able to show adaptive changes.

Threshold shifts in central structures involved in the thermoafferent system were observed in warm responsive neurons of rats, namely for the so-called switching-type neuron, meaning that up to a temperature threshold, these neurons have a very low basal firing rate which is increased abruptly to a maximal firing rate when the threshold is crossed. In studies with rats kept for 5 weeks at 3°C environmental temperature in our laboratory, the percentage of neurons with firing rates above the basal rate was significantly higher than in non-adapted rats. This holds for separate

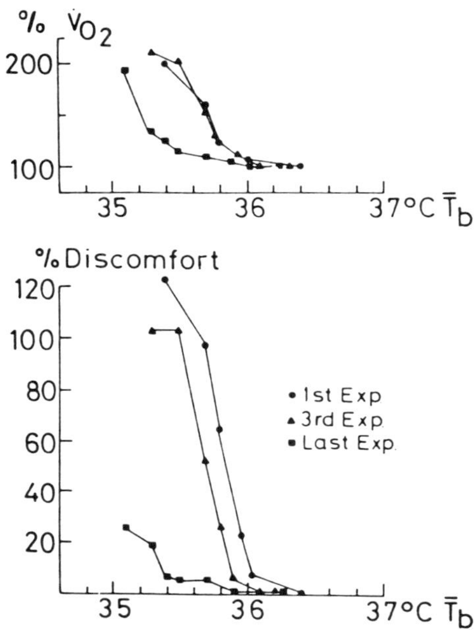

Fig. 3: Summary of a study of a subject, who was cold-exposed on 6 different days. The data from 3 cold-exposures are represented. Explanation of abbreviations: T_b means weighted body temperature. (●) 1st, (▲) 3rd, (■) final cold exposure. Oxygen uptake O_2, "100 %" = initial resting level. (From Brück et al.(5) with permission).

studies of thalamic, midbrain (nucleus raphe medialis et dorsalis) and preoptic neurons (8), (9): Fig. 4. It is concluded that after cold adaptation the increase of firing rate starts, on average, at a lower temperature.

Various responses of cold adaptation on cold and warm responsive units in guinea pigs were detected by Hinckel and co-workers, for a survey see e.g. (10): In contrast to adaptive changes in the subcoeruleus region where the preponderant cold responsive neurons decrease their peak activity, the warm responsive neurons in the pontine nucleus raphe magnus increase their peak activity significantly after long term cold exposure. Numerous lesioning experiments made possible further hypotheses about the origin of these adaptive changes.

Early evidence about the involvement of noradrenaline (NA) was found by Zeisberger and Brück (11): Micro-injections of NA into an area caudal to the preoptic area under thermoneutral conditions increased shivering or nonshivering thermogenesis or both in guinea pigs.

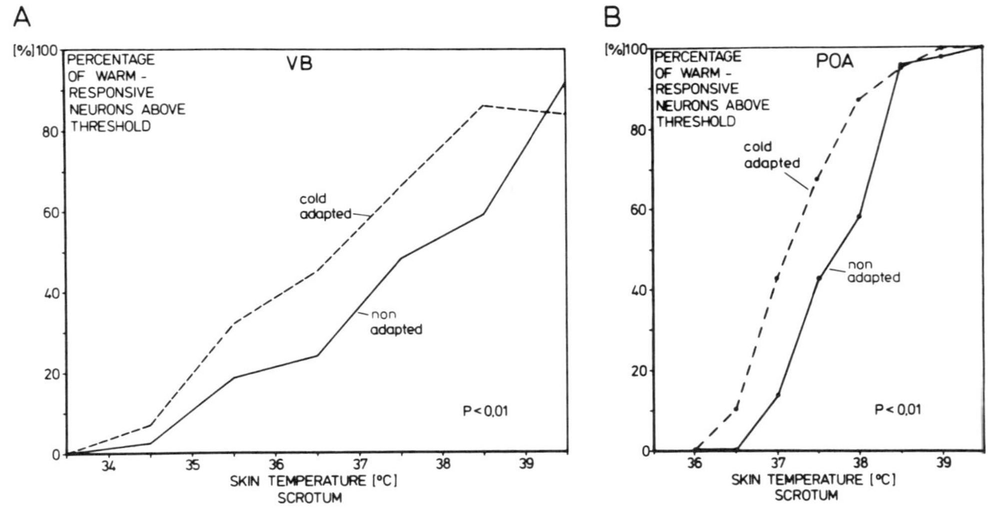

Fig. 4 A,B: Percentage of warm responsive switching-type neurons above threshold (firing at maximal discharge frequency). A: Ventrobasal complex of thalamus (VB), B: preoptic area (POA). (A from Werner et al. (8); B from Werner (34)).

These metabolic reactions could be suppressed by increasing skin and/or internal body temperature. It has been concluded that the threshold for the thermogenetic reactions is shifted to a higher level by intrahypothalamic injection of NA (Fig. 5). Phentolamine, an adrenergic a-blocking agent, has the reverse effect when given intrahypothalamically: the shivering threshold was decreased to values normally found in cold adapted and in newborn guinea pigs. The shift in threshold was combined with a change of gain of the thermogenetic response as a function of temperature (11, 12). A long-lasting shift of shivering threshold to a lower temperature level could be caused by peripheral NA infusion (13). In spite of lesioning of aminergic receptors (14) there was an unimpaired and adequate thermoregulatory reaction to peripheral cold stress in the chronic stage of the lesion, so that the transmitters involved (NA, 5-HT) appear to be predominantly modulatory in character. Another recent finding is the arginin vasopressin (AGV) release in cold adapted guinea pigs. This stressor hormone release obviously takes place to an extent, greatly exceeding the antidiuretic needs (15).

In contrast to the studies on humans mentioned above, where a lower mean body temperature T_b after cold adaptation was found, there was no significant change of T_b in studies on rabbits adapted to 10°C (16) and to 2°C (17). This is reasonable, as the results on humans suggest a predominantly functional adaptation, as explained in Fig. 2B (decreasing M and T_b), whereas the animal studies confirm both functional and morphological adaptation, the latter being obvious, because the fur insulation increases significantly. Fig 6 shows some results from the study of Werner and Graener (16), where metabolic heat production M decreased, insulation I increased and body temperature T_b did not change significantly during the seven week acclimation period. From Fig. 2 we know that neither morphological nor functional adaptation alone may explain such a result.

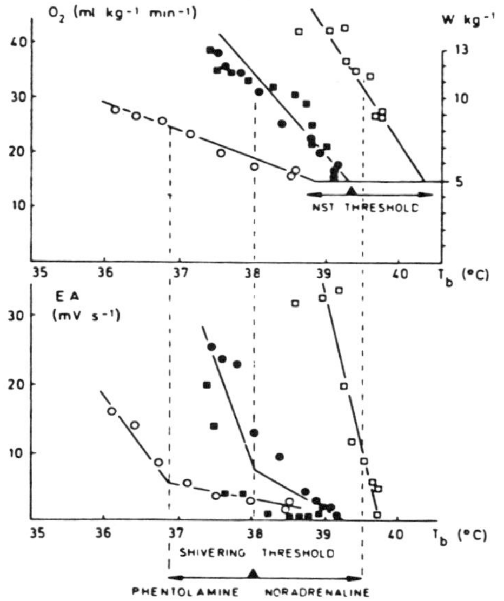

Fig. 5: Relations of metabolic rate (above) and electrical muscle activity (below) to the decrease in mean body temperature during an external cooling in young cold adapted guinea-pigs before (filled symbols) and after (open symbols) microinjection of noradrenaline or phentolamine into the hypothalamic area .(From Zeisberger (12), with permission).

Therefore in Fig. 7 we draw the status at time zero (as unbroken lines) and the status after seven weeks cold adaptation (as broken lines) in the form of a shift of the characteristic of the passive system PS (because of better insulation) and additionally of a shift of the characteristic of the active system AS. The postulated functional adaptation is in fact confirmed, as the point of intersection of the broken lines yields, as compared to the pre-adaptive state, no significant change of T_b in spite of a decreased M. From this system treatment it is clear that depending on the proportion of morphological to functional adaptation, the effects of a decreased or an increased T_b are also possible. The overall beneficial effect of cold adaptation is the conservation of energy during persisting cold stress.

When asking about detrimental effects of cold adaptation we should look primarily at its influences on heat defense mechanisms. It is self-evident that morphological cold adaptation in terms of improvement of insulation is detrimental for subsequent heat exposure. This also holds equally, if peripheral blood flow is reduced. A shift of the threshold for evaporative heat loss (REHL), was not found in guinea pigs (18), whereas in rabbits an upward shift was demonstrated (16), (17), an effect which is detrimental to an efficient heat defense.

Heat adaptation

To achieve adequate heat acclimation in humans, heat stress needs to be sufficient to cause skin wetness of greater than 50 % during 2 - 4 h/d for 7 - 14 days. In this way, an improved

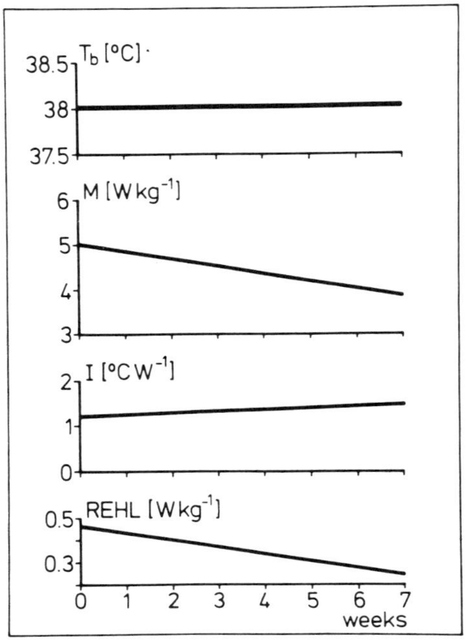

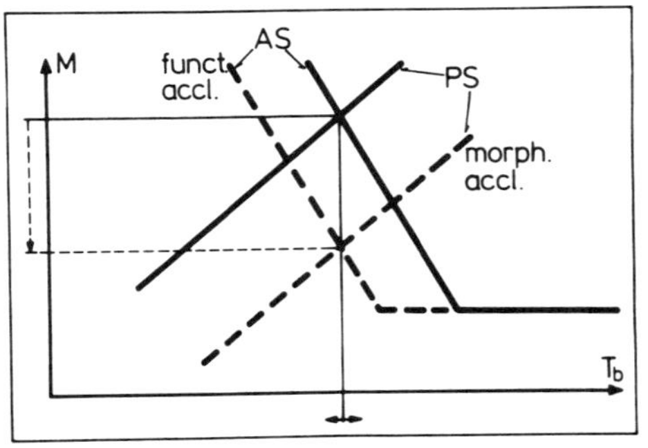

Fig. 7: Morphological and functional cold adaptation in rabbits. Relationships for the active system AS and for the passive system PS, unbroken lines for pre-adaptation state, broken lines for state after 7 weeks. Morphological plus functional adaptation explains the result of decreased metabolic heat production M and unchanged mean body temperature T_b.

Fig. 6: Cold adaptation in rabbits: mean body temperature T_b, metabolic heat production M, body insulation I, and respiratory evaporative heat loss REHL as a function of the 7-week adaptation process. (Data from Werner and Graener (16))

effectiveness of heat dissipation is achieved. The state of acclimation is charaterized by decreased core and skin temperature, decreased heat storage, increased sweating rate, increased sweat and cutaneous vasodilatation sensitivity, and a lower sweating, vasodilatation and shivering threshold (19, 20), see Fig. 8. The cardiovascular adjustments are represented by a lower heart rate, increased stroke volume, and increased plasma volume. The plasma volume expansion seems to be mediated by protein influx to the intravascular space and by sodium conservation (21, 22). However, in spite of new experimental evidence that plasma volume expansion is a decisive factor of heat acclimation (23), the detailed mechanisms do not seem to be finally resolved. For further details see e.g. (24).

The effects of training even in a cool environment on exercise-heat tolerance are generally well accepted (25). The mechanisms responsible for this phenomenon are plasma volume expansion (26), increased sensitivity of the sweating mechanisms, and a downward shift of the vasodilatatory and the sweating threshold to a lower body temperature (27, 28, 29). Additionally there seems to be evidence of a change in sweat gland morphology or sensitivity after acclimation (30). Often the ratio sweat production/heat production is increased to 2 or more, which if the water and salt loss is not adequately replaced, will amplify dehydration problems.

The positive effects of heat acclimation may be observed within two weeks. Rectal temperature and heart rate drop exponentially and, inversely, sweat loss increases asymptotically (31). A preceding increase was observed by Senay et al. (21) for plasma volume determined from

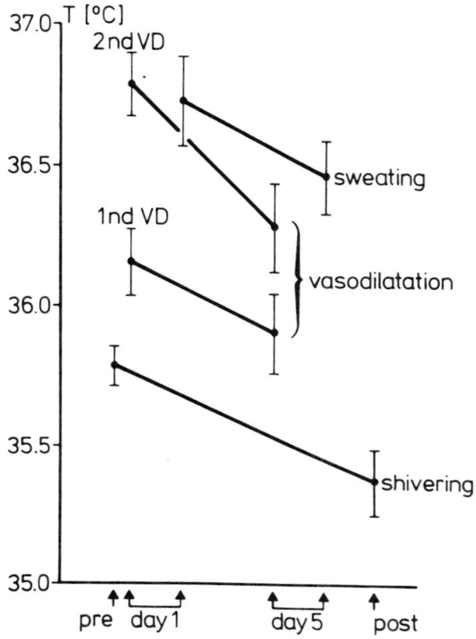

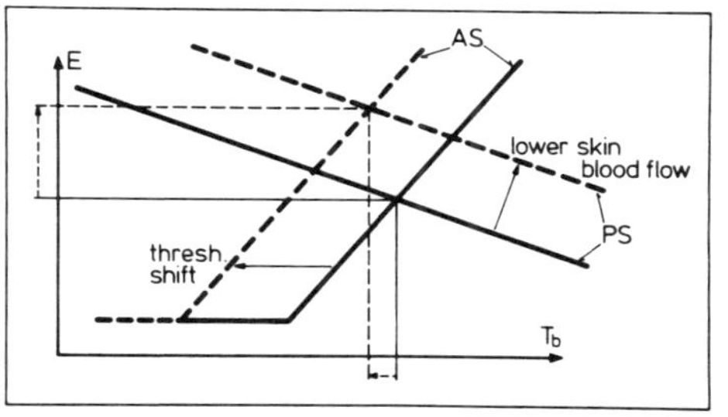

Fig.9: Heat adaptation in humans: Relationships for the active system AS and for the passive system PS, unbroken lines for pre-adaptation, broken lines for post-adaptation state. The two adaptive effects determine a higher evaporative heat loss E and a lower mean body temperature T_b.

Fig. 8: Concurrent shift of the threshold temperatures of all autonomic thermoregulatory effector reactions in the course of a 5-day heat adaptation series. Means ± SE of seven subjects. Shivering threshold was determined 1 day before and 1 day after the 5 heat exposure days. EMA Electrical muscle activity (mean from two muscle groups), O_2 oxygen uptake of the subjects; *1st* and *2nd* refer to the vasodilatation due to the sympathetic release and active vasodilatation, respectively (modified after Hessemer, Zeh & Brück (20), with permission).

hematocrit values prior to each exposure to 45°C dry-bulb and 32°C wet-bulb and working at 40 -50 % O_2 max for four hours each day. Detailed knowledge about cardiovascular heat adaptation in rats was recently presented by Horowitz and colleagues (32).

Fig. 9 summarizes the effects observed after heat adaptation of humans. As before in Fig. 7, the status before adaptation is given by the intersection of the unbroken characteristics PS and AS. The broken AS-line represents the downward shift of the sweating threshold, and the blood flow effect, for the sake of the simplicity, is represented by a shift of the PS-line. Higher blood flow would yield a shift to lower T_b, lower blood flow to higher T_b. The latter is shown in Fig. 9 as, in spite of a downward shift of the vasodilatation threshold, a lower skin blood flow is generally reported (because of the considerable decrease of body temperature). The point of intersection of the two broken lines confirms the experimental results: higher evaporative heat loss E and lower T_b. The overall beneficial effect is the lower thermal load in the body, and thereby a reduced strain of the circulatory system.

Is heat acclimation detrimental for a sudden cold exposure? Significant morphological effects of heat adaptation in terms of a reduction of insulation have not been reported, but the downward shift of the shivering threshold demonstrated in humans, see. Fig. 8 (20) and also in rabbits (33) may be maladaptive, as in (33) it is a comparable effect as that occuring in cold adaptation (17). Being an economic effect, it implies of course a lower body temperature.

Thermal adaptation and fever

The question of interaction of fever and acclimation was recently analysed in our laboratory (17). A global survey of the influence of cold acclimation on the total extent of temperature elevation in fever may be attained by the fever index, an integration of core temperature increase for a given time period after injection of pyrogens. The results are from experimental series, in which the interrelations were investigated by studies on different groups of rabbits acclimated for eight weeks to 2°C, 22°C, or 30°C. The fever course after injection of an exogenous pyrogen (LPS, 0.5µg/kg i.v.) was investigated both before and after adaptation. Half of the cold and of the heat adapted groups were taken for control experiments with an injection of isotonic NaCl-solution (n=10, each).

Fever after cold adaptation

The fever index for 300 minutes (FI_{300}, see Fig. 10) decreased significantly after cold adaptation for fever in a neutral or cold environment. This holds in comparison both to the same group before adaptation and to the thermoneutrally adapted control group. In a warm environment, however, FI_{300} was higher after cold adaptation.In a cold and in a neutral environment, where metabolic heat production is the dominant factor for the fever process, the effector activities during fever lead to a lower thermal strain than without cold adaptation. This can clearly be seen in Fig. 11: The pre-adaptive state is given by the unbroken PS-line and by the two unbroken AS-lines, the right one representing the upward shift to higher body temperatures induced by fever. This shift is reduced by the downward shift induced by cold adaptation (broken AS-line). Additionally, there is the shift of the PS-line because of better insulation. The post-adaptive febrile state is determined by the intersection of the two broken lines: M is reduced, leading to a body temperature lower than that for fever without cold adaptation, but still higher than in the afebrile state. Such a reduction of the fever index by cold adaptation is confirmed by Roth and Zeisberger (35).

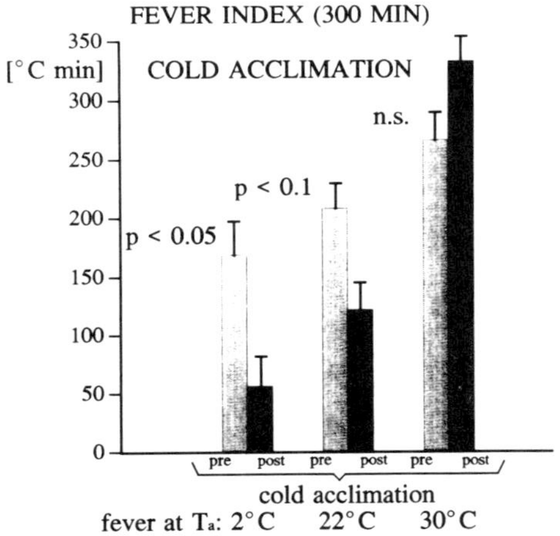

Fig. 10: Fever after cold adaptation in rabbits. Fever index at environmental temperatures of 2°C, 22°C, and 30°C before (shaded bars) and after (black bars) cold adaptation to 2°C. (Data from Beckmann and Werner (17)).Fever after cold adaptation

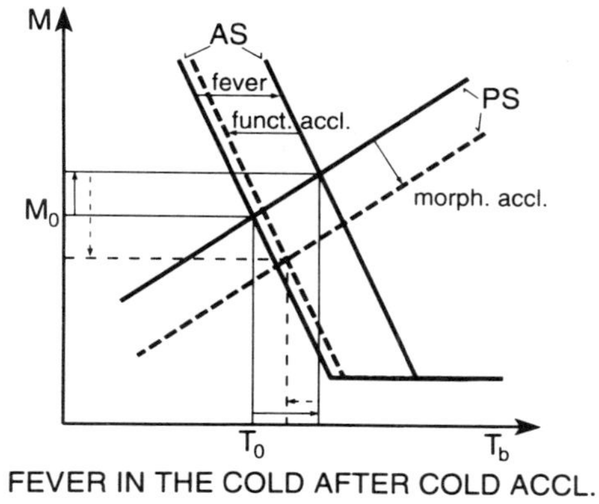

Fig. 11: Fever in the cold after cold adaptation. Relationship for the active system AS and for the passive system PS, unbroken lines for pre-adaptation state, broken lines for state after 7 weeks. A shift of AS-line to the right due to fever is partially compensated by a shift to the left due to adaptation. Functional plus morphological (shift of PS-line) adaptation result in reduction of metabolic heat production M and a mean body temperature T_b higher than in the afebrile state, but lower than in a febrile state without cold adaptation.

So cold adaptation is not detrimental in terms of an increase of the febrile response in a cold (and in a neutral) environment. This certainly does not hold for fever in the heat. The fever index was significantly higher as compared to the neutrally adapted group, (whereas as compared to the pre-adaptive state it could not be proved to be significant). A higher insulation undoubtedly is a detrimental effect in a hot environment, whether there is fever or not.

Fever after heat adaptation

After heat adaptation of rabbits the total level of temperature increase in fever is significantly reduced at all temperatures tested (2°C, 22°C, 30°C). The fever index FI_{300} (Fig.12) decreases and, in the cold, even becomes negative. The uniform reaction can be attributed both to a shift in the effector characteristic of metabolic heat production and an enhanced efficiency of heat loss mechanisms, which reduce the regulatory load-error during the dynamic fever process at high ambient temperatures. So in conclusion, it can be stated that heat adaptation is best suited to avoiding higher febrile temperatures, however, cold ambient temperatures may produce hypothermic problems even during fever.

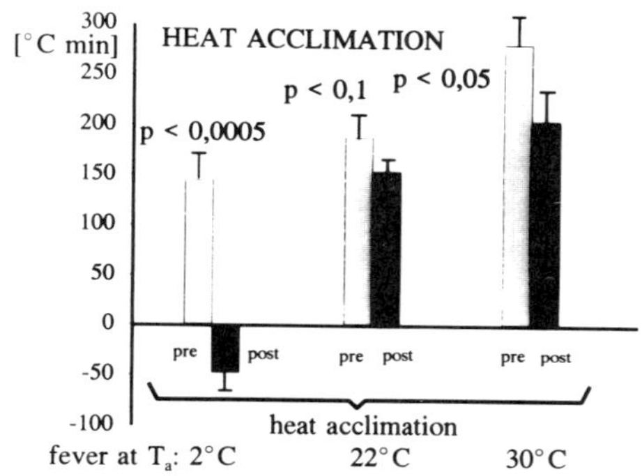

Fig. 12: Fever after heat adaptation in rabbits: Fever index at environmental temperature of 2°C, 22°C, and 30°C before (shaded bars) and after (black bars) heat adaptation to 30°C. (Data from Beckmann and Werner (17))

Final conclusions

Thermal adaptation of physiological (autonomic) control of body temperature improves the ability to cope with persisting thermal stressors. Cold adaptation aims primarily at a better economy, i.e. preservation of energy often at the cost of a lower mean body temperature *during cold stress,* whereas heat adaptation whether achieved by exposure to a hot environment or by endogenous heat produced by muscle exercise, often brings about a higher efficiency of control, i.e. a lower mean body temperature *during heat stress*, at the cost of a higher water loss. However, whether (as compared to the pre-adaptive state) a lower, an unaltered or a higher mean body temperature is achieved, depends on many factors, most particularly on the extent and the proportion of morphological and functional adaptation.

While cold adaptation is beneficial in a cold environment, it constitutes a detrimental factor for exposure to a hot environment, mainly because of morphological adaptation. Heat adaptation, on the other hand, is maladaptive for cold exposure, mainly because of functional adaptation.

Preceding heat adaptation clearly is best suited to avoiding higher body temperatures in the process of fever, no matter which environmental conditions prevail. On the other hand, cold adaptation is undoubtedly detrimental for coping with high febrile temperatures in a hot environment. Yet, in the cold, preceding cold adaptation may, because of reduced metabolic heat production, result in a lower febrile increase of body temperature.

So, to put it briefly, thermal adaptation is indeed suited to coping with the kind of climate which induced adaptation (whether the subject is febrile or not), and detrimental for the adverse climate, beyond that, preceding heat adaptation reduces the temperature increase during fever in the adverse (i.e. the cold) climate, even with the consequence of hypothermic stress.

Acknowledgements

The studies from my laboratories were supported by Deutsche Forschungsgemeinschaft.

References

1. Brück, K, Hinckel, P (1991): Thermoafferent networks and their adaptive modifications. In: Thermoregulation, Physiology and Biochemistry. Schönbaum, E, Lomax, P eds. New York: Pergamon Press, 129-152.

2. Brück, K (1990): Long-term and short-term adaptive phenomena in temperature regulation. In: Thermoreception and Temperature Regulation. Bligh, J, Voigt, K eds. Heidelberg: Springer-Verlag, 211-223.

3. Brück, K, Zeisberger, E (1991): Adaptive changes in thermoregulation and their neuropharmacological basis. In: Thermoregulation, Physiology and Biochemistry. Schönbaum, E, Lomax, P eds. New York: Pergamon Press, 255-307.

4. Heldmaier, G, Klaus, S, Wiesinger, H (1990): Seasonal adaptation of thermoregulatory heat production in small mammals. In: Thermoreception and Temperature Regulation. Bligh, J, Voigt, K eds. Heidelberg: Springer-Verlag, 235-243.

5. Brück, K, Baum, E, Schwennicke HP (1976): Cold-adaptive modifications in man induced by repeated short-term cold-exposures and during a 10-day and -night cold-exposure. Pflügers Arch 363, 125-133.

6. Bittel, JHM (1987): Heat debt as an index for cold adaptation in men. J Appl Physiol 62, 1627-1634.

7. Hensel, H, Schäfer, K (1982): Static and dynamic acitivity of cold receptors in cats after long-term exposure to various temperatures. Pflügers Arch 392, 291-294.

8. Werner, J, Schingnitz, G, Hensel, H (1981): Influence of cold adaptation on the activity of thermoresponsive neurons in thalamus and midbrain of the rat. Pflügers Arch 391, 327-330.

9. Werner, J (1991): Functional mechanisms of temperature regulation, adaptation and fever. In: Thermoregulation, Physiology and Biochemistry. Schönbaum, E, Lomax, P eds. New York: Pergamon Press, 185-208.

10. Hinckel, P, Perschel, WT (1987): Influence of cold and warm acclimation on neuronal responses in the lower brain stem. Can J Physiol Pharmacol 65, 1281-1289.

11. Zeisberger, E, Brück, K (1976): Alteration of shivering threshold in cold- and warm-adapted guinea pigs following intrahypothalamic injections of noradrenaline and of adrenergic alphareceptor blocking agent. Pflügers Arch 362, 113-119.

12. Zeisberger, E (1978): Role of a central catecholaminergic system in the thermoregulatory threshold deviation. In: New Trends in Thermal Physiology. Houdas, Y, Guieu, JD eds. Paris: Masson, 22-25.

13. Roth, J, Zeisberger, E, Schwandt, HJ (1988): Influence of increased catecholamine levels in blood plasma during cold-adaptation and intramuscular infusion on thresholds of thermoregulatory reactions in guinea-pigs. J Comp Physiol B 157, 855-863.

14. Behr, R, Zeisberger, E, Merker, G (1986): Shivering threshold changes after aminergic denervation of the anterior hypothalamus in normal and cold-adapted guinea-pigs. J. Aut. Nerv Syst Suppl , 553-559.

15. Zeisberger, E, Roth, J, Simon, E (1988): Changes in water balance and in release of arginine vasopressin during thermal adaptation in guinea pigs. Pflügers Arch 412, 285-291.

16. Werner, J, Graener, R (1986): Thermoregulatory responses to cold during a 7 week acclimatization process in rabbits. Pflügers Arch 406, 547-551.

17. Beckmann, U, Werner, J (1994): Interaction of acclimation and fever. In: Integrative and Cellular Aspects of Autonomic Functions. Pleschka, K, Gerstberger, R, Pierau, KF eds. London: John Libbey Eurotext Lim, 97-102.

18. Brück, K, Wünnenberg, W, Gallmeier, H, Ziehm, B (1970): Shift of threshold temperature for shivering and heat polypnea as a mode of thermal adaptation.. Pflügers Arch 321, 159-172.

19. Taylor, NAS (1986): Eccrine sweat glands adaptations to physical training and heat acclimation. Sports Med 3, 387-397.

20. Hessemer, V, Zeh, A, Brück, K (1986): Comparison of the effects of passive heat adaptation and moderate sweatless conditioning on responses to cold and heat. Eur J Appl Physiol 55, 281-289.

21. Senay, LC, Mitchell, D, Wyndham, CH (1976): Acclimatization in a hot, humid environment: body fluid adjustments. J Appl Physiol 40, 786-796.

22. Wyndham, CH, Rogers, GG, Senay, LC, Mitchell, D (1976): Acclimatization in a hot humid environment: cardiovascular adjustments. J Appl Physiol 40, 779-785

23. Sawka, MN, Hubbard, RW, Francesconi, RP, Horstman, DH (1983): Effects of acute plasma volume expansion on altering exercise-heat performance. Eur J Appl Physiol 51, 303-312.

24. Werner, J (1993): Temperature regulation during exercise: an overview. In: Exercise, Heat and Thermoregulation. Gisolfi, CV, Lamb, DR, Nadel, ER eds. New York: Brown & Benchmark, 49-84.

25. Gisolfi, CV (1973): Work-heat tolerance derived from interval training. J App Physiol 35, 349-354.

26. Convertino, VA, Mack, GW, Nadel, ER (1991): Elevated central venous pressure: a consequence of exercise training-induced hypervolemia? Am J Physiol 260, R273-R277.

27. Nadel, ER, Pandolf, KB, Roberts, MF, Stolwijk, JAJ (1972): Mechanisms of thermal acclimation to exercise in heat. J Appl Physiol 37, 515-520.

28. Roberts, MF, Wenger, CB, Stolwijk, JAJ, Nadel, ER (1977): Skin blood flow and sweating changes following exercise and heat acclimation. J Appl Physiol 43, 133-137.

29. Baum, E, Brück, K, Schwennicke, HP (1976): Adaptive modifications in the thermoregulatory system in long-distance runners. J Appl Physiol 40, 404-410.

30. Sato, F, Owen, M, Matthes, R, Sato, K, Gisolfi, CV (1990): Functional and morphological changes in the eccrine sweat gland with heat acclimation. J Appl Physiol 69, 232-236.

31. Mitchell, D, Senay, LC, Wyndham, CH, van Rensburg, AJ, Rogers, GG, Strydom, NB (1976): Acclimatization in a hot, humid environment: energy exchange, body temperatures, and sweating. J Appl Physiol 40, 768-778.

32. Horowitz; M, Haddad, W, Shochina, M, Meiri, U (1993): Long term heat acclimation: Acquired peripheral cardiovascular adaptations and their stabiliy under multifactorial stressors. In: Thermal Physiology 93, Milton, AS ed. University of Aberdeen, 52.

33. Beckmann, U, Werner J, (1993b): Effects of warm adaptation on fever. In: Thermal Physiology 93, Milton, AS ed. University of Aberdeen, 5.

34. Werner, J (1990): Models of cold and warm adaptation. In: Thermoreception and Temperature Regulation. Bligh, J, Voigt, K eds. Heidelberg: Springer-Verlag, 224-234.

35. Roth, J, Zeisberger, E (1992): Evidence for antipyretic vasopressinergic pathways and their modulation by noradrenergic afferents. Physiol Res 41, 49-55.

Thermal Balance in Health and Disease
Advances in Pharmacological Sciences

ALTERED RESPONSIVENESS OF COLD-ADAPTED RATS TO THERMAL STRESS

M. Székely, M.Balaskó and Z. Szelényi

Department of Pathophysiology, University Medical School of Pécs, H-7643 Pécs (Hungary)

Introduction

On cold-exposure an increase in the temperature difference between the body surface and the environment initiates an enhancement of heat loss of the organism. The fall of the temperature of the body shell induced by such direct heat loss is further accentuated by effector mechanisms such as peripheral vasoconstriction which, in turn, help the whole body limit the rise in heat loss. Both the increase of heat conservation developing in the cold and the augmentation of heat production are thought to be induced by some central nervous system mechanisms, the intensity of this action being influenced by the thermal input from centrally and peripherally located cold receptors. The weight and/or contribution of central (CNS, other body core sites) as well as peripheral (skin) thermoreceptors in determining thermal information on the body as a whole is not known but is believed to come more intensely from the deep thermoreceptors than from their peripheral counterparts, the latter set of information having been regarded as quantitatively important only in cases of extremely strong thermal stimuli causing a dynamic change in receptor functioning (see 6). It seems only too logical to suppose that a strong reduction in heat loss at the very beginning of cold-exposure might even result in a paradoxical increase in heat conservation and deep body temperature would rise under such circumstances, a phenomenon placing an unnecessary burden on temperature regulation since the thermal gradient to be maintained (mostly by an exaggerated and costly heat production) would be even greater than that at thermoneutrality or in cold without a paradoxical overshoot.

In earlier experiments carried out on cold-adapted rats (4, 9) we have observed a tendency of body core temperature to rise during a short cold exposure. In the present studies, therefore, we set out to analyse the same phenomenon in cold-adapted rats acutely exposed to cold lasting for one hour and compared their responses with those in rats adapted to the usual laboratory temperature.

Methods

Female Wistar rats with a starting body weight of 220 to 250 g were held singly either at an ambient temperature of 24-26 °C or at 3-4 °C for at least four weeks with laboratory chow and drinking water being available *ad libitum*. In slightly restrained rats habituated to rest in a brass cage within a metabolic chamber colonic temperature (Tc) and tail skin (Ts) temperature were measured with copper constantan thermocouples and metabolic rate (MR) was estimated by measuring carbon dioxide output using an open-circuit method (Kipp diaferometer). All parameters were recorded on a Beckman dynograph. Initially, the metabolic chamber was held in a water bath with a temperature which ensured a chamber air temperature just initiating tail skin vasodilatation, thus indicating an ambient temperature at the upper end of thermoneutrality. As expected, cold-adapted rats tended to have a thermoneutral temperature some 3 to 5°C lower than non-cold-adapted rats (24-25°C and 28-29°C, respectively) when determined on the basis of starting vasodilatation. After a period of some 60 to 90 minutes a standard cold-exposure lasting for 60 min was applied, the strength of which was an ambient temperature 15°C lower than thermoneutrality. Cold-exposure was then followed by a re-transfer of the metabolic chamber into a water-bath ensuring thermoneutrality. The dynamics of cold and thermoneutral exposures were such that they led to 80 per cent of the required change in air temperature in about 3 minutes. The strength of cold-exposure used in the two groups of rats with different thermal acclimation characteristics appeared to be comparable also in terms of the similar size of metabolic heat production when expressed as per cent change of resting MR.

Results

The results are summarized in Fig. 1. When exposed to their respective thermoneutral ambient temperatures, cold-adapted rats had significantly lower Tc than animals adapted to a higher ambient temperature while MR was about the same in the two groups. Cold-exposure led to an almost immediate fall in Tc of non-cold-adapted rats; this trend of fall in Tc lasting for the whole length of cold-exposure. Cold-adapted rats, however, exhibited an increase in Tc for the first half of cold-exposure and, later, Tc started to fall somewhat into the range of Tc of the other group. Still, at the end of cold-exposure Tc of cold-adapted rats was not lower than before cold-exposure. The rise of MR developed immediately in both groups and remained so throughout the whole period of cold-exposure. In cold-adapted rats MR was slightly higher than in the other group only in the first 10 minutes but later MR of the two groups ran similar courses. In fact, the extent of rise in MR was practically the same in the two experiments when expressed in per cent of the resting value (209 vs 220 per cent). On re-exposing the rats to thermoneutrality all changes were reversed; MR of cold-adapted animals fell immediately although Tc was still as high as at the end of cold-exposure; as a result of this Tc fell even below that measured at the beginning of the experiment. This paradoxical fall in Tc mirrored the rise observed at the beginning of cold-exposure. Tc of non-cold-adapted rats started to rise at thermoneutral ambient temperature as expected.

Discussion

Measured at their thermoneutral ambient temperatures, cold-adapted rats had a significantly lower Tc than non-cold-adapted rats. This phenomenon and even the size of the difference is very similar to that observed in guinea pigs (2). This so called "hypothermic" adaptation has been regarded as a useful mechanism saving some energy and - together with a better thermal insulation developing in various rodents - might serve a useful biological purpose (1). A transient increase in deep body temperature during the first period of a cold-exposure may find explanation in the speed with

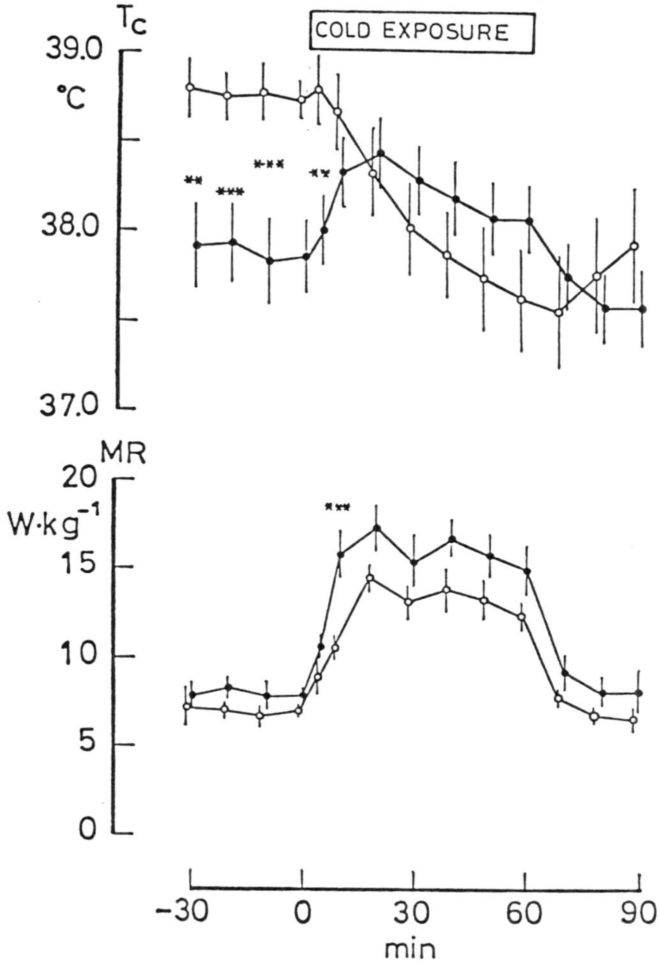

Fig. 1: Effects of cold-exposure on colonic temperature (Tc) and metabolic rate (MR) in cold-adapted rats (closed symbols) and rats adapted to 24-26 °C. Two or three astrerisks indicate significant differences between the two groups of rats ($p < 0.05$ or $p < 0.001$, respectively).

which ambient temperature has decreased. Such an overshoot in regulation should not last long and its effect on body temperature may well be minor (11). In the present experiments the rapidity of fall (or rise upon re-transfer) in air temperature within the metabolic chamber was not extreme and the paradoxical rise (or fall) in Tc remained for much of the thermal exposure. Also, the same extent of change in air temperature in non-cold-adapted rats failed to show this phenomenon. It should be emphasized that thermal exposure of the same type resulted in qualitatively different responses of body temperature depending on the type of thermal adaptation, thus indicating some basic modification in peripheral and/or central thermoregulatory mechanism(s) developing in the course of cold-adaptation.

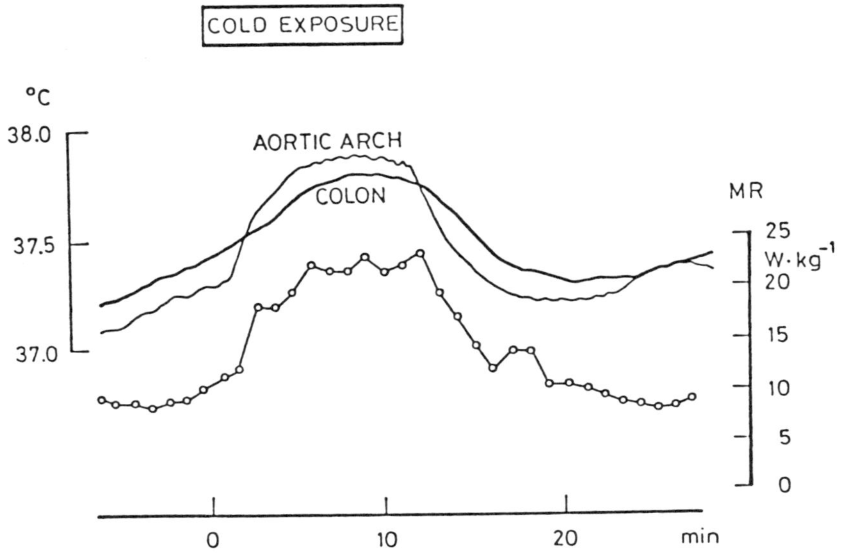

Fig. 2: Effects of an acute cold-exposure (from 28°C to 10°C) on temperatures of colon and of aortic arch (upper curves) and on metabolic rate (open circles) in a male Wistar rat adapted to 5-6°C for 30 days and implanted with a fine copper-constantan thermocouple reaching the aortic arch from one of the femoral arteries some 3 hours before 0 minute [modified from (9)].

It may be argued that Tc does not represent core temperature faithfully especially in a dynamic phase of thermal transients and that hypothalamic or arterial blood temperature might have been a better choice to study this phenomenon (11). Figure 2, taken from an earlier study, however, indicates that in cold-adapted rats the temperature of the aortic blood changes similarly to (or even more dynamically than) Tc, thus making the reality of the rise in core temperature at the beginning of a cold-exposure more likely than thought on looking at Tc alone. The rate of change in aortic temperature was, however, small when compared to that in an other experiment in which even a much more rapid change in core temperature induced in the goat failed to influence thermoregulatory effectors (8).

The apparent paradoxical behaviour of core temperature and of MR in cold-adapted rats on acute cold-exposure may be looked upon as a sign of positive feed-back regulation resulting from the role of tail-skin both as a site of heat loss with an ability to modify local blood flow between extremes of complete vasoconstriction and vasodilatation on central command (3) and as a possible site of cold-reception. There is some evidence that appendages such as the ear-pinna of the rabbit might possess also a thermoreceptor function (7) with the disadvantage that there might be - at least temporarily - a contradiction between these two thermoregulatory functions leading to changes in a fashion of a positive feed-back (vasoconstriction allowing heat conservation, in turn also a still stronger cold sensation, then more heat conservation, etc.) (5). If this would be valid for the rat tail, a strong vasoconstriction initiated by local as well as central mechanisms might allow such an apparently paradoxical regulatory behaviour. Why such a mechanism - if it occurs at all in the rat tail-skin - is present only in the cold-adapted state but not in non-adapted animals, is an open question, since no significant modification in peripheral thermosensitivity has been shown to develop in the course of thermal adaptation (6).

Changes of thermoregulatory effectors have traditionally been expressed in terms of some representative body temperatures in the hope to find (or at least to approach) the situation occurring in reality as to thermal afferentation. Depending on the species studied different weight has been assigned to the core and shell temperature and within the shell to different areas of the body surface (skin). In general the smaller a species, the higher weight is given to the surface, however, skin areas with sensitive vasomotorics (tail, ear-pinna, extremities) have rarely been considered to possess higher (or smaller) weight than the rest of the body surface. The tail of the rat (3), ear-pinna of the rabbit (7) and of the guinea-pig (10) have been shown to have an effector role to allow higher or smaller heat loss in animals exposed to hot or cold environments, respectively.

Most thermoregulatory studies carried out on small rodents so far have considered (mostly subcutaneous) trunk-skin temperature, only as that representing peripheral thermal input. Threshold mean body temperatures for shivering, rises in metabolic heat production or heat loss effectors have been utilized to characterize pharmacological effects on thermoregulation. Rapid changes in ambient temperature have been shown to induce an almost immediate rise in MR in guinea-pigs, a complication leading to difficulties in determining thermoregulatory thresholds, unless special attention is given to skin areas with rapidly changing temperature on thermal exposure (Szelényi & Hinckel, unpublished data). The present studies point to the possibility of exaggerated or even paradoxical thermoregulatory responses on thermal challenges in cold-adapted rats; a phenomenon the nature of which remains to be determined. The importance of temperature changes in skin areas with more rapid vasomotorics can be supposed to take part in this phenomenon . Experiments are in progress in this laboratory to see if thermal adaptation might also influence pharmacological effects of substances thought to modify central thermoregulatory processes.

Acknowledgements

This research is supported by the Hungarian National Foundation for Research (OTKA 472).

References

1. Brück, K (1981): Basic mechanisms in longtime thermal adaptation. In: Advances in Physiological Sciences, Vol. 32, Contributions to Thermal Physiology. Szelényi, Z, Székely, M eds. Pergamon Press, Oxford , 263-273.

2. Brück, K, Wünnenberg, W, Gallmeier, H, Ziehm, B (1970): Shift of threshold temperature for shivering and heat polypnea as a mode of thermal adaptation. Pflügers Arch 321, 159-172.

3. Dawson, NJ, Keber, AW (1979): Physiology of heat loss from an extremity: the tail of the rat. Clin Exper Phramacol Physiol 6, 69-80.

4. Donhoffer, Sz, Szelényi, Z (1967): The role of brown adipose tissue in thermoregulatory heat production in the warm- and cold-adapted adult rat. Acta Physiol Hung 32, 53-60.

5. Gordon, CJ, Heath, JE (1983): Reassessment of the neural control of body temperature: importance of oscillating neural and motor components. Comp Biochem Physiol 74A, 479-489.

6. Hensel, H (1981): Thermoreception and Temperature Regulation Monographs of the Physiological Society No. 38. Academic Press, London.

7. Kluger, MJ, Gonzalez, R, Mitchell, JW, Hardy, JD (1971): The rabbit ear as a temperature sensor. Life Sci 10, 895-899.

8. Roos, U, Jessen, C (1987): No dynamic effector responses to fast changes of core temperature at constant skin temperature. Can J Physiol Pharmacol 65, 1339-1346.

9. Székely, M, Kellermayer, M, Donhoffer, Sz (1971): Thermoregulatory role of periaortic brown adipose tissue in rats acclimated to different ambient temperatures. Acta Physiol Hung 40, 251-260.

10. Szelényi, Z, Hinckel, P (1987): Changes in cold- and heat-defence following electrolytic lesions of raphe nuclei in the guinea-pig. Pflügers Arch 409: 175-181.

11. Werner, J (1983): Influences of rate of temperature change on effector mechanisms in human thermoregulation. J therm Biol 8, 51-54.

Thermal Balance in Health and Disease
Advances in Pharmacological Sciences

ACCLIMATION OF AGED RATS TO COLD

Y. Asaki and M. Iriki*

Central Nervous System, Tokyo Metropolitan Institute of Gerontology, 35-2 Sakae-cho, Itabashi-ku, Tokyo-173, Japan. *Yamanashi Medical University, Tamaho, Nakakoma, Yamanashi-409-38, Japan

The ability to maintain thermal homeostasis decreases with age (1,2). Core temperature in the aged is liable to change in response to either heat or cold (3). Studies on a 3-hr cold (10°C) stress in mice suggested that adaptive thermoregulatory changes occurred by intermittent cold stress in different age groups from 8 to 22 months, but not in 30 months-old mice (4). Heat production in cold acclimated rats, which were 3, 12 and 24 months of age, was higher than that of non-acclimated groups during exposure to cold (5).

In the present study, young and aged rats were gradually acclimated to cold temperatures. To examine thermal homeostasis, rectal temperature in a freely moving rat was measured in a wide range of slowly and continuously changing ambient temperature. The aim of the present study was to elucidate whether through acclimation to cold, aged rats can achieve enhanced thermal homeostasis.

Methods

Wistar male rats aged 6 to 9 months and 24 to 30 months were used as young and old rats, respectively. Water and food were provided *ad libitum* and the ambient temperature in the animal colony of the Tokyo Metropolitan Institute of Gerontology was maintained at 22 ± 1°C with a light-dark cycle of 12 hr. The young and old rats were both divided into two groups. One, the Non-Acc group, was kept at an ambient temperature of 22°C. The other, the Acc group, was gradually subjected to cold temperatures. In the Acc group, the ambient temperature of breeding was changed from 22° to 20°C for the first week, to 15°C for the second week, and to 10 °C for the third to fifth week. Each group consisted of 6 to 8 rats.

The rectal temperature was measured by means of a thermistor thermometer. Twisting of the lead wire of the thermistor sensor was prevented by connecting it to the thermometer through the slip-ring (Airflyte Electronics CAY-675). Therefore, the animal could move freely. To evaluate thermal homeostasis, the ambient temperature was gradually changed from 0° to 30°C. Initially, the ambient temperature was maintained at 20°C for more than 30 min after the rats were placed in an artificial climatic chamber (Tabai Co. Ltd. PL4E). Next, the temperature was raised from 20° to 30°C, then dropped to 0°C, and then returned to 20°C at a constant rate of 10°C/hr. The cyclic interval was six hours. From this hysteresis experiment, the relation of the rectal temperature to one cycle of changing ambient temperature was determined. The first hysteresis experiment in each rat was used as a control experiment. The ambient temperatures of breeding for the Acc groups were lowered after the control experiment. After the control experiments, the hysteresis experiments on the four groups were repeated every weekend for five weeks.

Oxygen consumption was examined using an expired gas analyzer (NEC San-ei Co. Ltd. 1H26) over two hours from 1:00 to 3:00 p.m. Oxygen consumption (ml/g/hr) at an ambient temperature of 10 °C was estimated for both the Non-Acc and Acc groups.

Results and Discussion

Figure 1 shows typical results in the hysteresis experiments for the non-acclimated animals. A young rat is shown by the solid line and an old rat by the broken line. The relation of rectal temperature to ambient temperature in the old rat showed a remarkable decrease in rectal temperature and thermal hysteresis phenomenon during cold ambient temperature. The thermal hysteresis phenomenon was shown as the shape of a loop-like area in Fig. 1. To represent the results of the hysteresis experiment quantitatively, the following four indices were estimated: mean rectal temperatures and areas of thermal hysteresis loop in the thermal neutral zone of which ambient temperature is between 20° and 30°C, and in the cold zone between 0° and 10°C. They were respectively called rectal temperatures and Loop Areas in the neutral and the cold zone. In the neutral zone, there were no significant age differences in the rectal temperature and the Loop Area. Otherwise, in the cold zone, rectal temperature in the old group was significantly lower than that of the young group ($p<0.01$). Loop Area in the old group was significantly higher than that of the young ($p<0.01$).

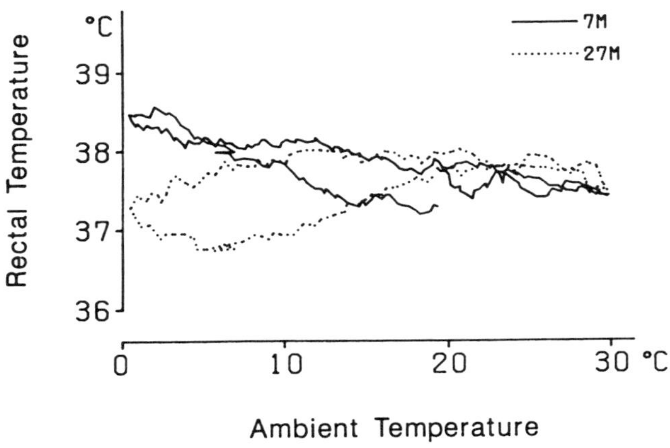

Fig. 1: Typical changes in the rectal temperature in response to ambient temperature change. The solid line and the dotted line show a young rat (7 months old) and an old rat (27 months old), respectively. Both rats were non-acclimated animals.

In the four groups, that is, the Non-Acc Young and Old groups and the Acc Young and Old groups, these indices were evaluated with regard to the hysteresis experiments of the control and the fifth week. The magnitude of the Loop Area was divided by the mean value of the area of the control in the Non-Acc Young group in the thermal neutral zone. In the neutral zone of ambient temperature, there were no significant differences for rectal temperatures and Loop Areas between the controls and the fifth week, respectively. In the thermal neutral zone, thermal homeostasis was thoroughly maintained in young and old rats regardless of cold acclimation. The thermal displacement, which makes a shift from a thermal balance condition, was not so large in the thermal neutral zone where the thermoregulatory system in even an old group could maintain thermal homeostasis.

Results of the hysteresis experiments in the cold zone are shown in Table 1. The rectal temperatures in the controls for the Non-Acc and Acc Old groups were significantly lower than those for the Non-Acc and Acc Young groups, respectively ($p<0.05$, $p<0.01$). As the Acc Young and Old groups in the control did not acclimate to cold, there were no differences in rectal temperature and Loop Area compared with the Non-Acc Young and Old groups, respectively. There was no significant difference in rectal temperature between the control and the fifth week for the Non-Acc Old group, but, in the Acc Old group, the rectal temperature at the fifth week was significantly higher than that of the control ($p<0.01$). In the Non-Acc and Acc Young groups, the rectal temperatures at the fifth week were significantly higher than those of the control, respectively ($p<0.05$, $p<0.001$). At the fifth week, the rectal temperature of the Acc Old group was significantly lower than that of the Acc Young group ($p<0.01$), but this difference was not significant in comparison with that of the Non-Acc Young group. After cold acclimation, the rectal temperature in the old rats in the cold zone was maintained without a decrease in temperature.

Table 1 Rectal temperature (Tr) and Loop Area (LA) in the Non-Acc and Acc Young groups and in the Non-Acc and Acc Old groups in the cold zone of ambient temperature from 0 to 10°C.

		Young			Old		
		Non-Acc		Acc	Non-Acc		Acc
					!		!!
Tr	C	38.24±0.06 (8)		38.26±0.09 (8)	37.40±0.22 (6)		36.97±0.30 (7)
		*		***	!!!		** !!
	5w	38.46±0.08 (8)	##	38.79±0.08 (8)	37.44±0.14 (6)	##	38.16±0.14 (7)
					!		!
LA	C	1.77±0.29 (8)		1.78±0.41 (8)	5.43±0.95 (6)		4.77±0.90 (7)
					!		**
	5w	1.55±0.36 (8)		1.79±0.38 (8)	3.58±0.67 (6)	#	1.26±0.39 (7)

The magnitude of the Loop Area was divided by the mean value of the area of the control in the Non-Acc Young group in the thermal neutral zone. The value shows mean ± SE (n).
C : the control, measured at the first experiment in each a group.
5w : the fifth week, measured at the fifth week after the first experiment.
Symbols (*, # and!) indicate significant differences.
*,** and ***: differences between the control and the fifth week ($p<0.05$, $p<0.01$ and $p<0.001$).
and ##: differences between the Non-Acc and the Acc groups ($p<0.05$ and $p<0.01$).
!, !! and !!!: age differences in Non-Acc groups or in Acc groups ($p<0.05$, $p<0.01$ and $p<0.001$).

The differences in Loop Areas between the control and the fifth week in both young groups and the Non-Acc Old group were insignificant. However, that in the Acc Old group at the fifth week was significantly lower than that of the control ($p<0.01$). The Loop Areas in the young groups did not differ between the Non-Acc and Acc groups at the fifth week, but those in the old groups differed significantly ($p<0.05$). The Loop Area in the Acc Old group at the fifth week was comparable to those of the young groups.

The Loop Area is the index of thermal hysteresis. The dead body of the rats, that can't maintain thermal homeostasis, remarkably displayed the thermal hysteresis even in the thermal neutral zone. Thermal hysteresis must indicate a thermal imbalance. If the thermoregulatory system maintains thermal homeostasis, core temperature is stable in response to ambient temperature changes, so that the Loop Area becomes a small value. As the decrease of the Loop Area in the cold acclimated old rats was clearly shown, their ability to regulate body temperature clearly improved.

The rectal temperature in cold acclimated old rats did not decline or exhibit large thermal hysteresis to exposure to cold. Old rats can acclimate to cold and maintain thermal homeostasis much like Non-Acc young rats.

Table 2 Oxygen consumption (ml/g/hr) in the Non-Acc and the Acc Young and Old groups at an ambient temperature of 10°C.

Young		Old	
Non-Acc	Acc	Non-Acc	Acc
1.605±0.060 (8)	1.700±0.026 (8)	1.438±0.021 (7)	1.729±0.054 (8)

Young Non-Acc vs Old Non-Acc: *; Old Non-Acc vs Old Acc: ***

The values represent means ± SEs (n).
* and *** indicate significant differences ($p<0.05$ and $p<0.001$).

Heat production during exposure to cold is one of the most important functions in maintaining thermal homeostasis. The oxygen consumption in ml/g/hr at 10°C in the four groups was measured using an expired gas analyzer. Oxygen consumption in the Non-Acc Old group was significantly lower than in the other groups (Table 2). Oxygen consumption in human beings (6) and rats (7) decreased with age. The old animal shows a decline of core temperature under a cold ambient temperature.

After cold acclimation, the oxygen consumption in young rats did not change, but in old rats it increased significantly ($p<0.001$). The increase of the oxygen consumption in acclimated old rats appeared to suppress the decrease of core temperature. On the other hand, in young groups, as even non-acclimated rats could maintain thermal homeostasis, no difference in heat production between acclimated and non-acclimated rats must be considered.

From these results, it can be seen that old rats can acclimate to cold and show a recovery in heat production. After cold acclimation, old rats can maintain thermal homeostasis to cold.

References

1. Collins, K. J. and Exton-Smith, A. N. (1983) : 1983 Henderson award lecture.Thermal homeostasis in old age. J. Am. Geriatr. Soc., 31 : 519-524.

2. Iriki, M. and Asaki, Y. (1987) : Decreased thermoregulatory function in aged rats. Proceedings of the third regional congress, Bangkok : 77-79

3. Cox, B., Lee, T., and Parkes, J. (1981) : Decreased ability to cope with heat and cold linked to a dysfunction in a central dopaminergic pathway in elderly rats. Life Sciences, 28 : 2039-2044.

4. Talan, M.I., Engel, B.T., and Whitaker, J.R. (1985) : A longitudinal study of tolerance to cold stress among C57BL/6J mice. J. Gerontol., 40 : 8-14.

5. Kiang-Ulrich, M. and Horvath, S. M. (1985) : Age-related differences in response to acute cold challenge (-10 °C) in male F344 rats. Exp. Gerontol., 20 : 201-209.

6. Shock, N.W. and Yiengst, M.J. (1955) : Age changes in basal respiratory measurements and metabolism in males. J. Gerontol., 10 : 31-40.

7. Balmagiya, T. and Rozovski, S.J. (1983) : Age-related changes in thermoregulation in male albino rats. Exp. Gerontol., 18 : 199-210.

Thermal Balance in Health and Disease
Advances in Pharmacological Sciences

AVIAN THERMOREGULATION DURING THE PERINATAL PERIOD

M. Nichelmann, B. Lange, R. Pirow, J. Langbein and S. Herrmann

Institut für Verhaltensbiologie und Zoologie der Humboldt-Universität zu Berlin, Invalidenstraße 43, D 10115 Berlin, Germany

Three different types of processes occur during the perinatal development of temperature regulation in avian species: (*a*) development of endothermy, (*b*) epigenetic determination of the thermoregulatory system, (*c*) thermoregulatory learning processes in the early postnatal phase.

The evolution of homeothermy (1) was made possible by developments in locomotion, respiration, the cardiovascular system and some aspects of cellular metabolism. Similarly, prerequisites of the ontogenetic development of endothermy in avian embryos are the development of (*a*) oxygen uptake through the egg shell, (*b*) transport of oxygen in the blood, (*c*) an appropriate cellular enzyme spectrum for heat production and (*d*) relevant nervous mechanisms for the thermoregulatory control.

According to Whittow and Tazawa (2) there are four perinatal stages in the development of endothermy: (*a*) an Arrhenius-limited stage in which the embryo's metabolic rate is directly related to its temperature; with a Q_{10} value of about 2 caused by a deficit of corresponding enzyme activity and/or by a lack of central nervous influences on heat production; (*b*) an oxygen-conductance-limited stage in which the oxygen uptake of the embryo is limited by the rate of diffusion of oxygen through the egg shell and the chorioallantois membrane, in relation to oxygen demand by the embryo; (*c*) a power-limited stage in which the embryo has a limited capacity to produce heat in response to cooling; (*d*) full-blown homeothermy.

Besides these developmental processes, epigenetic adaptation mechanisms occur during the perinatal period (3). Epigenetic adaptation is initiated by learning processes during the perinatal period.

The induced abilities can be termed innate if they are developed prenatally but they are not fixed genetically. Epigenetic adaptation preadapts the bird to an anticipated environment. It increases the probability of survival between hatching and the development of physiological adaptation. Also, it modifies the function of the control systems so that postnatal conditions are more readily tolerated. Experiments carried out in chickens (4) show that prenatal temperature experience in embryos of this species has a long-lasting influence on postnatal temperature regulation.

Lastly, we assume that in the new-hatched bird, as in mammalian species (5) learning processes are able to modify the thermoregulatory mechanisms and optimize thermoregulatory strategies.

We have chosen the Muscovy duck *Cairina moschata* as our experimental bird because the summit metabolism in 1-day-old Muscovy ducklings is 2 to 2.5 times higher than in other poultry species of the same age. Because of this and because non-shivering thermogenesis occurs in Muscovy ducks (6) we expected that metabolically relevant enzyme activity would be high enough in embryos of this species to overcome the power-limited stage of transition to endothermy.

Methods

Three types of experiments were carried out.

Firstly, during the second half of embryonic development the incubation temperature (ambient temperature, Ta) was changed from 37.5 °C (normal temperature, NT) to 34.5 °C (low temperature, LT) or to 39.5 °C (high temperature, HT). In some cases Ta was changed to 35.5; 39.0 or 40.5 °C. After 30 or 100 minutes of exposure, oxygen consumption was measured, using a modified Scholander respirometer. Additionally, using a non-invasive method (7; the tips of electrodes were inserted between shell and membrane) heart rate was measured continuously in NT-embryos from the 25th day of incubation until hatching. The heart rate was determined every second and averaged hourly.

Secondly, the temperature was lowered permanently to 34.5°C for the last 7 days of incubation. For the first 10 days after hatching, oxygen consumption and carbon dioxide loss were measured, heat production was calculated and colonic temperature (CT) was measured at ambient temperatures (Ta) ranging between 10 and 40 °C. By using a temperature channel, the preferred temperature (PT) was measured from additional birds (8).

Table 1: Environmental conditions of ducklings during the first 10 days of life

Series	Number of birds	Environmental conditions
1	8	temperature channel, 5 °C-steps between 20 and 40 °C,
2	8	constant 20 °C
3	8	constant 40 °C
4	8	voluntary choice between 25 and 35 °C
5	8	thermoneutral conditions

Thirdly, NT-incubated and normally hatched ducklings were kept under different conditions during the first 10 days of life (Table 1). HP and CT were estimated on day 11 and 12.

Results

Development of endothermy

Starting at the 24th day of incubation, a 30-min exposure to LT or HT increases or decreases the oxygen consumption, respectively, in comparison to the normal incubation temperature.

Table 2: Influence of Ta (duration of exposure 30 min) on heat production ($W \cdot kg^{-1}$ egg) of Muscovy duck embryos

	Ambient temperature (°C) 34.5	37.2	39.0
Age (days)			
24	1.50	1.26	1.19
31	1.94	1.68	1.46
32	2.19	2.09	1.96

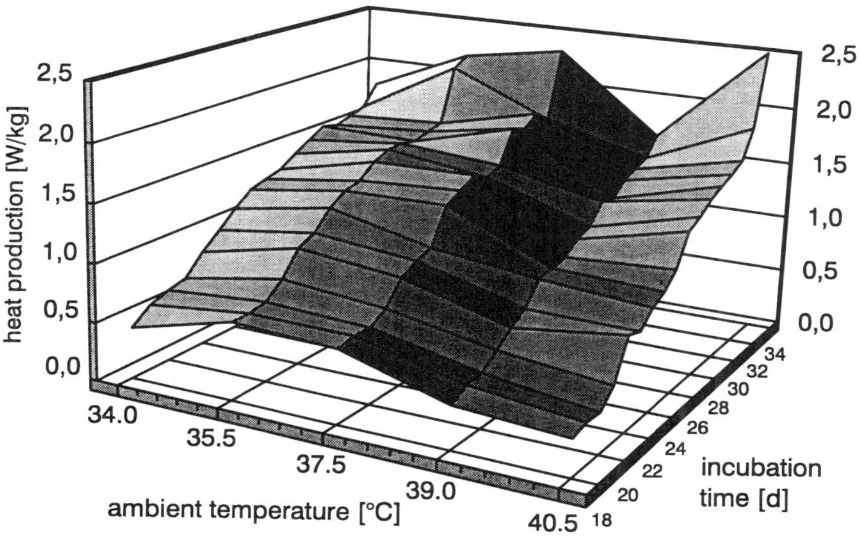

Fig. 1: Influence of incubation temperature on heat production in Muscovy duck embryos aged between 18 and 34 days.

In the 100-min exposure experiments, either an increase or a decrease in Ta resulted in decreased HP in 19- to 25-day-old embryos. For instance, on day 21 HP was 0.526 ± 0.104; 0.818 ±

0.098; 0.940 ± 0.102; 0.579 ± 0.049 and 0.478 ± 0.097 W•kg^{-1} at 34.0; 35.5; 37.5; 39.0 and 40.5 °C Ta, respectively. The relationship between HP and Ta could be described as a parabolic function with the extreme value y_{max} at 37.5 °C Ta (Figure 1). From the 26th day of incubation HP increased at Ta above 39.0°C (for instance from 1.189 ± 0.080 W•kg^{-1} to 1.375 ± 0.056 W•kg^{-1} at day 28 and from 1.403 ± 0.241 W•kg^{-1} to 1.642 ± 0.201 W•kg^{-1} at day 32).

These results demonstrate that embryos of muscovy ducks have a thermoneutral temperature (TNT) of 40.5 °C up to the 25th day of incubation, when it decreases to 39.0 °C. The maximum values of HP were measured at NT (37.5 °C) (summit metabolism); below this Ta, HP began to decrease, probably because of hypothermia.

Ta affected the onset and duration of the plateau phase, the period of stagnation of oxygen consumption during final embryonic development. The plateau phase occurred from days 26 - 31 at Ta 27.5 °C, days 28 - 32 at 35.5 °C and days 30 - 32 at 40.5 °C.HP during the plateau phase was 1.611 ± 0,165 W•kg^{-1} at 37.5 °C; 1.572 ± 0.113 W•kg^{-1} at 35.5°C and 1.581 ± 0.165 W•kg^{-1} at 40.5 °C ($p<0.01$); Ta had an influence on the duration of the plateau phase but not on the level of HP during this period. This result supports the hypothesis that a limited rate of diffusion through the egg shell and shell membranes causes the beginning of the plateau phase.

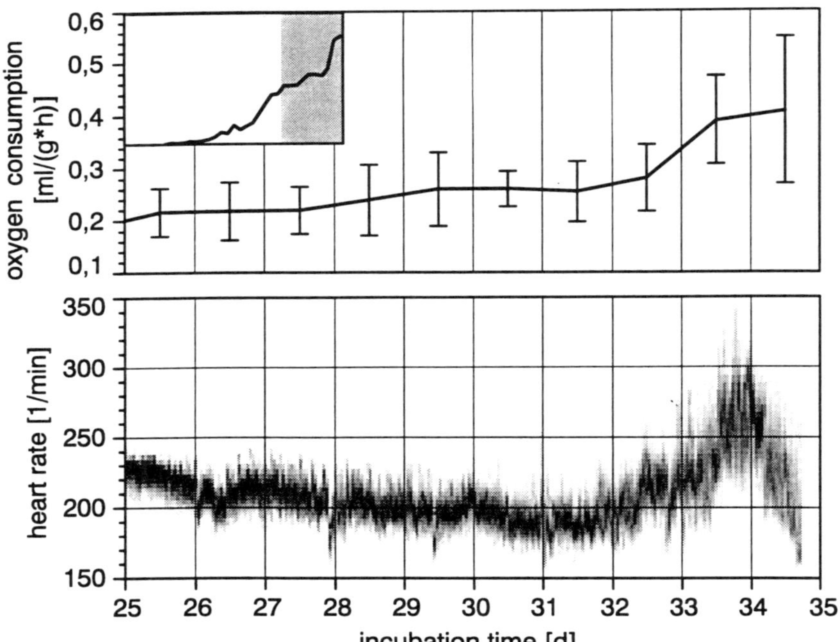

Fig. 2: Time course of oxygen consumption (upper part) and heart rate (lower part) in Muscovy duck embryos. Both parameters are very closely related.

Because the development of endothermy is connected with the development of the cardiovascular system, the heart rate of embryos was estimated from the 25th day of incubation. The experiments gave the following results.
* Latest on the 26th day of incubation the heart activity is under the influence of the autonomic nervous system (Nichelmann 1993).
* The relationships between age and heart rate show a typical course.
* Oxygen consumption and heart rate are very closely related (Figure 2).
* Heart rate is lower under LT conditions at all developmental stages except shortly before hatching.

Epigenetic determination of the thermoregulatory system

Lowering the Ta during the last 7 days of incubation to LT resulted in an increase in the cold resistance of ducklings after hatching. LT-birds treated in this way had a higher HP and a higher CT after hatching than NT-birds. As an example, on the 1st day of life the CT decreased from 42.1 ± 0.6 °C at 40.0 °C Ta to 37.5 ± 2.9 °C at 10 °C Ta in NT-birds and from 41.7 ± 0.6 °C to 39.5 ± 1.1 °C in LT-birds. Especially during the first 4 days of life, CT was higher in the LT-birds at low and moderate Ta and higher in the NT birds at high Ta. At low Ta (10 and 15 °C) the HP of the LT-birds were higher (28.2 ± 3.9 and 22.6 ± 3.6 W·kg, respectively) than those of NT-birds (18.1 ± 2.4 and 18.0 ± 1.3 $W \cdot kg^{-1}$). During the first days of life, the PT of the LT-birds was lower than the PT of the NT-birds. The corresponding values of the LT-birds on the 1st, 5th and 10th day were 38.7 ± 1.4; 34.1 ± 3.4; 33.6 ± 3.7 and of the NT-birds 39.1 ± 2.5; 36.0 ± 3.4; 33.6 ± 3.7 °C, respectively.

Postnatal thermoregulatory experiences

Ducklings, kept under different environmental conditions for the first 10 days of life (Table 1) had different cold resistances on days 11 and 12 because of different HP rates at low Ta. (Table 3)

Birds of the first 3 series were kept under conditions which activate heat loss and/or HP, whereas birds of the last series were kept at PT (series 4) or under thermoneutral conditions (series 5), so that the autonomic mechanisms of HP or heat loss where either not or only slightly activated to attain compensated heat balance.

Table 3: Influence of early environmental conditions on CT (° C) and HP ($W \cdot kg^{-1}$) of 11- or 12-day-old Muscovy ducklings (for series number see Table 1) at 10 °C Ta.

Series	CT	HP
1	41.11 ± 0.11	22.9 ± 1.8
2	40.95 ± 0.07	19.9 ± 1.0
3	41.17 ± 0.15	22.2 ± 1.6
4	40.69 ± 0.16	18.4 ± 0.9
5	40.54 ± 0.19	17.5 ± 1.1

Discussion

The results of our experiments show that the efficiency of the thermoregulatory system changes during the perinatal period and may be influenced by environmental factors. Three topics should be discussed.

1. Starting from the 18th day of incubation short term cold exposure increases oxygen uptake and HP. This opposes the statement of Whittow and Tazawa (2) that, prior to pipping, the avian embryo is in the Arrhenius-limited stage of development and its metabolic rate is directly related to its temperature. Whittow and Tazawa (2) based this calculation mainly on experiments carried out in domestic fowl embryos. The summit metabolism in newly hatched Muscovy ducklings is 2.0 to 2.5 times higher than in other poultry species (9) and they can produce heat by non-shivering thermogenesis (6); it is probable that the enzymes associated with heat production are developed very early in the prenatal period, so that the embryos are able to produce additional heat in the course of cooling. Of course, this is only possible if the neural mechanisms of thermoregulation are sufficiently developed at this time. Further research is needed to answer these questions. Our experiments show that the cardiovascular system is capable to supporting the development of endothermy. In the long term experiments heat loss was higher than HP. Therefore, the body temperature of the embryos dropped and consequently the HP decreased, too. It is noteworthy that Muscovy duck embryos have a thermoneutral temperature between 39.0 and 40.5 °C depending on age, whereas the normal incubation temperature is the temperature of summit metabolism. The decrease in HP between NT and thermoneutral temperature must be produced by active regulatory mechanisms, because it is against the van't Hoff rule.

2. The results support the hypothesis that a prenatal thermal load increases the sensitivity of the heat production mechanisms. In 1-day-old LT-birds HP at 10 °C Ta was about 35% higher than in NT-birds. Because of this, the CT did not decrease at 10 °C in LT-birds. Also, the higher HP explains the higher CT of the LT-birds in the Ta-zone between 10 and 30 °C. Furthermore, the results are in agreement with the hypothesis that avian prenatal cold experiences lead to a downward shift of the thermoregulatory set-point. The higher CT of LT-birds in the first day of life is due to the greater sensitivity of the HP mechanisms. Because of this, the actual body temperature can be controlled at the set-point, unlike that of the NT-birds, which have a lower HP. At day 5, the difference between actual body temperature and the thermoregulatory set point is extremely small in NT-birds because of the increasing HP. Finally, at the 10th day, CT in NT-birds is higher than in LT- birds: actual body temperature and thermoregulatory set-point are totally adjusted (10). LT-birds have a lower PT than NT-birds. This result also supports the hypothesis of a prenatally adjustable set-point. The lower thermoregulatory set-point seems to be a special strategy of epigenetical adaptation to cold conditions in the prenatal period (3, 4, 11).

3. The thermoregulatory system of Muscovy ducklings is influenced by thermoregulatory experience in the first 10 days of life. Birds which activate physiological thermoregulatory mechanisms (e.g. shivering or panting) in this phase of development, also use these mechanisms more readily later in extreme ambient situations (hot, cold, wet, high wind speed etc.). Conversely, birds controlling CT in the first few days mainly by behavioural thermoregulation (huddling together in low temperatures, enlarging the body surface in high temperatures, selecting optimal ambient temperatures etc.) prefer these adaptive behavioural strategies later (12). If birds are kept under constant optimal temperatures, thermoregulatory mechanisms develop imperfectly (13).

References

1. Duncker, H-R (1991): The evolutionary biology of homoiothermic vertebrates: the analysis of complexity as a specific task of morphology. In: Verhandlungen der Deutschen Zoologischen Gesellschaft. Pfannenstiel, H-D eds. 84, 39 - 60, Stuttgart-Jena-New York: Gustav Fischer Verlag.

2. Whittow, GC, Tazawa, H (1991): The early development of thermoregulation in birds. Physiol Zool 64, 1371 - 1390

3. Nichelmann, M (1992): Verhaltensbiologische Probleme im perinatalen Zeitraum. In: Verhaltensentwicklung. Nichelmann, M, Tembrock, G eds. Berlin: Akademie Verlag, 7 - 24.

4. Minne, B, Decuypere, E. (1984): Effects of late prenatal temperatures on some thermoregulatory aspects in young chickens. Arch exper Vet Med 38, 374 - 383

5. Schmidt, I (1987): Vergleichende Untersuchungen autonomer und verhaltensthermoregulatorischer Reaktionen endothermer Tiere. Habilschrift, Marburg

6. Barré, H, Geloen, A, Chatonnet, J, Dittmar, A, Rouanet, J-L (1985): Potentiated muscular thermogenesis in cold-acclimated Muscovy ducklings. Am J Physiol 249, (Regul Inter Comp Physiol 18) R533 - R 538

7. Pirow, R, Nichelmann, M (1993): Wiederkehrende Muster im Verlauf der Herzperiodendauer bei Embryonen von Cairina moschata. In: Verhandlungen der Deutschen Zoologischen Gesellschaft. Pfannenstiel H-D eds. 86, 268, Stuttgart-Jena-New York, Gustav Fischer Verlag

8. Nichelmann, M (1993): Early experience and adaptation. In: Proceedings of the Fourth European Symposium on Poultry Welfare. Savory, CJ, Hughes, BO, eds. Potters Bar: Universities Federation for Animal Welfare, 8 - 18

9. Tzschentke, B, Nichelmann, M, Burmeister, A, Modrey, P, Blank, A, Herrmann, S (1993) Development of temperature preference in birds during the first 10 days of life. In: Proceedings of the International Congress on Applied Ethology. Nichelmann, M, Wierenga, WK, Braun, S eds. Berlin: Humboldt-Universität zu Berlin, 524 - 526

10. Herrmann, S, Nichelmann, M (1991): Einfluß der Bruttemperatur auf physiologische und ethologische Temperaturregulations-mechanismen bei Moschusentenküken (Cairina moschata). In: Verhandlungen der Deutschen Zoologischen Gesellschaft. Pfannenstiel H-D eds. 84, 309 - 310, Stuttgart-Jena-New York: Gustav Fischer Verlag

11. Decuypere, E (1984): Incubation temperature in relation to postnatal performance in chickens. Arch exper Vet med 38, 439 - 449

12. Langbein, J, Nichelmann, M (1991): Die postnatale Ontogenese des Temperaturregulationssystems bei der Moschusente Cairina moschata. Zool Jb Physiol 95, 173 - 183

13. Nichelmann, M, Tzschentke, B (1991): Thermoregulatorische Präferenzen: Sind sie ein Maß für die Optimierung des Stallklimas? In: Artgemäße Tierhaltung 1990. Darmstadt, KTBL, 184 - 195

Thermal Balance in Health and Disease
Advances in Pharmacological Sciences

NOREPINEPHRINE-INDUCED PHASE SHIFTS OF THE JUVENILE CIRCADIAN CORE TEMPERATURE RHYTHM IN RAT PUPS

B. Nuesslein-Hildesheim and I. Schmidt

Max-Planck-Institut, W. G. Kerckhoff-Institut, Parkstraße 1, D-61231 Bad Nauheim, Germany

Within the first week of life, rat pups artificially reared in constant dim light and at ambient temperatures (Ta) below their thermoneutral zone develop an endogenous circadian rhythm of core temperature (Tc) and metabolic rate (MR). The circadian rhythms of these individually housed pups are in phase with the light-dark cycle in the breeding colony (5,6,11). The entrainment of the juvenile suprachiasmatic nucleus (SCN), considered to function as a "biological clock", seems normally to occur through maternal influences before the 19th day of gestation (10). Moreover, crossfostering experiments have shown that the phase of the circadian rhythm of rat pups can also be influenced by nursing mothers during the first week of life (4,5,7,10).

The pre- and postnatal maternal signals that entrain a rat pup's developing circadian system are unknown. Lesioning of the maternal SCN prevented entrainment of pups, whereas the surgical removal of various maternal endocrine organs (pituitary, adrenals, thyroid-parathyroids, ovaries, or pineal) did not disrupt maternal-fetal communication (10). This commumication has also been investigated by restoring rhythmicity to SCN-lesioned dams, for example by restricted daily feeding (references in 4,5,10) or by periodically injecting melatonin (2). Rhythmic changes in maternal sympathetic activity have not yet been considered as a zeitgeber for the juvenile rat`s Tc rhythm. Such changes seem, however, likely candidates for the maternal-fetal signal transfer in view of the above findings as well as because of the marked circadian changes in the pups' nonshivering thermogenesis that develop shortly after birth (8,9). In a first attempt to investigate a possible zeitgeber effect of a sympathetically mediated signal, we artificially reared rat pups for 12 days, while giving them periodic 12-h infusions of norepinephrine (NE) from the first through the sixth postnatal day. The effect of this treatment was evaluated by testing in the second postnatal week for a possible phase difference between the Tc rhythm of the treated pups and that of pups artificially reared without NE infusions.

Methods

Animals. We used lean (Fa/-) Zucker rats for these experiments because the juvenile circadian Tc rhythm has been most thoroughly characterized in this strain (5-9,11). Pups were derived from two colonies housed in different rooms, each maintained at 22°C and 55% humidity. One room was illuminated from 7:00 to 19:00 (LD colony) and the other from 19:00 to 7:00 (DL colony). The colonies were tended at the onset of the daily activity phase.

Artificial rearing. One-day-old pups (day 0=day of birth) were removed from the breeding colony and artificially reared in a way similar to that described for older pups (6). The esophageal catheters, though, were made from especially thin - 0.3x0.5 mm (i.d.xo.d.) - Silastic tubing (Dow Corning, special order). Synthetic rat milk was fed at rates increased according to body size and the degree of stomach filling from 1.0 ml/d on day 1 to 4.4 ml/d on day 12. To avoid digestive failure associated with hypothermia, Ta had to be kept at 33-32°C during the first 4 days. Pups could thereafter be exposed to Ta similar to those in our previous studies (6); that is, to cold loads resulting during the second postnatal week in a MR 50-100% greater than the MR at thermoneutrality. As in our earlier studies, 24-h cues during artifical rearing were avoided by tending the pups at 16-h intervals (at 6:00 and 22:00 on even-numbered days and at 14:00 on odd-numbered days).

Procedure. For each of the 6 experiments, pups from the LD and the DL colony were removed from their mothers simultaneously and then artificially reared with synthetic milk until postnatal day 12 in continuous light. Eighteen pups from the LD colony and 22 pups from the DL colony were used, and 6 pups from both groups received milk supplemented with D(-)NE for 12 h a day from day 1 to day 6. Starting at the time of lights-out in the colony of origin, the treated pups were given NE at doses increasing from 300 $\mu g \cdot h^{-1} \cdot kg^{-1}$ for 1- and 2-day-old pups to 500 $\mu g \cdot h^{-1} \cdot kg^{-1}$ for 3- to 6-day-old pups. Colonic Tc was measured continuously as previously described (6), except that the thermocouple ends had to be shielded in the first 4 days with the 0.3x0.5-mm Silastic tubing. An accurate record of deep body temperature was ensured by gradually increasing insertion depth of the thermocouples throughout the experiment from 1.5 to 2.2 cm.

Evaluation and Statistics. We evaluated the data of only those 11 NE-treated and 19 untreated pups that survived the experiment in good health. The hourly mean Tc for each of these pups was calculated from the data stored every 12 s. The period length and acrophase of the Tc rhythm were determined by cosinor analysis (3) for days 1-6 and for days 8-12. Cosinor parameters were included in the evaluation, only when the "percentage of variability accounted" (PVA), an index for the goodness of curve fitting, was ≥ 20. Acrophase on a given day was calculated by taking into account the individual period length. Mean phase angles of samples were evaluated by vector addition, and only those whose synchronization was judged significant ($p \leq 0.05$) by Rayleigh testing (1) are presented here. The degree of synchronization is expressed as the length of the mean vector (r). The difference between the mean acrophases of groups was tested for significance ($p \leq 0.05$) by Mardia-Watson-Wheeler testing (1), and differences between the period lengths of groups were tested for significance ($2p \leq 0.05$) by Mann-Whitney u-testing. Values are presented as means ±SE.

Results

Periodic Tc changes during NE treatment. Periodic 12-h NE infusions into 1- to 6-day-old artificially reared rat pups increased Tc during the infusions to maximum hourly mean values between 36 and 39°C, and resulted in daily Tc changes with amplitudes of 1-3°C (Fig. 1, top). Cosinor analysis of data gathered during days 1-6 yielded well-fitting sine curves (PVA≥20) for the periodic Tc changes of all NE-treated pups. The mean period length of this exogenously driven Tc rhythm was 24.0±0.2 h (N=11). The treated pups were synchronized with each other (r=0.89) and the mean phase angle for the acrophases on day 1 was 8.5 h (0 h = beginning of dark phase in the colony of origin). The untreated pups showed only a weak circadian periodicity in the first days of the experiment, but the difference between their daily minimum and maximum Tc gradually increased to about 1°C by the end of the first postnatal week.

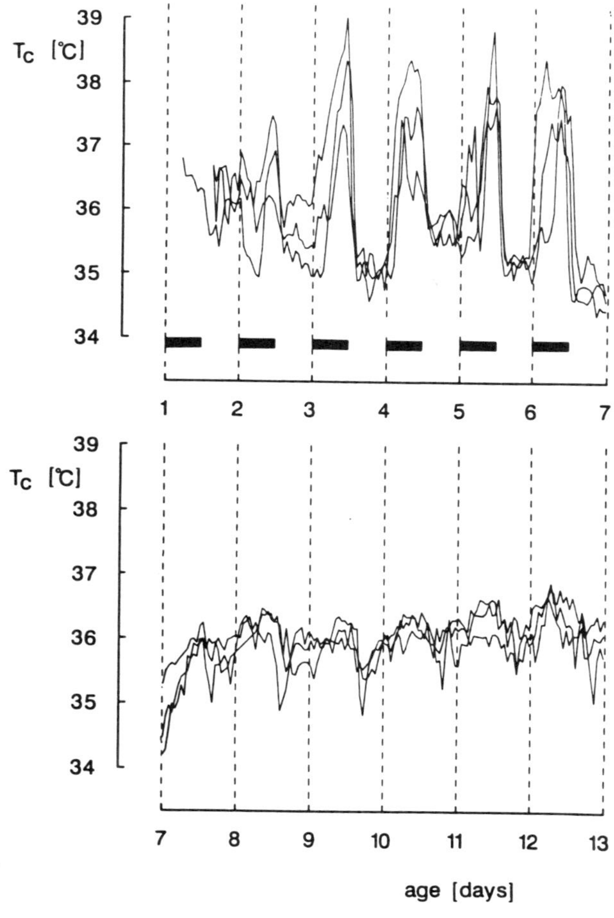

Fig. 1. Core temperature (Tc) of 3 pups artificially reared in continuous light. Black bars on days 1-6 denote 12-h norepinephrine infusions during the time the lights were out in the colony of origin. Dashed lines mark the beginning of the dark phase in the colony of origin (7:00).

Circadian rhythm of NE-treated and control pups during the second week of artificial rearing. The treated pups continued to show periodic changes of Tc after the periodic NE-infusions were discontinued (Fig. 1, bottom), but cosinor analysis for days 8-12 yielded well-fitting sine curves (PVA≥20) for only ≈60% of the NE-treated pups, whereas the PVA was ≥20 for ≈95% of the control pups. As shown in Fig. 2, the circadian Tc rhythms of the control pups with PVA≥20 were still synchronized with each other after artificial rearing up to 12 days of age (N=18, r=0.76). The mean phase angles of the acrophases on postnatal days 8 and 12 were respectively 2.6 h and 3.7 h. The rhythm freeran with a period length of 24.7±0.4 h, but the phase shift over these 4 days was too small to reach statistical significance. The rhythms of the NE-treated pups with a PVA≥20 were also still synchronized with each other at the end of the artificial rearing (N=7, r=0.87). The mean phase angle of their acrophases on postnatal days 8 and 12, however, were respectively 6.8 and 8.7 h. The mean period length of their Tc rhythm (24.5±0.2 h) was not significantly different from that of the control pups. Correspondingly, their phase shift from day 8 to 12 was also not significant. In contrast, the 5-h difference observed six days after the end of the treatment in the mean phase angles of the acrophases in the treated and control pups (Fig. 2) was significant, thus indicating a lasting phase-shifting effect of the periodic NE-treatment.

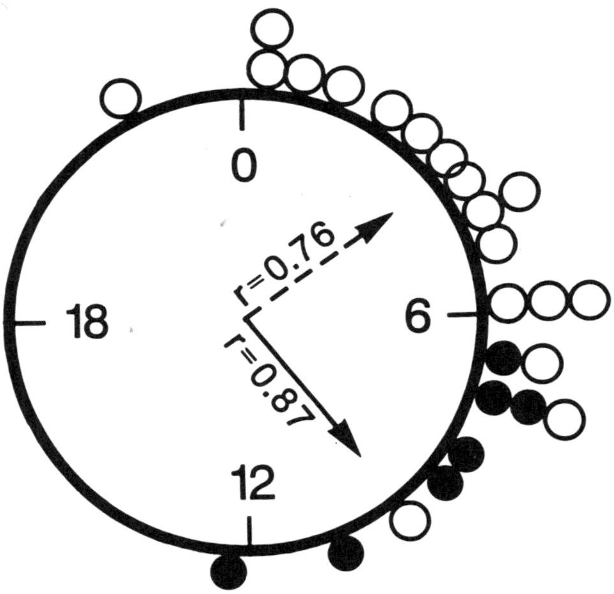

Fig. 2. Acrophase of individual norepinephrine (NE)-treated and control pups shown around a circle representing 24 h on postnatal day 12. Length of the mean vectors (r) represents the synchronization in the group receiving NE (solid line, closed symbols) and in the control group (dashed line, open symbols).

Discussion

This study demonstrates that the endogenous circadian Tc rhythm of rat pups can be entrained by periodic NE-infusions during the first week of life. During these infusions, the Tc of the treated pups was about 2 C higher than that of the control pups, and their MR reached about 185% the average daily MR (8-10 W/kg) of the control pups (5). Although fewer treated than control pups showed circadian rhythms that permitted the calculation of well-fitting sine curves during the second week of life, more than half of the pups which had been given the periodic NE-infusions, continued to show a clear circadian Tc rhythm which was about 5 h phase shifted in comparison to the controls at the end of the experiment. Because the period length was the same for treated and control pups, and the phase of the treated pups' rhythm did not shift significantly between days 8 and 12, this difference cannot merely reflect a gradual return of the rhythm of the treated pups to the original phase angle. In continuous light, during postnatal days 1-6 periodic 12-h infusions of NE starting at the beginning of the dark phase in the breeding colony can thus induce a persistent phase difference between the Tc rhythm of treated pups and the free-running rhythms of pups reared in the absence of 24-h cues. Pilot studies have indicated that pups receiving the same 12-h NE-treatment starting in the middle of the dark phase in the breeding colony undergo an even more pronounced phase shift (5). These results thus indicate that signals that cause a periodic 12-h increase of MR and Tc during the daily dark phase could be involved in setting the phase of the juvenile rat's circadian Tc cycle.

Acknowledgements

This study was supported by the DFG (Schm 680/1). We thank our NESP (Native English Speaking Person), Randy Kaul, for his advice.

References

1. Batschelet, E (1981): Circular Statistics in Biology. London: Academic Press.

2. Davis, FC, Mannion, J (1988): Entrainment of hamster pup circadian rhythms by prenatal melatonin injections to the mother. Am J Physiol 255, R439-R448.

3. Halberg, F, Tong, YL, Johnson, EA (1967): Circadian system phase - an aspect of temporal morphology; procedures and illustrative examples. In: The Cellular Aspects of Biorhythms. von Mayersbach, H ed. Berlin, Heidelberg: Springer-Verlag, 20-48.

4. Hiroshige, T, Honma, K-I, Watanabe, K (1982): Prenatal onset and maternal modifications of the circadian rhythm of plasma corticosterone in blind infantile rats. J Physiol 325, 521-532.

5. Nuesslein B (1993) Der juvenile circadiane Kerntemperatur-Rhythmus von Ratten: Physiologische Grundlagen und synchronisierende Faktoren. Theses, Univerität Marburg

6. Nuesslein, B, and Schmidt, I (1990a): Development of circadian cycle of core temperature in juvenile rats. Am J Physiol 259, R270-R276.

7. Nuesslein, B, and Schmidt, I (1990 b): Phase setting of juvenile circadian core temperature rhythm in rats: A pre- or postnatal event? J Interdis Cycle Res 21, 224-226.

8. Nuesslein, B, Petrova, O, Schildhauer, K, and Schmidt, I (1993): Morning depression of cold defense in juvenile rats. In: Integrative and Cellular Aspects of Autonomic Functions: Temperature- and Osmoregulation. Pleschka, K, Gerstberger, R, Pierau, F-K eds. London, Paris: John Libbey Eurotext, 295-304.

9. Redlin, U, Nuesslein, B, Schmidt, I (1992): Circadian changes of brown adipose tissue thermogenesis in juvenile rats. Am J Physiol 262, R504-R508.

10. Reppert, SM, Waever, DR, Rivkees, SA (1988): Maternal communication of the circadian phase to the developing mammal. Psychoneuroendocrinol 13, 63-78.

11. Schmidt, I, Markewicz, B, Nuesslein, B (1990): Endogenous circadian rhythm of metabolic rate in periodically cold exposed week-old rat pups. J Interdis Cycle Res 21, 241-243

Thermal Balance in Health and Disease
Advances in Pharmacological Sciences

THE EFFECTS OF CIRCADIAN RHYTHM AND CHANGE IN AMBIENT TEMPERATURE ON BODY TEMPERATURE IN AGED MALE RATS

D. Shemi and J. Kaplanski

Department of Clinical Pharmacology, Ben Gurion University of the Negev, Beer Sheva, Israel

Introduction

In homotherms, body temperature is one of the most vital, constant, regulated parameters. As it has been shown in many investigations, the ability to maintain body temperature, to cope with variations in ambient temperature and to develop fever is compromised with aging (1,2). Most of the data indicate a deficiency in aging to cope with a cold environment (3,4,5,6). Another change occurring with aging is a reduced amplitude of body temperature oscillation in animals and humans (7,8,9,10). It is the purpose of the present investigation to elucidate the relationship between body temperature and aging in rats at different ambient temperature and different light schedules. Moreover, we tried to correlate hypothalamic, prostaglandin E_2 (PGE_2) production with aging and body temperature.

Methods

Old male Sprague-Dawley rats (22 months) weighing between 450-500 g and young (7 months) weighing between 350-450 g were maintained in a control (ctl) room (12L 12D) kept at 22±1°C and relative humidity of 35-55% for at least 10 weeks before commencing the experiments. Some of the animals were maintained in a "reverse" (rev) room (12D 12L) for at least 2 weeks before beginning the experiments. The rats were housed on wood shavings, 2 per cage with water and food ad libitum. The experiments started at 8:00 and were carried out at an ambient temperature of 22±1°C (ctl room and rev room) for the circadian temperature measurements; at 4±1°C for the cold exposure experiments and at 34±1°C for the heat exposure. Rectal temperature was recorded at different times of the day by a thermocouple probe, (Anritherm, HL 600, Anritsu Meter Corp. Japan) inserted 5 cm into the rectum. Following the final rectal temperature measurements in ctl room and rev room, rats were decapitated and the anterior hypothalami were excised. One hypothalamus was incubated per vial for 3 hrs in Krebs-Henseleit incubation medium containing 0.2% glucose, in an environment of 95% O_2 and 5% CO_2 at 37°C. The medium was replaced every 60 min. The medium samples were frozen and assayed at a later time for PGE_2 content by radio-immunoassay. Statistical evaluation was carried out using Student's t test (2 tailed) and fractional analysis.

Results

The rectal temperature of the rats in the rev room at 8:00 and at 12:00 (peak of the circadian rhythm) were significantly higher ($P<0.05$) than the rectal temperature of the rats in the ctl room at 8:00 and at 12:00 (trough of the circadian rhythm). This was observed in the young and aged rats (fig 1). Rectal temperatures of aged rats were significantly lower ($P<0.01$) at 16:00 in the ctl room and at 8:00 and 12:00 in the rev room (fig 1). Two hours after exposure of the rats to cold, rectal temperatures of the aged rats were significantly reduced ($P<0.05$) compared to that of control aged rats and to the rectal temperature of cold young exposed rats ($P<0.01$) (fig 2). Thirty min after exposure of the rats to heat (hot room) rectal temperatures of aged rats were lowered by about 0.5°C ($P<0.05$). Three hrs after commencing the experiment, rectal temperatures of the aged rats were significantly higher ($P<0.05$) than the rectal temperatures of young rats by about 1°C (fig 3). In this experiment, 2 aged rats died during the heat exposure. The *ex vivo* experiments showed that the hypothalamic enzymatic activity resulting in PGE_2 synthesis in the young and aged rats in ctl and rev rooms were similar (2nd hr ctl room young 31.1±3.12, aged 35.2±3.75; 2nd hr rev room young 26.7±3.91 aged 27.2±2.12 pg/mg tissue/hr).

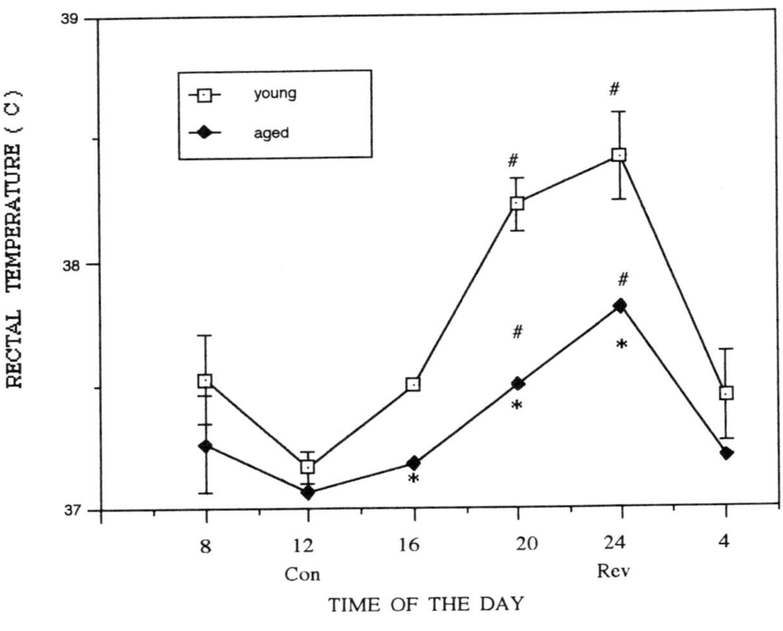

Fig. 1: Circadian rectal temperature of aged and young male rats. * $p < 0.05$ aged vs young; # $p< 0.05$ Rev vs Con.

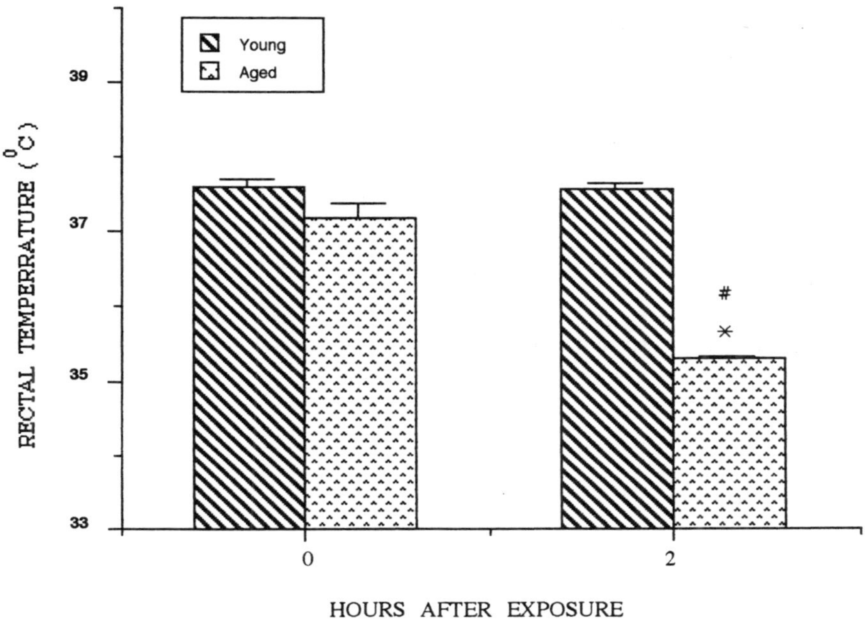

Fig. 2: Rectal temperature of aged and young male rats after 2 hrs cold exposure. * $p < 0.05$ aged vs young; # $p < 0.05$ 2 hrs exposure vs 0 time

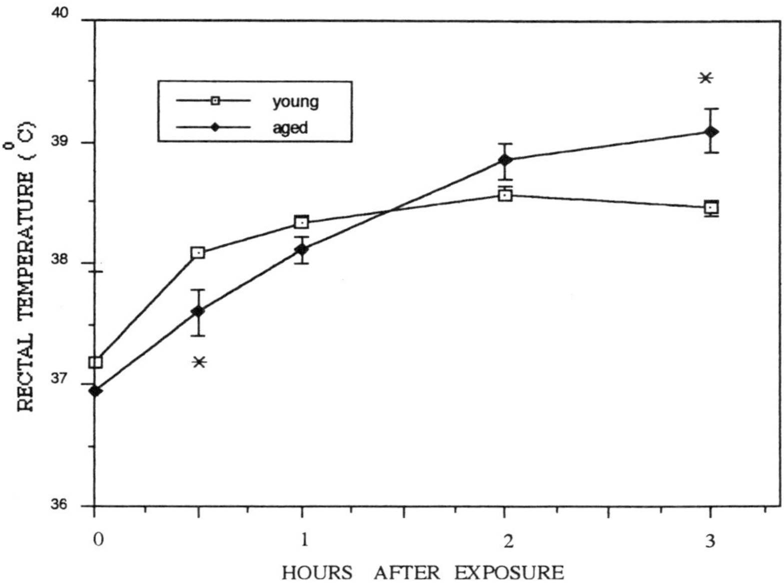

Fig. 3: Rectal temperature of aged and young male rats during 3 hrs heat exposure. * $p < 0.05$ aged vs young.

Discussion

The results of our study confirm some other observations that aging reduces the amplitude of the circadian rhythm of body temperature (11). Our data are also in agreement with studies which report lower resistance to cold stress in old age (4,12). We have found also a poor adaptation to heat in the aged rats. Male aged rats have difficulties to cope with the high ambient temperature. In this environment these aged rats do not have the capability to maintain body temperature in normal range, like the young rats. In this situation a high incidence of death occurred in the old rats and body temperature of the surviving rats is increased much more than in the young. The basal hypothalamic PGE_2 production of the old rats measured at 2 points is of the same magnitude as of the young rats. It could be concluded that old rats have a poor tolerance to cold and hot environments and a weakening of the circadian rhythm of body temperature. The exact mechanism of these deteriorating processes in the thermoregulatory mechanisms have to be elucidated.

References

1. Werner, J (1980): The concept of regulation for human body temperature. J Therm Biol 5, 75-82.
2. Shido, O, Yoneda, Y, Nagasaka, T (1989): Changes in body temperature of rats acclimated to heat with difficult acclimation schedules. J Appl Physiol 67, 2154-2157.
3. Estler, CJ (1971): Efficiency of thermoregulation in acutely cold-exposed young and old mice. Life Sci 10, 1291-1298.
4. Kiang-Ulrich, M, Horvath, SM (1985): Age-related differences in response to acute cold challenge (-10^{0}C) in male F344 rats. Exp Gerontol 20, 201-209.
5. McDonald, RB, Horowitz, BA, Stern, JS (1988): Cold-induced thermogenesis in younger and older Fischer 344 rats following exercise training. Am J Physiol 254, R908-R916.
6. Wagner, JA, Horvath, SM (1985): Influences of age and gender on human thermoregulatory responses to cold exposure. J Appl Physiol 58, 180-186.
7. Aschoff, J (1966): Circadian activity pattern with two peaks. Ecology 47, 657-662.
8. Aschoff, J (1983): Circadian control of body temperature. J Therm Biol 8, 143-147.
9. Thornhill, JA, Hirst, M, Gowdey CW (1978): Measurement of diurnal core temperatures of rats in operant cages by AM telemetry. Can J Physiol Pharmacol 56, 1047-1050.
10. Fioretti, MC, Riccurda, C, Menoconi, E, Martini, L (1974): Control of the circadian rhythm of body temperature in the rat. Life Sci 14, 2111-2119.
11. Vitiello, MV, Smallwood, RG, Avery, DH, Pascualy, RA, Martin, DC, Prinz, PN (1986): Circadian temperature rhythms in young adult and aged men. Neurobiol Aging 7, 97-100.
12. Talan, M (1984): Body temperature of C57BL/6J mice with age. Exp Gerontol 19, 25-29.

Thermal Balance in Health and Disease
Advances in Pharmacological Sciences

DIURNAL VARIATIONS IN RESPONSE OF GOLDEN SPINY MICE *(Acomys russatus)* TO NORADRENALINE INJECTION.

N. Kronfeld[1], N. Zisapel[2] and A. Haim[3]

[1]Depts. of Zoology and [2]Biochemistry, Tel-Aviv University, Tel-Aviv 69978, Israel. [3]Dept. of Biology, University of Haifa at Oranim, Kiryat Tivon 36910, Israel.

Introduction

Non-shivering thermogenesis (NST) is an important mechanism for rapidly increasing heat production in winter-acclimatized, as well as in cold- and long-scotophase acclimated small mammals (1,2,3). The great advantage of regulatory NST as a thermogenic mechanism is that when a rodent is not exposed to low ambient temperatures, metabolism can be kept at a relatively low rate; during exposure to cold, heat production is increased within a relatively short period of time (4). Therefore, this mechanism enables rodents to conserve energy.

It is commonly considered that NST capacity for cold acclimated mammals is a function of body mass (2). However, recent studies indicate that capacity for NST is not only allometrically determined. It appears that natural selection acting through habitat or habits may play an important role. For example, it has been demonstrated that NST capacity in desert rodents that are characterized by low RMR's (resting metabolic rate) is relatively high (5).

The golden spiny mouse, *Acomys russatus* is distributed in Israel in the most extreme hot and arid rocky desert regions . Its activity pattern under natural conditions, in the presence of the common spiny mouse *A.cahirinus,* is that of diurnal species. Like other desert rodents, the golden spiny mice have lower RMR than expected (6). Elimination of the sympatric species, the common spiny mouse *A.cahirinus* from the common habitat results in nocturnal activity of *A. russatus* (7). Accordingly, when kept in captivity without individuals of *A. cahirinus*, the golden spiny mouse exhibits a nocturnal activity pattern (8). Rubal *et al.* (8) demonstrated that in the golden spiny mouse, the increase in oxygen consumption precedes the increase in body temperature by *ca.* 3 h. This result suggests that the golden spiny mouse behaves like the pouched mouse (*Saccostomus campertris*), in which the VO_2 rhythm rather than the overall thermal conductance rhythm was found to be the driving force in the T_b rhythm (9).

The low RMR of the golden spiny mouse, coupled with the daily temporal order in VO_2 and T_b rhythms, suggest that NST may be a major driving force in the T_b rhythm. In such case it should be assumed that NST would exhibit diurnal variations. In order to check this hypothesis we tested to what extent the thermogenic state of BAT (Brown adipose tissue) changes in response to NA (Noradrenaline) injection during the maximum and minimum phases of the T_b and VO_2 daily cycle in *A.russatus*.

Materials and methods

Golden spiny mice (8 females, mean body mass 77.8 ±12.8 gr.) from a colony in the University of Haifa at Oranim (that were collected at Ein-Gedi on the Dead sea shore, and bred in Oranim) were used in the experiment. Prior to the experiment, the mice were individually caged with a photoperiod regime of 12L:12D (lights on 06:00 - 18:00 h) at T_a = 28±1°C. Rat pellets and carrot were supplied *ad lib.*

The diurnal variations in the response of the golden spiny mice to noradrenaline (NA) injection were measured in one individual every 3 h in a metabolic chamber at T_a=31±1°C, i.e. their lower critical point (6). Oxygen consumption (VO_2) and body (rectal) temperature (VO_{2min}, T_{bmin}) were recorded and immediately after that, NA was injected (1.5 mg/kg body mass), and VO_2 measurements were resumed for at least 30 min (VO_{2NA}). T_b was measured once again at the end of this period (T_{bNA}).

NST capacity was calculated from the net increase in O_2 consumption and in T_b due to the NA injections. O_2 consumption was measured in an open-circuit system (10,11) using flow of dried air at a rate of 480 ml/min. T_b was measured using a thermometer inserted 3 cm into the rectum of the tested mice.

Results

The results show clear and significant diurnal variation in the response of the animals VO_2 and T_b to NA injections. Maximal T_{bNA} and VO_{2NA} Values were recorded in the middle of the dark (at 24:00 h) and the light periods (at 12:00 h). The minimum values were recorded at the beginning and end of the dark period (6:00 and 18:00 h). The metabolic rate and body temperature measured in this and other studies, under the same conditions (8), indicate maximal T_b and VO_2 at the beginning and end of the dark period and minimal values in the middle of the dark and light periods (fig. 1,2). Here the ratio between T_{bNA}/T_b, VO_{2NA}/VO_2 and NST capacity are highest in the middle of the dark and light periods, and lowest at the beginning and end of the dark period (fig. 1,2).

Discussion

Our results clearly show daily variation in the response of the golden spiny mice to NA injections. This variation may be the result of a daily rhythm in the availability of NA receptors of BAT, or a result of the endogenous secretion of NA which occupies the available receptors. It is widely

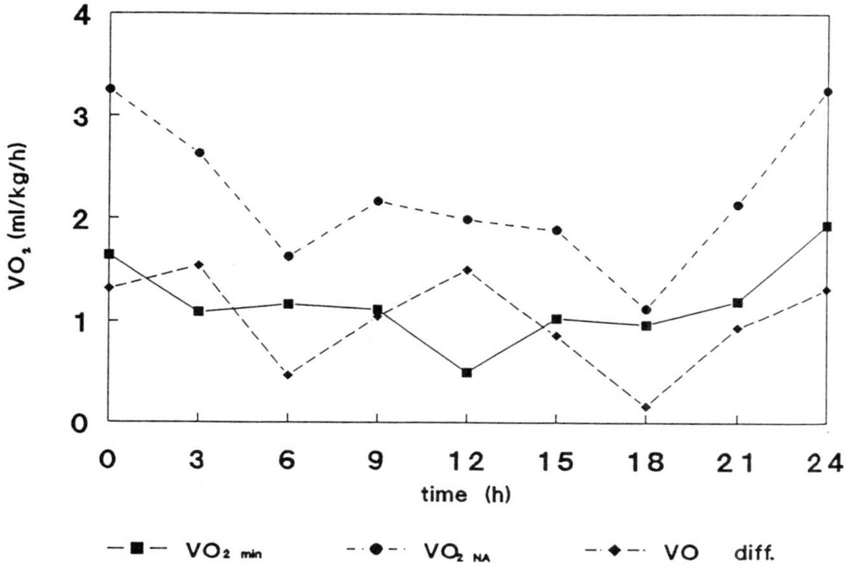

Fig. 1: The daily variation of the effect of NA injections on VO_2 in the golden spiny mouse.

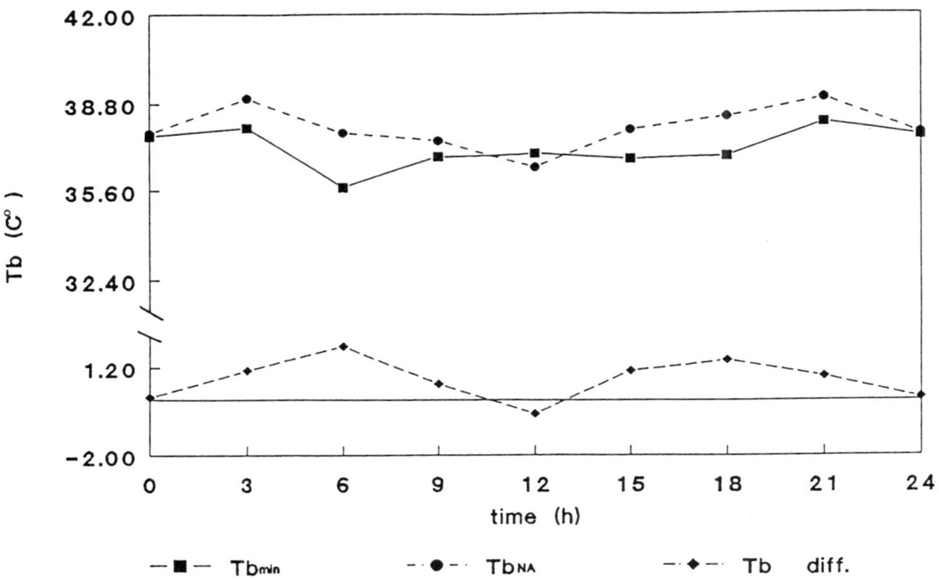

Fig. 2: The daily variation of the effect of NA injections on T_b in the golden spiny mouse.

accepted that GDP-binding sites on the mitochondrial uncoupling protein (UCP) are masked and unmasked for short-term regulation of NST (12). On the other hand, propranolol, (which is known to block BAT thermogenesis) had a big effect on thermogenesis when injected to juvenile rats when Vo_2 and Tb where close to their circadian maximum, and a negligible effect when Vo_2 and Tb where close to their circadian minimum (12). The fact that the maximal NST capacity was observed when RMR and Tb were at their minimum supports the endogenous secretion of NA (which occupies the available receptors) hypothesis.

The daily variation found in the response of the golden spiny mouse to NA injections suggests that this species is using NST for short time thermoregulation. The capacity to increase heat production rapidly from a low metabolic rate just before leaving the den for activity, enables this species to have a low RMR values (44% than predicted for their body mass at the thermoneutral zone or lower critical temperature (6)), and so to decrease the energy allocated for thermoregulation. This ability is advantageous in ecosystems like deserts, where food and water are limiting factors, so energy conservation is critical.

Acknowledgments

The authors thank Mr. H.Kosik for the maintenance of the colony. The research was supported by a BSF grant No. 91-00248.

References

1. Lynch, GR (1970): Effect of photoperiod and cold acclimation non-shivering thermogenesis in *peromyscus leucopus*. Am. Zool. 10, 308.

2. Heldmaier, G (1972): Cold-adaptive changes of heat production in mammals. In Bioenergetics international symposium on environmental physiology (smith, R.E. ed.) pp 79-81. Federeration of the American Society of Experimental Biology, Bethesda, Md.

3. Haim, A (1982): Effects of long scotophase and cold acclimation on heat production in two diurnal rodents. J. comp. Physiol. 148B, 77-81.

4. Wunder, BA (1985): Energetics and thermoregulation. In Biology of new world Microtus (Tamarin, R.H. ed.), pp. 812-844. Special publication of the American Society of mammalogists, No. 8.

5. Haim, A, and Izahaki, I (1993): The ecological significance of resting metabolic rate and non-shivering thermogenesis for rodents. J. therm. biol.18, 71-81.

6. Haim, A, Borut, A (1981): Heat production and dissipation in Golden spiny mouse *Acomys russatus* from two extreme habitats. J. comp. Physiol. 142, 445-450.

7. Shkolnik, A (1971): Diurnal activity in small desert rodent. Int. J. Biometeor. 15, 115-120.

8. Rubal, A, Choshniak, I, Haim, A (1992): Daily rhythms of metabolic rate and body temperature of two murids from extremely different habitats. Chronobiol. Int. Vol.9, No. 5, 341-349.

9. Haim, A, Ellison,GTH, Skinner, JD (1988): Thermoregulatory circadian rhythms in the Pouched mouse Saccostomys campastris. Comp. Biochem. Physiol. 91A, 123-127.

10. Depocas,F, Hart, JS (1957): Use of the Pauling oxygen analyzer for measurments of oxygen consumption of mammals in open circuit system and in short lag, closed circuit apparatus. J. appl. Physiol. 20, 388-392.

11. Hill, JS (1972): Determination of oxygen consumption by use of the paramagnetic oxygen analyzer. J. appl. Physiol. 33, 261-263.

12. Redlin, U, Nuesslein, B, Schmidt, I (1992): Circadian changes of brown adipose tissue thermogenesis in juvenile rats. Am. J. Physiol. 262 (Regulatory integrative comp. Physiol. 31), R504-R508.

DAILY RHYTHMS OF BODY TEMPERATURE IN COEXISTING RODENTS OF THE GENUS ACOMYS ACCLIMATED TO LONG-PHOTOPERIOD: EFFECTS OF ETHANOL AND MELATONIN

A. Haim[1] and N. Zisapel[2]

[1]Department of Biology, University of Haifa at Oranim, P.O. Kiryat Tivon 36910 and [2]Department of Biochemistry, Tel Aviv University, Tel Aviv 69978, Israel

Introduction

The daily rhythm of body temperature in endothermic animals is the product of two different rhythms, those of heat production and heat dissipation (2). While nocturnal species exhibit a daily rhythm of body temperature with maximal body temperature (Tb) values during the scotophase, diurnal ones increase their Tb during day time with maximal values at mid-day. Such a daily rhythm ensures that during their activity time, body temperature (Tb) is high while, during the resting phase, it is kept at relatively lower values, due to a decrease in heat production. Using such a "strategy", reduces daily energy expenditure (8).

An important behavioural adaptation in small mammals, such as rodents inhabiting arid environments, is their nocturnal activity. An exception to this activity pattern is found in the golden spiny mouse (Acomys russatus). This species, which is distributed throughout the extremely dry and hot regions of the Israeli Rift Valley, is diurnal (6). The diurnal activity of this species in these places is due to its exclusion from nocturnal activity by the common spiny mouse A. cahirinus. The removal of the latter from the common habitat results in nocturnal activity of the former (12). However, under controlled laboratory conditions, indirect introduction through an air inflow of diluted chemical signals released by A. cahirinus to nocturnal A. russatus resulted in a phase shift of 2h (3). Direct contact with chemical signals (urine and faeces) released by A. cahirinus advanced onset of activity of A. russatus by about 6h (4).

The effects of daily injections of melatonin and 7% ethanol in saline (used as a vehicle) on daily rhythms of Tb were studied previously in A. russatus, acclimated to a long photoperiod (16L:8D) at an ambient temperature (Ta) of 28°C (5). The results of that study showed, that under 16L:8D A. russatus exhibited circadian rhythmicity of Tb of a nocturnal species with maximal values recorded at 24:00h and minimal ones recorded at 12:00h. Daily s.c. injections (at 16:00h) for three days of 0.2ml of 7% ethanol in saline (vehicle) caused hyperthermia with maximal Tb values recorded at 12:00h. Similar injections of 50µg melatonin in 0.2ml vehicle for eight days abolished the daily Tb rhythm (Fig. 2). Daily injections of 0.2ml saline used as control had no effect on the daily rhythm of Tb in long photoperiod acclimated A. russatus.

The aim of the present research was to study the responses of long photoperiod (16L:8D) acclimated A. cahirinus (the nocturnal species) to ethanol and melatonin, injection and compare them with those of A. russatus.

Material and Methods

Seven A. cahirinus (males, mean body mass 41.5±3.2g) from our breeding colony (University of Haifa at Oranim) were acclimated for 3-5 weeks to a long photoperiod 16L:8D (lights on between 06:00-22:00h) at Ta=28±1°C, (Forma Scientific). During the photophase cool white fluorescent illumination and red dim light were used while during scotophase only the red dim light was used. Each individual was kept in a separate cage and was supplied with rat pellets and fresh carrots as a water source ad lib. Sawdust was used for bedding. Body (rectal) temperature Tb was measured by using a copper-constantan thermocouple, and monitored on a TH-65 Wescor thermometer. The thermocouple was inserted to a depth of 3cm for no more than 30sec. Preliminary tests had shown that a daily injection of 0.2ml saline given at 16:00h did not cause any change in the daily rhythm of Tb.

After acclimation, the daily rhythm of Tb in such mice was recorded by measuring Tb at intervals of 6h for 72h. The same mice were injected daily, for three days s.c., with a vehicle (0.2ml of 7% ethanol in saline) at 16:00h, 6h before lights went off. Tb in such mice was then recorded at 6h intervals over 48h. Finally, the mice were injected daily, at 16:00h for eight days, with 50μg melatonin in 0.2ml vehicle and Tb was then recorded every 6h for 72h.

All results are given as mean ±SD. Student's t-test was used for statistical analysis.

Results

Common spiny mice A. cahirinus acclimated to a long photophase showed a daily rhythm of Tb. Maximal Tb values (37.8±0.3°C) were recorded at 24:00h, while minimal Tb values (36.4±0.4 and 36.5±0.3°C) were recorded at 12:00 and 18:00h respectivly. Daily injections 0f 7% ethanol for three days caused a transient hypothermia in the first 24h after the treatment. However, Tb values returned to control values within 48h after the treatment. The melatonin treatment had no effect on the phase of the Tb daily rhythm of A. cahirinus, but increased the amplitude by lowering the minimal Tb value at 18:00h. As a result the difference between the maximal Tb values obtained at 24:00 and 06:00h and the minimal value recorded at 18:00h, in the melatonin injected mice (2.4°C) was higher than that in the control (1.4) (Fig 1). The daily rhythms of Tb of the golden spiny mouse A. russatus under the same treatments are shown in Fig. 2.

Discussion

A daily pattern of Tb variations is common among endothermic vertabrates. High Tbs are maintained during activity, while low values are recorded during periods of rest (9). Interestingly,

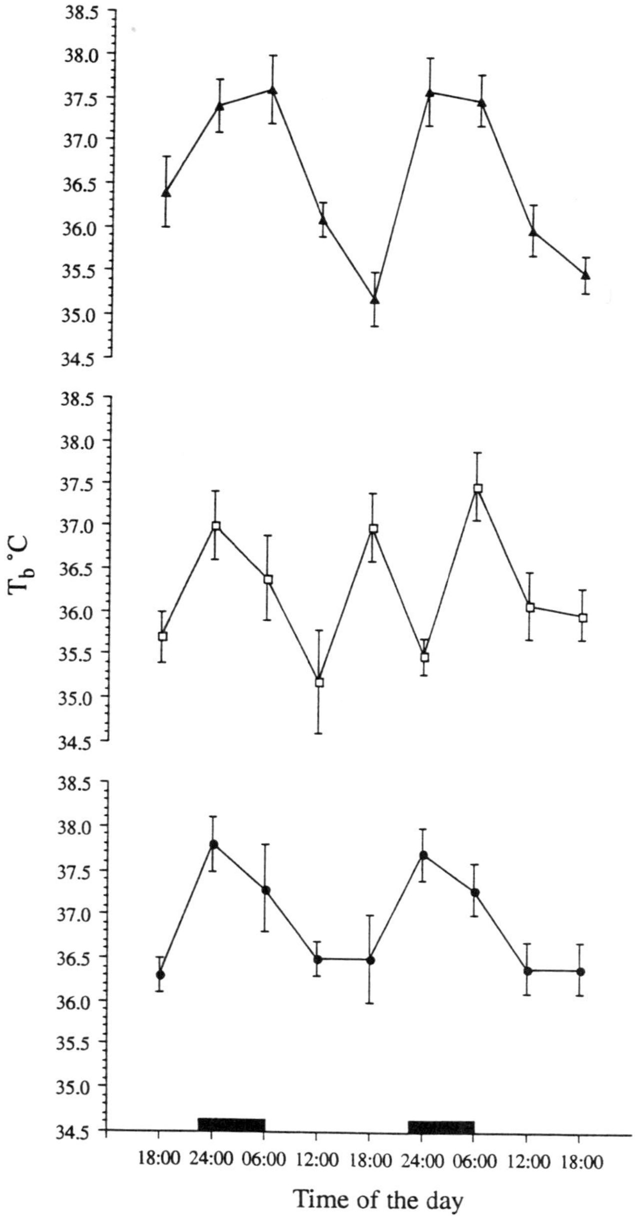

Fig. 1: Daily rhythms of body temperature (Tb°C) in the common spiny mouse Acomys cahirinus. Closed circles - mice acclimated to an ambient temperature of 28±1°C and a photoperiod regime of 16L:8D. Open circles - mice treated with 7% ethanol in saline (vehicle). Triangles - mice treated with 50 μg melatonin in vehicle. Each point is a mean of n=7 ± SD.

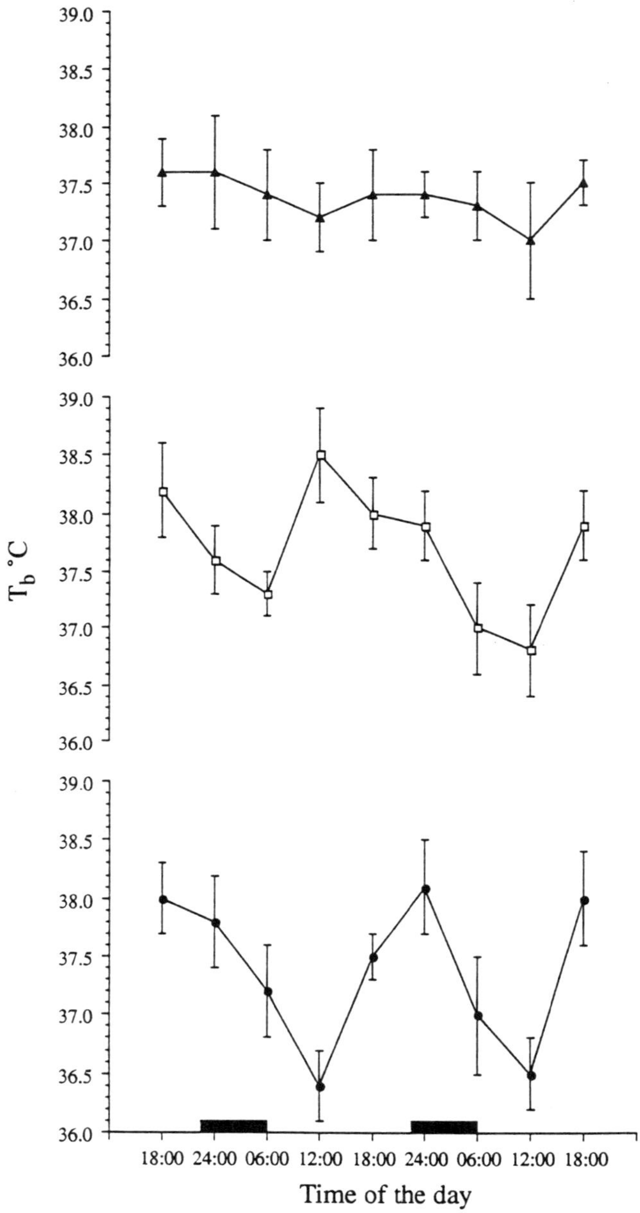

Fig. 2: Daily rhythms of body temperature (Tb°C) in the golden spiny mouse Acomys russatus. Symbols are the same as in figure 1.

the results of this study (Fig 1) show that, as in other nocturnal species, an increase in photoperiod causes the maximal Tb values to be maintained for a period of 6h with a daily amplitude of only 1.4°C. Djungarian hamsters, Phodopus sungorus, acclimated to a long photoperiod regulate Tb only with a small daily variation, while in hamsters acclimated to a short photoperiod, daily variations in Tb increased by 3-4°C (8).

In mice ethanol administration has been shown to decrease Tb. This response was found to depend upon time of administration (7). The maximal response was noted when ethanol was administered just before the beginning of the animals' activity phase. Recently the effects of chronic ethanol administration on the pineal N-acetyltransferase (NAT) were studied in the rat (10). In ethanol treated rats it was noted that the onset of nocturnal NAT activity as well as its nocturnal peak were delayed by 2h when compared with control rats. In humans, ethanol ingestion was found to decrease Tb significantly (11).

In A. cahirinus the administration of ethanol 6h before onset of darkness caused an hypothermia and a change in the daily rhythm of Tb. Minimal values of Tb were recorded at 24:00h on the first day of recovery from the drug. These responses are compatible with those in rats. Comparing the results of A. cahirinus with those of the golden spiny mouse A. russatus (5) a great difference is noted in the response to ethanol (Fig 2). When A. russatus is kept separately from A. cahirinus it behaves as a nocturnal species with a maximal Tb value recorded at 02:00h and a daily Tb amplituded of only 1°C. In contrast to A. cahirinus when treated with the vehicle (7% ethanol in saline), A. russatus changed its Tb daily rhythm pattern and became diurnal with a maximal Tb recorded at 12:00h. They also increased their Tb in contrast to the hypothermia recorded in A. cahirinus.

The effects of melatonin on the mammalian circadian clock have also been studied (1) and it was shown that in mammals, daily melatonin administration could entrain the circadian clock. The response of the two species to melatonin is also different. In A. cahirinus the administration of melatonin in 7% ethanol (vehicle) increased the daily Tb amplitude to 2.4°C, as reported for short photoperiod acclimated hamsters (8). This response may represent the sum of effects of ethanol and melatonin, since at long photoperiod the amplitude of Tb rhythm in A. cahirinus was only 1.4°C. In A. russatus in contrast to A. cahirinus a daily injection of melatonin abolished the Tb daily rhythm.

These results show that, under long photoperiod acclimation, the response of the daily Tb rhythm in these two sympatric species to ethanol and to melatonin is different. This difference may emerge from the lability of the biological clock in A. russatus, a species that can change its daily rhythm both in natural conditions and in the laboratory in response to the presence of A. cahirinus or even to chemical signals released by it into its environment. These relations are currently investigated in our laboratories.

Acknowledgements

The authors thank Mr. H. Kosik for the maintenance of the colonies. We also thank Dr. J. Hoffeman for her remarks on this manuscript. This research was supported by a BSF grant No. 91-00248.

References

1. Armstrong, SM, Cassone, VM, Chesworth, MJ, Redman, J, Short, RV (1986): Synchronization of mammalian circadian rhythms by melatonin. J Neural Trans (Suppl) 21, 375-398.

2. Aschoff, J (1982): The circadian rhythm of body temperature as a function of body size. In: A Companion to Animal Physiology. Taylor, CR, Johanson, K, Bolis, L eds. Cambridge: Cambridge University Press, 173-188.

3. Fluxman, S, Haim, A (1993): Daily rhythms of body temperature in Acomys russatus: the response to chemical signals released by Acomys cahirinus. Chronobiol Int 10, 159-164.

4. Haim, A, Rozenfeld, F (1993): Temporal segregation in coexisting Acomys species: the role of odour. Physiol and Behav (in press).

5. Haim, A, Zisapel, N (1992): Body temperature rhythms in long photoperiodacclimated spiny mice (Acomys russatus): Effects of melatonin. J Interdiscipl Cycle Res 23, 220-222.

6. Harrison, DL, Bates, PJJ (1991): The Mammals of Arabia. Kent: Harrison Zoological Museum Publication.

7. Haus, E, Halberg, F (1959): 24-hour rhythm in susceptibility of C mice to a toxic dose of ethanol. J appl Physiol 14, 878-880.

8. Heldmaier, G, Steinlechner, S, Ruf, T, Wiesinger, H, Klingenspor, M (1989): Photoperiod and thermoregulation in vertebrates: Body temperature rhythms and thermogenic acclimation. J Biol Rhythms 4, 251-265.

9. Honma, KI, Hiroshige, T (1987): Internal synchronization among several circadian rhythms in rats under constant light. Am J Physiol 235, R243-R249.

10. Jimenez, J, Osuna, C, Rubio, A, Guerrero, JM (1993): Effect of chronic ethanol administration on the rat pineal N-acetyltransferase and thyroxine type II 5'-deiodinase activities. Biosci Rep 13, 91-98.

11. Reinberg, A, Smolensky, M, (1990): Chronobiology and thermoregulation. In: Thermoregulation Physiology and Biochemistry. Schonbaum, E, Lomax, P eds. New York: Pergamon Press, 61-100.

12. Shkolnik, A (1971): Diurnal activity in a small desert rodent. Int J Biometeor 15, 115-120.

Thermal Balance in Health and Disease
Advances in Pharmacological Sciences

CIRCADIAN RHYTHM IN THE SNAIL *HELIX POMATIA* L. DURING ESTIVATION AND HIBERNATION

W. Wünnenberg

Zoologisches Institut, Zoophysiologie, Olshausenstr. 40, 24098 Kiel, F.R.G.

Under natural conditions numerous life processes of animals show oscillations corresponding to nychthemeral fluctuations of environmental factors, such as light, ambient temperature (T_a), humidity (r.h.) etc.. Some of these diurnal biological cycles are directly caused by environmental changes - they disappear when the animals are kept under constant environmental conditions. In contrast to such exogenously driven biological cycles endogenous oscillations were found the cyclicity of which persists under constant environmental conditions with a period length of approximately 24 h (circadian rhythms).

In general, circadian rhythms are supposed to reflect an adaptational response to the daily recurrent environmental fluctuations. Furthermore, it was discussed that they play an important role in the control of mammalian hibernation and it was suggested that a bout of hibernation is an extended sleep phase of the circadian cycle (1). The prediction generated by this hypothesis is that the circadian period is stretched to the length of hibernation bouts as a result of a temperature effect on the circadian oscillator.

Likewise in snails endogenous rhythms might be involved in the timing of arousal from dormancy. It was shown that in the estivating snail *Otala lactea* exogenous stimuli may not be sufficient to induce arousal indicating that both exogenous and endogenous cues are involved in the control of arousal from dormancy (2).

In the pulmonate snail *Helix pomatia* kept at constant T_a, constant r.h., and a constant photoperiod (LD 12:12; photophase from 8:00 h to 20.00 h) heart rates showed a diurnal rhythm with minimum values at 12:00 h and maximum values at 24:00 h (3). This diurnal rhythm persisted when estivation was induced by exposing the snails to dehydrating conditions (T_a=25±0.5°C, r.h.=10%). In the estivating animals, however, maxima of heart rates were shifted towards the beginning of the scotophase.

The existence of circadian rhythms in snails is well documented for *Aplysia californica* (4). There are, however, only a few studies on land snails in which periodic fluctuations of environmental factors were excluded. One of these studies indicates that locomotor behaviour of *Limax maximus* and *Limax flavus* exhibits a circadian rhythm with a period that deviates slightly from 24 h (5).

The present study was performed to investigate the circadian rhythm of heart rate in *Helix pomatia* exposed to constant T_a, constant r.h., constant light (LL) or constant darkness (DD). Of peculiar interest was the question whether circadian rhythms persist when the animals are in dormant states such as estivation and hibernation. An additional aim was to obtain information on the quality of the temperature compensation mechanism of the endogenous pacemaker.

Material and Methods

Experiments were carried out in adult specimens of *Helix pomatia* collected in the field. Whole body mass (including shell) ranged between 18 and 22 g. Snails were maintained in the laboratory at T_a=20-22°C and a 12:12 LD cycle and fed on lettuce. Egg-shell fragments were provided as a calcium source.

Heart rate was determined by registration of the electrocardiogram (ECG) which was picked up by means of fine stainless steel electrodes as described previously (3). For implantation of the electrodes the snails were immobilized by light anaesthesia. A small hole (diameter 3-4 mm) was drilled into the shell and the electrodes were attached to the epithelium above the heart region by a tissue adhesive. Dental cement was used to close the hole in the shell and to fix the electrodes. The leads connecting the electrodes with the preamplifier during the experiment were flexible and light so that the behaviour of the snail was obviously not disturbed.

One week following the implantation of the electrodes the snails were transferred into a climatic chamber. T_a and r.h. in this chamber oscillated in a range of 0.5°C and 5%, respectively, with a periodicity of 40 min. Hibernating animals were kept at constant darkness, the non-dormant and estivating snails at constant light (approximately 150 lux). With the exception of the experiments on estivating snails, the animals were kept in individual Plexiglas cages the floor of which was covered with soil, leaves, and moss pads. In all experiments on non-dormant snails food was provided ad libitum.

Estivation was induced by starvation and dehydration. Snails were placed into Plexiglas boxes which were perfused by air that was dried by silicate gel.

Hibernation was evoked by exposure to cold (+5±0.5°C) and constant darkness. In this series of experiments implantation of the electrodes was performed 4 to 6 weeks after the onset of hibernation, i.e. in snails that had already secreted a thick calcareous epiphragm across the shell aperture. Implantation of electrodes in hibernating snails, however, induced arousal from hibernation in nearly 30 to 50% of the animals.

ECG was registered at 2h intervals for periods of 10 min. Identification of circadian rhythms and estimation of period length was performed by the following procedure. Daily maxima of heart rates were determined for each experimental day. As the course of heart rates of an individual snail did not correspond to a sine curve usually 1 to 3 peak values per day were detected. The times of the day when these peak values occurred were plotted versus the consecutive experimental days in the conventional manner (Fig. 2). For these data regression lines were calculated the slopes of which indicated the periods of the freerunning rhythms.

Where appropriate, data are presented as mean values ±SD. The level of significance was set to 0.05.

Results

In all snails used in the present study heart rates could be determined by registration of the ECG. Usually an undisturbed record of ECG was possible over periods of 6 to 8 weeks and up to 5 months in the hibernating snails.

Mean heart rates of non-dormant and dormant snails exposed to different temperatures are shown in Fig. 1. In the non-dormant animals heart rates of 8.4±1.65 min^{-1} were registered at T_a=5°C and of 43.5±5.9 min^{-1} at T_a=25°C corresponding to Q_{10} values of 2.3 to 2.8. In hibernating snails heart rates (4.53±0.76 min^{-1}) were significant lower ($P<0.001$) as compared with non-dormant animals exposed to the same T_a. In animals which were exposed to starvation and dehydrating conditions a gradual decrease of heart rates from 43.5±5.9 min^{-1} to 30.5±5.3 min^{-1} was observed during the first week of estivation.

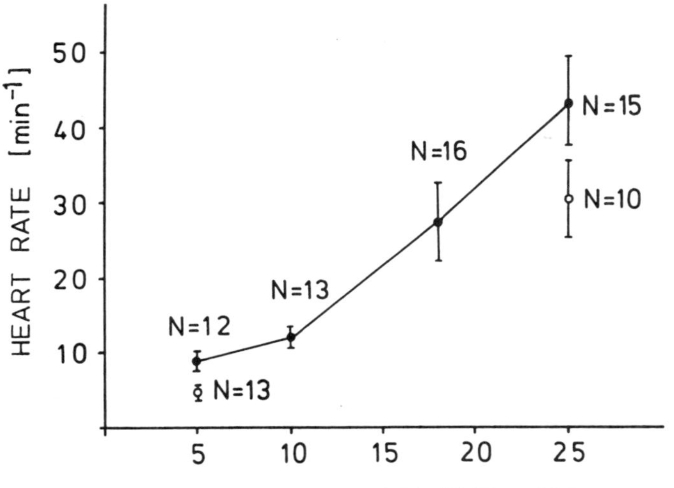

Fig. 1: Mean heart rate (±S.D.) of dormant (o) and non-dormant snails (•) kept at various ambient temperatures. N=number of animals.

Circadian rhythms were recorded in seven hibernating snails exposed to T_a=5±0.5°C, r.h.=80±5%, and DD. An example of this type of experiments is shown in Fig. 2. The times at which the daily peak values of heart rate occurred are plotted versus the consecutive experimental days. Regression lines fitted to these data indicate a period length of 23.1 h and 23.26 h, respectively. The mean duration of freerunning periods of all hibernating snails amounted to 23.15±0.19 h.

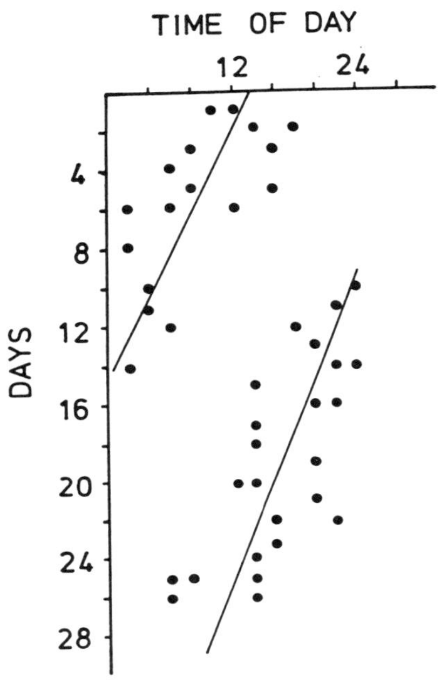

Fig. 2: Freerunning circadian rhythm of heart rate maxima in a single snail hibernating at T_a=5°C, r.h.=80%, and DD. Regression lines indicate a period length of 23.1 h and 23.26 h, respectively.

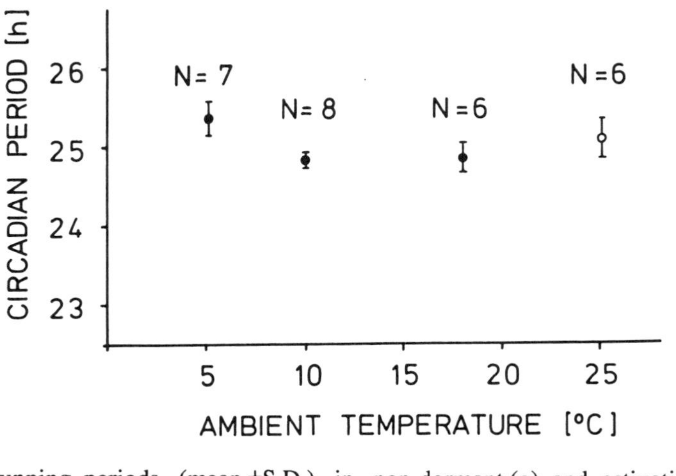

Fig. 3: Freerunning periods (mean ±S.D.) in non-dormant (•) and estivating snails (o) as a function of ambient temperature. All animals were exposed to constant light. N=number of animals.

Unlike the hibernating animals the non-dormant as well as the estivating snails developed circadian rhythms with periods exceeding 24 h when kept at constant light (Fig. 3). Figure 3 further shows that periods of circadian rhythms are temperature compensated, exhibiting a Q_{10} near 1 in the temperature range from 10 to 18°C. Likewise in snails estivating at T_a=25°C mean period length did not differ significantly from periods of non-dormant animals exposed to T_a=10°C and T_a=18°C, respectively. However, in non-dormant snails exposed to T_a=5°C mean periods of freerunning rhythms are significantly longer ($p=0.001$) than in animals exposed to T_a=10°C.

Discussion

The present study shows that registration of heart rates is a suitable method for investigations of circadian rhythms. Compared with other biological functions, such as locomotor activity, the registration of heart rates has the advantage that it can also be used for studies in dormant snails - even in animals that are digged into the soil.

Mean heart rates of dormant snails are 30 to 50% lower than heart rates of non-dormant animals exposed to the same T_a (Fig. 1). Like heart frequency, a parameter thought to reflect metabolic rate, oxygen consumption of estivating snails drops to 10 to 30% of resting rates measured in non-dormant animals at the same T_a (6). As shown by experiments in the snail *Otala lactea* dormancy is characterized by hypercapnia and respiratory acidosis - a condition that strongly depresses metabolic rate (7).

This mechanism of metabolic rate depression by hypercapnia as discussed for dormant snails resembles the situation of mammalian hibernation. In hibernating mammals blood acid-base state corresponds to a strong respiratory acidosis (8). This pH change is regarded to contribute to the metabolic depression that occurs during hibernation, in addition to the temperature-related reduction of metabolism (8). Thus, a common molecular basis for metabolic depression in hibernating mammals and dormant land snails is possible (6).

As shown by the present experiments circadian rhythms persist during dormant states. In the hibernating animals kept at DD the period length of the freerunning rhythms was significantly shorter than 24 h reflecting the well-known finding that in nocturnal animals decreasing light intensity leads to a shortening of period duration (9). Even in hibernating snails exposed to a 8:16 LD cycle freerunning periods of less than 24 h were found (3). Obviously light perception in hibernating snails that are more or less digged into the substrate is not sufficient for photoentrainment.

Unlike the hibernating animals the estivating snails developed freerunning circadian rhythms not discernible from those found in non-dormant animals. Furthermore, it was shown that in estivating snails which are completely withdrawn into their shells rhythms can be synchronized by the photoperiod (3). The mechanisms mediating photoentrainment in *Helix pomatia* have not been defined. Experiments in blinded garden slugs (*Limax maximus*, *Limax flavus*) indicate that the eyes are not necessary for either maintenance of rhythmicity or for LD entrainment (10). The results of our experiments in the estivating animals lead to the assumption that extraocular photoreception also exists in *Helix pomatia*.

In accordance to studies in other invertebrates as well as in vertebrates (cf. 11) the present experiments in *Helix pomatia* show that over a wide range of T_a the circadian rhythm is completely temperature compensated (Fig. 3). Although this data indicate that a transfer of the animals from T_a=10°C to T_a=5°C lengthens the circadian period significantly, the influence of T_a is rather small and corresponds to a Q_{10}=1,04 only. More pronounced temperature effects would have to be expected if the processes underlying circadian rhythms showed a temperature dependence of the magnitude postulated for mammalian hibernators (1).

References

1. Heller, HC, Krilowicz, B, Kilduff, TS (1989): Neural mechanism controlling hibernation. In: Living in the cold II. Malan, A, Canguilhem, B, eds. Montrouge: John Libbey Eurotext, 447-459.

2. Herreid, CF, Rokita, MA (1976): Environmental stimuli for arousal from dormancy in the land snail *Otala lactea* (Müller). Physiol. Zool. 49, 181-190.

3. Wünnenberg, W (1991): Diurnal rhythms of heart rate in the snail *Helix pomatia* L. Comp. Biochem. Physiol. 99A, 415-417.

4. Lickey, ME, Block, GD, Hudson, DJ, Smith, JT (1976): Circadian oscillators and photoreceptors in the gastropod *Aplysia*. Photochem. Photobiol. 23, 253-273.

5. Sokolove, PG, Beiswanger, CM, Prior, DJ, Gelperin, A (1977): A circadian rhythm in the locomotor behavior of *Limax maximus*. J. exp. Biol. 66, 47-64.

6. Storey, KB, Storey, JM (1990): Metabolic rate depression and biochemical adaptation in anaerobiosis, hibernation and estivation. Q. Rev. Biol. 65, 145-174.

7. Barnhart, MD, McMahon, BR (1987): Discontinuous carbon dioxide release and metabolic depression in dormant land snails. J. exp. Biol. 128, 123-138.

8. Malan, A (1978): Hibernation as a model for studies on thermogenesis and its control. In: Effectors of thermogenesis. Girardier, L, Seydoux, J, eds. Basel: Birkhäuser, 303-314.

9. Aschoff, J (1960): Exogenous and endogenous components in circadian rhythms. Cold Spring Harbor Symp. quant. Biol. 25, 11-28.

10. Beiswanger, CM, Sokolove, PG, Prior, DJ (1981): Extraocular photoentrainment of the circadian locomotor rhythm of the garden slug *Limax*. J. exp. Zool. 216, 13-23.

11. Bünning, E (1977): Die physiologische Uhr. Berlin, Heidelberg, New York: Springer-Verlag.

Thermal Balance in Health and Disease
Advances in Pharmacological Sciences

HEAT ACCLIMATION ALTERS PORTAL BLOOD FLOW DURING HEAT STRESS

W. Haddad and M. Horowitz

Division of Physiology, Hadassah School of Dental Medicine, The Hebrew University, Jerusalem, Israel

Introduction

Previous studies performed in our laboratory on desert and comfort climate dweller rodents have shown that acclimation to heat involves cellular and/or metabolic changes in the peripheral effector organs. These changes seem to modify the mode of thermoregulatory responses during heat stress in the acclimated body (1).

In the cardiovascular system, heat acclimation increases the share of cardiac output distribution to the splanchnic organs and improves responsiveness of heat-induced vasomotor responses. Thus, greater heat-induced splanchnic vasoconstriction allows greater thermoregulatory blood flow in cutaneous vascular beds coincidentally with the maintenance of adequate splanchnic blood flow (2). This significantly increases endurance to heat stress. Presently accumulated findings indicate that in heat acclimated rats membranal or cellular changes in the vasculature play a role in modifying splanchnic vasomotor responses (3,4). For example, heat stressed α adrenergic blocked heat acclimated rats show splanchnic vasodilatation. This vasodilatation is not observed before acclimation and may reflect either a change in adrenergic receptor density or an increase in the receptor-mediated vasodilatation when acclimation has occurred. The observation of a greater increase in arterio-venous anastomotic blood flow in heat-stressed, heat-acclimated rats, compared to non-acclimated, following a adrenergic blockade further supports this view. Additional evidence for altered vascular responsiveness to various transmitters following acclimation to heat is given by the observation that heat acclimation attenuates heat-induced plasma volume expansion (4). A number of data infer that this acute response is mediated by the ratio between α and β_2 activated receptors, possibly at a precapillary site, leading to transcapillary fluid absorption. The pronounced splanchnic vasodilatation occurring during heat stress in heat acclimated a blocked rats draws our attention to this vascular bed as a possible active site for fluid absorption, differing in its magnitude before and after heat acclimation.

The present investigation was conducted to gain further insight into the role played by the blood vessels vs central thermoregulatory responses in modifying splanchnic circulation during heat stress following heat acclimation. We studied the portal blood flow and the pharmacological properties of the portal vasculature in heat acclimated rats. For comparison, the pharmacological properties of the tail artery were studied and discussed in respect to tail blood flow.

Materials and Methods

Male, 3 week old albino rats (Rattus norvegicus, Zabar strain) of initial body weight (BW) 80-90 grams were used in this study. They were fed on Ambar laboratory chow and water ad lib. The animals were divided into two groups: untreated controls (C) and heat acclimated (AC). The control group was held at an ambient temperature of 24±1°C. Heat acclimation was attained by continuous exposure to 34±1°C for 2, 5 and 30 days.

Experiments on conscious rats

Portal blood flow (PBF), a representative measurement of the splanchnic blood flow draining into the liver, was measured using a TSI blood flow meter in intact and chemically denervated rats. Chemical denervation was attained by 60 mg/kg body wt 6 hydroxydopamine administered IP in the following order: day 1 - 10mg/kg; days 2-4 -5mg/kg twice/day (AM and PM); day 8 - 10mg/kg twice/day. Experiments were performed on the 9th day. Successful denervation was confirmed by measurement of systemic blood pressure prior to and following IV administration of 1mg/kg tyramine-HCL (Sigma) according to Hales, 1984 (5). On the day of the experiment, rats were anesthetized with 60mg/kg Na-pentobarbital. A rectal thermistor (YSI no 402) was inserted 7cm deep, beyond the anal sphincter; the portal vein was exposed and the TSI flow probe (2S) was placed around the vessel and exteriorized at the animal's back between the scapulae. The animals were allowed to recover from the anesthesia for several hours. Following normothermic-control blood flow monitoring, the rats, lightly restrained, were placed in a temperature controlled chamber at 42°C. PBF and rectal temperature (T_{re}) were then monitored in the heat stressed animals. In another series of experiments, the effect of adrenaline (3µg/kg) administered IV on PBF at each rectal temperature was determined. All experiments were terminated at T_{re} 42°C.

Experiments on isolated blood vessels

Experiments were performed on portal vein and tail artery rings from C and AC rats. The cumulative dose response curves to noradrenaline in the presence and absence of propranolol (10^{-6}) were determined. For these experiments blood vessels were placed in an organ bath of Krebs Henseleit bicarbonate buffer aerated with 95%O_2-5%CO_2, at 37°C according to Horowitz et al., 1994 (3). Isometric tension in response to the drug (administered directly to the organ bath) was measured using a Statham LC2 force transducer.

Statistics

To evaluate the significance of differences obtained between the groups one way analysis of variance and Student's t test were used. $p < 0.05$ were considered significant.

Results

Experiments on conscious rats

The portal blood flow in heat stressed control and 30 days AC rats before and after sympathectomy is shown in Fig. 1.

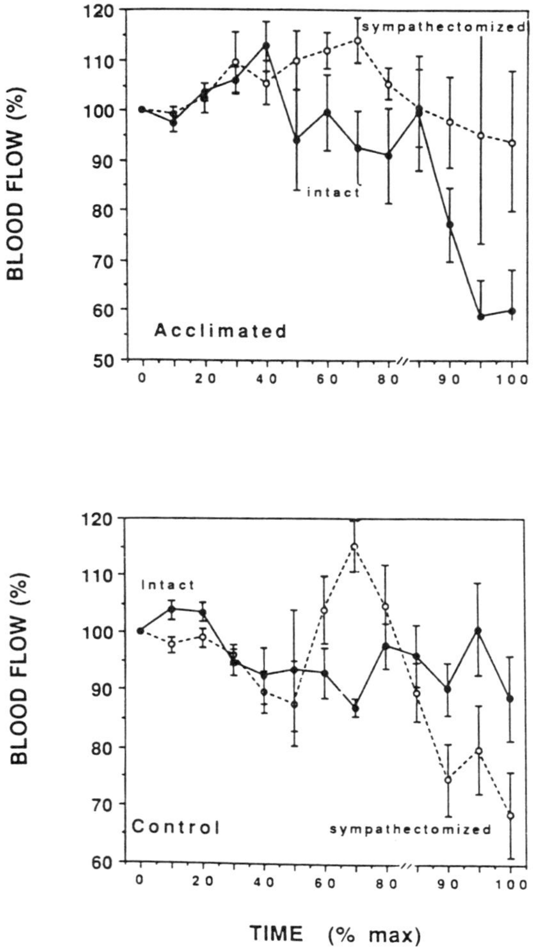

Fig. 1: Portal blood flow in non-acclimated and heat acclimated, heat-stressed rats before and after sympathectomy. Data (M ± SEM) are presented as percentage of the initial flow. The time (min) to 42°C was taken as 100 % (n = 6-8 rats).

It is evident that during heat stress acclimated rats maintained their PBF unchanged for a longer period than non-acclimated rats. The temperature threshold (Tsh) for splanchnic vasoconstriction was higher in AC than in C rats and the magnitude of splanchnic vasoconstriction was greater. During severe hyperthermia splanchnic vasodilatation took place in the controls but not in the acclimated rats. After sympathectomy, significant vasodilatation occurred in AC but not in C heat stressed rats. In contrast to the 30 days acclimated rats, short term heat acclimation (2 days, data not shown) did not produce significant changes in PBF compared to the non-acclimated rats during heat stress.

Adrenaline administration (Fig. 2) resulted in increased PBF. This increase was more pronounced in C than in the 30 days AC rats. However, in C rats, responsiveness to adrenaline decreased with the rise in T_{re}, whereas in AC rats an increase in body temperature from 37°C to 39°C improved responsiveness to the drug. Above 39°C, responsiveness to the drug decreased similarly to that observed for the non-acclimated rats.

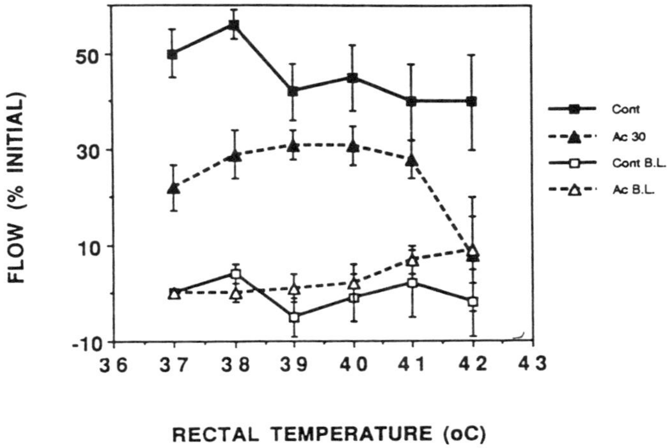

Fig. 2: Portal blood flow (M ± SEM) during rectal temperature rise before and after administration of 3 µg/kg adrenaline. Empty symbols represent baseline flow and solid symbols represent adrenaline effect (n = 6-7 rats).

Isolated blood vessels

In the portal vein, the largest adrenergic receptor population are β_2 adrenergic receptors. Thus, the activity of the a receptors can only be studied following β adrenergic blockade. Force generation in β adrenergic blocked (with 10^{-6}M propranolol) portal vein rings from AC rats in response to noradrenaline (10^{-4} - 10^{-9} M) stimulation was stronger than that measured in C rats . It is noteworthy, however, that maximal effect of the drug was not achieved in the acclimated rings. It is also evident that force generation was higher on day 2 than on day 30 of the acclimation period. Force generation in response to a adrenergic stimulation on day 5 of the acclimation did not differ significantly from that generated by C blood vessels (Fig.3).

In contrast to the portal vein, in the tail artery rings (Fig.3), propranolol produced a significant decrease in noradrenaline-induced force generation. This decrease was most pronounced on day

30 and least pronounced on days 2 and 5 of heat acclimation. Force generation in response to noradrenaline solely was almost similar in C and 30 days AC rings but was lower in rings excised from short term AC rats. This suggests that aadrenergic responsiveness was lowest on day 2 and maximal on day 30 of acclimation. Similarly to the portal vein, maximal response was not achieved with the noradrenaline concentrations used.

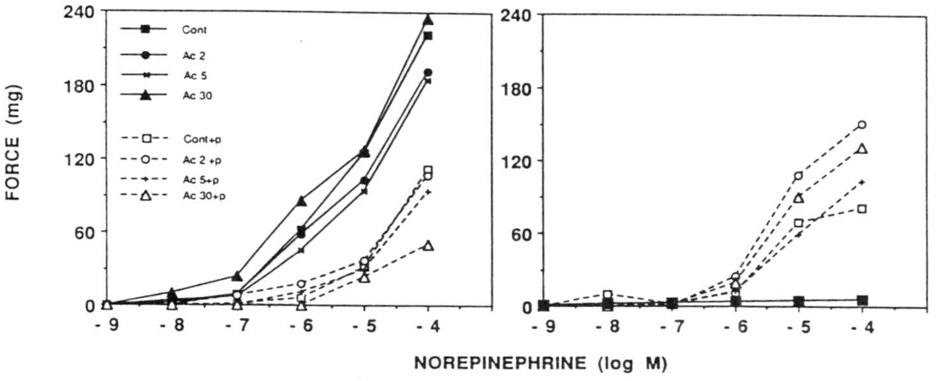

Fig. 3: Norepinephrine-induced force generation in the tail artery (left) and in the partal vein (right) of the acclimating rats (n = 6-10 rats).

Discussion

Our observations support the postulate that heat acclimation produces changes in the vasculature, possibly at the membranal or intracellular level. The data suggest that in conscious heat stressed rats, a vasodilatating component compromises thermally induced vasoconstriction to augment PBF. Our data also imply an apparent delay in the vasoconstriction temperature threshold in AC rats. However, due to differences between C and AC rats in the response of PBF to adrenaline (improved flow in AC rats vs reduced flow in C rats in response to T_{re} rise) this apparent delayed threshold may be different from the temperature threshold for vasoconstriction in arterial vessels in the splanchnic circulation. Unfortunately, continuous arterial blood flow measurements in the splanchnic area have not been performed in acclimated animals. Temperature threshold for skin vasodilatation is lower in acclimated rats (6), as in other animal species and humans (7).

Experiments performed on chemically sympathectomized rats and on isolated blood vessels imply that changes in β and α adrenergic signalling responsiveness occur during acclimation. These changes, however, are not uniform for all vascular beds. The present investigation shows that while in the portal vein the α adrenergic response was most pronounced on day 2 of the acclimation, in the tail artery it was lowest on day 2. In both cases, 30 days acclimation produces more pronounced α adrenergic activity than in non-acclimated vessels. Although the mode of the changes in adrenergic transduction pathways on acclimation has not yet been completely elucidated, our data supply strong evidence for temporal variation in the adrenergic transduction pathway with acclimation.

Our observation that in the tail artery noradrenaline-induced force generation is reduced in the presence of propranolol, implies the existence of β receptors other than β_2. Purdy et al., 1988 (8) and Toda and Okamura, 1990 (9) provided evidence for the presence of β_1 adrenergic receptors in coronary vasculature. Activation of this receptor produces force generation. Our data therefore suggest that in the tail artery both α and β_1 adrenergic receptors play a role in force generation. However, the contribution of α receptors is more pronounced in non-acclimated blood vessels In the long term acclimated blood vessels β1 receptors play the major role in pressure generation.

References

1. Horowitz, M: (1994): Heat stress and heat acclimation: the cellular response modifier of autonomic control. In: Integrative and cellular aspects of autonomic functions. Pleschka, K, Gerstberger, K (eds), John Libbey & Co. London (in Press).

2. Horowitz, M, Samueloff, S (1988): Cardiac output distribution in thermally dehydrated rodents. Am J Physiol 254, R109-R116.

3. Horowitz, M, Haddad, W, Shochina, M, Meiri, U (1994): Long term heat acclimation: acquired peripheral cardiovascular adatpations and their stability under multifuctorial stressors. In: Temperature regulation: recent physiological and pharmacological advances. Milton, E (ed) Birkhaeuser Verlage, Basel, pp 291-296.

4. Horowitz, M, Givol, N (1989): Heat acclimation and heat stress: cardiac output distribution, plasma volume expansion and the involvement of the adrenergic pathway. In: Thermoregulation: Research and Clinical Applications. Lomax, P, Schönbaum, E. (eds) Karger, Basel pp 204-207.

5. Hales, JRS (1984): Pharmacological evidence against sympathetic nervous control of regional blood flow during heat stress. In: Thermal Physiol 1984 Hales JRS (ed) pp 267-270.

6. Horowitz, M, Meiri, U (1985): Thermoregulatory activity in the rat: effects of hypohydration, hypovolemia and hypertonicity and their interaction with short-term heat acclimation. Comp Biochem Physiol 82A, 577-582.

7. Rowell LB (1983): Cardiovascular adjustment to thermal stress. In: Handbook of Physiology, The cardiovascular system Am Physiol Soc, Sec 2, vol 2, pp,967-1025.

8. Purdy, R, Stupecky, G, Coulombe, P (1988): Further evidence for a homogeneous population of β_1 adrenoceptors in bovine coronary artery. J Pharmacol Exp Ther 245, 67-71.

9. Toda, N, Okamura, T (1990): Beta adrenoceptor subtype in isolated human monkey and dog epicardialncoronary arteries. J Pharmacol Exp Ther 253, 518-524.

FUNCTIONAL EVIDENCE OF VASOMOTOR CONTROL BY TACHYKININS IN THE HEAT DISSIPATING TISSUES OF THE CANINE NOSE AND FACE

T. Ikeda and K. Pleschka

Max-Planck-Institut für Physiologische und Klinische Forschung, W. G. Kerckhoff-Institut, 61231 Bad Nauheim, F.R.G.

In contrast to man and non-furred mammals, in which excessive body heat is dissipated by evaporation of the skin, furred mammals mainly depend on evaporative heat loss via the ventilated mucous membranes of mouth, tongue and nose. Considering that this respiratory mechanism also serves to condition the inspired air, a precise neurogenic vasomotor control of convective heat transfer from the core to these mucous surfaces (especially the nose) is a prerequisite for the provision of adequate perfusion rates over a wide range. While the role of sympathetic vasoconstrictor and parasympathetic vasodilator efferents in nasal blood flow control has been clarified, a local vasomotor control by afferent fibers, is still debated. Nevertheless, the innervation of the nasal and facial vasculature by substance P (SP) - immunoreactive afferent nerve fibers, their relationship with the sensory, in particular the trigeminal system, and the relaxing potency of SP on vasomotor tone, suggest the involvement of this peptide in locally induced vasomotor changes. In addition to local axon reflex SP-release, SP-fibers may influence parasympathetic efferents at the sphenopalatine ganglionic level. This type of vasomotor control is assumed to be important in connection with allergic and hyperreactive disorders of the nose (1). In recent studies it has been demonstrated that the various requirements for the adjustments of nasal blood flow under physiological conditions are met by the arteriovenous anastomoses (AVA) at relatively stable capillary flow (CAP-FLOW) (2). Physiological control of vasomotor tone of these vessels is not solely governed by symapthetic adrenergic but also by parasympathetic cholinergic efferents (3). Since the latter are associated with SP fibers, it appears reasonable also to suggest their co-action, at least in a modulatory way. To asses the extent to which SP modulates blood flow partitioning through AVA and capillaries in conditions with unimpaired sympathetic and parasympathetic inputs to the nose and face, total blood flow (IMA-FLOW) and blood pressure of the internal maxillary artery (BP-IMA) were recorded bilaterally during unilateral infusion of SP at different doses in anesthetized, paralyzed and artificially ventilated dogs. Distribution of IMA-FLOW to precapillaries (CAP-FLOW) and AVA (AVA-FLOW) was determined by the tracer microsphere technique at three doses of SP infusion and the corresponding resting conditions. In addition, the vasodilator potencies of SP agonists on IMA-FLOW and its vascular compartments were also determined to evaluate which of the established three neurokinin receptors (NK1, NK2, NK3) are involved in neurotransmission.

Methods

Animals: Thirty-three beagle dogs of either sex and with an average body weight of 20.05 ± 1.51 kg (mean ± SD) were used in this study. Anesthesia was induced with a combination of ketamine hydrochloride (Ketavet, Parke-Davis, Berlin, F.R.G., 12 mg/kg) and acepromazine maleate (Vetranquil, Albrecht, Aulendorf, F.R.G., 1.6 mg/kg) followed by an intravenous injection of sodium pentobarbitone (Nembutal, CEVA, Hannover, F.R.G., 15 mg/kg). Anesthesia was maintained by intravenous (i.v.) infusion of sodium pentobarbitone at a rate of 2 mg/kg/h. Following the surgical procedure, the animals were paralyzed with pancuronium bromide (Organon, 0.08 mg/kg priming dose followed by a constant infusion of 2 mg/h) and artificially ventilated through a cannula previously inserted into the trachea.
Surgical procedures: Cannulae (Portex, England) were inserted into the aorta via the femoral artery to monitor systemic blood pressure (BP) and into the vena cava via the femoral vein. The tip of the latter catheter was placed at the level of the right heart for drug administration and blood sampling. Both IMAs were exposed and electromagnetic flow probes were placed for continuous recording of blood flow. For measurement of blood pressure of both IMAs (BP-IMA) and close arterial application of drugs and radioactive microspheres, cannulae were placed into the pterygoid branches.
Experimental protocol: The concentration-response curve for the different neuro-peptides was obtained by infusing the corresponding drug in rising concentrations. Each stimulation period was interrupted by a control period in which isotonic saline was infused. After all variables had reached stable levels, SP was infused into the IMA at doses of 1, 2, 5, 10, 20, and 50 pmol/min. CAP-FLOW and AVA-FLOW were determined at three SP doses (5, 10, 20 pmol/min) and the corresponding resting conditions. The same protocol was used for SP agonists SP-methylester (SP-M), septide (SEP), neurokinin A (N-A), neurokinin B (N-B), and senktide (SEN).
Physiological measurements: BP and BP-IMA were monitored with pressure transducers (Statham, P23Db) connected to pressure amplifiers (Gould, Cleveland, Ohio,USA). Blood flow in both IMAs was continuously monitored with flow probes connected to a 2 - channel electromagnetic flowmeter (Statham SP 2202). BP, BP-IMA, and IMA-FLOW, were recorded directly on an 8 - channel Gould polygraph and stored in a microcomputer. The quotient of BP-IMA and IMA-FLOW served as an estimate of resistance to IMA-FLOW. The norepinephrine, epinephrine and dopamine concentrations of venous blood samples were measured by high performance liquid chromatography (Waters, Milford,USA) with electrochemical detection.
Measurement of tissue blood flow: Estimations of blood flow in various tissue compartments were based on determination of the proportional tissue distribution of isotope-labeled microspheres (15.0 ± 1.5 μm in diameter) injected into the IMA. Briefly, radioactive microspheres were injected into the IMA and either trapped by the smaller CAP of the tissues or drained by the larger AVA into the venous system and finally trapped by the capillaries of the lung. Thus, the local tissue activity indicates the CAP blood flow, while the lung activity represents the blood flow through the AVA. Microspheres used in this study were labeled with ^{141}Ce, ^{51}Cr, ^{113}Sn, ^{103}Rn, ^{95}Nb and ^{46}Sc (NEN, Dreieich, F.R.G.), suspended in a solution containing 10% Dextran and 0.01% Tween 80 and sonicated for 30 s prior to injection. Injections were made against the direction of the blood stream to promote turbulence and thereby enhance mixing. Each isotope served as a marker of separate events. At the end of the experiment, the animals were sacrificed with an anesthetic overdose and tissue specimens of anatomical defined compartments were reovered for analysis by a gamma-spectrometry (Nuclear Data, ND-62) as previously described (2).
Drugs: The drugs tested in this study were substance P, SP-methyl ester, septide, neurokinin A, neurokinin B, and senktide from Bachem, Swizerland.
Statistics: Mean values and standard deviations were calculated for all values during steady-states. Differences between paired control and experimental periods were tested for statistical significance using the Student's *t*-test for paired data and incorporating the Bonferroni-Holm technique to correct for multiple comparisons. Probability levels of 0.05 or less were selected for determining statistical significance.

Results

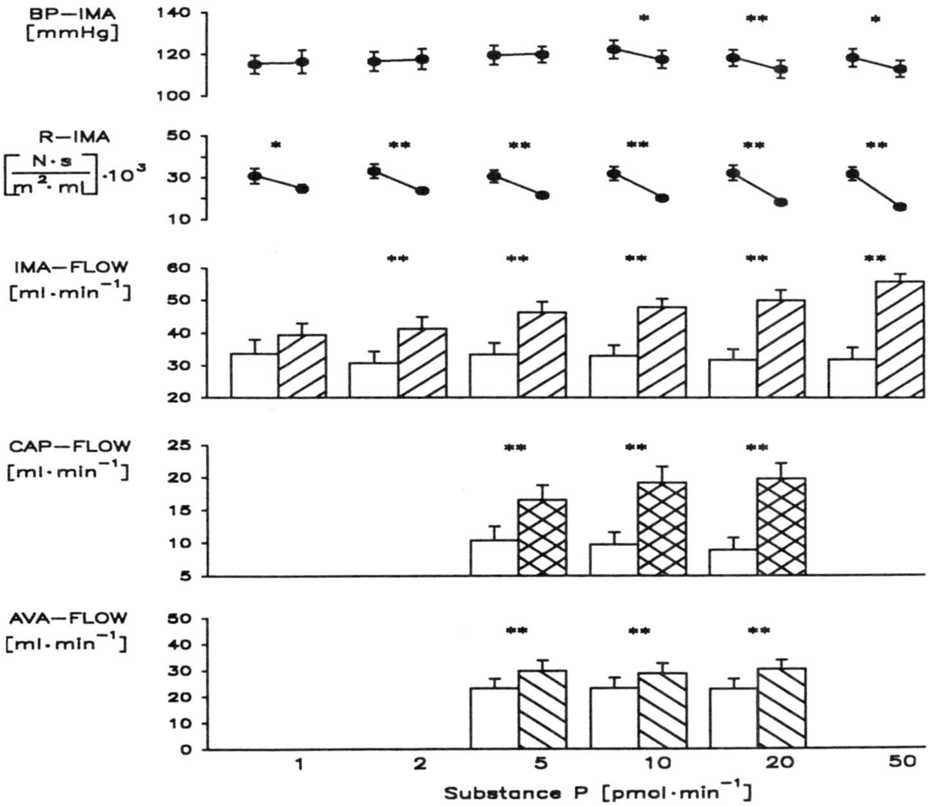

Fig. 1: Effects of substance P on the hemodynamics of the internal maxillary artery (IMA). Blood pressure (BP-IMA), resistance to flow (R-IMA), blood flow (IMA-FLOW) and IMA-FLOW distribution to precapillaries (CAP-FLOW) and arteriovenous anastomoses (AVA-FLOW) are displayed during control periods (open bars) and during substance P infusion (hatched bars). Values are expressed as means ± SE for 9 dogs. Significant differences between control values and those after subsequent treatments are designated by asterisks (* $P < 0.05$, ** $P < 0.001$).

As demonstrated in Figure 1, unilateral infusion of SP caused dose-dependent ipsilateral IMA-FLOW increases which were due to significant decreases in R-IMA. The SP threshold dose for a vasodilatatory response was determined as 1 pmol/min. It is also apperent that the increase in IMA-FLOW in response to SP was in fact due to increases in both AVA-FLOW and CAP-FLOW, but with a larger proportional contribution from CAP-FLOW to the total flow. Plasma catecholamines (pg/ml),which were 225.3 ± 18.0 for norepinephrine, 269.2 ± 29.9 for epinephrine and 39.1 ± 4.3 for dopamine remained nearly unaffected during the experimental course. Figure 2 shows the potencies of SP agonists. It indicates that with the exceptions of neurokinin B (N-B) and Senktide (SEN), all other tachykinins dose-dependently decreased R-IMA and increased unilateral IMA-FLOW during their infusion in a dose dependent manner.

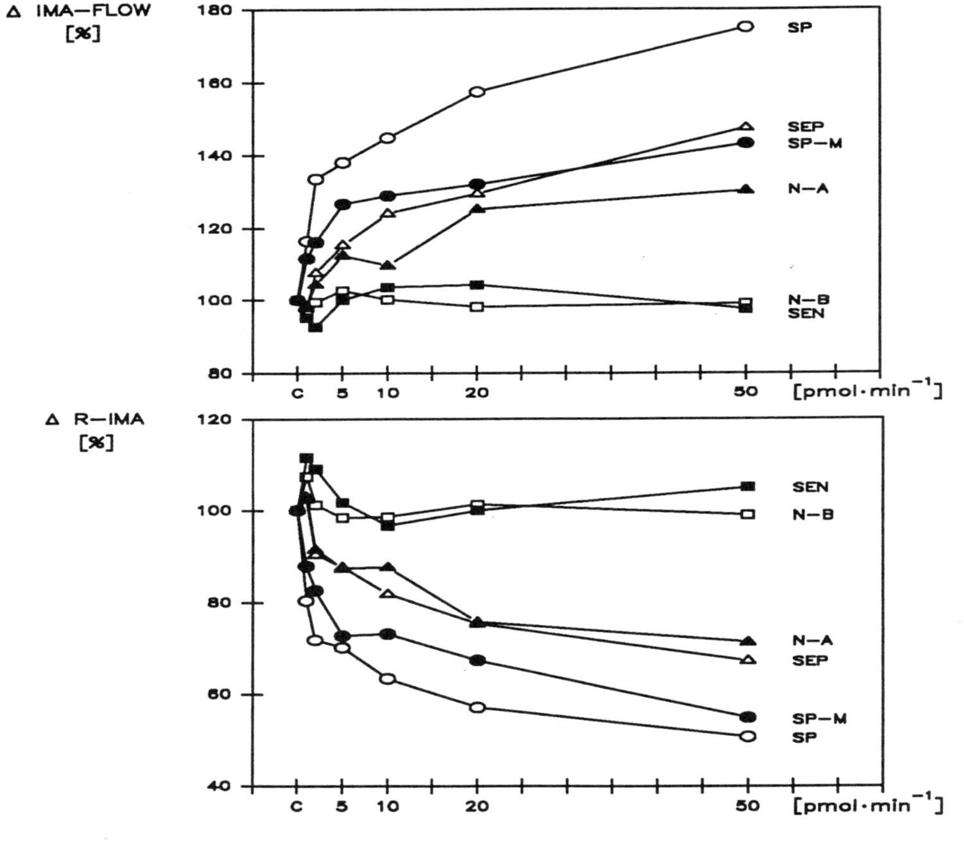

Fig. 2: Effects of substance P (SP) and its agonists on blood flow (IMA-FLOW) and resistance to flow (R-IMA) of the internal maxillary artery in % of the value obtained at control (C). The agonists include: substance P-methylester (SP-M), septide (SEP), neurokinin A (N-A), neurokinin B (N-B) and senktide (SEN). Values are expressed as means of multiple measurements: n = 9 for SP; n = 6 for SP-M, SEP, N-A; n = 3 for N-B, SEN.

The dose-dependence of AVA-FLOW and CAP-FLOW to infused tachykinins is shown in Figure 3. In general, the CAP-FLOW values continue to increase with the tested doses, whereas the AVA-FLOW levelled off at the lowest dose of agonists infused. According to their dilatory potencies as determined by individual magnitudes of change in CAP-FLOW with dose, the following rank order was obtained: SP > SP-M > SEP > N-A.

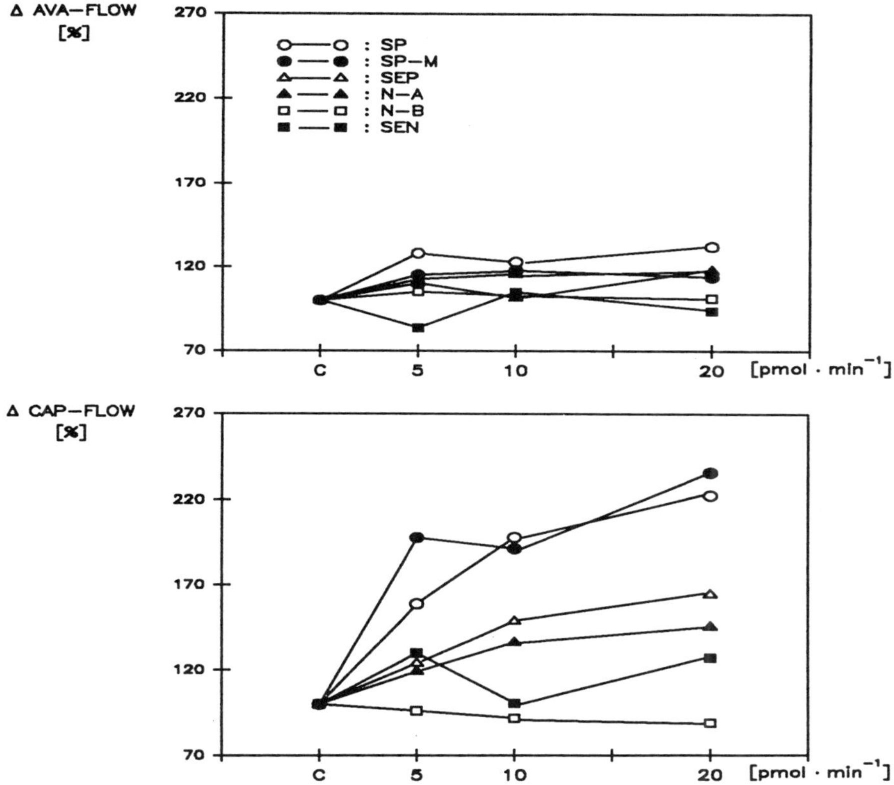

Fig. 3: Effects of substance P (SP) and of its agonists substance P - methylester (SP-M), septide (SEP), neurokinin A (N-A), neurokinin B (N-B) and senktide (SEN) on IMA-FLOW distribution to flow through arteriovenous anastomoses (AVA-FLOW) and precapillaries (CAP-FLOW) in % of the value obtained at control (C). Values are expressed as means of multiple measurements: n = 9 for SP; n = 6 for SP-M, SEP, N-A; n = 3 for N-B, SEN.

Conclusions

The results from this study show that the dilatory effects of substance P in nasal and facial tissues are more pronounced in the precapillaries than in the AVA. Thus, the AVA which are indirectly affected by vasomotor efferent control of the IMA vascular bed, particularly under thermoregulatory requirements (4), are essentially uninfluenced by substance P in physiological conditions. But the distinct capillary flow increase caused by substance P, similiar to that of histamine vasodilatation (5), is compatible with the proposed importance of this peptide as a modulator in local vascular control (e.g. inflammation). The mediation of the vasodilator response induced by substance P and its agonists suggest that the neurokinin receptors NK1 and NK2 are the ones most likely involved in neurogenic afferent vasomotor control of facial and nasal tissues.

Acknowledgements

The authors are grateful to Dr. Charles Webber Jr. (Loyola University, Chicago) for reviewing the manuscript and to Mrs. I. Lürkens and Mrs. C. Kammer for their excellent technical assistance.

References

1. Änggård, A, Lundberg, JM, Lundberg, L (1983) Nasal autonomic innervation with special reference to peptidergic nerves. Eur J Respir Dis 64 (suppl 128), 143 - 148.

2. Hashimoto, M, Sommerlad, U, Pleschka, K (1987) Sympathetic control of blood flow to AVAs and capillaries in nasal and facial tissues supplied by the internal maxillary artery in dogs. Pflügers Arch. 410, 589 - 595.

3. Sugahara, M, Pleschka, K (1992) Nutrient and shunt flow responses to vidian nerve stimulation in nasal and facial tissues of the dog. Eur Arch Otorhinolaryngol 249, 79 - 84.

4. Pleschka, K, Sugahara, M, Hashimoto, M, Sommerlad, U, Lürkens, I, Ernst C (1987) Local circulatory control in thermal stress. In: Comparative physiology of environmental adaptations, vol 2. Dejours, P ed. Basel: Karger, 107 - 122.

5. Bari, F, Ariwodola, JO, Pleschka K (1993) Histamine responsiveness of the various vascular beds of facial and nasal tissues in the dog. J Vasc Res, 30 - 37.

Thermal Balance in Health and Disease
Advances in Pharmacological Sciences

VASCULAR RECEPTORS INVOLVED IN EFFECT OF NEUROPEPTIDE-Y (NPY) ON PERIPHERAL BLOOD FLOW IN THE RAT FOOT AND TAIL

M. E. Heath

Thermal Stress/Adaptation, Code 53, Naval Medical Research Institute, 8901 Wisconsin Avenue, Bethesda, Maryland USA 20889-5607

Neuropeptide-Y (NPY) is a 36 amino acid peptide. It is occurs, in separate vesicles, in the same peripheral sympathetic adrenergic fibers as norepinephrine (NE) (1,2), and co-exists with VIP in peripheral non-noradrenergic neurons (3). NPY also occurs in the adrenal medulla and is one of the most abundant of all neuropeptides found in the brain (4). Three NPY receptor sub-types have been identified. Wahlestedt et al (5), demonstrated the existence of two sub-types of receptors by studying the effects of C-terminal peptide fragments of NPY at the level of the synaptic cleft. One receptor (Y_1) is post-synaptic and occurs in the vascular smooth muscle. The other receptor (Y_2) is pre-synaptic and inhibits norepinephrine release. Both Y_1 and Y_2 receptors bind with NPY and PYY with similar affinity. More recent studies of rat cardiac ventricular membranes has revealed a NPY-binding site with different affinities for these two neuropeptides, that is thought to be a third (Y_3) receptor sub-type (6).

NPY, like NE, is highly vasoactive. Both central and peripheral administration of NPY modify cardiovascular function (4). This includes increased blood pressure, reduced heart rate, and increased resistance in the vascular beds of many peripheral tissues and organs (7). In vivo studies have demonstrated marked reductions in blood flow in the human forearm, and in the brain, heart, kidney, pancreas, thyroid, spleen, skeletal muscle, and intestine, in humans and in other mammals, in response to exogenous NPY (4, 8). In comparison to the vascular constriction induced by NE, that caused by NPY is described as slower in onset, but more prolonged (8).

Some studies have shown that NPY, like NE, is released during exposure to severe cold in rats (9, unpublished observation this laboratory). Since our laboratory is directed at understanding the effects of non-freezing cold injury, it was of interest to assess the effects of peripherally administered exogenous NPY on peripheral blood flow and body temperatures. Heath and Thomas (10) were the first to demonstrate a pronounced increase in blood flow in a peripheral tissue in response to exogenous NPY. Using an in vivo preparation in rats (11), we showed that NPY caused an increase in total blood flow (BF_T) and blood volume in the tail of the rat that was concurrent with a decrease in superficial cutaneous microcirculatory blood flow (BF_{sk}) in the tail and foot pad (10). The purpose of the present study was to: (a) compare the effects of NPY to the effects of NE on BF_T in the rat tail, on BF_{sk} in the tail and foot pad, and on core temperature (T_c) and skin temperature (T_{sk}) on the foot and tail; and, (b) determine the relative importance or contribution of the postjunctional Y_1 and prejunctional Y_2 receptor populations in these responses to NPY.

Methods

Animals. Eight male Long-Evans rats of 290-310 g were studied. At least one week prior to experiments, rats were equipped with a cannula (ITTC Life Science, S-26) in a jugular vein while anesthetized with ~50 mg/kg sodium pentobarbital. Five rats were also equipped with a reentrant tube sutured to the muscle adjacent to the carotid artery for measurement of T_c. Sterile surgical technique was used. In addition, hair was removed from the tail using a depilatory so it would not interfere with laser Doppler blood flow measurements.

Measurements. Air (T_a), T_c, and T_{sk} on the tail and foot were measured with thermocouples. BFT was measured in the tail by venous occlusion plethysmography using mercury-in-silastic strain gauges (D.E. Hokanson Inc., 4.5 cm, wrapped twice around tail) connected to an electronic plethysmograph (D.E. Hokanson Inc., model EC-4; see ref. 11). BF_{sk} of the tail and foot pad was monitored with laser Doppler flowmeters (TSI Laserflo, Model BPM 403A; see ref. 11).

Preparation for experiments. Rats were anesthetized with ~50 mg/kg sodium pentobarbital, and gently introduced into a cylindrical plexiglas rat restrainer. The restrainer allowed free access to the hind leg and tail and to the end of the cannula and reentrant tube located between the shoulders. A laser Doppler flow probe and a thermocouple were positioned on the skin adjacent to each other mid-way along the tail. A pneumatic cuff at the base of the tail, and a mercury-in-silastic strain gauge ~2 cm distal to the laser Doppler probe, were used to measure tail BFT. A second laser Doppler flow probe was placed on the right foot and a thermocouple was taped in the same position on the left foot. All instrumentation was connected to a IBM compatible computer via an A/D converter (Keithley). Data collection was directed by LabTech Notebook software (Laboratory Technologies Corp.). Thermocouple and laser Doppler channels were sampled at 1 s intervals and averaged for 20 s. The venous occlusion plethysmographic measure of tail BF_T was done at 20 s intervals. The occlusion cuff was inflated to 50-55 mmHg for 5 s and tail BF_T assessed between the 2nd and 4th s of the cuff inflation.

Protocols. Fully instrumented rats rested in the restrainer for 15-30 min before the intravenous administration of NPY (16, 32, 64 or 128 μg/kg), NE (25, 50, 100, 400 μg/kg), the Y_1 receptor agonist: [Leu^{31} Pro^{34}]NPY (63.25 μg/kg), the Y_2 receptor agonist: NPY[13-36] (44.7 μg/kg), or saline control. The amounts of [Leu^{31} Pro^{34}]NPY (63.25 μg/kg) and NPY[13-36] (44.7 μg/kg) administered were equivalent to 64 μg/kg of NPY. The volume of all injections was 300 μl. Baseline recording was begun 5 min preceding the injection, and measurements were continued for at least 35 min post-injection. All experiments were done at T_a of 24-26°C.

Results

NE and NPY. Figure 1 illustrates the effect of NE (left) and NPY (right) on BF_T, BF_{sk} and T_{sk} in the tail and foot pad. NE causes an immediate marked decrease in tail BF_T, tail volume, and BF_{sk} in the tail and foot, but no significant change in body temperatures. The magnitude of change in

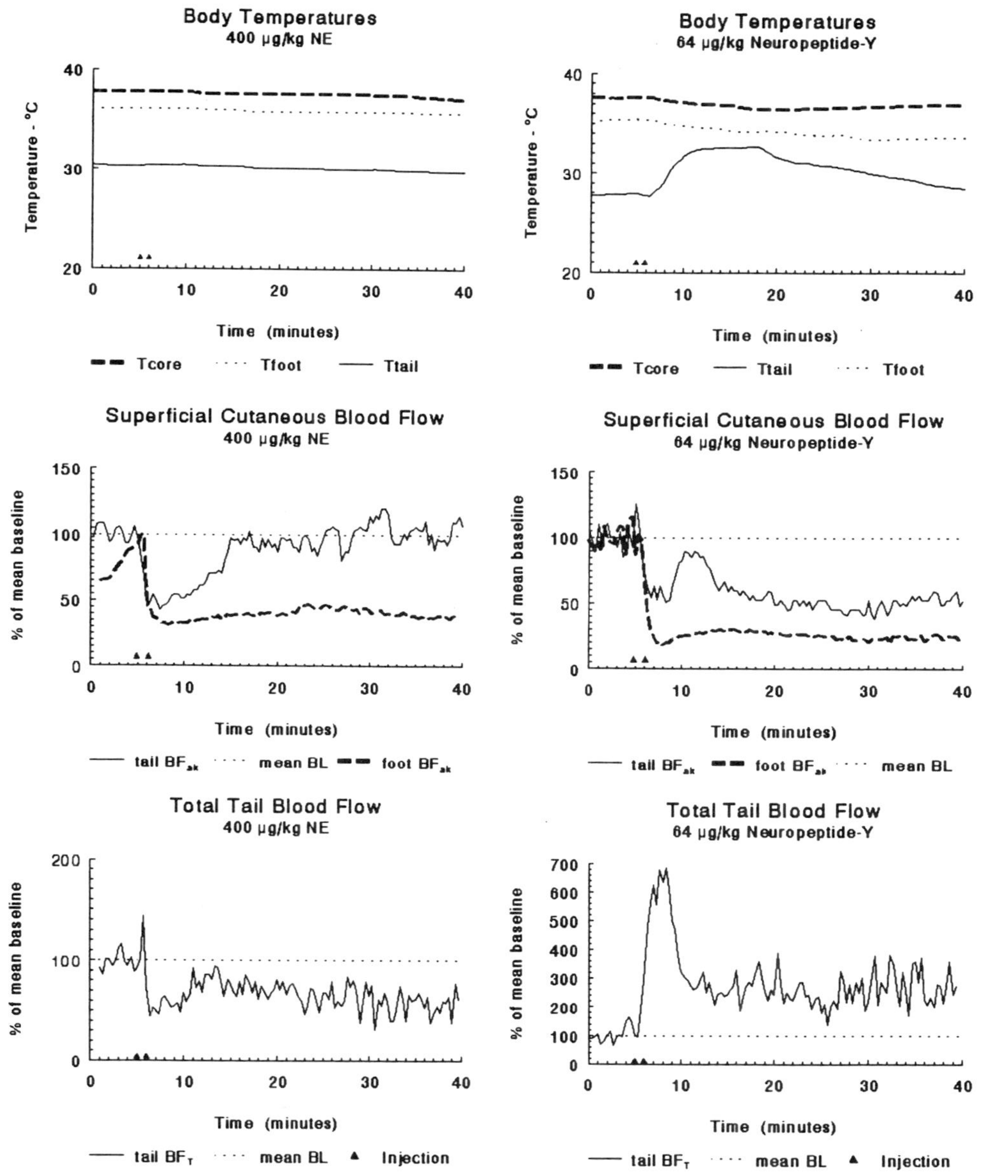

Fig. 1: Effect of NE (left) and of NPY (right), on tail BF_T , BF_{sk} and T_{sk} on the tail and foot pad of the rat.

BF, and the duration of the constriction was dose dependent. The effect of NPY on these same parameters is markedly different. NPY has an immediate *dynamic* effect that is of short duration and is followed by a prolonged *static* effect. It causes an immediate increase in tail BF_T and tail volume (not shown), that peaks within 1-3 min. Concurrent with the increase in BF_T, is an increase in tail (but not foot) T_{sk}. In marked contrast is the decrease in BF_{sk} in the tail and foot to <50% of baseline within 1-3 min. Following these dynamic changes, BF values begin to return toward baseline level, without fully acheiving it. Both the magnitude of the dynamic changes in BF, and the level of BF during the prolonged static phase are dose dependent. The largest response occured for 64 μg/kg NPY, and was not increased when 128 μg/kg NPY was given. It is clear that while exogenous NE causes the expected vasoconstriction in foot and tail blood flow, exogenous NPY causes a vasodilation in the tail.

NPY[13-36] and [Leu³¹ Pro³⁴]NPY. The effects of NPY[13-36], the Y_2 (prejunctional) receptor agonist, and [Leu³¹ Pro³⁴]NPY, the Y_1 (postjunctional) receptor agonist, on tail BF_T and BF_{sk} in the tail and foot pad, are shown in Figure 2.

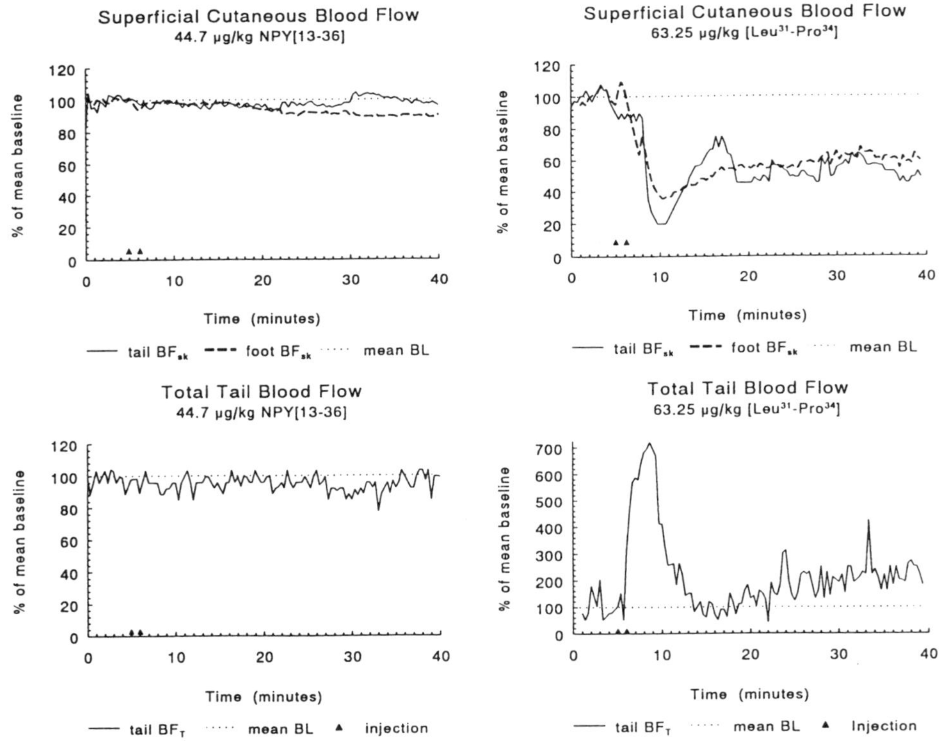

Fig. 2: Effect of the Y_2 receptor agonist NPY[13-36] (left), and of the Y_1 receptor agonist [Leu³¹Pro³⁴]NPY (right), on tail BF_T and on BF_{sk} in the tail and foot pad of the rat.

The Y2 receptor agonist (left) has no effect on tail BF_T or BF_{sk}. In marked contrast, the Y_1 receptor agonist (right) has an effect that is nearly identical to that of NPY. These results suggest that the increase in BF_T and concurrent decreases in BF_{sk} may be mediated by postjunctional Y_1

receptors in the vascular smooth muscle. The observation that the Y_2 receptor agonist has no effect on BF values indicates that the increase in BF_T in response to NPY is in no way due to inhibition of norepinephrine release.

Discussion

The most important observations of this study are: (a) that although NE caused the expected vasoconstriction in the tail and foot, NPY caused an increase in tail BF_T, tail blood volume, and tail T_{sk}, rather than a decrease, and (b) that this vasodilation effect of NPY on tail BF is apparently mediated by Y_1 rather than Y_2 receptors.

We decided to study the effects of NPY on peripheral BF because of our interests in peripheral BF during cold exposure, and in vasoactive agents that might contribute to non-freezing cold injury. The effects of exogenous NE were compared with the effects of exogenous NPY on foot and tail BF, because both neurotransmitters are located in the same sympathetic adrenergic nerve terminals and are often co-released (1,2), because both are released during severe cold exposure (9), and because of the well known vasoconstriction effects of NE, and the growing literature on the vasoconstriction effects of NPY (4). The observation that NE caused reductions in every BF parameter measured, was expected. Although the present observation that NPY causes vasodilation in the rat tail reiterates the findings of our previous study (first report: Heath and Thomas, 1993), it is contrary to observations made in most other tissues (12).

The concurrent increase in tail BF_T and blood volume indicates that some vessels in the tail are dilating in response to NPY. In vitro studies of the rat tail artery found that NPY caused a slow depolarization and concentration-dependent contraction via a direct effect on smooth muscle cells. This vasoconstrictor effect of NPY on rat tail arteries, along with our own observations that microvascular flow is diminished by NPY, limits the possible sites of vasodilation to arterioles, venules, veins, and arteriovenous anastomose (AVAs). As suggested by Heath and Thomas (11), the AVAs are the likely candidates for several reasons. They occur in large numbers in the rat tail and play a role in thermoregulatory heat loss. The observed increase in tail T_{sk} , even while superficial microcirculatory BF_{sk} was diminished (Figure 1), lends strong support to the suggestion. AVAs can also divert blood away from the capillaries as the blood flows via the path of least resistance. AVAs also allow for a much higher rate of BF through tissue than do capillaries, and when open, increase the vascular space and total blood volume in the tissue.

The Y_2 prejunctional receptor agonist, NPY[13-36], had no effect on tail BF_T and blood volume, or on BF_{sk} and T_{sk} in the tail and foot pad. This indicates that NPY is not acting on this prejunctional receptor, which has been shown to inhibit release of NE. Thus, the increase in tail BF_T is not due to inhibition of NE release.

The Y_1 post-junctional receptor agonist, [Leu^{31} Pro^{34}]NPY, had essentially the same effect on all measured parameters as NPY had. That is, tail BF_T and blood volume, and T_{sk} increased while BF_{sk} in the foot and tail decreased. The amount of [Leu^{31} Pro^{34}]NPY administered was equivalent to 64 μg/kg NPY, and the magnitude and range of responses caused by both were similar. This suggests that NPY is acting on Y_1 receptors in the vascular smooth muscle. An alternative explanation is that, both NPY and the Y_1 post-junctional receptor agonist, [Leu^{31}

Pro^{34}]NPY, are having the same effect on some other receptor type, be it Y_3 receptors, adrenergic receptors, or a receptor type as yet unidentified.

References

1. Lundberg, JM, Hökfelt, T (1983): Co-existence of peptides and classical neurotransmitters. Trends Neurosci 6, 325-333.

2. Ekblad, E, Edvinsson, L, Wahlestedt, C, Uddman, R, Häkanson, R, Sundler, F (1984): Neuropeptide Y coexists and cooperates with noradrenaline in perivascular nerve fibers. Regul Pept 8, 225-235.

3. Kong, JY, Thureson-Klein, AK, Klein, RL (1990): Are NPY and enkephalins costored in the same noradrenergic neurons and vesicles? Peptides 11, 565-575.

4. Walker, P, Grouzmann, E, Burnier, M, Waeber, B, (1991): The role of neuropeptide Y in cardiovascular regulation. Trends-Pharmacol-Sci 12, 111-115.

5. Wahlestedt, C, Yanaihara, N, Häkanson, R (1986): Evidence for different pre- and post-junctional receptors for neuropeptide Y and related peptides. Regul Pept 13, 307-318.

6. Balasubramaniam, A, Sheriff, S, Rigel, DF, Fisher, JE (1990): Characterization of neuropeptide Y binding sites in rat cardiac ventricualr membranes. Peptides 11, 545-550.

7. Waeber, B, Aubert, JF, Corder, R, Evequoz, P, Nussberger, J, Gaillard, R, Brunner, HR (1988): Cardiovascular effects of neuropeptide Y. Am J Hypertension 1, 193-199.

8. Lundberg, JM, Pernow, J, Locroix, S, Franco-Cereceda, A. Rudehill, A, Hokfelt, T (1989): Neuropeptide tyrosine (NPY) and sympathetic cardiovascular control. In: Neuropeptide Y. Mutt, V, Fuxe, K, Hokfelt, T, Lundberg, JM eds. New York:Raven Press, 191-200.

9. Zukowska, G-Z, Vaz, AC (1988): Role of neuropeptide Y (NPY) in cardiovascular responses to stress. Synapse 1988, 293-298.

10. Heath, ME, Thomas, JR (1994): Neuropeptide-Y (NPY) reduces cutaneous microcirculatory blood flow and increases total blood flow in the rat tail. In: Temperature Regulation: Recent Physiological and Pharmacological Advances, Milton, A ed. Basel Switzerland:Birkhauser, 171-175.

11. Heath, ME, Shelton, J, Thomas, JR (1993): A whole animal model for in vivo studies of the effects of environmental (thermal) stress and vasoactive substances on peripheral blood flow. Technical Report No. 93-07, Naval Medical Research Institute, 1-15.

12. Sheikh, SP (1991): Neuropeptide Y and peptide YY: major modulators of gastrointestinal blood flow and function. Am J Physio 261, G701-G715.

Thermal Balance in Health and Disease
Advances in Pharmacological Sciences

COMPARISON OF CORTISOL PRODUCTION BY ISOLATED ADRENAL CELLS FROM SYRIAN HAMSTERS AND GUINEA-PIGS AT DIFFERENT INCUBATION TEMPERATURES

R. Werner and R. Reimer

Zoologisches Institut, Zoophysiologie, Olshausenstr. 40, D-24098 Kiel, Germany

During acute cold exposure the cortisol production of guinea-pigs and Syrian hamsters is increased, as indicated by elevated plasma cortisol concentrations (1, 2). The augmented formation of corticosteroids obviously favours the resistance to cold. In this respect an intensifying effect of corticosteroids on the calorigenic action of catecholamines has been suggested (2). In the guinea-pig, a non-hibernator, corticosteroidogenesis naturally occurs at a body temperature (T_b) above 36°C. In the Syrian hamster, however, a T_b far below 30°C is not unusual during hibernation. Thus the question arises if there are interspecies differences in corticosteroidogenesis at low body temperatures between the hibernator and the nonhibernator. In order to answer this question cells from adrenals of both species were isolated and stimulated by adrenocorticotropic hormone (ACTH). In addition, the effect of cold acclimation on steroidogenesis of adrenal cells was investigated.

Methods

Adrenal cells were obtained from guinea-pigs *(Cavia aperea porcellus)* weighing 486 ± 41 g (mean ± S.E.M.; n=24, 12m, 12f) and male Syrian hamsters *(Mesocricetus auratus)* weighing 98 ± 6 g (n=86), kept at 20°C or 10°C, respectively, and a 14L/10D light-dark cycle for at least four weeks. After halothane anaesthesia the animals were killed by decapitation and the adrenals were removed quickly and trimmed free of adherent fat and external connective tissue. Adrenals of one guinea-pig or four Syrian hamsters were used for one preparation. Adrenal cells were isolated using a modified method derived from (3). Adrenals were minced with scissors into small pieces (<0.5 mm^3), washed with ice-cold Earle´s balanced salt solution (EBSS, pH=7.4) and placed in EBSS containing 5 $mg \cdot ml^{-1}$ collagenase (Sigma: Type I), 0.1 $mg \cdot ml^{-1}$ DNase and 1 $mg \cdot ml^{-1}$ bovine serum albumin (BSA). The mixture was incubated for 30 min at 37°C with shaking at 60 $cycles \cdot min^{-1}$. This, like all other incubations was performed under a 95% O_2/5% CO_2 atmosphere. The enzyme-treated tissue was dispersed by repeated aspiration through 1 ml pipette tips with different diameters. The suspension was filtered through a steel mesh (100 µm) into conical polystyrol centrifuge tubes, which were centrifuged at 100g for 10 minutes at 4°C. The cell pellets were suspended in fresh medium 199 (M 199, pH=7.4) containing 5 mg·ml^{-1} BSA and washed again. The suspended cells were incubated for 30 minutes at 37°C. After this preincubation the cell suspensions were centrifuged and the cell pellets were suspended in fresh M 199. Cell viability, as determined by trypan blue exclusion, was >80 %. The total numbers of cells obtained from one pair of adrenals were on the average 1335000 in guinea-pigs and 105000 in Syrian hamsters.

The incubations were performed in polystyrol tubes containing 1 ml of the suspension with a final cell concentration of about 30000 cells·ml^{-1}. The cells were incubated in absence or presence of porcine ACTH (10^{-8} - 100 *IU* ACTH · l^{-1}) at temperatures between 0°C and 37°C for 2.5 h. Incubations were terminated by transferring the tubes to an ice bath. Then the tubes were centrifuged at 300 g for 10 min at 0°C. The supernatant solution was stored at -20°C until analysis. For cortisol determination 200µl aliquots of incubation medium were extracted with dichloromethane (recovery > 90%). Cortisol was assayed by RIA using a highly specific cortisol antiserum with a specified cross-reactivity: cortisol 100, 11-deoxycortisol 2.7, cortisone 2.6, 11-deoxycorticostrone <0.2, 17α-OH progesterone, corticosterone and progesterone <0.1% (Guildhay, Surrey, England). [1,2,6,7-^{3}H]Cortisol was obtained from Amersham Buchler (Braunschweig, Germany).

Results

Guinea-pigs

The maximum steroid response of adrenal cells from guinea-pigs maintained at an ambient temperature (T_a) of 22°C was measured at concentrations between 10^{-2} and 10 *IU* ACTH · l^{-1} (=ED_{100}; Fig. 1). The amount of the ACTH-stimulated cortisol production is reduced with decreasing incubation temperature (T_i) without altering the sensitivity of the cells, *i.e.* the median effective dose (ED_{50}). ED_{50} was between 10^{-4} and 10^{-3} *IU* ACTH · l^{-1}. The basal cortisol synthesis is also reduced at low T_i. The maximal cortisol production stimulated by 10 *IU* ACTH · l^{-1} (=ED_{100}) was 17 pmol ·30000 cells^{-1} · 2.5 h^{-1} at T_i=37°C (Fig. 2). This value was set 100%. The cortisol production was 32% at T_i=30°C, 16% at 25°C, 5% at 20°C and 3.5% at 10°C and 0°C. A temperature coefficient (Q_{10}) of 5 is calculated for the drop in temperature from T_i=37°C down to 30°C and a Q_{10} of 6.5 for a drop from 30°C to 20°C. Significant increases of the *in vitro* cortisol production due to ACTH were observed at T_i=37°C, 30°C and 25°C (Fig. 2; non-hatched columns). Cold acclimation resulted in a considerable increase of adrenal weight and increase of the cellular ACTH-stimulated cortisol production. The wet weight of a single adrenal was 195 ± 14 mg (mean ± S.E.M.) in guinea-pigs maintained at 22°C and 279 ± 33 mg in those maintained at 10°C. The weight of one pair of adrenals related to body weight was 708 ± 33 mg · kg^{-1} (22°C) and 1200 ± 98 mg · kg^{-1} (10°C). The *in vitro* cortisol production by adrenal cells from guinea-pigs acclimated to 10°C (Fig. 2; hatched columns) was augmented to about 165 % (at T_i=37°C) compared to that of guinea-pigs kept at T_a=22°C. The cortisol production declined with decreasing T_i. No significant increase was observed at T_i=20°C in both acclimation groups.

Syrian hamsters

Adrenal cells from Syrian hamsters acclimated to T_a=22°C show a maximum steroid response at concentrations between 10^{-1} and 10 *IU* ACTH · l^{-1} (=ED_{100}). The maximum amount of ACTH-

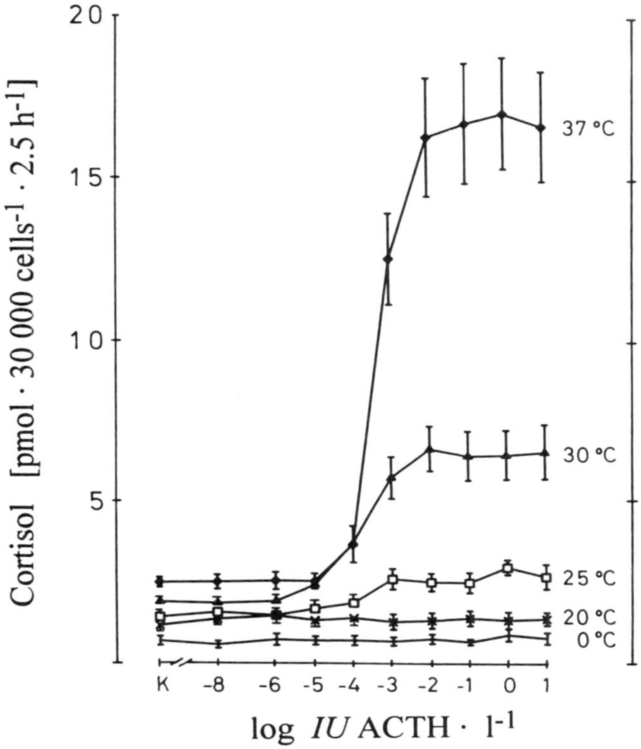

Fig. 1: Cortisol production by isolated adrenal cells stimulated by 10^{-8} up to 10 *IU* ACTH · l^{-1} (K = unstimulated). Dose-response curves (mean ± S.E.M.) at incubation temperatures of 37 °C (n=6), 30°C (n=5), 25°C (n=3), 20°C (n=5), and 0°C (n=8). Guinea-pigs were acclimated to 22°C

stimulated cortisol production of cells from Syrian hamsters is 1.34 pmol · 30000 cells^{-1} · 2.5 h^{-1} at T_i=37°C and less than 1/10 that of guinea-pigs. Setting this value 100% the ACTH-stimulated cortisol production was 76% at T_i=30°C, 46% at 20°C, 26% at 10°C and 25 % at 0°C in the Syrian hamster (Fig. 3). A Q_{10} of 1.5 results from the drop in temperature of T_i=37°C down to 30°C and Q_{10} of 1.6 for 30°C down to 20°C. Like in guinea-pigs the amount of the ACTH-stimulated cortisol production is reduced with decreasing T_i without significant altering the sensitivity of the cells. ED_{50} was between 10^{-3} and 10^{-2} *IU* ACTH · l^{-1}. In Syrian hamsters a significant ACTH-stimulated increase of the cortisol synthesis was observed at Ti ≥ 20°C.

Cold acclimation resulted in a considerable decrease of ACTH-stimulated cortisol production (Fig. 3; hatched columns). At T_i=37°C the *in vitro* cortisol production by adrenal cells from Syrian hamsters acclimated to 10°C (Fig. 3) was only about 61 % compared to hamsters kept at T_a=22°C. The ACTH-stimulated and basal cortisol production declined with decreasing T_i. However, ED_{50} and adrenal weight remained unchanged. The wet weight of a single adrenal was 12.2 ± 1.6 mg (mean ± S.E.M.) in both groups. In Syrian hamsters the increase of ACTH-stimulated cortisol synthesis was significant at T_i ≥ 20°C.

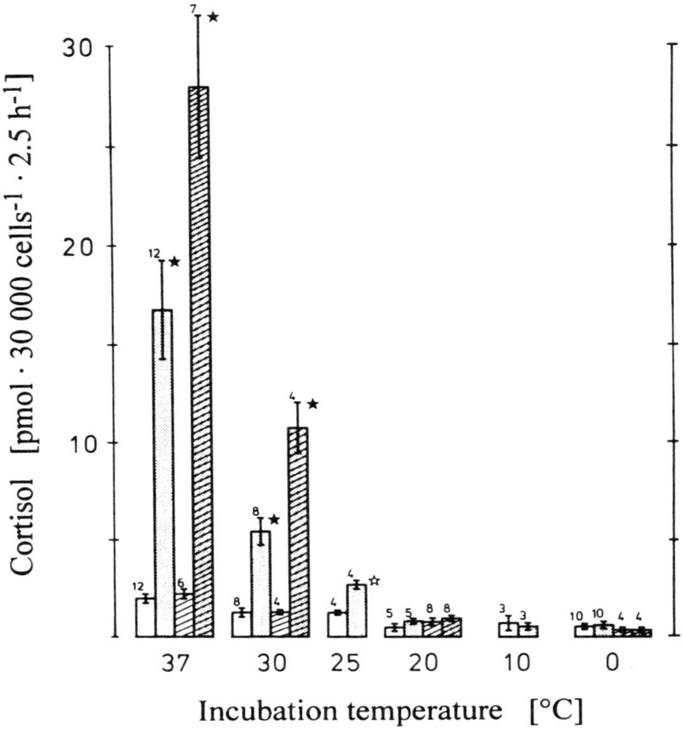

Fig. 2: Cortisol production (mean ± S.E.M.) by isolated adrenal cells from guinea-pigs acclimated to an ambient temperature of 22°C (non-hatched) or 10°C (hatched) in absence (non-shaded) or presence of 10 *IU* ACTH · l^{-1} (shaded) at different incubation temperatures.

Discussion

Little was known so far concerning the effect of incubation temperature or acclimation temperature on the rate of cortisol synthesis. The high rate of cortisol synthesis in guinea-pigs is consistent with high cortisol concentrations determined in plasma (1). On the other hand a low rate of cortisol production is in accordance with comparatively low plasma cortisol levels in Syrian hamsters (2). The amount of the ACTH-stimulated cortisol production is reduced with decreasing T_i in both species. As derived from the temperature coefficient (Q_{10}=5) the considerable larger amount of the cortisol synthesis of guinea-pigs is affected much more severely by low T_i compared to Syrian hamsters (Q_{10}=1.5). A further difference was found in the lower limit of an ACTH-stimulated increase in cortisol production. An increase was still evident at T_i=20°C in the hibernator, while no increase was observed in the non-hibernator at that temperature. In addition, the basal (*e.g.* unstimulated) cortisol production is reduced with decreasing T_i. However, as derived from dose-response curves, the median effective dose (ED_{50}) is unchanged in both species. That implies that the sensitivity of the adrenal cells to ACTH is not altered by low T_i, but the capacity for cortisol production.

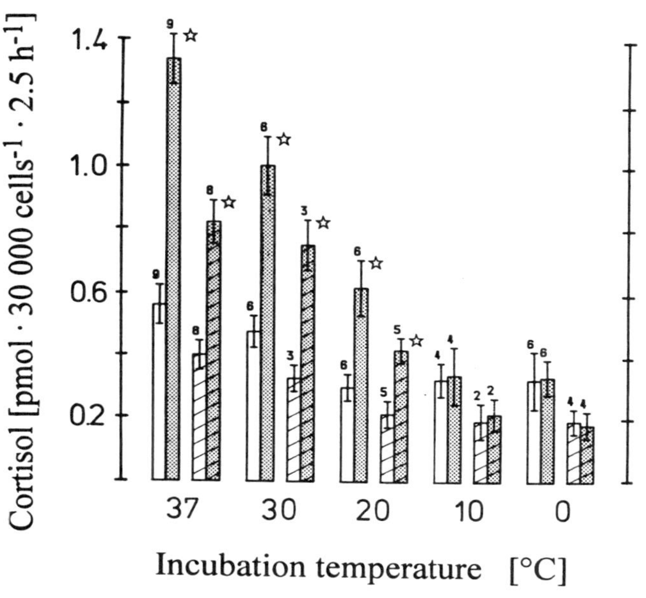

Fig. 3: Cortisol production (mean ± S.E.M.) by isolated adrenal cells from Syrian hamsters acclimated to an ambient temperature of 22°C (non-hatched) or 10°C (hatched) in absence (non-shaded) or presence of 10 *IU* ACTH · l^{-1} (shaded) at different incubation temperatures.

No effect of cold acclimation on the sensitivity of adrenal cells to ACTH was identified in the hibernator or the non-hibernator. The adrenal weight gain determined in guinea-pigs is consistent with a hypertrophy of the adrenal cortex found in rats (4). Glucocorticoids restore thermogenic responses and cold tolerance in hypophysectomized and adrenalectomized animals and are important for survival in the cold (5). In this connection the augmentation of the ACTH-stimulated cellular cortisol production may be the result of the acclimation of guinea-pigs to cold.

On the contrary the adrenal weight remained unchanged in Syrian hamsters and cold acclimation resulted in a decrease of ACTH-stimulated and basal cortisol production. Concerning the reduction of cortisol synthesis it should be taken into account that the Syrian hamster is a hibernator. During hibernation the sympathetic activity and particularly the adrenocortical activity is depressed. In contrast to the arousal or euthermic state metabolic heat production is reduced during hibernation and there is no special requirement for a permissive effect of corticosteroids on the calorigenic action of catecholamines. Compared to euthermic hamsters at an environmental temperature of 8°C or 28°C the cortisol concentration in the hibernating state was only 1.5% or 4%, respectively (6). In Syrian hamsters corticosteroids are important for the control of carbohydrate metabolism during hypothermia and the spontaneous rewarming from artificial hypothermia (7).

The hypothalamo-pituitary-adrenocortical axis is activated during acute cold exposure and cold-induced increases of the cortisol level and of heat production were observed during moderate hypothermia (1, 2). Further experiments, using an adrenocorticostatic agent, corroborate the hypothesis that corticosteroids intensify the calorigenic action of catecholamines in a permissive manner (2). Corticosteroids are potent to exert rapid and non-genomic effects (8, 9). Furthermore, acute effects of corticosteroids on the action of catecholamines were demonstrated, *i.e.* at the β-receptor level by affecting the coupling mechanisms between the receptor and the N-protein and presumably the catalytic moiety of the adenylate cyclase (8, 9). At T_i=20°C there is still a significant increase of the cortisol production of adrenal cells from Syrian hamsters. However, if the hypothesis applies to arousal from hibernation corticosteroido-genesis should be increased even at lower temperatures in order to intensify the calorigenic effect of catecholamines. The interscapular brown adipose tissue is a well known source of nonshivering thermogenesis (10). In addition, by warming the adrenals, the periadrenal brown fat found in the Syrian hamster is apparently an important calorigenic tissue to support corticosteroidogenesis during arousal from hibernation. By means of such a positive feedback mechanism an increased production of corticosteroids may exert a permissive, intensifying effect on the calorigenic action of catecholamines even during early arousal from hibernation.

References

1. Werner, R (1988): Effect of metopiron-ditartrate on thermogenesis in the guinea-pig. Comp Biochem Physiol 90 C, 445-450.

2. Werner, R (1992): Sympathoadrenal activity during helox-cold induced hypothermia in Syrian hamsters. Comp Biochem Physiol 103 A, 137-143.

3. Sala, GB, Hayashi, K, Catt, KJ, Dufau, ML (1979): Adrenocorticotropin action in isolated adrenal cells. J Biol Chem 254, 3861-3865.

4. Deane, HW, Lyman, CP (1954): Body temperature, thyroid and adrenal cortex of hamsters during cold exposure and hibernation, with comparison to rats. Endocrinology 55, 300-315.

5. Palazzolo, DL, Quadri, SK (1987): Plasma thyroxine and cortisol under basal conditions and during cold stress in the aging dog. Proc Soc Exp Biol Med 185, 305-311.

6. Werner, R (1991): Adrenocortical and sympathetic activity during hibernation in Syrian hamsters. Acta endocrinol 124, Suppl. 1, 51.

7. Musacchia, XJ, Deavers, RD (1978): Glucocorticoids and carbohydrate metabolism in hypothermic and hibernating hamsters. In: Effectors of Thermogenesis. Girardier, L, Seydoux, J, eds. Basel, Stuttgart, Birkhäuser Verlag, Experientia Suppl 32, 247-258.

8. Duval, D, Durant, S, Homo-Delarche, F (1983): Non-genomic effects of steroids. Interactions of steroid molecules with membrane structures and functions. Biochim Biophys Acta 737, 409-442.

9. Stiles, GL, Caron, MG, Lefkowitz, RJ (1984): α-Adrenergic receptors: Biochemical mechanisms of physiological regulation. Physiol Rev, 64: 661-743.

10. Jansky, L (1973): Non-shivering thermogenesis and its thermoregulatory significance. Biol Rev 48, 85-132.

Thermal Balance in Health and Disease
Advances in Pharmacological Sciences

HIBERNATION AND COLD ADAPTATION INDUCE DIFFERENT CHANGES IN THE BODY TEMPERATURE REGULATION IN THE GOLDEN HAMSTER

B. Štefl

Department of Physiology and Developmental Biology, Faculty of Science, Charles University, Vinicna 7, Prague 2, 128 00 CZ

Introduction

In the golden hamster the entrance to hibernation torpor is possible under artificial conditions (lower ambient temperature, short light period of a day, changes in food supply, etc.) after a preparatory period lasting about 2 months. Adaptation changes occurring during this pre-hibernation period are connected with changes in hormonal status of the hibernator (1). The adaptation changes studied most often concern mechanisms of heat production, especially the total capacity of non-shivering thermogenesis (NST) and parallel events on the organ, cellular, and subcellular level (e.g. 2, 3, 4, 5). This study aims to reveal eventual adaptation changes in body temperature control of golden hamsters that might be induced through a pre-hibernating period (i.e. by the readiness to hibernate) or simply through cold-acclimation contrary to warm-acclimated animals.

Methods

The dynamics of thermoregulatory responses in the awake state was studied in three groups of golden hamsters during the controlled decrease of ambient temperature from 30 °C to 19 °C in the metabolic chamber. The three groups were: warm- acclimated hamsters (WA) kept at an ambient temperature 25 °C; cold-acclimated hamsters (CA), kept at least 4 weeks at 6 °C (both groups were held under artificial light:dark conditions 14:10); hibernating hamsters (HI), kept in a dark quiet room at an ambient temperature 8 to 10 °C. Hibernating animals were forced to awake from torpor 5 hours before the start of the experiment, so that they were fully awake and euthermic at the beginning. At least one hour before cooling the resting values of total oxygen consumption (converted to the total metabolic rate, MR, using an energetic equivalent 20.95 kJ/l O_2), deep body temperature (measured in colon, Tc), and peripheral body temperature (measured subcutaneously above the sacral bone, Ts) were measured. During the ramp cooling lasting 30 min the threshold temperatures for the onset of a thermogenic response were determined and the dynamic changes of MR, Tc, and Ts were recorded. From this, the conclusions are drawn of plausible changes in the body temperature control elicited by cold-acclimation or by hibernation of the golden hamster.

Results

The time record of averaged experiments performed on the CA group (n = 15) is presented as an example in Fig. 1. Zero time indicates the moment when the threshold temperatures for the onset of cold thermogenesis were reached. From this time on, the total heat production increases as a consequence of body temperature fall, until it reaches a steady level. Then the increasing of MR ceases despite of further possible decrease in body temperatures.

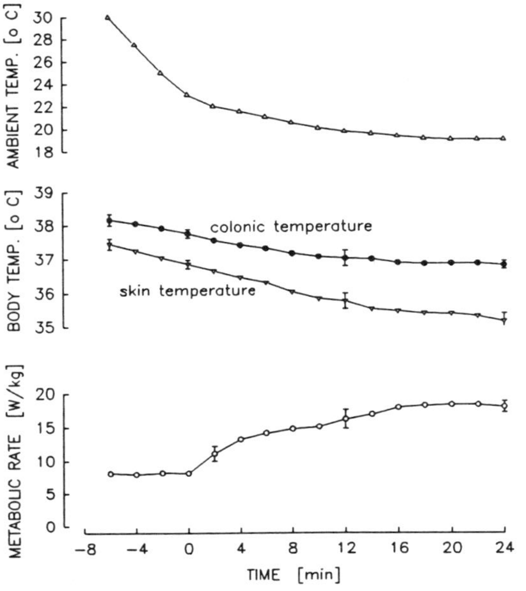

Fig. 1: Time courses of ambient temperature, total metabolic rate, and body temperatures of cold-acclimated golden hamsters during external cooling in the metabolic chamber (means, n = 15, vertical bars indicate S.E.M.)

Results from the three experimental groups are summarized in Tab. 1. At the thermoneutral zone (30 °C) WA and CA hamsters differ only in the resting metabolic rate, which is in CA group (and HI group, too) significantly higher than in WA. Resting values of deep and peripheral body temperature, as well as thresholds of cold thermogenesis do not differ between CA and WA hamsters. HI hamsters have significantly lower both resting and threshold temperatures than CA or WA ones. As a consequence of (equally intensive and equally lasting) cold stimulus, Tc and Ts decrease and MR increases. The relative changes of all measured variables from resting levels are in all 3 experimental groups of the equal size, except of Tc. Deep body temperature of HI hamsters decreases significantly less ($P < 0.05$) than in CA or WA hamsters.

The dynamics of changes of body temperatures and MR stands out when MR is plotted against Tc or Ts (Fig. 2). Metabolic rate and body temperatures do not correlate at rest and at the beginning of cooling, until the thresholds are reached. After this, an approximately linear relationship between MR and Tc, or MR and Ts takes place during the increasing phase of metabolic rate (till the steady level of heat production is attained). Linear regression coefficients of those relationships may serve as arbitrary characteristics of the body temperature controller (6). They represent apparent starting sensitivities of the regulator to the change of the particular input signal (i.e. body core temperature or body shell temperature). When determined from averaged values of variables in individual

Table 1: A comparison of resting levels at thermoneutral zone, threshold temperatures for the onset of cold thermogenesis and relative changes of the total metabolic rate (MR), body core temperature (Tc) and peripheral body temperature (Ts) after 30 min of external cooling of hibernation-aroused (HI), cold- (CA) and warm-acclimated (WA) golden hamsters.
* - differences between HI and WA or CA groups are significant ($P < 0.05$)
\# - differences between WA and CA or HI groups are significant ($P < 0.05$)

	resting levels			thresholds		cooling levels		
	MR [W/kg]	Tc [°C]	Ts	Tc [°C]	Ts	MR [W/kg]	Tc [°C]	Ts
HI n = 5	7.3	36.78 *	35.62 *	36.48 *	35.34 *	9.7	-0.66 *	-2.04
± S.E.M.	0.6	0.19	0.16	0.30	0.28	0.5	0.12	0.26
CA n = 15	8.2	38.17	37.45	37.76	36.85	9.3	-1.31	-1.99
± S.E.M.	0.3	0.17	0.16	0.13	0.13	0.5	0.12	0.15
WA n = 9	5.8 #	37.83	37.23	37.41	36.79	10.4	-1.51	-2.20
± S.E.M.	0.4	0.18	0.20	0.18	0.12	0.4	0.31	0.28

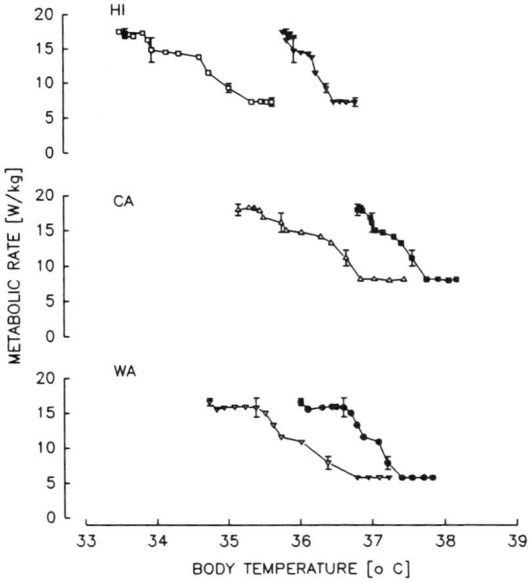

Fig. 2: Plots of total metabolic rate versus body temperature derived from averaged values of corresponding groups; HI - hibernating aroused, CA - cold-acclimated, WA - warm-acclimated. Full symbols are for deep body temperatures.

experimental groups, these linear coefficients of the dependence of MR on Tc are in [W/kg x °C] -13.8, -10.1, and -12.4, whereas the coefficients of MR versus Ts -5.5, -5.6, and -7.2, for HI, CA, and WA animals, respectively. The ratio of the apparent sensitivities to central versus peripheral input indicates the relative importance of the signals from the appropriate body part in thermoregulation. These ratios (sensitivity to central input divided by sensitivity to peripheral input) were calculated from averaged values of particular variables of given experimental groups. The results are 1.7 : 1 for WA, 1.8 : 1 for CA, and 2.5 : 1 for HI hamsters. This indicates, the importance of temperature input signals from body core is higher (or the importance of peripheral input decreased) in awake golden hamsters between hibernating bouts, in comparison with situation in cold- or warm-acclimated hamsters.

Discussion

Cold acclimation of normothermic mammals, e.g. guinea pigs, besides of development of NST leads to the shift of threshold temperature for the induction of shivering (7). Adaptive relevancy of these changes are seen in lower heat loss to surroundings (as a consequence of lower body temperature) and in greater efficiency of thermogenic mechanisms (NST in comparison to shivering). Cold acclimation of the golden hamster does not lead to similar events. Unlike normothermic rodents (rat, guinea pig, etc.) in golden hamsters brown adipose tissue (BAT) and NST persist throughout the life (3). In this sense hamsters are ever cold adapted, though also BAT of hamsters is subjected to adaptation changes in the cold (8). Cold acclimation rather results in higher metabolic rate at thermoneutral zone comparing to warm-acclimated individuals. The reason of this could lie at least partly in higher activity of thyroid gland (9). On the other hand, awake hamsters between hibernating bouts (adapted "to hibernate") have significantly lower body temperature, as well as the thresholds for cold thermogenesis. Thus, from this point of view the acclimation to cold and the adaptation to hibernate are not the same.

The apparent sensitivity to central input signal, i.e. the increase of MR calculated on the unit change of Tc, is the highest in HI group. Needless to say, this arbitrary characteristic must not be confused with real sensitivity of the regulator (the gain) that can only be measured when feedback loops from parallel inputs are opened (10), which is extremely difficult in living animals. Nevertheless, it can serve as easily available working parameter to compare the qualities of body temperature regulation under well defined experimental conditions (6). As in this experiments none of the body temperatures was manipulated independently, the mutual ratio of the apparent sensitivities to central over sensitivity to peripheral input is used as a characteristic of thermoregulation. This is actually a measure, what is the change in body shell temperature related to unit change of body core temperature. The higher ratio thus reflects the fact, that on the cold stimulus of the same intensity the deep body temperature of HI animals falls significantly less than that of CA or WA hamsters. This indicates the adaptation to hibernation increase somehow the efficiency of thermoregulatory responses (as the increase in MR is of the same magnitude in the 3 groups).

Underlying mechanisms could consist in the changes of passive components of thermoregulation (changes in heat conductivity of tissues, better insulation), and in the development of vasomotoric reactions allowing selective maintenance of high body core temperature. The biological significance of lower body temperature in animals between hibernating bouts is conceivable as an energy saving mechanism. The endeavour to keep deep body temperature on the desired height (at least for some period) is comprehensible, as well. It seems, the mechanisms assuring this are in animals adapted to hibernate more effective than in cold-acclimated or warm-acclimated golden hamsters.

References

1. Janský, L, Haddad, G, Kahlerová, Z, Nedoma, J (1984): Effect of external factors on hibernation of golden hamsters. J Comp Physiol B 154, 427-433.

2. Pohl, H (1965): Temperature regulation and cold acclimation in the golden hamster. J Appl Physiol 20, 405-410.

3. Vybiral, S, Janský, L (1974): Non-shivering thermogenesis in the golden hamster. Physiol bohemoslov 23, 235-243.

4. Abbotts, B, Wang, LCH (1980): Seasonal thermogenic capacity in a hibernator, Spermophilus richardsonii. J Comp Physiol 140, 235-240.

5. Heldmaier, G, Boeckler, H, Buchberger, A, Klaus, S, Puchalski, W, Steinlechner, S, Wiesinger, H (1986): Seasonal variation of thermogenesis. In: Living in the Cold. Heller, HC, Musacchia, XJ, Wang, LCH, eds. New York-Amsterdam-London: Elsevier, 361-372.

6. Štefl, B (1989): Dynamic effects of acute cooling in body temperature regulation of the euthermic warm-acclimated golden hamster. Physiol bohemoslov 38, 171-178.

7. Brück, K, Wünnenberg, W, Gallmeier, H, Ziehm, B (1970): Shift of threshold temperature for shivering and heat polypnea as a mode of thermal adaptation. Pflügers Arch 321, 159-172.

8. Himms-Hagen, J (1986): Cold- versus diet-induced thermogenesis in brown adipose tissue: different strategies in different species. In: Living in the Cold. Heller, HC, Musacchia, XJ, Wang, LCH, eds. New York-Amsterdam-London: Elsevier, 93-100.

9. Hoffman, RA (1968): Hibernation and effect of low temperature. In: The Golden Hamster.Its Biology and Use in Medicinal Research. Hoffman, RA, Robinson, PF, Magalhaes, H, eds. Ames Iowa: The Iowa State Univ Press, 25-39.

10. Brown, AC, Brengelmann, GL (1970): The interaction of peripheral and central inputs in the temperature regulation system. In: Physiological and Behavioral Temperature Regulation. Hardy, JD, Gagge, AP, Stolwijk, JAJ, eds. Springfield, Illinois: Thomas, 684-702.

Section IV:

THERMAL BALANCE IN THE CLINIC

BASIC ASPECTS OF FETAL THERMAL HOMEOSTASIS

H. J. Schröder and G. G. Power*

Abteilung für experimentelle Medizin Universitäts-Frauenklinik, Martinistr. 52, 20246 Hamburg, FRG, *Division of Perinatal Biology, Loma Linda University Medical School, Loma Linda, Ca. 92350, USA

Introduction

The fetus in utero generates heat that has to be dissipated via the mother. Immediately after birth, the fetus must be capable above all of responding quickly and efficiently to heat loss to cooler surroundings. Fetal, perinatal and neonatal heat production and temperature regulation have been the focus of experimental research during the last three decades (1-3). Our work has concentrated on the sheep fetus, and thus the following brief overview will outline the principles of fetal thermal homeostasis. Most animal studies in this field have been performed on chronically instrumented, unanesthetized fetal sheep at gestational ages from 120 to 139 days (term 145 days). Fetal weight during this period is in the range of 2.5 to 4 kg.

The fetus in utero

The temperature T of a body is determined by its heat content and specific heat (SH) as T = heat content/specific heat. Heat content is measured in Joules g^{-1} (or cal g^{-1}, 1 cal = 4.18 J), temperature in centigrade (°C) and specific heat in J $°C^{-1}$ g^{-1}. Temperature is stable when heat transfer into the body (heat production) equals heat transfer out of the body (dissipation). Heat transfer is measured in Watt (J sec^{-1}; 1 W = 14.4 cal min^{-1}).

These principles are valid for each of the fetal compartments, including the placenta, and for the fetus as a whole. It is likely that heat content, specific heat, heat transfer (per unit weight) and therefore temperatures are different in the fetal compartments. These temperature variations will be small because the temperature of the fetal environment is similar (within 0.3 to 1 °C) to the fetal temperature. The term "fetal temperature" applies to the fetal arterial blood temperature if not otherwise indicated.

The fetal-maternal temperature difference . It has been known for more than a century that the fetal tissues are somewhat warmer than the maternal tissues. Fetal temperature typical exceeds maternal temperature by less than 1°C in most mammalian species including man (2, 4-7). A major determinant of fetal temperature is the maternal temperature: if maternal temperature increases, fetal temperature will increase also (Figure 1). The maternal-fetal temperature difference is maintained, provided that the maternal temperature increases slowly (in fetal sheep <0.2 °C/h). If maternal temperature increases rapidly, however, the temperature gap will become smaller or even reverse due to thermal inertia (6, 7). The characteristic finding of a positive maternal-fetal temperature difference proves that the fetus generates heat, and that resistance to heat flux between maternal and fetal tissues is not zero.

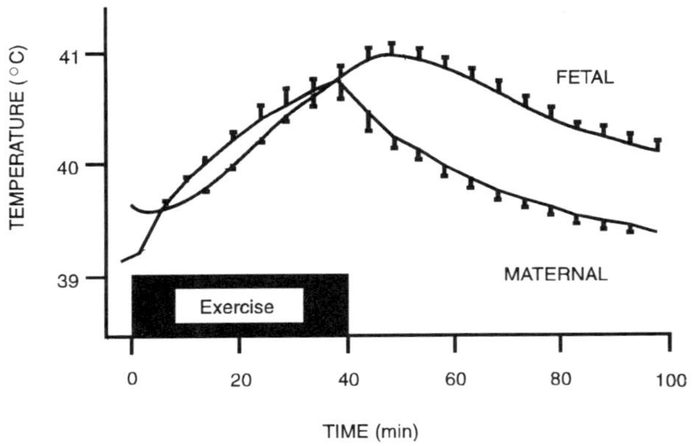

Fig. 1: Time course of maternal and fetal temperatures during maternal exercise. Fetal temperature lags behind the rapidly rising maternal temperature, and the usual maternal-fetal temperature difference is transiently reversed. Modified from (6), with permission.

Fetal heat production

The amount of heat the fetus generates per unit time can be deduced from fetal oxygen consumption (indirect calorimetry) and more directly by "differential calorimetry". The results of these fundamentally different techniques agree quite well for fetal sheep in late gestation: both indicate a heat production of 3-4 W kg^{-1} fetal body weight.

Fetal oxygen consumption. Indirect calorimetry assumes that a specified amount of heat will be produced for each ml of oxygen consumed.The value of the "energy equivalent" is about 5 cal ml^{-1} STPD O_2 (21 J ml^{-1} or 468 J $mmol^{-1}$ O_2), it depends moderately on the respiratory quotient and thus on the contribution of different substrates to total fetal metabolism (1).

Fetal oxygen consumption has frequently been measured in a variety of species and ranges from about 4 ml min^{-1} kg^{-1} body weight to 12 ml min^{-1} kg^{-1} (8-10), and 10 ml (~8.3 ml STPD) min^{-1} kg^{-1} may be accepted as average. Fetal heat production H_{fetus} thus varies from 1.2 to 3.5 W kg^{-1} with 2.9 W kg^{-1} as estimated mean. The value does not change appreciably during the last third of gestation or among different species that have been studied to date. The error introduced by fetal growth (up to 1%/d in humans) may be neglected.

Differential calorimetry: principle and measurements . In fetal sheep a small electrical nichrome heating element may be chronically introduced into the inferior vena cava (10). When heated, fetal temperature starts to rise as the energy H_{heater} is introduced, and eventually a new steady state is approached during which the maternal-fetal temperature difference T_{m-f} has widened by a new additional difference dT_{m-f}. It is known, then, from the ratio of dT_{m-f} to H_{heater} how large fetal heat production must be to generate a given maternal-fetal temperature difference. Dividing the spontaneous T_{m-f} by the above ratio therefore yields fetal heat production.

This principle of differential calorimetry has been applied to chronically instrumented fetal sheep (10) ranging in gestational age from 120 to 135 days. The average value of fetal heat production was 3.3± 0.3 W kg^{-1} (n=7, fetal weight 3.3±0.4 kg) which agrees favorably with the results determined in other fetal sheep by indirect calorimetry. Several caveats apply. Maternal and fetal temperatures must reach a steady state, the heat conductances between mother and fetus must not change between basal observations of T_{m-f} and those made during heating, and there must be no external heat source for the fetus. In other words, when fetal heat production stops, the maternal-fetal temperature difference should become zero. Because the placenta probably directs part of the heat it produces towards the fetus, the last assumption will be violated (see below).

Fetal heat dissipation

From the maternal-fetal temperature difference (~0.5°C) and the fetal heat production (3.5 W kg^{-1}) the overall heat conductance between fetus and mother can be calculated as 7 W kg^{-1} $°C^{-1}$. Two pathways are open for fetal heat to move towards the maternal tissues: a convective pathway following the umbilical circulation towards the placenta, and a conductive pathway from fetal surfaces, through the surrounding fluids, towards the uterine wall. The allocation of heat dissipation between these two pathways largely determines the freedom the fetus has to control heat dissipation without interfering with oxygen and carbon dioxide exchange.

The convective pathway. According to Fick's principle, heat loss from the umbilical circulation H_{umb} will equal

$$H_{umb} = dT_{a-v} \times Q_{umb} \times SH_{blood}$$

where dT_{a-v} is the temperature difference between umbilical arteries and vein (veins), and Q_{umb} is fetal placental blood flow (in fetal sheep typically 3.3 ml sec^{-1} kg^{-1} in late gestation). Specific heat for blood is 3.6 J ml^{-1} $°C^{-1}$ (or 85% of water). A first estimate of the expected umbilical arterial-venous temperature difference therefore yields about 0.3 °C for fetal sheep, which obviously assumes that fetal heat is lost exclusively through the umbilical circulation (see below). Attempts have been made to measure the umbilical av-temperature difference in fetal sheep, and the results are in the range of 0.1 to 0.4°C (5, 11). It should be noted that determining the true difference (and therefore placental heat exchange) with accuracy poses a considerable technical problem.

Heat conductance of the human umbilical cord. Heat conductances were measured in isolated human umbilical cord segments (12), and estimated (for total cord length and normal blood flow rates) to be 4 W $°C^{-1}$ between the vessels and the bath, and as 0.4 W $°C^{-1}$ between the arteries and the vein. From these data it may be concluded that only a fraction less than 10% would leave or enter the umbilical vessels. Furthermore, the normal flow arrangement in umbilical arteries and veins is not such an efficient countercurrent system that it forms a "heat trap" keeping heat within the fetal body.

Placental heat exchange . Heat exchange across the placenta should be subjected to the same physical rules governing placental exchange in general (13). In this system, the maternal-fetal temperature difference, the placental permeability and the arrangement of blood streams (e.g. countercurrent or concurrent exchange system) are the predominant factors controlling placental transfer. Heat conductances in water approximates 9×10^{-2} cm^2 min^{-1} (13) which is about 40 times the diffusion coefficient of water in water (at 39°C). The placental permeability for heat transfer in sheep has been estimated to be 0.8×10^6 W °C^{-1} kg^{-1}. It is expected, therefore, that the placenta does not measurably limit the diffusion of heat between fetal and maternal blood streams, and that any limitation of the convective pathway is due to the limited heat transfer capacity of the umbilical circulation.

The placental transfer of heat has been directly compared with the diffusive transfer of labeled water using an isolated placenta preparation (14). Most unexpectedly, the transfer efficiency for heat was distinctly less than measured simultaneously for water. This apparent paradox can be resolved by assuming that heat flux in the placenta is not confined to the exchange area but also occurs between small umbilical vessels as they enter or leave the placenta. Such a heat shunt, bypassing the exchange region, gains importance because of the remarkably high diffusivity of heat. This does not imply that the placenta limits heat dissipation but rather that placental heat exchange is less than it could be if heat kept to the usual pathways. Data from other placentas do not exist.

The conductive pathway: conductances of fetal and uterine surfaces . Two heat conductances, G, affect the conductive pathway: the conductance of the fetal surfaces G_{surf} controls heat transfer from the fetus into the amniotic/allantoic fluid, and the conductance of the uterine wall G_{ut} influences heat flux from the extrafetal fluids into the wall of the uterus. Estimates of both conductances have been obtained in fetal sheep (15) by observing fetal and amniotic fluid temperatures after injection of cold saline into the amniotic cavity. With support of a computer model which simulated the time courses of temperature changes, estimates of G_{surf} and G_{ut} were calculated. On the average, in ten fetal sheep G_{surf} and G_{ut} were 2.7±0.3 W °C^{-1} kg^{-1} and 1.8±0.2 W °C^{-1} kg^{-1}, respectively. Thus total conductance of the conductive pathway equals 1.1 W °C^{-1} kg^{-1}, which leaves a balance of 6-7 W °C^{-1} kg^{-1} (15) to the placental "conductance". However, the conductance values should be regarded as rough estimates only because unstirred layers of fluid will cause the conductances to be underestimated.

Heat flux from the uterus . Fetal heat will finally reach the uterine tissues by both the conductive and the convective pathways. It will leave the uterus either with the blood streams from the placenta and the uterine tissues, or directly from the uterine surface to the neighboring organs. Applying Fick's principle, heat efflux with the uterine venous blood has been measured as 2.0±0.15 W kg^{-1} total uterine mass (16) and 1.9±0.2 W kg^{-1} (8), which includes fetal and placental heat production, and heat production of the uterine tissues as well. An additional 5 to 10% has been found to leave the uterine surface directly as measured by a heat flux transducer that was attached to the uterine surface (16).

Placental heat production

Metabolically the placenta is a very active organ with an oxygen consumption ranging from 6.7 to more than 14 ml O_2 min^{-1} kg^{-1}. Accepting 10 ml (~8.3 STPD) min^{-1} as a representative value (8, 9) it therefore generates about 1.5 W (placental weight in sheep is 0.5 kg). It is not known how the heat leaves the placenta. Presumably, most will be carried away by the maternal and fetal blood streams, but local conduction to the uterine tissues (decidua, muscle) and the amniotic/allantoic fluid seems possible also.

In model calculations (see below) it has been assumed that placental heat exits to the maternal organism and the fetus in proportion to their respective placental blood flow rates. Because the ratio of both flow rates is close to one, it follows that less than 10% (0.8 W from the placenta, 10-11 W from the fetus) of H_{fetus} enters the fetal circulation. About 0.07°C of the fetal-maternal temperature difference is due to placental heat production (17).

An attempt to directly measure uteroplacental heat production from uterine blood flow rates and av-temperature differences after killing the fetus did not agree with the oxygen consumption measurements (8). It was pointed out that fetal heat may be dissipated to neighboring maternal structures bypassing the maternal circulation. The possibility then that the fetus might have direct access to the environment through the maternal abdominal wall, for example, has not been explored further.

Combining data: a model of fetal-maternal heat exchange

Thermoregulation requires the control of heat influx and heat efflux to and from the body to maintain its heat content or temperature. The limits in between which the thermoregulatory efforts of the fetus can be effective, depend on its control of heat production and of heat dissipation. Inevitably the elimination of heat is tied to the placental transport of oxygen, carbon dioxide and metabolic substrates. The situation is comparable to that of gill fish which cannot be shielded from the surrounding water temperature if they are to maintain adequate gas exchange.

An elementary computer model was constructed (17) to explore the limits of fetal thermoregulation and the influence of single components of the dissipative pathways on fetal body temperature.

In the model, heat production, blood flow rates and conductances were varied from 10 to 300% of values taken to be representative as controls, and the effect on steady-state temperatures in the various compartments was calculated. Changing any heat conductance was found to influence fetal temperature at most 0.1°C, and the "heat trap" (see above) of the umbilical cord was insignificant. Major fetal temperature changes (more than 0.5°C) were only possible with pronounced changes of heat production or blood flow rates. The effects of fetal cardiac output (which greatly influences fetal placental blood flow) and heat production, two variables under possible fetal control for thermoregulation, are displayed in Figure 2.

A rise of maternal temperature larger than the usual maternal-fetal temperature difference must increase fetal temperature under all circumstances, because the maternal-fetal difference cannot become less than zero.

Decreases of cardiac output and raise of heat production (of unrealistic magnitude) might lead to fetal temperature increases of close to 5°C. It should be noted, however, that this requires an impossible combination of placental blood flow(-90%) and oxygen consumption (+200%) changes.

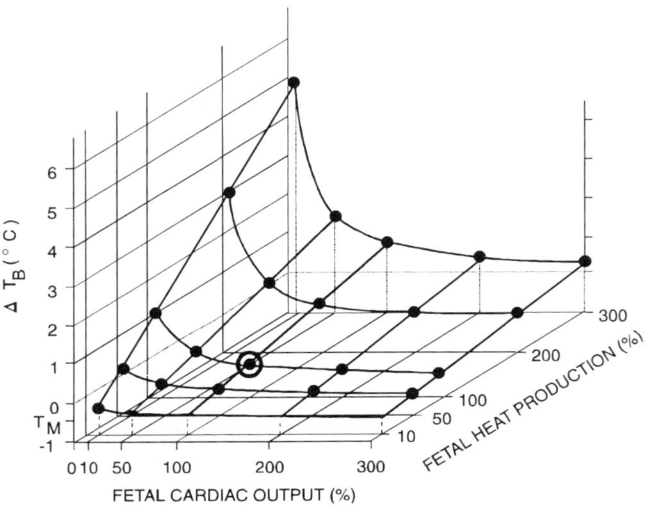

Fig. 2: Changes of fetal arterial blood temperature T_B with changes of fetal cardiac output and fetal heat production (or oxygen consumption). "100" indicates normal values. Note that fetal temperature can be decreased at most by 0.5°C, but might increase up to 5°C; this, however, would require a combination of cardiac output and oxygen consumption values which are impossible to obtain in reality. The fetus thus can do little to fight changes of its temperature. Encircled data point represents "normal" values. T_M is maternal temperature. From (7), with permission.

An example (17) illustrates the limited effect of fetal thermoregulatory efforts. Without thermoregulatory responses, the fetus would follow a maternal temperature drop of 2°C. If fetal cardiac output increases 50%, fetal oxygen consumption increases 100% and fetal skin blood flow drops to zero, fetal temperature would decrease 1.7°C instead of 2°C, a rather modest result of a strenuous fetal effort.

The tight coupling of heat exchange and oxygen supply in the placenta has been demonstrated experimentally in fetal sheep. Reduction of fetal placental blood flow to 25% of normal led to a fetal temperature increase of 0.13±0.02 °C, a small value understandable only if the reduced oxygen delivery and consumption and hence heat production, is taken into account (18).

We conclude that by means of the placenta the fetal temperature is clamped to the maternal temperature. An effort of the fetus to avoid changes of its own temperature will be mostly futile. Teleological thinking suggests that mimicking an ectothermic (poikilothermic) being in utero might be a good strategy for survival; it certainly prevents the rash burning of "fuel" (brown fat) which is urgently needed after birth (19). The following sections demonstrate that indeed effective fetal thermoregulatory actions do not occur in utero. The problem then arises how this system is subdued in utero and activated rapidly during birth.

Fetal thermoregulation

To study fetal responses to temperature changes, preparations have been developed in which coils of plastic tubing have been placed around the fetal body (4) or in the amniotic fluid (7). The tubes then can be perfused with cold water. The cooling power of this device is at least 50 W (7), it can be used for heating experiments also. To increase fetal pO_2, some fetuses have

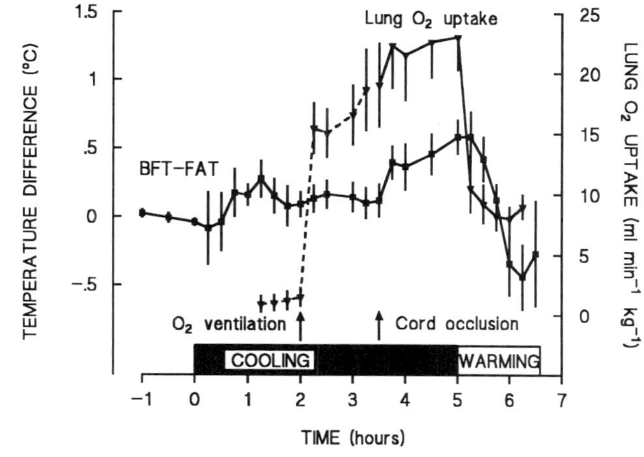

Fig. 3: Two indices of thermogenic activity during simulated birth of sheep in utero. The extent to which brown fat temperature exceeds arterial blood temperature (BFT-FAT), shown as squares, rises after cord occlusion. Pulmonary O_2 uptake, equal to whole body O_2 use after cord snare (triangles), increases threefold from baseline indicating powerful thermogenic responses. Both indices return rapidly to normal after cooling ceases. Average results ± SEM from 6 neart term fetal sheep.

been equipped with a tracheal tube which allowed artificial lung ventilation from the outside, and with a cord snare to interrupt umbilical blood flow. Combining cooling, ventilation and cord obstruction is known as "birth in utero" (20). Besides the increase of oxygen consumption (Figure 3), two other changes are taken to represent fetal thermogenesis: the increase of plasma free fatty acid or glycerol levels (see Figure 4), and, less compelling, the widening of the temperature difference between brown adipose tissue and fetal arterial blood during cooling (Figure 3).

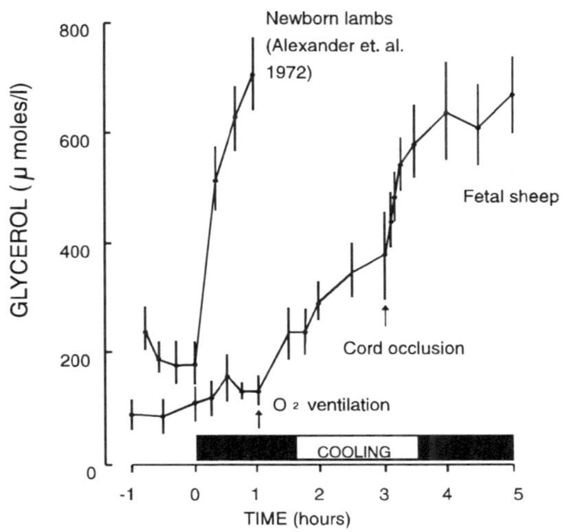

Fig. 4: Plasma glycerol, an indicator of nonshivering thermogenesis, rises rapidly and markedly in newborn lambs exposed to cold. However, it rises only negligibly in cooled fetal sheep, and only moderately after supplemental oxygen is given. Only after cord occlusion do responses approach newborn levels., a result that suggests an intrauterine inhibitor, possibly of placental origin.

Thermogenic responses to cooling

Thermogenic responses are brought about by shivering in skeletal muscle, by nonshivering thermogenesis (NST) in brown adipose tissue, and perhaps by other untested mechanisms. Their importance and strength vary widely in different species (21). In the newborn lamb nonshivering thermogenesis is a major contributor to heat production, accounting for roughly one half of the total response (22; Power, unpublished). In humans NST is active in the immediate newborn period. The contribution of shivering and nonshivering thermogenesis depends on species (piglets, for example, lack brown fat completely), maturity at birth, age, and possibly a number of other factors. The biochemical mechanism and physiology of these various means of facultative heat production are fundamentally different.

Shivering thermogenesis . During mild cold exposure muscle tone is increased, which changes during more severe cold stress to frank isotonic muscle contractions. When shivering is maximal, whole body oxygen consumption increases several fold, and heat production rises commensurately. Thus in adults a decrease of core temperature by 1 °C will increase heat production by 2 to 5 W kg^{-1} (23).

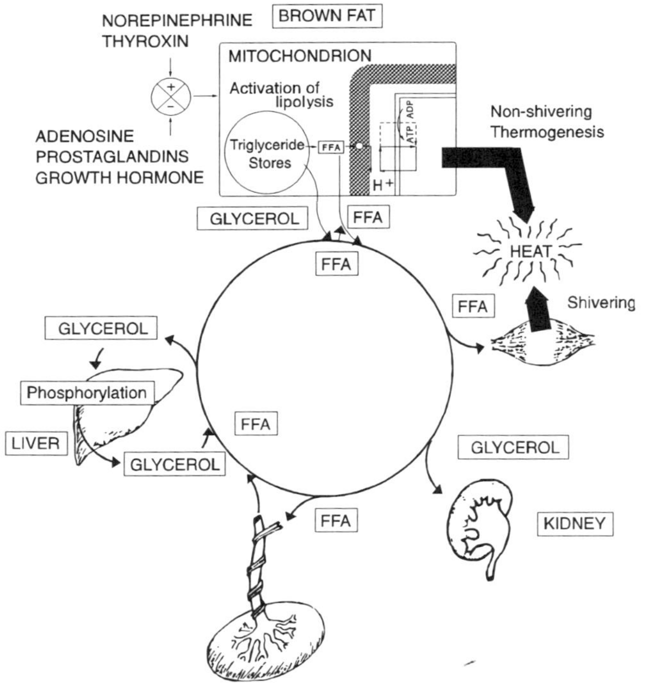

Fig. 5: An overview of thermogenic responses producing heat in the fetal sheep. Potential stimuli and facilitators of nonshivering thermogenesis are shown to be counterbalanced by a number of intrauterine inhibitors of lipolysis. When lipid stores are broken down, glycerol and fatty acids are released which provide substrate for increased oxygenation not only in brown fat, but also in shivering skeletal muscle. See Figure 3 and 4 for plasma levels of glycerol, oxygen consumption and temperature changes during cooling.

Nonshivering thermogenesis . Oxidising of brown tissue lipids is the most powerful mechanism for NST. Brown fat is located typically in perirenal and periaortic tissues, and between the scapulae. When fully stimulated by exposure to cold in the newborn period, brown adipose tissue, which makes up only about 1.5 per cent of body mass (sheep, humans) can release two- to fivefold more heat than the basal heat production of the entire body. Brown fat may release up to 0.5 W per gram (24), more power than any other tissue in the body can generate. When thermogenically active, brown fat receives a relatively large blood flow, about 22% of cardiac output (25), cf. Figure 6. This flow provides oxygen and substrate to the fat and carries away heat and some of the fatty acids which are released from lipid stores. As lipolysis proceeds some of the fatty acids released are oxidised within the brown fat itself, but others (an estimated 30-70 %) enter the blood and are transported throughout the body where they provide substrate for other organs. The dispersal of fatty acids makes energy reserves available for shivering as well as nonshivering thermogenesis.These metabolic shifts are summarised schematically in Figure 5.

Cooling is the principal stimulus that initiates nonshivering thermogenesis in the newborn. Increased sympathetic activity causes the release of norepinephrine from nerve endings that acts via adrenergic receptors of adipocytes, largely of type $ß_{3\text{-}a}$ newly discovered class of receptors that is less inhibited by propranolol than other ß-types (21), but also via α_1 receptors. The latter account for 10 to 40% of the total response (21; Power unpublished). Both $ß_3$ and α_1 responses are stimulatory, though they act by different intracellular pathways. The essential element of heat release in brown fat is the presence of a unique uncoupling protein found only in the inner mitochondrial membrane of brown fat cells. This protein, called thermogenin, permits protons to pass across the mitochondrial membrane without the need to conserve energy in the formation of high-energy chemical bounds. Thermogenin is not continuously active, rather it appears normally to be kept in an inhibited state by high cytosolic concentrations of ATP. Thus activation of nonshivering thermogenesis appears to depend on removal of the normal inhibition, but how this is accomplished is not known. There is evidence that the fatty acids themselves induce a shift in thermogenesis from the inhibited to the activated state (21).

Other mechanisms . Futile ion cycling in skeletal muscle, the mechanism that is responsible for malignant hyperthermia in adult humans, could play a role in newborn also. The basis of this mechanism is ATPase-dependent cycling of Ca2+ at the sarcoplasmic reticulum (27) with the consequent release of heat. Another possibility is heat production by smooth muscle contraction in blood vessel wall. Such a mechanism was recently demonstrated in the adult (28), but its role in the newborn period is unknown.

Control of heat dissipation . The effective mechanism of heat dissipation by evaporation is unavailable to the fetus. During cooling or warming, placental blood flow remains essentially constant (Figure 6).

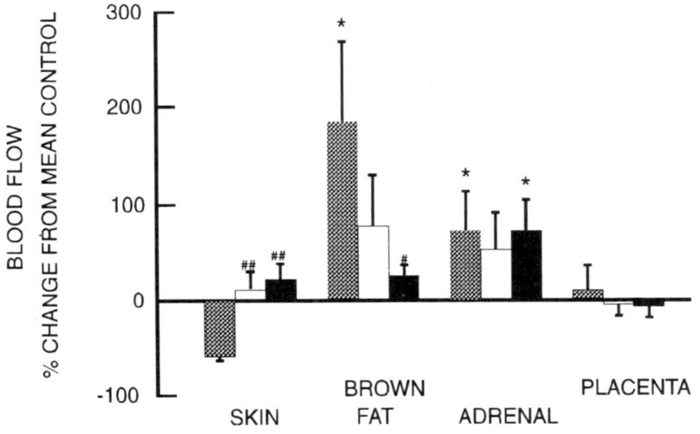

Fig. 6: Blood flow to various organs in 8 near-term fetal sheep during cooling (stippled bars), recovery from cooling (open bars), and warming (closed bars). Values are expressed as percent changes from mean control flows. Significant differences from control are indicated by *, and significant differences from cooling by # or ##.
From (25), with permission.

Model calculations show that variations of skin blood flow (Figure 6) have very little influence on fetal body temperature (0.03 °C, 17). Thus the fetus seems unable to control heat dissipation effectively.

Thermogenic responses to cooling after birth

When the newborn of a number of species including man is exposed to cold, strong thermogenic responses are evoked (1, 3, 22). In the newborn lamb, for instance, oxygen consumption and heat production typically increase three- to fivefold (22). These responses have been termed *summit metabolism*. Typically they begin within a few minutes after birth, reach their peak in 30 min, and persists for several hours if necessary, until energy reserves are depleted (22). They are essential for survival in species such as the sheep which bear their young in winter and early spring. Thermogenesis in the newborn period is impaired by hypoxia (29) and birth injury (30).

Lack of fetal thermoregulatory responses to intrauterine cooling

In marked contrast to newborn responses, fetal thermogenic responses to cooling are minimal. Even when the fetal temperature is decreased 2 °C using the cooling coil described earlier, the fetus produces little additional heat and NST is not induced (7, 31). This difference between newborn and fetus is shown strikingly in Figure 4. Nor does the fetus respond to ß-agonists or antagonist during intrauterine cooling (7), agents which are powerful stimuli and inhibitors of NST after birth.

Effective shivering is also minimal before birth because fetal temperature responses are not altered appreciably by neuromuscular blocking agents such as gallamine (4, 7), and neuromuscular blockade only reduces total heat production by about 10% in the cooled, hyperoxygenated fetal sheep (Power, unpublished).

It appears therefore that both shivering and nonshivering thermogenesis are largely quiescent in utero. The fetus will passively accept maternal temperature decreases as well as increases (Figure 1).

Why is fetal heat production impaired in utero?

The same late trimester sheep fetus that fails to respond to cooling when in utero will demonstrate intense thermogenic reactions when prematurely delivered. The puzzle has not yet been completely resolved, but explanations may follow two lines: one is that the specific intrauterine situation limits thermogenic responses, the other that these responses are actively subdued in utero.

Possible limitations of intrauterine thermogenesis . Immaturity of the sensory system or of the hypothalamic axis does not appear to be present because cooling leads to prompt cardiovascular responses, such as increases in blood pressure and increased sympathetic nerve activity elsewhere in the body (32). Nor does it appear to be inadequate differentiation of the necessary enzymes and biochemical machinery in fetal brown fat. This is clearly shown by the observation that samples of ovine brown fat do have the capacity for marked (about tenfold) increases in metabolic rate when stimulated with norepinephrine outside the body (33).

Brown fat metabolism has long been known to be suppressed at low oxygen levels, and to be stimulated by increased oxygen levels. For example, as arterial pO_2 falls below 30 mmHg (which is the upper limit in fetal sheep), nonshivering thermogenesis is impaired in infants and puppies (34, 35). However, when the fetal sheep is cooled and then ventilated with pure oxygen in utero, thermogenic responses (rise of plasma glycerol) are still only modest (about one-half maximal), as shown in Figure 4. They remain moderate despite the success of ventilation in raising fetal pO_2 to levels higher than are normally achieved in the newborn period. Yet after cord occlusion, thermogenesis and oxygen consumption do increase to higher levels, despite no further change in oxygenation. These observations suggest that increased oxygenation is necessary for maximal nonshivering thermogenesis during cooling but is not sufficient to account for the onset of maximal thermogenesis at birth.

A high blood flow is necessary to supply oxygen and to carry away heat produced by the brown fat, and thus blood flow may limit heat production. However, blood flow to brown fat does increase about threefold during intrauterine cooling in the fetal sheep (Figure 6), yet thermogenesis remains virtually undetectable. It seems likely therefore that blood flow responds to brown fat activity, rather than acting as a controlling factor.

Another possibility is lack of hormonal stimulation to brown fat, but it appears to be more than sufficient before birth. Intrauterine cooling, for example, leads to marked elevations in plasma norepinephrine (31, 32) and thyroxin (36) in the fetal sheep, yet brown adipose tissue remains largely quiescent. Even large infusions of T_3 into cooled and well-oxygenated fetal sheep do not initiate thermogenesis (37).

In conclusion, because there are appropriate responses in the fetal cardiovascular and endocrine systems in utero, and because the sensory and hypothalamic pathways for sensing and integration are differentiated well before birth, it appears that the thermogenic organs are receiving appropriate stimuli. Rather than being developmentally immature and unable to initiate thermogenesis, it appears that the fetus is actively prevented from doing so.

Possible intrauterine inhibitors of thermogenesis . To date, adenosine and the prostaglandins have been proposed as possible intrauterine inhibitors of thermogenesis. Both are powerful inhibitors of lipolysis in vitro and hence suppressers of nonshivering thermogenesis. Moreover, their concentrations are remarkably high before birth, and fall rapidly during the first 15 min after birth coincident with the onset of nonshivering thermogenesis.

Adenosine is an ubiquitous nucleotide with powerful antilipolytic properties. Its concentration in the fetal sheep is typically in the range of 1 to 1.5 uM, and well into range of antilipolytic (and cardiovascular) actions. Its concentration falls about 50% after cord occlusion in sheep whose birth is simulated in utero (38). This result suggests the placenta is a major source of adenosine. It supports but does not establish the hypothesis that declining adenosine concentrations enable the onset of NST in the immediate newborn period. Prostaglandins such as PGE_2 have antilipolytic actions also. Their concentration is elevated before birth and, like adenosine, decreases rapidly after birth. Significantly, NST can be induced in the cooled oxygenated fetal sheep if the synthesis of the prostaglandins is blocked by prolonged treatment with indomethacin (19). This result strongly supports an inhibitory role for the prostaglandins. Again, however, because additional responses occur after cord occlusion, the prostaglandins are not likely to be the only inhibitors of lipolysis before birth.

Whether other hormones also play a role in neonatal thermogenesis is largely unexplored. Other serious candidates may be growth hormone, the corticosteroids, and melatonin. Indeed, much redundancy in the control of thermogenesis seems likely, as is true for other systems essential for survival, such as blood pressure control. Redundancy among multiple stimuli and inhibitors may account for the rapid, sustainable, and powerful responses observed.

Conclusions

For the fetus, the protective maternal environment turns into a trap when the maternal body temperature changes. During maternal hyperthermia (fever, exercise, malignant hyperthermia, sauna bathing) the fetal temperature will inevitably rise, and the fetus may suffer from its own temperature increases as well as from the reactions of the maternal organism. The consequences will depend on species, gestational age (e.g. increased risk of malformations during organogenesis), duration and amount of temperature rises. The upper limit for short term heat exposure will be in the range of 43°C (39).

The predominant use of the placenta as "radiator" for the fetus implies that heat exchange is tightly coupled with other placental transfer functions. To avoid heat losses during maternal hypothermia (e.g. boat accidents) a distinct reduction of fetal placental blood flow would be necessary which at the same time would compromise the supply of oxygen.

Experiments mostly in fetal sheep have shown that fetal hypothermia does not induce thermoregulatory actions well known from the newborn, even when prematurely delivered. If this is true for the human fetus also, cold stress would be of minor importance because the fetus would not suffer from the consequences of a strong rise of its own metabolism which is so deleterious in adults (maternal reactions, however, may be dangerous for the fetus also). This might be reassuring in certain clinical situations, and for fetal surgery.

It is likely that a combination of factors prevents the activation of thermoregulatory responses in utero. The minimum requirement for activation after birth seems to be the separation from the placenta (with subsequent reduction of antilipolytic substances like adenosine of prostaglandins), the increase of catecholamines and the rise of fetal arterial pO_2 following the first breaths.

In 1885 Preyer (39) suggested that "no embryo possesses a thermoregulatory mechanism" which he thought would develop slowly after birth in "idiothermic" animals. Experimental evidence now demonstrates that he was largely correct.

Acknowledgements

The skilful assistance of R.Kirby and R.Christel in preparing the manuscript is gratefully acknowledged.

References

1. Brück K (1978): Heat production and temperature regulation. In: Perinatal Physiology, Stave U, ed. New York London:Plenum Publishing Corporation, 455-498.

2. Adamsons K Jr. (1966): The role of thermal factors in fetal and neonatal life. Pediatr Clin North Am 13, 599-619.

3. Alexander G, Nicol D, Thorburn G (1973): Thermogenesis in prematurely delivered lambs. In: Fetal and Neonatal Physiology, Comline R, Cross K, Dawes G, Nathanielsz P, eds. Cambridge UK:Cambridge University Press, 410-417.

4. Gunn TR, Gluckman PD (1983): Development of temperature regulation in the fetal sheep. J Develop Physiol 5, 167-179.

5. Cefalo RC, Hellegers AE (1978): The effects of maternal hyperthermia on maternal and fetal cardiovascular and respiratory function. Amer J Obst Gynec 131, 687-694.

6. Lotgering FK, Gilbert RD, Longo LD (1983): Exercise response in pregnant sheep: blood gases, temperatures, and fetal cardiovascular system. J Appl Physiol 55, 842-850.

7. Schröder HJ, Hüneke B, Klug A, Stegner H, Carstensen M, Leichtweiss H-P (1987): Fetal sheep temperatures in utero during cooling and application of triiodothyronine, norepinephrine, propranolol and suxamethonium. Pflügers Arch 410, 376-384.

8. Gilbert RD, Power GG (1986): Fetal and uteroplacental heat production in sheep. J Appl Physiol 61, 2018-2022.

9. Harned HS (1978): Respiration and the respiratory system. In: Perinatal Physiology, Stave U, ed. New York London:Plenum Publishing Corporation, 53-101.

10. Power GG, Schröder H, Gilbert RD (1984): Measurement of fetal heat production using differential calorimetry. J Appl Physiol 57, 917-922.

11. Abrams R, Caton D, Clapp J, Barron DH (1970): Thermal and metabolic features of life in utero. Clin Obstet Gynecol 13, 549-564.

12. Schröder HJ, Nissen B (1992): Heat conductances of human umbilical cord blood vessels. Placenta 13, A56.(Abstract)

13. Faber JJ, Thornburg KL (1983): Placental Physiology. New York:Raven Press.

14. Schröder HJ, Hatano H, Leichtweiss H-P (1989): Wärmeaustausch in der isolierten Meerschweinchenplacenta. Hamb Ärzteblatt 43, 410-412.

15. Gilbert RD, Schröder H, Kawamura T, Dale PS, Power GG (1985): Heat transfer pathways between fetal lamb and ewe. J Appl Physiol 59, 634-638.

16. Rudelstorfer R, Tabsh K, Khoury A, Nuwayhid B, Brinkmann III CR, Assali NS (1986): Heat flux and oxygen consumption of the pregnant uterus. Amer J Obst Gynec 154, 462-470.

17. Schröder H, Gilbert RD, Power GG (1988): Computer model of fetal-maternal heat exchange in sheep. J Appl Physiol 65, 460-468.

18. Schröder HJ, Power GG (1993): Raise of fetal core temperature by reduction of umbilical blood flow in instrumented fetal sheep. Pflügers Arch 422 (suppl. 1), R85.(Abstract)

19. Gunn TR, Ball KT, Gluckman PD (1993): Withdrawal of placental prostaglandins permits thermogenic responses in fetal sheep brown adipose tissue. J Appl Physiol 74, 998-1004.

20. Power GG, Kawamura T, Dale PS, Schröder H, Gilbert RD (1986): Temperature responses following ventilation of the fetal sheep in utero. J Develop Physiol 8, 477-484.

21. Nedergaard J, Cannon B (1992): Brown adipose tissue: development and function. In: Fetal and Neonatal Physiology, Polin RA, Fox WW, eds. Philadelphia:Saunders.

22. Alexander G, Williams D (1968): Shivering and non-shivering thermogenesis during summit metabolism in young lamb. J Physiol Lond 198, 251-276.

23. Necker R (1984): Central thermosensitivity: CNS and extra-CNS. In: Thermal Physiology, Hales JRS, ed. New York:Raven Press, 53-61.

24. Rothwell NJ, Stock MJ (1984): Nonshivering and diet-induced thermogenesis: the role of brown adipose tissue. In: Thermal Physiology, Hales JRS, ed. New York:Raven Press, 145-153.

25. Kawamura T, Gilbert RD, Power GG (1986): Effect of cooling and heating on the regional distribution of blood flow in fetal sheep. J Develop Physiol 8, 11-21.

26. Cannon B, Nedergaard J (1982): The function and properties of brown adipose tissue in the newborn. In: The biochemical development of the fetus and neonate, Jones CT, ed. Amsterdam New York:Elsevier, 697-730.

27. Block BA (1987): Billfish brain and eye heater: a new look at nonshivering heat production. NIPS 2, 208-213.

28. Colquhoun EQ, Hettiarachi M, Ye J-M, Rattigan S, Clark MG (1990): Inhibition by vasodilators of noradrenaline and vasoconstrictor-mediated, but not skeletal muscle contraction-induced uptake in the perfused rat hindlimb: implications for non-shivering thermogenesis in muscle tissue. Gen Pharmac 21, 141-148.

29. Hill JR (1959): The oxygen consumption of newborn and adult mammals. J Physiol (Lond) 149, 346-373.

30. Scopes JE, Ahmed I (1966): Indirect assessment of oxygen requirements in newborn babies by monitoring deep body temperature. Arch Dis Childh 41, 25-33.

31. Power GG, Gunn TR, Johnston BM, Gluckman PD (1987): Oxygen supply and the placenta limit thermogenic responses in fetal sheep. J Appl Physiol 63, 1896-1901.

32. Gunn TR, Johnston BM, Iwamoto HS, Fraser M, Nicholls MG, Gluckman PD (1985): Haemodynamic and catecholamine responses to hypothermia in the fetal sheep in utero. J Develop Physiol 7, 241-249.

33. Klein AH, Reviczky A, Padbury JF (1984): Thyroid hormones augment catecholamine-stimulated brown adipose tissue thermogenesis in the ovine fetus. Endocrinology 114, 1065-1069.

34. Baum D, Anthony CL Jr, Stowers C (1971): Impairment of cold-stimulated lipolysis by acute hypoxia. Am J Dis Child 121, 115-119.

35. Silva JE, Larsen PR (1986): Interrelationship among thyroxine, growth hormone, and the sympathetic nervous system in the regulation of 5'-iodothyronine deiodonase in rat brown adipose tissue. J Clin Invest 77, 1214-1223.

36. Fraser M, Gunn TR, Butler JH, Johnston BM, Gluckman PD (1985): Circulating thyrotropin (TSH) in the ovine fetus: evidence for pulsatile release and the effect of hypothermia in utero. Pediatr Res 19, 208-212.

37. Power GG, Gunn TR, Johnston BM, Nichols G, Gluckman PD (1989): Umbilical cord occlusion but not increased plasma T3 or norepinephrine stimulate brown adipose tissue thermogenesis in the fetal sheep. J Develop Physiol 11, 171-177.

38. Sawa R, Asakura H, Power GG (1991): Changes in plasma adenosine during simulated birth of fetal sheep. J Appl Physiol 70, 1524-1528.

39. Preyer WT (1885): Specielle Physiologie des Embryo. Leipzig:Th.Grieben (L.Fernau).

Thermal Balance in Health and Disease
Advances in Pharmacological Sciences

THERMOREGULATION AND HEAT BALANCE: GENERAL ANESTHESIA

D.I. Sessler

Thermoregulation Research Laboratory, University of California, San Francisco, California 94143-0648.

The thermoregulatory system usually maintains core body temperature within 0.2°C of "normal," which is near 37°C in humans. Yet, perioperative hypothermia is common due to anesthetic-induced inhibition of thermoregulation combined with exposure to a cold operating room environment. Hypothermia causes a number of complications including shivering, prolonged duration of drug action, coagulopathies, cardiac ischemia, and decreased resistance to surgical wound infections. It can be prevented and treated in a variety of ways. *All data are presented as means ±SD.*

DEFINITIONS AND THERMOREGULATORY MODEL

Temperature is regulated by central structures which compare integrated thermal inputs from the skin surface, neuroaxis, and deep tissues with reference temperatures for each thermoregulatory response (1). Input temperature outside these values initiate regulatory responses. The core temperature triggering an afferent thermoregulatory responses defines the *threshold* for that parameter. The slope of response intensity *vs.* the difference between thermal input and the threshold temperature is the *gain* of that response. Response intensity that no longer increases with further deviation in core temperature identifies the *maximum intensity*.

Control of autonomic responses is roughly 70-80 percent determined by thermal input from core structures (2). In contrast, a large fraction of the input controlling behavioral responses is derived from the skin surface. The *interthreshold range* (core temperatures *not* triggering autonomic thermoregulatory responses) is bounded by the sweating threshold at its upper end and by vasoconstriction at the lower end. In humans the range is only 0.2°C (fig. 1) (3).

THERMOREGULATION DURING GENERAL ANESTHESIA

Behavioral regulation is not relevant during general anesthesia since patients are unconscious and frequently paralyzed. All general anesthetics so far tested markedly impair normal autonomic thermoregulatory control.

Cold responses

Vasoconstriction: Halothane (4) and enflurane (5) decrease the vasoconstriction threshold 2-4°C from its normal value near 37°C. Isoflurane also decreases the threshold, and the decrease is proportional to end-tidal concentration: Threshold (°C) = 37.1 - 3.1 [Isoflurane] (6). Nitrous oxide/fentanyl anesthesia (7) and nitrous oxide/propofol (8) anesthesia also significantly decrease the vasoconstriction threshold.

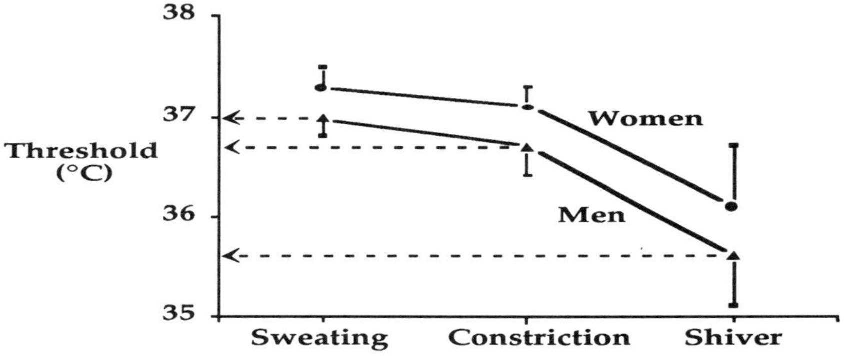

Fig. 1: The thresholds for sweating, vasoconstriction, and shivering in unanesthetized men and women. Skin temperature was kept constant throughout, and the core was cooled by central venous infusion of cold fluid. Temperatures between sweating and vasoconstriction define the intertheshold range, which is ≈0.2°C. All the thresholds were ≈0.3°C greater in women than in the men. [Data from (3), with permission.]

Thermoregulatory vasoconstriction is comparably impaired in infants, children, and adults given halothane (9) or isoflurane (10) anesthesia. In contrast, the vasoconstriction threshold is about 1°C less in patients aged 60-80 years than in those between 30 and 50 years old (fig. 2) (11). Painful stimulation slightly increases vasoconstriction thresholds (5); consequently, thresholds will be somewhat less when surgical pain is prevented by simultaneous local or regional anesthesia. Vasoconstriction is more-or-less an "on/off" response, making it difficult to evaluate gain. Maximum response intensity, however, is well preserved during isoflurane anesthesia (12).

Nonshivering Thermogenesis: An increase in metabolic rate independent from muscular activity does not occur in anesthetized adults (13), which is not terribly surprising because this response is not particularly important in unanesthetized adults (14). In anesthetized infants, however, nonshivering thermogenesis remains an important protection against hypothermia and can double metabolic heat production.

Shivering: Shivering is rare during surgical doses of general anesthesia, suggesting that general anesthesia decreases the shivering threshold far more than the vasoconstriction threshold. Consistent with this theory, low concentrations of nitrous oxide significantly decrease the shivering threshold (15). Shivering is common, however, during postoperative recovery when brain anesthetic concentrations decrease rapidly (16).

Two electromyographic patterns contribute to involuntary postoperative muscular activity: 1) a tonic pattern resembling normal shivering, typically having a 4-8 cycle/minute waxing-and-waning component; and, 2) a phasic, 5-7 hertz bursting pattern resembling pathological clonus. Both are thermoregulatory, that is, always preceded by core hypothermia and arterio-venous shunt vasoconstriction (17). The tonic pattern consistently demonstrates the 4-8 cycle/minute waxing-and-waning pattern of normal shivering (18), and apparently is a simple thermoregulatory response to intraoperative hypothermia. In contrast, the clonic pattern is not a normal component of thermoregulatory shivering and appears specific to recovery from volatile anesthetics. While the precise etiology of this tremor pattern remains unknown, it likely results from anesthetic-induced disinhibition of normal descending control over spinal reflexes.

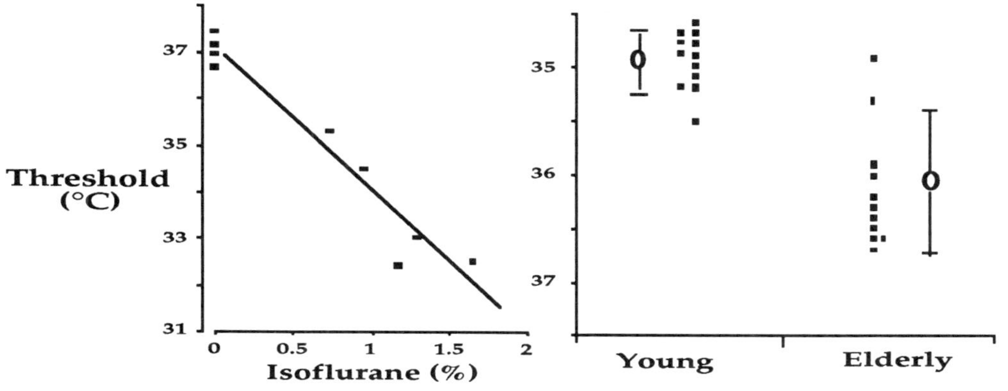

Fig. 2: Left: The thermoregulatory threshold in four unanesthetized volunteers and five surgical patients given isoflurane/air/oxygen anesthesia. The threshold was inversely correlated with end-tidal isoflurane concentration. [From (6), with permission.]
Right: The vasoconstriction threshold was significantly less in the elderly (33.9 ± 0.6°C) than in younger patients (35.1 ± 0.3°C). Filled squares indicate the vasoconstriction threshold in each patient; the open ovals show the mean and standard deviations in each group. [From (11), with permission.]

Maintaining normothermia will prevent most postanesthetic tremor. Additionally, shivering can be treated by skin-surface warming because the regulatory system tolerates more core hypothermia when cutaneous warm input is augmented (19). And finally, postanesthetic shivering also can be treated using a variety of drugs, including meperidine (25 mg, iv), clonidine (75 µg, iv), and ketanserin (10 mg, iv) (20). All opioids probably impair thermoregulation to some extent.

However, meperidine reportedly is considerably more effective in treating shivering than equi-analgesic doses of other μ agonists. The efficacy of meperidine may be mediated in part by its kappa agonist activity (21). The specific mechanism by which ketanserin stops shivering remains unknown. Similarly, how clonidine arrests shivering also remains unknown, but the drug comparably reduces the vasoconstriction and shivering thresholds (22), suggesting it acts on the central thermoregulatory system, rather than peripherally preventing shivering.

Warm responses

Sweating: The sweating threshold is increased approximately 1°C by one MAC of isoflurane, and the dose-response curve is linear (23). Sweating is comparably impaired by enflurane anesthesia and there is no direction-dependent hysteresis (24). That is, sweating rate during enflurane anesthesia depends only on skin and core temperature, not the direction in which core temperature is changing. Interestingly, the sweating threshold remains greater in women than men, even during anesthesia (23). The gain and maximum intensity of sweating are well preserved during both isoflurane and enflurane anesthesia (fig. 3).

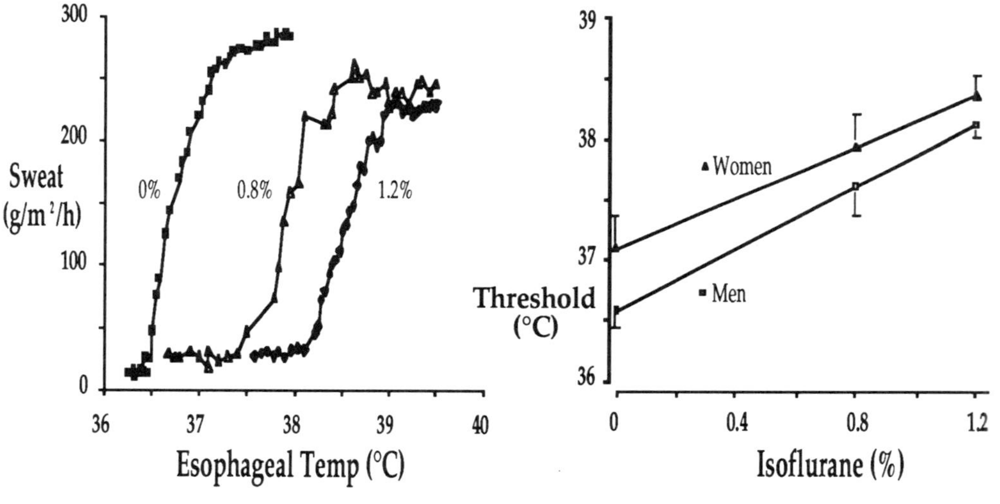

Fig. 3: Sweating rate as a function of core temperature in one man is shown on the left. Although the sweating threshold was increased by isoflurane, gain and maximum response intensity remained relatively well preserved. The threshold for sweating was greater in women than men, and the difference was preserved during anesthesia. Isoflurane increases the sweating threshold only one third as much as it decreases the vasoconstriction threshold. [From (23), with permission.]

Active Vasodilation: The threshold for active vasodilation also is increased by general anesthesia, and the increase appears to be similar to that of sweating (24,25). As expected from their mechanisms, both sweating and active vasodilation are prevented by nerve block.

Overall effects of general anesthesia

The mechanism(s) by which drugs produce anesthesia remains disputed (as does even the definition of "anesthesia"). The volatile anesthetics (*e.g.*, halothane, isoflurane) are chemically similar. However, other anesthetics (*e.g.*, ketamine, propofol) differ significantly. Nonetheless, the thermoregulatory effects of various general anesthetics are more notable for being similar than different. For example, routine surgical doses of most anesthetics reduce the vasoconstriction threshold from a normal value near 37°C to ≈33-35°C.

Anesthetic-induced impairment has a specific form: warm-response thresholds are elevated whereas cold-response thresholds are reduced. Consequently, the interthreshold range is increased from its normal values near 0.2°C to ≈4°C (fig. 4). The gain of most responses, however, remains normal, as do maximum response intensities. The result is a broad range of temperatures over which anesthetized patients are poikilothermic: within this range core temperature will vary passively depending largely on the environment. For additional detail and references, see (26).

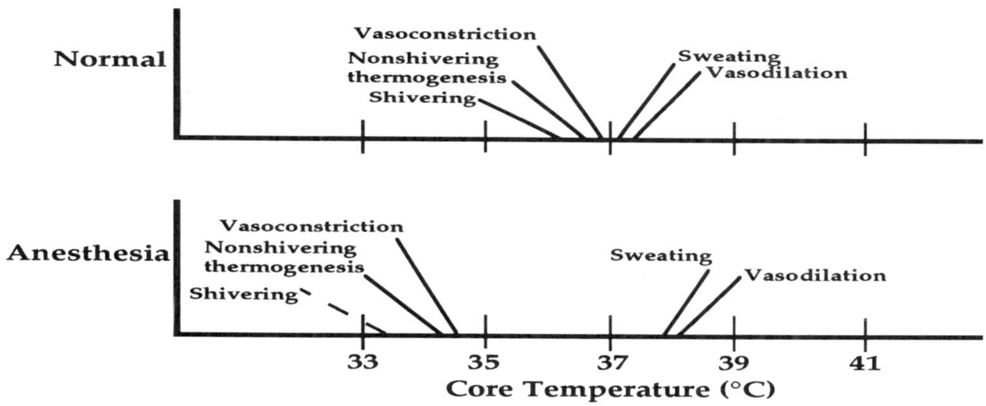

Fig. 4: The slanted lines represent different regulatory responses and the dark horizontal lines show core body temperature; the intersection of each line with the temperature scale is the threshold. The interthreshold range is shown as the distance between the first cold response (vasoconstriction) and the first warm response (sweating). During general anesthesia (bottom part of figure), the thresholds for vasoconstriction and nonshivering thermogenesis are shifted down to ≈ 34.5°C (depending on anesthetic type and dose). Similarly, the thresholds for active precapillary vasodilation and sweating are increased ≈1°C. The interthreshold range thus increases from ≈0.2°C to ≈4°C. [From Temperature Monitoring, In Anesthesia, 4rd edition. Miller RD Ed. New York, Churchill Livingstone, in press, with permission.]

INADVERTENT HYPOTHERMIA DURING GENERAL ANESTHESIA

Inadvertent hypothermia during anesthesia is by far the most common perioperative thermal disturbance. Hypothermia results from a combination of anesthetic-impaired thermoregulation and exposure to a cold operating room environment.

Patterns of intraoperative hypothermia

Hypothermia during general anesthesia develops with a characteristic pattern. An initial rapid decrease in core temperature is followed by a slow, linear reduction in core temperature. Finally, core temperature stabilizes, and subsequently remains virtually unchanged; each phase has its own etiology.

Redistribution Hypothermia: Core temperature usually decreases 0.5-1.5°C during the first hour after induction of general anesthesia (27). Induction of anesthesia decreases metabolic heat production only about 20 percent, a reduction insufficient to explain the observed core hypothermia. Most general anesthetics are direct vasodilators and all impair central thermoregulatory control, thus inhibiting normal tonic thermoregulatory vasoconstriction (4-8). Nonetheless, anesthetic-induced vasodilation only slightly increases cutaneous heat loss (28), suggesting that increased heat loss is not the major cause of hypothermia immediately following induction of anesthesia (fig. 5).

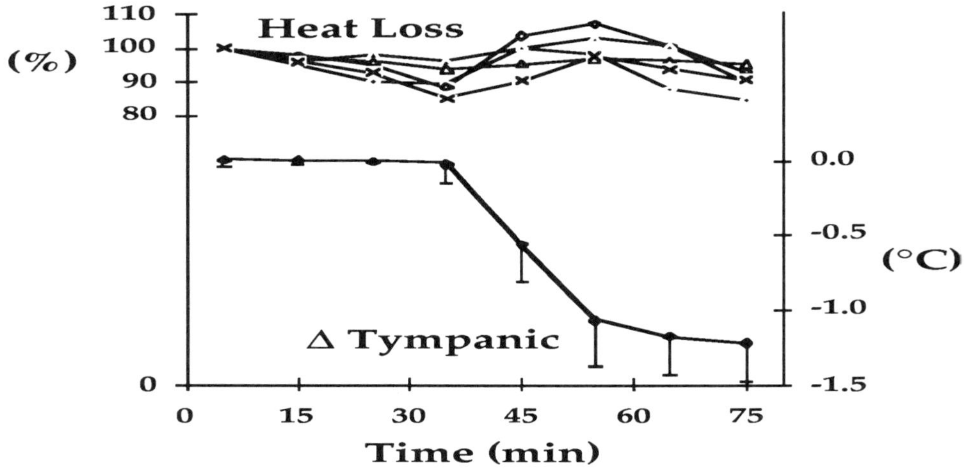

Fig. 5: Tympanic membrane temperature remained constant during the control period, but then decreased 1.2 ± 0.2°C during the 50 minutes after induction of anesthesia. Changes in total cutaneous heat loss were minimal during the control period (0-30 minutes) and during isoflurane administration (30-80 minutes). It already is established that induction of isoflurane anesthesia decreases metabolic heat production only ≈20 percent. Accordingly, these data indicate that the initial core hypothermia following induction of anesthesia results from a core-to-peripheral, internal redistribution of body heat. [From (28), with permission.]

This apparent paradox — dramatic core hypothermia *not* resulting only from decreased metabolic heat production or increased cutaneous heat loss — is explained by altered distribution of heat within the body. Heat usually is not evenly distributed within the body because tonic thermoregulatory vasoconstriction typically maintains a significant core-to-peripheral temperature gradient. Although induction of general anesthesia only slightly increases cutaneous heat loss, anesthetic-induced vasodilation does allow redistribution of heat from the core thermal compartment to peripheral tissues (28). This process markedly decreases core temperature, but *mean* body temperature and body heat content remain unchanged.

Linear Second Phase: Following initial redistribution hypothermia, core temperature usually decreases in a slow, linear fashion for 2-3 hours. This reduction results simply from heat loss exceeding metabolic heat production (27).

Core Temperature Plateau: After 3-4 hours of anesthesia, core temperature usually reaches a plateau and remains virtually constant for the duration of surgery. The core temperature plateau may simply represent a thermal steady state (heat production equaling heat loss) in patients remaining relatively warm. But perhaps more commonly, the plateau phase is associated with peripheral thermoregulatory vasoconstriction triggered by core temperatures 33-35°C. Thermoregulatory vasoconstriction during anesthesia significantly decreases cutaneous heat loss (12), but the decrease alone may be insufficient to produce a thermal steady state. Evidence suggests that an additional factor is constraint of metabolic heat to the core thermal compartment. In this scenario, distribution of metabolic heat (which largely is produced centrally) is restricted to the core compartment, maintaining its temperature. Peripheral tissue temperature, in contrast, continues to decrease because it is no longer being supplied with sufficient heat from the core (fig. 6) (29).

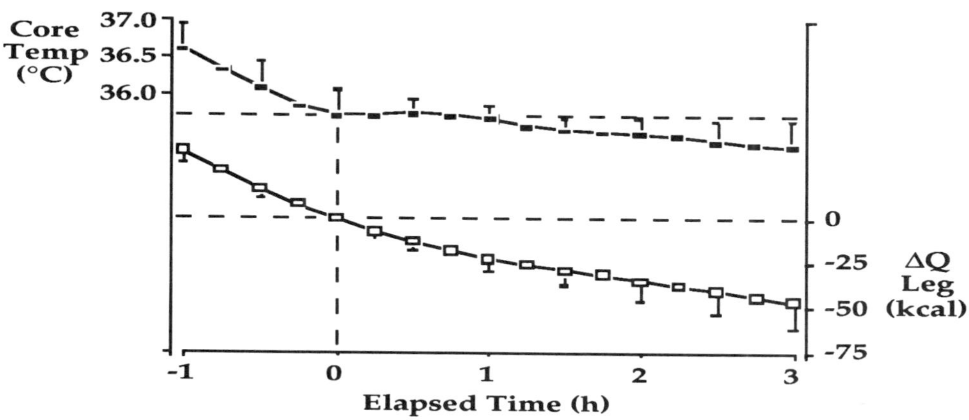

Fig. 6: Core temperature, which was decreasing 1°C/hour before thermoregulatory vasoconstriction, decreased only 0.4°C in the subsequent three hours. Leg heat content, however, continued to decrease throughout the core temperature plateau, indicating that body heat content continued to decrease. These data indicate that the core temperature plateau associated with thermoregulatory vasoconstriction results both from decreased cutaneous heat loss and from re-establishment of the normal core-to-peripheral tissue temperature gradient. Zero elapsed hours indicates the time of vasoconstriction. [From (29), with permission.]

A core temperature plateau resulting from thermoregulatory vasoconstriction thus is not a thermal steady state because body heat content continues to decrease.

Consequences of mild intraoperative hypothermia

Benefits: Substantial protection against ischemia and hypoxia is provided by just 2-3°C hypothermia (30). Protection initially was thought to result from the ≈8 percent/°C reduction in tissue metabolic rate. However, the efficacy of mild hypothermia far exceeds that of treatments such a high-dose isoflurane or barbiturate coma which comparably reduce metabolic rate (31). These data suggest that other actions (*e.g.*, decreased release of excitatory amino acids) explain the protective action of hypothermia. An additional benefit of mild hypothermia is that it makes acute malignant hyperthermia more difficult to trigger. Furthermore, once triggered, the syndrome is less severe (32).

Adverse Effects: Drug metabolism is markedly decreased by perioperative hypothermia. The duration of action of vecuronium is more than doubled by 2°C of core hypothermia, and the prolongation is a pharmacokinetic effect, not a pharmacodynamic one (33). Atracurium duration is less dependent on core temperature: a 3°C reduction in core temperature increases the duration of muscle relaxation only by 60 percent. During a constant infusion of propofol, plasma concentration is ≈30 percent greater than normal when individuals are 3°C hypothermic.
Coagulation is impaired by mild hypothermia. The most important factor appears to be a cold-induced defect in platelet function (34). Interestingly, the platelet function defect is related to local skin temperature, not core temperature. Enzymatic clotting probably also is highly temperature dependent. However, the importance of impaired coagulation has yet to be tested in surgical patients.

Wound infections are the most common serious complication of anesthesia and surgery, probably causing more morbidity than all anesthetic complications combined. Hypothermia can contribute to wound infections both by directly impairing immune function (35) and by triggering thermoregulatory vasoconstriction that in turn decreases wound oxygen delivery (36). Mild hypothermia, maintained only during anesthesia, impairs resistance to the both *E. coli* and *S. Aureus* dermal infections in guinea pigs (37). Preliminary data indicate that mild intraoperative hypothermia also aggravates clinical wound infections and delays hospital discharge in patients undergoing colon surgery. Consistent with poor wound healing, urinary nitrogen excretion remains elevated for several postoperative days in patients allowed to become hypothermic during surgery (38). Thermal comfort is markedly impaired by postoperative hypothermia (17). Patients, asked years after surgery, often identify feeling cold in the immediate postoperative period the worst part of their hospitalization — sometimes rating it worse than surgical pain. It is likely that postoperative thermal discomfort is also physiologically stressful, elevating blood pressure, heart rate, and plasma catecholamine concentrations. Consistent with increased autonomic nervous system activation, mild intraoperative hypothermia appears to increase the incidence of postoperative myocardial ischemia (39).

Maintaining Intraoperative Normothermia

Most patients are poikilothermic during surgery because they do not become sufficiently hypothermic to trigger thermoregulatory responses (4-8). Minimizing intraoperative hypothermia therefore depends on techniques that limit cutaneous heat loss to the environment, evaporation from surgical incisions, and conductive cooling produced by administration of cold intravenous fluids.

Preventing redistribution hypothermia

Redistribution is difficult to treat (27), but can be prevented. Skin-surface warming before induction of anesthesia only slightly alters core temperature (which remains well regulated), but does increase body heat content and reduce the normal core-to-peripheral temperature gradient. When peripheral tissue temperature is sufficiently increased, subsequent inhibition of normal tonic thermoregulatory vasoconstriction produces little redistribution hypothermia because heat only can flow down a temperature gradient (40,41).

An alternative to pre-warming is pharmacological vasodilation induced well before induction of anesthesia. Vasodilation does redistribute heat from the core to peripheral tissues (reducing the core-to-peripheral temperature gradient), but core temperature remains well regulated in unanesthetized patients. Subsequent induction of anesthesia then provokes little redistribution hypothermia (fig. 7) (42).

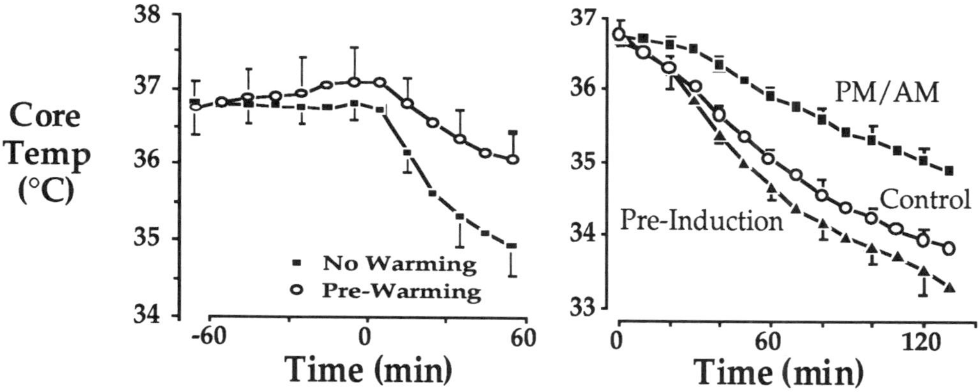

Fig. 7: Left: During the pre-induction period (-120 to 0 minutes), volunteers were either actively warmed or passively cooled (no warming). At induction of anesthesia (time = 0 minutes), active warming was discontinued and volunteers were exposed to the ambient environment. During the 60 minutes following induction of anesthesia, core temperature decreased significantly less when volunteers were pre-warmed. ($\Delta T = -1.1 \pm 0.3$°C) compared to when the same volunteers were not warmed ($\Delta T = -1.9 \pm 0.3$°C). [From (40), with permission.]
Right: Intraoperative core temperature decreased most in patients given nifedipine only immediately before induction of anesthesia (Pre-Induction) and least in those given nifedipine both the night before and the morning of surgery (PM/AM). [From (42), with permission.]

Inspiratory gas and intravenous fluid heating

Simple thermodynamic calculations indicate that less than 10 percent of metabolic heat production is lost *via* the respiratory tract. Because little heat is lost *via* respiration, even active airway heating and humidification minimally influences core temperature (27). Respiratory heat transfer maintains core temperature slightly better than mean body temperature because the heat is transferred directly into the core thermal compartment. However, the influence of respiratory gas heating and humidification at steady state remains trivial.

Heat loss due to cold intravenous fluids becomes significant when large amounts of crystalloid solution or blood are administered. One unit of refrigerated blood or one liter of crystalloid solution administered at room temperature decreases mean body temperature ≈0.25°C; warming fluids prevents these losses.

Cutaneous warming

The easiest method of decreasing cutaneous heat loss is to apply passive insulation to the skin surface. Insulators readily available in most operating rooms include cotton blankets, surgical drapes, plastic sheeting, and reflective composites ("space blankets"). A single layer of each reduces heat loss approximately 30 percent, and there are no clinically important differences among the insulation types (43). Adding additional layers of insulation further reduces heat loss only slightly. For example, one cotton blanket reduces heat loss about 30 percent but three cotton blankets reduces heat loss only by 50 percent (44).

Passive insulation alone rarely is sufficient to maintain normothermia in patients undergoing large operations: active warming will be required in those cases. Since about 90 percent of metabolic heat is lost *via* the skin surface, only cutaneous warming will transfer sufficient heat to prevent hypothermia. Circulating-water mattresses are nearly ineffective (45), presumably because little heat is lost from the back into the foam insulation covering most operating room tables. The most effective perianesthetic warming system is forced-air (46). Among the commercially available systems, the Bair Hugger® (Augustine Medical, Inc.) is best. Forced-air warming has been shown to maintain normothermia even during the largest operations and is superior to circulating-water (fig. 8) (47).

Just as thermoregulatory vasoconstriction is effective in preventing loss of centrally generated metabolic heat to peripheral tissues, it also decreases transfer of cutaneously applied heat to the core (48). Skin-surface warming is, therefore, far more effective when patients are vasodilated. From a practical point of view, this means that it is easier to maintain intraoperative normothermia (when most patients are vasodilated) than rewarm them postoperatively (when virtually all hypothermic patients are vasoconstricted).

SUMMARY

Thermoregulatory impairment induced by general anesthesia has a specific form: warm-response thresholds are elevated (23,24) whereas cold-response thresholds are reduced (4-8). Consequently, the interthreshold range is increased from its normal value near 0.2°C (3), to ≈4°C. Hypothermia during general anesthesia develops with a characteristic pattern. An initial rapid decrease in core temperature results from an internal core-to-peripheral redistribution of body heat (28). Subsequently, heat loss exceeding metabolic heat production produces a slow, linear reduction in core temperature (27). Finally, core temperature stabilizes, and subsequently remains virtually unchanged. This core temperature plateau usually results when thermoregulatory vasoconstriction decreases cutaneous heat loss (12) and constrains metabolic heat to the core thermal compartment (29).

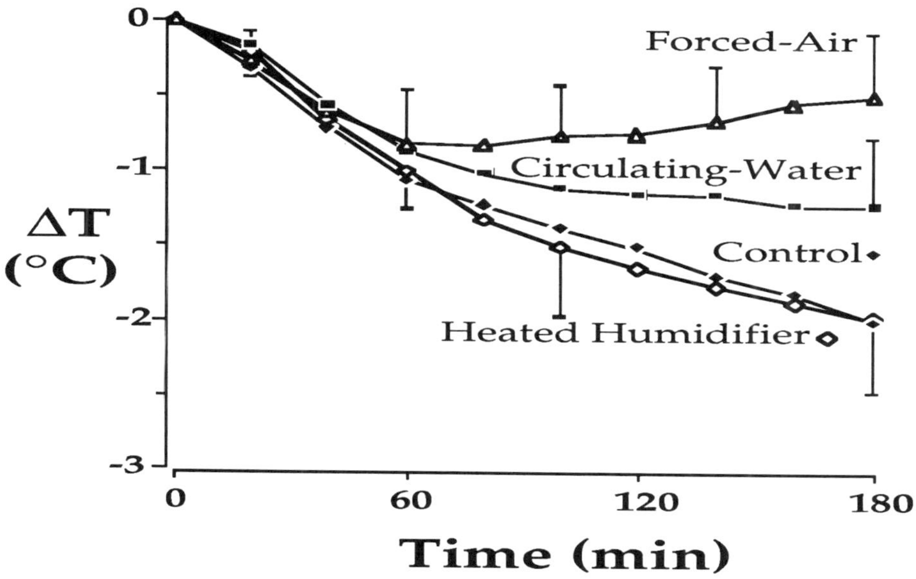

Fig. 8: Three different active warming devices were applied immediately following induction of anesthesia. Redistribution uniformly decreased tympanic membrane temperature during the first hour after induction of anesthesia. Tympanic membrane temperature increased over the remaining two hours in patients receiving forced-air warming, but remained nearly constant in those lying on a warmed circulating-water blanket. Patients in the control and heated humidifier groups continued to become more hypothermic throughout the operation. [From (27), with permission.]

Major adverse effects of intraoperative hypothermia include prolonged duration of drug action (33), impaired coagulation (34), reduced resistance to surgical wound infections (37), postoperative shivering (16,17), myocardial ischemia (39), and decreased postoperative thermal comfort (17).

An effective method of reducing intraoperative hypothermia is to minimize redistribution by pre-warming (40,41) or drug-induced vasodilation (42). Most clinically available passive insulators reduce cutaneous heat loss comparably, by ≈30% (43) and adding additional layers helps only little (43). Among the clinically available active systems, forced-air is by far the most effective (44) and can maintain normothermia even during the largest operations (47).

ACKNOWLEDGMENTS

This review is based on Dr. Sessler's chapter in *Anesthesia* (4rd edition. Miller RD Ed. New York, Churchill Livingstone, in press). Most of the described studies were supported by the Augustine Medical, Inc., and National Institutes of Health grants GM 39723, and GM 49670.

REFERENCES

1. Simon, E, Pierau, F, Taylor, DCM (1986): Central and peripheral thermal control of effectors in homeothermic temperature regulation. Physiol Rev 66, 235-300.
2. Tam, H, Darling, RC, Cheh, H, Downey, JA (1978): The dead zone of thermoregulation in normal and paraplegic man. Can J Physiol Pharmacol 56, 976-983.
3. Lopez, M, Sessler, DI, Walter, K, Emerick, T, Ozaki, M Rate- and gender-dependence of the sweating, vasoconstriction, and shivering thresholds in humans. Anesthesiology, in press
4. Sessler, DI, Olofsson, CI, Rubinstein, EH, Beebe, JJ (1988): The thermoregulatory threshold in humans during halothane anesthesia. Anesthesiology 68, 836-842.
5. Washington, DE, Sessler, DI, McGuire, J, Hynson, J, Schroeder, M, Moayeri, A (1992): Painful stimulation minimally increases the thermoregulatory threshold for vasoconstriction during enflurane anesthesia in humans. Anesthesiology 77, 286-290.
6. Støen, R, Sessler, DI (1990): The thermoregulatory threshold is inversely proportional to isoflurane concentration. Anesthesiology 72, 822-827.
7. Sessler, DI, Olofsson, CI, Rubinstein, EH (1988): The thermoregulatory threshold in humans during nitrous oxide-fentanyl anesthesia. Anesthesiology 69, 357-364.
8. Hynson, JM, Sessler, DI, Belani, K, Washington, D, McGuire, J, Merrifield, B, Schroeder, M, Moayeri, A, Crankshaw, D, Hudson, S (1992): Thermoregulatory vasoconstriction during propofol/nitrous oxide anesthesia in humans: Threshold and Sp_{O_2}. Anesth Analg 75, 947-952.
9. Bissonnette, B, Sessler, DI (1992): Thermoregulatory thresholds for vasoconstriction in pediatric patients anesthetized with halothane or halothane and caudal bupivacaine. Anesthesiology 76, 387-392.
10. Bissonnette, B, Sessler, DI (1990): The thermoregulatory threshold in infants and children anesthetized with isoflurane and caudal bupivacaine. Anesthesiology 73, 1114-1118.
11. Kurz, A, Plattner, O, Sessler, DI, Huemer, G, Redl, G, Lackner, F (1993): The threshold for thermoregulatory vasoconstriction during nitrous oxide/isoflurane anesthesia is lower in elderly than young patients. Anesthesiology 79, 465-469.
12. Sessler, DI, Hynson, J, McGuire, J, Moayeri, A, Heier, T (1992): Thermoregulatory vasoconstriction during isoflurane anesthesia minimally decreases heat loss. Anesthesiology 76, 670-675.
13. Hynson, JM, Sessler, DI, Moayeri, A, McGuire, J (1993): Absence of nonshivering thermogenesis in anesthetized humans. Anesthesiology 79, 695-703.
14. Jessen, K (1980): An assessment of human regulatory nonshivering thermogenesis. Acta Anaesthesiol Scand 24, 138-143.
15. Passias, TC, Mekjavic, IB, Eiken, O (1992): The effect of 30% nitrous oxide on thermoregulatory responses in humans during hypothermia. Anesthesiology 76, 550-559.

16. Just, B, Delva, E, Camus, Y, Lienhart, A (1992): Oxygen uptake during recovery following naloxone. Anesthesiology 76, 60-64.

17. Sessler, DI, Rubinstein, EH, Moayeri, A (1991): Physiological responses to mild perianesthetic hypothermia in humans. Anesthesiology 75, 594-610.

18. Israel, DJ, Pozos, RS (1989): Synchronized slow-amplitude modulations in the electromyograms of shivering muscles. J Appl Physiol 66, 2358-2363.

19. Sharkey, A, Lipton, JM, Murphy, MT, Giesecke, AH (1987): Inhibition of postanesthetic shivering with radiant heat. Anesthesiology 66, 249-252.

20. Joris, J, Banache, M, Bonnet, F, Sessler, DI, Lamy, M (1993): Clonidine and ketanserin both are effective treatments for postanesthetic shivering. Anesthesiology 79, 532-539.

21. Kurz, M, Belani, K, Sessler, DI, Kurz, A, Larson, M, Blanchard, D, Schroeder, M Naloxone, meperidine, and shivering. Anesthesiology, in press

22. Delaunay, L, Bonnet, F, Liu, N, Beydon, L, Catoire, P, Sessler, DI (1993): Clonidine comparably decreases the thermoregulatory thresholds for vasoconstriction and shivering in humans. Anesthesiology 79, 470-474.

23. Washington, D, Sessler, DI, Moayeri, A, Merrifield, B, Prager, M, McGuire, J, Belani, K, Hudson, S, Schroeder, M (1993): Thermoregulatory responses to hyperthermia during isoflurane anesthesia in humans. J Appl Physiol 74, 82-87.

24. Lopez, M, Ozaki, M, Sessler, DI, Valdes, M (1993): Physiological responses to hyperthermia during epidural anesthesia and combined epidural/enflurane anesthesia in women. Anesthesiology 78, 1046-1054.

25. DuBois, D, DuBois, EF (1916): A formula to estimate the approximate surface area if height and weight be known. Arch Intern Med 17, 863-871.

26. Sessler, DI (1993): Perianesthetic thermoregulation and heat balance in humans. FASEB J 7, 638-644.

27. Hynson, J, Sessler, DI (1992): Intraoperative warming therapies: A comparison of three devices. J Clin Anesth 4, 194-199.

28. Sessler, DI, McGuire, J, Moayeri, A, Hynson, J (1991): Isoflurane-induced vasodilation minimally increases cutaneous heat loss. Anesthesiology 74, 226-232.

29. Belani, K, Sessler, DI, Sessler, AM, Schroeder, M, McGuire, J, Washington, D, Moayeri, A (1993): Leg heat content continues to decrease during the core temperature plateau in humans. Anesthesiology 78, 856-863.

30. Busto, R, Globus, MY, Dietrich, WD, Martinez, E, Valdes, I, Ginsberg, M (1989): Effect of mild hypothermia on ischemia-induced release of neurotransmitters and free fatty acids in rat brain. Stroke 20, 904-910.

31. Todd, MM, Warner, DS (1992): A comfortable hypothesis reevaluated: Cerebral metabolic depression and brain protection during ischemia (editorial). Anesthesiology 76, 161-164.

32. Nelson, TE (1990): Porcine malignant hyperthermia: Critical temperatures for *in vivo* and *in vitro* responses. Anesthesiology 73, 449-454.

33. Heier, T, Caldwell, JE, Sessler, DI, Miller, RD (1991): Mild intraoperative hypothermia increases duration of action and spontaneous recovery of vecuronium blockade during nitrous oxide-isoflurane anesthesia in humans. Anesthesiology 74, 815-819.

34. Valeri, CR, Khabbaz, K, Khuri, SF, Marquardt, C, Ragno, G, Feinhold, H, Gray, AD, Axford, T (1992): Effect of skin temperature on platelet function in patients undergoing extracorporeal bypass. J Thorac Cardiovasc Surg 104, 108-116.

35. Van Oss, CJ, Absolam, DR, Moore, LL, Park, BH, Humbert J R (1980): Effect of temperature on the chemotaxos, phagocytic engulfment, digestion and 0_2 consumption of human polymorphonuclear leukocytes. J Reticuloendothel Soc 27, 561-565.

36. Sheffield, CW, Hopf, HW, Sessler, DI, Schroeder, M, Hunt, TK, West, J (1993): Local heat reverses the decrease in subcutaneous oxygen tension produced by thermoregulatory vasoconstriction (abstract). Anesth Analg 76, S389.

37. Sheffield, CW, Sessler, DI, Hunt, TK, Scheuenstuhl, H Mild hypothermia during halothane anesthesia decreases resistance to *S. Aureus* dermal infection in guinea pigs. Journal of Wound Healing and Regeneration, in press

38. Carli, F, Emery, PW, Freemantle, CAJ (1989): Effect of peroperative normothermia on postoperative protein metabolism in elderly patients undergoing hip arthroplasty. Br J Anaesth 63, 276-282.

39. Frank, SM, Beattie, C, Christopherson, R, Norris, EJ, Perler, BA, Williams, GM, Gottlieb, SO (1993): Unintentional hypothermia is associated with postoperative myocardial ischemia. Anesthesiology 78, 468-476.

40. Hynson, JM, Sessler, DI, Moayeri, A, McGuire, J, Schroeder, M (1993): The effects of pre-induction warming on temperature and blood pressure during propofol/nitrous oxide anesthesia. Anesthesiology 79, 219-228.

41. Just, B, Trévien, V, Delva, E, Lienhart, A (1993): Prevention of intraoperative hypothermia by preoperative skin-surface warming. Anesthesiology 79, 214-218.

42. Vassilieff, N, Rosencher, N, Sessler, DI, Conseiller, C, Lienhart, A Nifedipine and intraoperative core body temperature in humans. Anesthesiology, in press

43. Sessler, DI, McGuire, J, Sessler, AM (1991): Perioperative thermal insulation. Anesthesiology 74, 875-879.

44. Sessler, DI, Schroeder, M (1993): Heat loss in humans covered with cotton hospital blankets. Anesth Analg 77, 73-77.

45. Morris, RH, Kumar, A (1972): The effect of warming blankets on maintenance of body temperature of the anesthetized, paralyzed adult patient. Anesthesiology 36, 408-411.

46. Sessler, DI, Moayeri, A (1990): Skin-surface warming: Heat flux and central temperature. Anesthesiology 73, 218-224.

47. Kurz, A, Kurz, M, Poeschl, G, Faryniak, B, Redl, G, Hackl, W (1993): Forced-air warming maintains intraoperative normothermia better than circulating-water mattresses. Anesth Analg 77, 89-95.

48. Ereth, MH, Lennon, R, Sessler, DI (1992): Isolation of peripheral and central thermal compartments in vasoconstricted patients. Aviat Space Environ Med 63, 1065-1069.

Thermal Balance in Health and Disease
Advances in Pharmacological Sciences

ALTERATION OF HEAT DISSIPATION BY DIABETES INSIPIDUS IN HUMANS

R. Behr[1], Ch. Dietrich[2] and K. Brück[2]

[1]Department of Neurosurgery, University of Würzburg, Josef-Schneider-Str. 11, 97080 Würzburg, Germany, [2]Institute of Physiology, University of Giessen, Aulweg 129, 35392 Giessen, Germany

Introduction

Antidiuretic Hormone (ADH) is known to act in the control of water balance. Further it has a great potential as endogenous antipyrogen (1) and is implicated in learning processes. In patients with diencephalic lesions, diabetes insipidus (DI) is a common symptom, which can easily be treated by Minirin[R] (desmopressinacetate) substitution. However, a disturbed water balance with hypohydration does not only affect circulation and the homeostasis of blood electrolytes, but also heat dissipation processes (2), which are clinically difficult to diagnose and to describe. The aim of our study was, to investigate the thermoregulatory capabilities of patients with central DI in a hot environment. Besides physiological interests, especially the clinician must be aware of all the functional aspects of a lacking ADH when treating or operating upon these patients.

Methods

Patients and Control Group: 16 patients (10 male and 6 female), five of whom developed DI postoperativly, five spontaneously as symptom of a cranial tumor, and six some time after an operation, were studied. In total, 29 thermoregulatory examinations were performed. Six of the patients suffered from suprasellar pituitary adenomas, four from craniopharyngeomas, two from hypothalamic gliomas, one from a teratoma, one from a primary central lymphoma, one from a metastasis of a sarcoma in the hypothalamus and one from a primary central sarcoidosis.Thirteen healthy subjects (9 male and 4 female) who were neither thermally adapted by regular heat or cold exposure nor due to endurance training served as controls. As in patients with such pathology who did not develop DI, changes of thermoregulatory threshold temperatures were described (3), a second control group was defined. This group consisted of 36 patients with comparable pathology, e.g. suprasellar pituitary adenomas, but without DI.

Climatic Chamber: Patients and controls were examined in a climatic chamber. During the examination the temperature in the chamber could be varied between 5 and 58 °C. The radiant temperature of the walls was kept close to the air temperature by thin aluminum plates placed in front of the walls and intensively flushed with room air. Air movement in the chamber was less than 0.5 m/sec. Ambient humidity was controlled by regulation of the dew point and was maintained at 50% relative humidity. Before the examinations the chamber was adjusted to 30 °C (3).

In the chamber the subjects could sit comfortably on a plastic mesh reclinging couch. The patients were observed by means of an intercom system. Data were recorded every second outside the chamber by an electronic data recording and analysis system.

Measurements

Temperatures: All temperatures were measured with copper-constantan thermocouples with an accuracy of better than 0.1 °C. The air temperature in the climatic chamber was measured with a free-hanging dry temperature element, which was thermally insulated.

Tympanal temperature was measured with a thermocouple in the outer ear near the eardrum, which was held in place with a plastic ear plug. The ear was also covered with cotton wool and with a cotton-filled hearing protection headphone for further insulation. Tympanal temperature rather than esophageal temperature was measured because many patients were resistant to the esophageal probe, and in some postoperative cases, especially after transnasal approach to the tumour, it would have been impossible to introduce a probe through the nose even if the patients had consented.

Skin temperatures were measured at four locations, chest (T_A), upper arm (T_B), thigh (T_C) and lower leg (T_D) (4). Mean skin temperature was calculated from the four measurments using Ramanathan's equation (4):

$$\overline{T}_{sk} = 0.3\,(T_A + T_B) + 0.2\,(T_C + T_D)$$

The weighted mean body temperature was calculated using the following formula:

$$\overline{T}_b = 0.87 \times T_{ty} + 0.13 \times \overline{T}_{sk}$$

The coefficents 0.87 and 0.13 reflect the relative sensitivity of the thermolegulatory system to changes in the core and mean skin temperatures; they do not describe the heat content of the body core and shell (5). The thermoregulatory threshold temperatures were defined by plotting the mean body temperature aginst the data of each thermophysiological effector, e.g. sweat rate (SR), and skin blood flow (SBF). A linear regression line was calculated for the resting period and for the period of activation. The point of intersection of the two regression lines determins the threshold temperature.

Sweat Secretion Rate (SR): To determine the sweat secretion rate, a plastic capsule which covered a skin area of 7 cm^2 was fastened air-tight with a rubber band over the right thorax on the medioclavicular line. Warm dry air flowed at a constant rate through the capsule. The difference in the humidity of the entering and leaving air was measured with an Uras 2T, Hartman & Braun infared absorption meter.

Skin blood flow: Heat transport was measured with the fluvographic method as a standard for local skin circulation. A 6-plate element (6) was used. It was set in a Plexiglas holder which allowed a largely constant pressure against the skin, and the holder was fixed to the inner side of the lower arm with adhesive tape. Zero blood circulation (λ_0) was defined by placing a blood pressure measuring cuff pumped to 220 mmHg on the upper arm for 2-3 minutes. This procedure was repeated during the neutral phase and before the warming phase. The base line drift was electronically corrected between the various values. The increase in heat transfer ($\lambda - \lambda_0$) is approximately linearly proportional to plethysmographically determined increase in skin circulation. Two phases of vasodilation in the lower arm are usualy seperated. The first, quantitatively smaller dilation takes place in conjunction with the relaxation of vasoconstriction when previously vasoconstricted persons are warmed (7,8). The second vasodilation phase occurs simultaniously with the increase in sweating and is an active, quantitatively significantly stronger process. It is based on the sympathetic innervation of the skin (8) and connected to an intact sweat gland function (9). Here only the second, active phase was considered for the determination of the threshold temperature.

Water-Electrolyte Balance: Urine osmolality, serum osmolality, serum sodium and the urine volume/24 hours were measured. Osmolality was measured by observing the lowering of freezing temperatures (Osmomat 030, Gonotec). Electrolyte values were determined using the flame-photometric method.

Statistics: The data were statistically analyzed with a personal computer running the program "PC-Statistic". The distribution of the measured and calculated values was determined using the Kolmogorow-Smirnow test. Since normal distribution was shown, for further analysis the analysis of variance (ANOVA) und the Student's t-test were employed.

Ethics: The study protocol was submitted to the Ethics Commission of the Justus-Liebig-University of Giessen, which had no reservations, provided the patients were in a physically stable condition and gave their informed consent.

Results

Thermoregulatory Findings: Since patients with pituitary adenomas, craniopharyngeomas and hypothalamic gliomas, but without diabetes insipidus (DI), also show elevated heat dissipation threshold temperatures, as previously demonstrated (3), a mixed group of patients with these diagnoses but without DI served as an inter-patient control in addition to the healthy control group. Analysis of variance of the results of the control group, the patient group without DI and the patients with DI who showed a heat dissipation threshold temperature - it should be mentioned here that in some DI cases no increase in sweat secretion on exposure to heat occurred - showed significant differences, $p < 0.0001$ (ANOVA & Kruskal-Wallis Test). The comparison of means for the sweating threshold showed a significantly higher threshold temperature for the patients with DI (37.25 °C $\bar{T}_b$), in comparison to 37.02 °C T_b for patients with the same diagnoses but without DI and 36.66 °C T_b for the healthy controls ($p < 0.001$). The threshold temperature for skin blood flow was identical (36.91 °C $\bar{T}_b$) in both patient groups and significantly increased in comparison to the controls (36.59 °C $\bar{T}_b$). The inital slope of the sweat secretion rate for patients with DI was not significantly reduced in comparison to the controls. The maximal sweating rate differance (max. sweating rate minus evaporation in a thermoneutral environment) was, however, lowered by 0.33 mg H_2O/cm^2*min in comparison to the controls ($p < 0.01$). The raised heat dissipation thresholds of the DI patients led to significantly increased maximal tympanal temperatures (37.71 °C) and maximal mean skin temperatures (38.95 °C), $p < 0.01$, when compaired to controls. In the neutral phase of the examination, however, the body temperatures of the several groups did not differ significantly and the maximal ambient temperature during heat load was almost equal (56.24 °C and 55.72 °C). Table 1 presents the most important thermophysiological data.

Table 1: Thermoregulatory data of controls (CON) and patients (DI) during thermoneutral ambient temperatures and heat load. * = $p < 0.01$

T_a	= ambient temperature
T_{ty}	= tympanic temperature
$\bar{T}_{sk}$	= mean skin temperature
$\bar{T}_b$	= mean body temperature
MR	= metabolic rate
MR_o	= metabolic rate adjusted to body surface
x-SR	= temperatures at sweating threshold
x-λ	= temperatures at vasodilation threshold
x-max	= maximal observed temperatures during heat load
Δ-SR_{max}	= maximal measured sweat rate minus evaporation at thermoneutral ambient temperatures

Neutral ambient temperature

	CON	N	DI	N	p
T_a	30.14 (0.44)	13	29.62 (0.19)	24	
T_{ty}	36.78 (0.06)	13	36.74 (0.17)	24	
$\bar{T}_{sk}$	34.45 (0.20)	13	34.10 (0.45)	24	
MR	91.93 (3.93)	13	79.81 (3.89)	20	+
MR_o	48.09 (1.63)	13	40.58 (1.86)	20	*

Heat load

	CON	N	DI	N	p
T_a	53.57 (1.00)	13	51.71 (1.40)	15	
T_{ty}-SR	36.49 (0.05)	13	37.11 (0.11)	12	*
$\bar{T}_{sk}$-SR	37.70 (0.27)	13	38.02 (0.41)	12	
$\bar{T}_b$-SR	**36.66 (0.08)**	**13**	**37.25 (0.09)**	**12**	*
Δ-SR_{max}	**0.55 (0.09)**	**13**	**0.22 (0.03)**	**24**	*
T_{ty}--λ	36.49 (0.04)	13	36.81 (0.08)	12	*
$\bar{T}_{sk}$-λ	37.30 (0.35)	13	37.56 (0.42)	12	
$\bar{T}_b$-λ	**36.59 (0.06)**	**13**	**36.91 (0.08)**	**12**	*
Δ–λ	0.05 (0.01)	13	0.03 (0.00)	12	
T_a-max	56.24 (0.59)	13	55.72 (0.53)	23	
T_{ty}-max	36.74 (0.07)	13	37.71 (0.13)	24	*
$\bar{T}_{sk}$-max	38.17 (0.18)	13	38.95 (0.20)	24	*

As mentioned above, in some measurements (n=12) no sweating threshold was discernible; even at very high body temperatures no increase in sweat secretion rate could be observed. The maximal tympanal temperature in these cases was 37.94 °C and the maximal $\bar{T}_{sk}$ 39.43 °C, both significantly ($p < 0.01$) above the values for patients with thresholds. The maximal environmental temperature for these cases was also significantly higher than that for the patients with thresholds, and the exposure time (time from T_a= 40 °C to the end of examination) was 12 minutes longer. In a thermoneutral environment the body temperatures were not different. The mean maximal sweating rate difference for the group without thresholds (0.08 mg H_2O/cm^2*min) was lower than that for the patient group with thresholds (0.36 mg H_2O/cm^2*min), $p < 0.01$, and the healthy control group (0.55 mg H_2O/cm^2*min), $p < 0.05$.

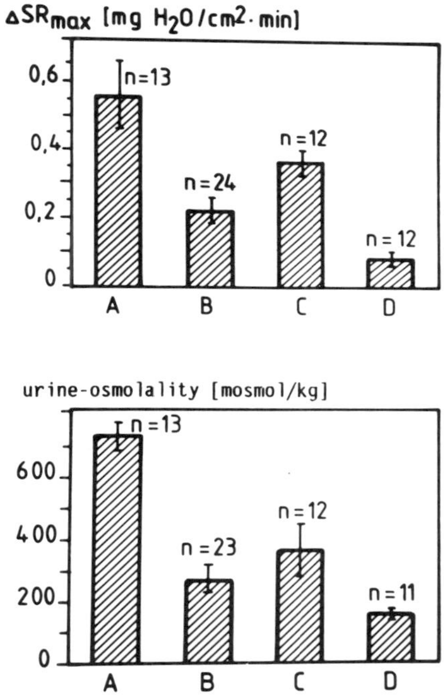

Fig. 1: left: Comparison of the maximal sweating rate difference (Δ-SR_{max}) and urine osmolality between controls and patients with DI. A = controls, B = total group of DI patients, C = patients with sweating thresholds, D = patients without sweating thresholds. The non-sweaters (D) have the lowest urine-osmolality. A/B $p < 0.01$, C/D $p < 0.05$, t-Test & Wilcoxon.

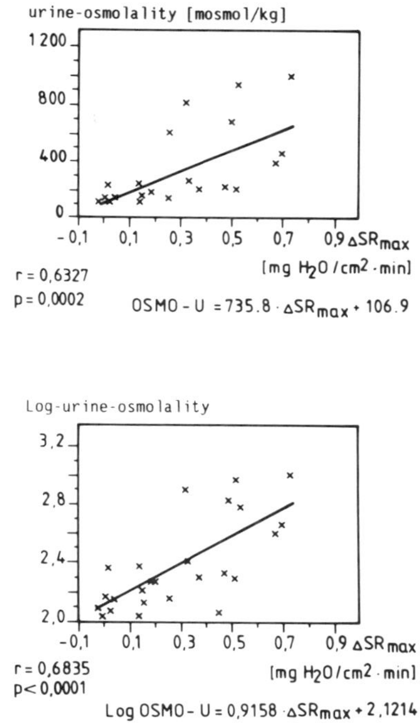

Fig. 1: right: Linear and logarithmic regression line and correlation of Δ-SRmax and urine osmolality. The relation is significant with a coefficient of 0.63 and 0.68 respectively.

Laboratory Findings and Correlation with Thermophysiological Data A shortage of antidiuretic hormone leads to a reduction in the concentrating ability of the kidneys which can be measured as reduction in urine osmolality. If water loss is not adequately compensated, hypertonic dehydration with increased serum osmolality and serum sodium follows. In the total DI patient group the urine osmolality was significantly lowered, to 266 mosmol/kg in comparison to 736 mosmol/kg in the healthy control group. When the total DI-patient group is divided into patients with and without thresholds, the patients without thresholds also show a significantly lower urine osmolality (158.8 mosmol/kg),$p < 0.05$, and higher serum sodium (148.5 mval/ml),$p < 0.01$, than the DI-patients with thresholds (364.2 mosmol/kg, 139.8 mval/ml). Plotting the data of urine osmolality or the logarithmic values against the maximal sweat rate (Fig. 1) a significant relationship can be demonstrated.

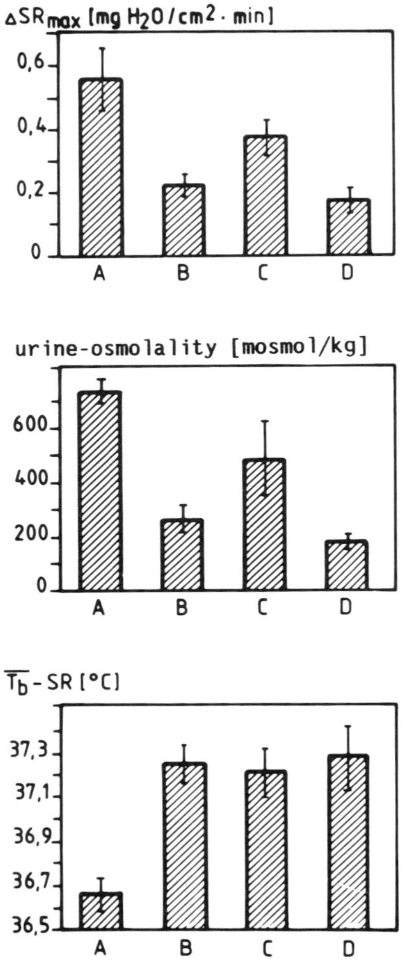

Fig. 2: Comparison of the maximal sweating rate (Δ-SR_{max}), the urine osmolality and the threshold temperatures ($\bar{T}_b$-SR) between controls (A), the total group of DI patients (B), DI patients with Minirin^R therapy (C) and DI patients without Minirin^R (D). $\bar{T}_b$-SR remains unchanged by Minirin^R, while the sweating rate and the urine osmolality are significantly higher compared to the untreated group (D) ($p < 0.01$).

After descriping the relationship between the sweat secretion rate and urine osmolality, the effect of substitution therapy with Minirin^R should be mentioned. Six patients received, due to pronounced DI, Minirin^R in doses from 0.005 to 0.015 mg per day intranasally. Dosage was adapted to the individual cases and, in view of the levels of urine osmolality, rather under- than overdosed. This was deliberate, as too high a dosage can easily lead to excessive water retention with attendant pulmonary and cardiocirculatory problems. In comparison to all DI patients who received no Minirin^R, there was a significantly higher urine osmolality and also a significantly greater sweat secretion rate in the substituted patients (Fig.2). Serum sodium levels were significantly higher in patients who did not receive Minirin^R, the serum osmolality also tended to be higher (not statistically significant). As Fig.2 shows, the peripheral, renal effects of ADH deficiency can be compensated by substitution therapy with Minirin^R and the maximal sweating rate also increases significantly. However, the centrally regulated sweating threshold is not affected.

According to Fig.2, in patients receiving Minirin^R an increase in maximal sweating rate occurred, that should enhance the cooling effect of sweating. And in fact, although they had effectively the same T_{sk} (respectively 34.12 and 34.19 °C) in a thermoneutral environment, the maximal $\bar{T}_{sk}$ for the Minirin^R group was 38.62 °C, as opposed to 39.06 °C for the patients without Minirin^R ($p < 0.05$). The maximal tympanal temperature was also, even after somewhat higher tympanal values in a thermoneutral environment (36.88 °C compared to 36.69 °C), significantly lower for the Minirin^R-treated group (37.35 °C compared to 37.84 °C $p < 0.05$) in comparable hot environmental temperature (54.89 °C for the Minirin^R group compared to 55.93 °C). This lower core temperature occurred despite a longer heat exposure time (52.2 min.compared to 48.8 min, $p < 0.05$) for the Minirin^R group. Due to wide distribution of values, no change in the intial steepness of the curve of the sweat secretion rate could be determined. The values for the Minirin^R group tended to be somewhat higher. As mentioned above, these findings apply if all the patients who received no Minirin are considered. However, when the patient group without Minirin^R is divided into patients with sweating thresholds and patients without sweating thresholds, no significant difference appears between the untreated group with thresholds and the Minirin^R group. On the other hand, the differences between patients who showed no thresholds and were not treated with Minirin^R and the Minirin^R group were quite noticeable. With almost the same exposure time (52.2 compared with 54.6 min.) and very slightly lower T_a-max (54.89 °C compared with 56.97 °C, $p < 0.05$), the Minirin^R patients had significantly lower tympanal and mean skin temperatures than the non-sweaters. They also showed significantly higher urine osmolality and sweat secretion. The body temperatures of these two groups in a thermoneutral environment were not significantly different.

From a thermoregulatory point of view, Minirin^R therapy should be particularly profitable for patients who have severe DI and show no sweating threshold temperature.

Comparison of Pre- and Postoperative Findings: Five patients, four men and one woman, who did not have DI preoperatively, were examined postoperatively after the development of DI and served as their own controls. The histological diagnoses were three cases of craniopharyngeomas, one pituitary adenoma and one teratoma. There were no differences in the pre- and postoperative anthropomorphic data.

The preoperative heat production thresholds of the patients were lowered on the average to $\bar{T}_b$-MR (metabolic rate) 35.60 °C (SEM 0.14) and $\bar{T}_b$-EMA (electrical muscle activity) 35.65 °C (SEM 0.15), $p < 0.05$. Postoperatively these thresholds were normalized ($\bar{T}_b$-MR 35.96 °C, (SEM 0.02) and $\bar{T}_b$-EMA 35.91 °C (SEM 0.05). No significant changes in the slope of the curves for heat production mechanisms were observed.

The sweating thresholds were 0.39 °C higher post- than pre-operatively; one patient with a craniopharyngeoma showed no threshold postoperatively. As could be expected, the maximal sweating rate was significantly lowered postoperatively (from 0.61 to 0.32 mg H_2O/mg*min), and the urine osmolality was clearly, although not significantly reduced.

Discussion

The results of the thermoregulatory investigations of patients with central DI focus the discussion on the function of ADH in the thermoregulatory process on two points. The first is, to what extent sweat rate capacity is influenced by lack of ADH. Clinical experience shows, that a moderate or latent DI without severe alterations of serum osmolality can be compensated by an increased water ingestion. In these cases, however, urine osmolality is always reduced. Severe DI, with an increase of serum sodium, was combined with the lowest values of urine osmolality and without increase of sweat rate during strong heat exposure. In all patients with DI, the maximal observed local sweat rate was significantly reduced. The patients with the lowest urine osmolality and elevated serum sodium levels depicted the lowest increment of heat-induced sweat rate. In this group serum osmolality at rest was not significantly increased, which may be due to the relatively high standard deviation. In this study we were not able to measure the whole body fluid content in order to determine the degree of dehydration. However, if one assumes that the urine osmolality measured from urine samples collected during normal daily activity of patients without water deprivation or forced ingestion reflects to some extent the degree of dehydration, one may state that in our patients the reduced hydration/urine osmolality was significantly related to a reduced local sweat rate. Similar results were reported in healthy subjects (10). After dehydration by diuretics, exercise led to an increased body core temperature compared to normal hydrated controls and to a reduced whole body sweating rate as well as a lowered local sweat rate. In a series of 13 patients with aquired DI, in most cases the local sweat rate was also significantly reduced (11). The role of hydration for sweat production rate is stressed by the effects of substitution therapy. Continous Minirin^R treatment caused a significant increase of urine osmolality, a sign. decrease of serum Na^+ and a reduction of serum osmolality. The local sweat rate of the treated patients was significantly higher than in the untreated group. The single intravenous injection of 10 I.E. Pitressin^R (argipressine) immediately before heat exposure did not change the local sweat rate, nor the sweating threshold temperature. The plasma level of arginine-vasopressin (AVP) was increased by this injection from below 1 to 2.4 pg/ml. This indicates that an acut increase of AVP does not alter thermoregulatory capacities. However, this was proved only in one single examination. The chronic effect of Minirin^R, which allows an augmentation of body water content is thought to be the reason for the improved local sweat rate.

The second point is, to what extent the thermoregulatory threshold temperatures are affected by a lack of ADH. The results show very clearly, that compared to healthy controls, the sweating thresholds are highly significant elevated. This holds also true for the vasodilation threshold. However, due to previous evidence in patients with diencephalic lesions without DI (3), in whom an elevation of heat dissipation thresholds was found, one must ask, whether the disturbed sweating threshold in DI patients is caused by the lesion, e.g. the tumor itself, or by the disturbed ADH system. Therefore a group of 36 patients with comparable diagnosis, mostly suprasellar pituitary adenomas and craniopharyngeomas, was investigated, who had elevated heat loss thresholds but no DI. In DI patients the sweating threshold was significantly rised compared to the other patient group. This indicated, that DI had an additional effect on the evaporation threshold. Further support was given by a very rare case of primary central sarcoidosis. The single symptom of this female patient was a DI. Magnetic resonance tomography revealed a thickening of the pituitary stalk without any hypothalamic involvement. Heat production thresholds were completly normal in two external cold examinations. Sweating thresholds were elevated to 37.22 and 37.38 °C T_b (two examinations). In contrast to the peripheral benefit of Minirin^R, with improvement of hydration and sweat rate, the centrally governed threshold temperatures were not affected at all, which depends probably on the inability of Minirin^R to penetrate the blood brain barrier. In healthy subjects, who were dehydrated by diuretics, no change of the sweating thresholds was observed (10). This indicates, that an intact central ADH balance plays an important role in the regulation of the evaporation thresholds. It remains the question, however, which state of balance is responsible for the increased thresholds. Is it a complete lack, a reduction or even an augmentation of central ADH. A complete lack is not possible, because it would suppose a destruction of the ADH producing nuclei. This is clinically not possible exept a huge destruction of hypothalamic tissue with fatal consequences. Due to the pathology of our patients it is much more likely, that the site of the lesion is the pituitary stalk or the posterior lobe of the pituitary. This would explain a reduced peripheral ADH level. In the cerebro-spinal fluid (CSF) of DI patients a significantly increased level of ADH was measured compared to plasma levels (12). If we take these findings into consideration, we must conclude, that the increased heat dissipation thresholds in DI are caused by an increased central activation of ADH. The benefit for the organism is, that in chronic hypohydration a delayed onset of sweating is an additional water saving mechanism, which is realised in heat-adapted animals, e.g. the camel. For the clinician it is of interest, that DI patients are subject to a higher risk of hyperthermia due to a passive or active heat load which can not be sufficiently counteracted and may result in heat stroke.

Acknowledgements

Supported by DFG Grant: Be 1015/1-3

References

1. Kasting, N W (1982): Vasopressin: A homeostatic effector in the febrile process. Neurosci. & Biobehavioral Reviews, Vol. 6, 215-222

2. Senay, L C (1979): Temperature regulation and hypohydration: A singular view. J. Appl. Physiol. 47, 1, 1-7

3. Behr, R, Hildebrandt, G, Koca, M, Brück, K (1991): Modifications of thermoregulatory characteristics in patients with suprasellar pituitary adenomas. Brain 114, II, 697-708

4. Ramanathan, N L (1964): A new weighting system for mean surface temperatures of the human body. J. Appl. Physiol. 19, 531-533

5. Brück, K, Zeisberger, E (1990): Adaptive changes in thermoregulation and their neuropharmacological basis. In: Thermoregulation Physiology and Biochemistry, International Encyclopedia of Pharmacology and Therapeutics, 255-307. Eds: Schönbaum, E, Lomax, P, Pergamon Press, 1990

6. Hensel, H, Brand, S (1977): 6-Plate element for recording of cutaneous blood flow. Pflügers Arch., 368, 165-167

7. Hessemer, V, Brück, K (1985): Influence of menstrual cycle on shivering, skin blood flow and sweating responses measured at night. J. Appl. Physiol. 59, 6, 1902-1910

8. Roddie, I C (1983): Circulation to skin and adipose tissue. In: Handbook of Physiology, Vol. 3, Sec. 2, 285-317. Hrsg.: Shepherd, J T, Abboud, F M, Geiger, S R, Am. Physiol. Soc., Bethesda, 1983

9. Brengelmann, G L, Freund, P R, Rowell, L B, Oleru J E, Kraning, K K (1981): Absence of active cutaneous vasodilation associated with congenital absence of sweat glands in humans. Am. J. Physiol., 240, H 571-H 575

10. Fortney, S M, Nadel, E R, Wenger, C B, Bove, J R (1981): Effect of blood volume on sweating rate and body fluids in exercising humans. J. Appl. Physiol.: Resp. Envir. Exercise Physiol. 51, 1594-1600

11. Taussig, L M, Braunstein, G D (1973): Effects of vasopressin on sweat rate and composition in patients with diabetes insipidus and normal controls. L. Investig. Dermatol. 60, 4, 197-202

12. Luerssen, T G, Shelton, R L, Robertson, G L (1977): Evidence for seperate origin of plasma and cerebrospinal fluid vasopressin. Cin. Res. 25, 14.

DOES INTRAOPERATIVE CORE HYPOTHERMIA INCREASE THE INCIDENCE OF SURGICAL WOUND INFECTIONS AND PROLONG HOSPITALIZATION?

A. Kurz, D. I. Sessler, R. Lenhardt, E. Narzt, G. Huemer and F. Lackner

Department of Anesthesia and Intensive Care Medicine, University of Vienna, Austria, and the Thermoregulation Research Laboratory, University of California, San Francisco, California 94143-0648.

Wound infections are among the most common serious complications of anesthesia and surgery. The reported incidence of wound infections in patients having two or more factors increasing risk of infection (*e.g.*, abdominal surgery, surgery lasting more than 2 hours, "contaminated" or "dirty-infected" wounds, or three or more underlying diagnoses), ranges from 8 to 27% (1).

Development of wound infections is influenced by various factors including: 1) character and magnitude of contamination; 2) effects of hemostasis, foreign bodies, damaged tissues, *etc.* on the local milieu; 3) wound perfusion, which delivers (and in some cases disperses) immune components such as oxygen, inflammatory cells, growth factors, cytokines, *etc.*; 5) antibiotic administration; and, 6) immune function (2,3). Non-specific or natural immunity is the most important host defense following acute bacterial contamination. It includes granulocyte demargination, diapedesis, phagocytoses, and oxygen-dependent bacterial killing and non-oxidative killing. The first few hours following bacterial contamination constitute a *decisive period*, during which infection is established (4). The effects of antibiotic administration and of hypoperfusion are especially important during this period.

Numerous techniques allow anesthesiologists and surgeons to maintain core normothermia in their patients (5,6). Although consequences of intraoperative hypothermia have been identified (*e.g.*, impaired coagulation, prolonged drug action, postoperative shivering), not even a single prospective, interventional study has demonstrated adverse *outcomes* from mild hypothermia during surgery. Consequently, most patients do become 1-3°C hypothermic because preventive measures rarely are instituted.

Previous work and our preliminary studies suggest that core hypothermia impairs immune function, decreases subcutaneous oxygen tension, and reduces resistance to test dermal infections. Accordingly, we tested the hypothesis that 2-3°C core hypothermia during elective colon surgery increases the incidence of clinically important wound infections and prolongs the duration of hospitalization.

Methods

We studied 110 ASA Physical Status 1-3 patients undergoing elective colon resection. We restricted our study population to patients undergoing colon resection because the incidence of wound infection in such patients is ≈15% (1). None had a history of thyroid disease, dysautonomia, Raynaud's syndrome, or malignant hyperthermia.

Protocol: The patients were anesthetized with isoflurane, nitrous oxide (60%), and fentanyl. They then were randomly assigned to normal thermal management (n = 54, hypothermic group) or additional warming (n = 56, normothermic group). Patients in the hypothermic group were covered with a single layer of surgical drape and given unwarmed iv fluids; those assigned to normothermia were given iv fluids warmed to 37°C and covered with an upper-body forced-air warmer (Bair Hugger, Augustine Medical, Inc. Eden Prairie, MN) (5,6).

Per surgical routine, antibiotics were administered starting at the time of surgery and maintained for two postoperative days. Intravenous fluids were administered intraoperatively at a basal rate of 10-15 $ml \cdot kg^{-1} \cdot h^{-1}$. Additionally, 4 ml of crystalloid was given to compensate for each estimated ml of intraoperative blood loss. Blood was administered as necessary to maintain a hematocrit ≥25-30% (leukocyte filters were not used).

Factors known to influence the incidence of wound infections and rates of healing were managed by protocol. Postoperatively, supplemental oxygen was administered at a rate of 6 L/minute *via* nasal prongs for the first six postoperative hours. Additional oxygen was given as necessary to maintain oxygen saturation (S_pO_2) ≥ 95%. Postoperative surgical pain was treated using patient-controlled analgesia (PCA).

Measurements: Morphometric characteristics of each treatment group were tabulated. Similarly, detailed records of anesthetic and fluid management were kept. Blood pressures were measured oscillometrically to avoid vasoconstriction-induced artifactual changes in radial artery pressure End-tidal isoflurane and carbon dioxide concentrations were measured with an infrared detector. Since blood transfusions decrease immune function (transfused leukocytes decrease killer T-cell function), the number and type of blood transfusions were recorded.

Core temperatures were measured at the tympanic membrane. Appropriate probe placement was confirmed when volunteers easily detected a gentle rubbing of the attached wire. The aural canal was occluded with cotton, the probe securely taped in place, and a gauze bandage positioned over the external ear. Tympanic membrane temperatures correlate well with distal esophageal temperatures in the perianesthetic period (7). Additionally, we recorded distal esophageal temperature intraoperatively.

During the first six hours of postoperative recovery, tympanic membrane temperatures, and hemodynamic responses were measured in a subgroup of 53 patients. The patients' subjective level of pain and degree of thermal discomfort was evaluated using 100-mm-long visual analog scales. All measurements were taken at 10-minute intervals intraoperatively, and at 20-minute intervals during recovery.

An observer strictly blinded to group assignment and intraoperative and postoperative core temperatures evaluated the wounds on a daily basis until discharge. The date of discharge was determined by the attending surgeon.

Statistical analysis: Morphometric data, anesthetic parameters, and temperatures were compared using unpaired *t*-tests. Wound infections were compared using Chi square analysis. All values are expressed as means ± SD; differences were considered significant when $P < 0.01$.

Results

Morphometric characteristics, hemodynamic values, anesthetic management, and fluid management were similar in the patients assigned to each treatment (table 1).

	normothermia	**hypothermia**
Age (yr)	59 ± 17	60 ± 15
Weight (kg)	73 ± 16	70 ± 13
Height (cm)	170 ± 10	168 ± 11
Duration of Surgery (min)	203 ± 65	213 ± 88
Final Intraoperative Core Temperature (°C)	37 ± 0.3	34.2 ± 0.4*
Administered Fentanyl (μg)	710 ± 215	650 ± 230
End-Tidal Isoflurane (%)	0.5 ± 0.2	0.6 ± 0.3
Transfused Blood (units)	2 ± 2	1 ± 2

Table 1: Morphometric characteristics, anesthetic management and intraoperative temperatures in 110 patients assigned to hypothermia and normothermia. Only core temperatures differed significantly between the two groups.

Core temperatures at the end of surgery in the hypothermic group were significantly less than those in the warmed patients (34.2 ± 0.4 *vs.* 37.0 ± 0.3°C). Initial postoperative temperatures in the warmed patients were 36.7 ± 0.4°C, and increased only slightly during the six-h recovery period. In contrast initial core temperatures were 34.7 ± 0.6°C in those allowed to become hypothermic, and increased at a rate of 0.5°C/h for the first three hours of recovery and at 0.3°C for the subsequent two hours.

Only a few of the warmed patients shivered at any time during recovery and the normothermic patients reported feeling significantly warmer than those allowed to become hypothermic. In contrast, shivering was common in the hypothermic patients. By the end of the second elapsed hour, however shivering was rare even in the hypothermic patients. Pain scores and the amount of opioid administered in the two groups were virtually identical at every postoperative measurement interval. Heart rates and blood pressures were similar in the normothermic and hypothermic patients (figure 1).

The overall incidence of surgical wound infections was 17%. There were five surgical wound infections in the normothermic patients and fourteen in those allowed to become hypothermic

($P = 0.1$). The duration of hospitalization was 10 ± 3 days in the normothermic patients and 14 ± 7 in the hypothermic ones ($P < 0.01$).

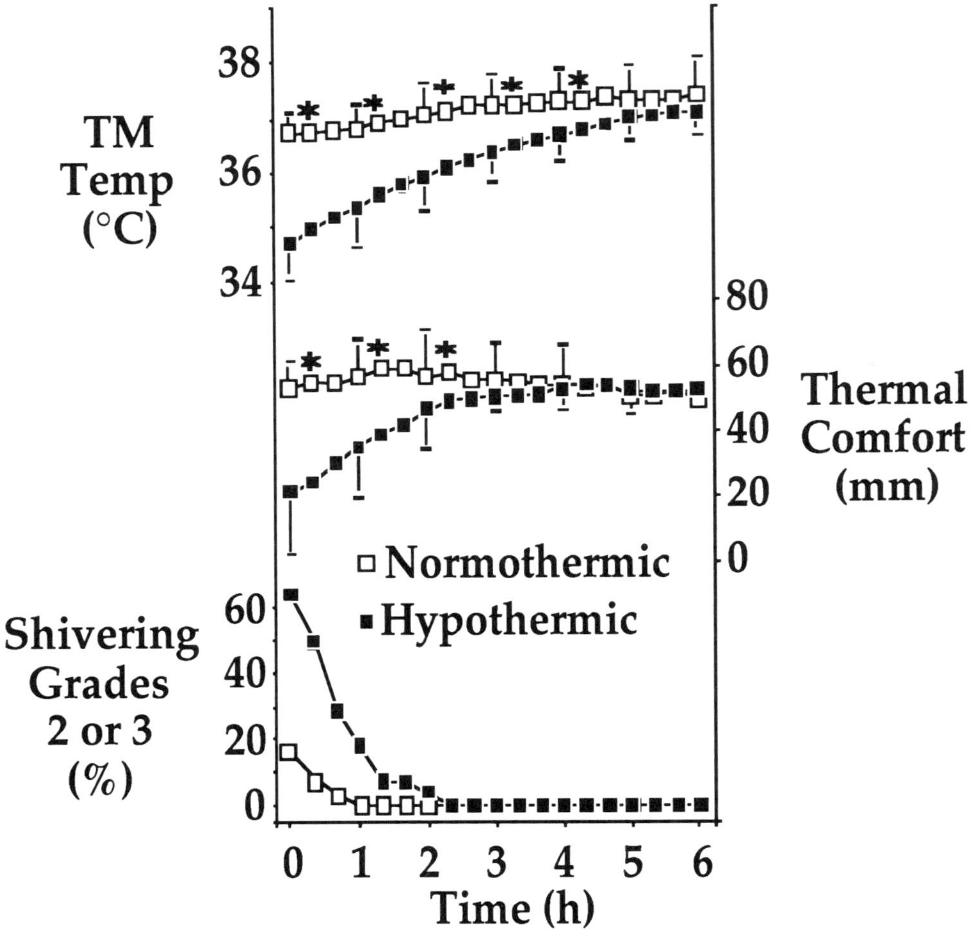

Fig. 1. Initial postoperative core temperatures differed by ≈2.0°C in the normothermic and hypothermic patients, and remained significantly different until more than 4 hours after surgery. The hypothermic patients reported feeling unpleasantly cold, and the sensation persisted for more than two hours. Only a few of the warmed patients shivering at any time during the recovery period. In contrast, most of the hypothermic patients demonstrated grades two or three tremor during the first postoperative hour. By the end of the first elapsed hour, however, shivering was rare, even in the hypothermic patients. Asterisks (*) indicate statistically significant differences between the groups.

Discussion

As expected, forced-air warming maintained normothermia in all patients assigned to intraoperative heating (5,6). In contrast patients of the control group became 2-3°C hypothermic. Interestingly, hypothermia persisted for more than four hours. Factors which may have contributed to prolonged postoperative thermoregulatory inhibition include: residual volatile anesthetic, opioids given to treat surgical pain, and a prolonged effect of intraoperative core hypothermia.

Hemodynamic responses were similar in the two groups. Little or no change in arterial blood pressure during thermoregulatory vasoconstriction is consistent with the relatively small fraction of the cardiac output traversing arterio-venous shunts. Our hypothermic and normothermic patients had comparable pain scores and required similar amounts of opioids. However it remains likely that hypothermia delayed opioid clearance and that plasma concentrations were greater in the hypothermic patients. Thus hypothermia may have directly aggrevated surgical pain, or have increased it *via* shivering (which strains surgical incisions).

Mild perioperative core hypothermia (*e.g.* 3°C below normal core temperature) is common, though preventable (5,6). There are at least two reasons to suspect that hypothermia facilitates development of postoperative wound infections. First, normal core body temperature is maintained by thermoregulatory responses, including cutaneous vasoconstriction (8). Although arteriovenous shunt vasoconstriction (in fingers and toes) is most dramatic (9), capillary flow in non-acral skin also decreases (10). Low tissue oxygen values are associated with impaired resistance to infection (unpublished data), and thermoregulatory vasoconstriction decreases subcutaneous oxygen tension ($P_{sq}O_2$) in humans (11).

Second, mild hypothermia impairs various immune functions, including leukocyte mobility and phagocytosis. Mild hypothermia also may impair non-specific ("natural") immune functions including chemotactic migration of leukocytes, and phagocytosis (2). Although general anesthesia as well as blood transfusions impair immune function, isoflurane concentrations and the amount of blood administered were virtually identical in the two groups, suggesting that the observed differences in the incidence of infection resulted specifically from the differences in core temperatures. And finally, tissue hypothermia produces substantial platelet dysfunction (12); an important component of inflammation is the activation of platelets and the coagulation cascade with release of immune mediators including kinins, fibrin split products, and growth factors.

The morbidity (and related cost) associated with surgical infections is considerable; estimates of prolonged hospitalization vary from 5 to 20 days per infection (1,13). Our data suggest that simply maintaining intraoperative normothermia in patients having colon surgery not only reduces the incidence of surgical wound infections by roughly a factor of three, but also significantly reduces the duration of hospitalization.

Acknowledgments

These studies were supported by National Institutes of Health grants GM 39723 and GM 49670.

References

1. Haley, RW, Culver, DH, Morgan, WM, White, JW, Emori, TG, Hooton, TM (1985): Identifying patients at high risk of surgical wound infection: A simple multivariate index of patient susceptibility and wound contamination. Am J Epidemiol 121, 206-215.

2. Van Oss, CJ, Absolam, DR, Moore, LL, Park, BH, Humbert J R (1980): Effect of temperature on the chemotaxis, phagocytic engulfment, digestion and O_2 consumption of human polymorphonuclear leukocytes. J Reticuloendothel Soc 27, 561-565.

3. Burke, JF (1961): The effective period of preventive antibiotic action in experimental incisions and dermal lesions. Surgery 50, 161-168.

4. Miles, AA, Miles, EM, Burke, J (1957): The value and duration of defence reactions of the skin to the primary lodgement of bacteria. Br J of Exp Path 38, 79-97.

5. Hynson, J, Sessler, DI (1992): Intraoperative warming therapies: A comparison of three devices. J Clin Anesth 4, 194-199.

6. Kurz, A, Kurz, M, Poeschl, G, Faryniak, B, Redl, G, Hackl, W (1993): Forced-air warming maintains intraoperative normothermia better than circulating-water mattresses. Anesth Analg 77, 89-95.

7. Cork, RC, Vaughan, RW, Humphrey, LS (1983): Precision and accuracy of intraoperative temperature monitoring. Anesth Analg 62, 211-214.

8. Sessler, DI (1991): Temperature disturbances. In: Critical Care Practice. Hoyt J, Allen S, Tennesen A eds. Philadelphia: WB Saunders, 400-412.

9. Hales, JRS (1985): Skin arteriovenous anastomoses, their control and role in thermoregulation. In: Cardiovascular Shunts: Phylogenetic, Ontogenetic and Clinical Aspects. Johansen, K, Burggren, W eds. Copenhagen: Munksgaard, 433-451.

10. Sessler, DI, Moayeri, A, Støen, R, Glosten, B, Hynson, J, McGuire, J (1990): Thermoregulatory vasoconstriction decreases cutaneous heat loss. Anesthesiology 73, 656-660.

11. Sheffield, CW, Hopf, HW, Sessler, DI, Schroeder, M, Hunt, TK, West, J (1993): Local heat reverses the decrease in subcutaneous oxygen tension produced by thermoregulatory vasoconstriction (abstract). Anesth Analg 76, S389.

12. Valeri, RC, Cassidy, G, Khuri, S, Feingold, H, Ragno, G, Altschule, MD (1987): Hypothermia-induced reversible platelet dysfunction. Ann Surg 205, 175-181.

13. Bremmelgaard, A, Raahave, D, Beir-Holgersen, R, Pedersen, JV, Andersen, S, Sorensen, AI (1989): Computer-aided surveillance of surgical infections and identification of risk factors. J Hosp Infect 13, 18.

Thermal Balance in Health and Disease
Advances in Pharmacological Sciences

HYPERTHERMIA FOLLOWING HYPOTHERMIC CARDIOPULMONARY BYPASS DOES NOT CORRELATE WITH POSTOPERATIVE ENDOTOXEMIA

H. Schmidt, H. Böhrer, J. Motsch, R. Urbaschek* and E. Martin

Department of Anesthesiology and *Department of Med. Microbiology (Mannheim) University of Heidelberg, 69120 Heidelberg, INF 110, Germany

Introduction

Hypothermic extracorporeal circulation in open-heart surgery is usually followed by a phase of postoperative hyperthermia. This hyperthermia is part of a systemic inflammatory response, which can lead to organ dysfunction and postoperative morbidity. In recent years, it has been found that endotoxin is released during cardiopulmonary bypass (1,2,3). Endotoxin (bacterial lipopolysaccharide) is known as a potent pyrogenic substance and triggers the release of several fever-inducing cytokines including interleukin-6 (IL-6) (4). In human volunteers, administration of endotoxin produces pathophysiologic changes similar to those observed in patients after cardiopulmonary bypass (4). Aim of this study was to detect a correlation between endotoxin and interleukin-6 release after cardiopulmonary bypass and the degree of postoperative hyperthermia.

Methods

After institutional approval and informed consent, 14 consecutive adult patients undergoing cardiac surgery were studied for postoperative endotoxin levels. Additionally, interleukin-6 levels were determined in 12 patients two hours after cardiopulmonary bypass (CPB). Mean patient age was 60 years (range: 36 to 77 years). Preoperatively, there were no biochemical or clinical signs of infection in any of the patients. Operative procedures consisted of coronary bypass grafting (CABG) or replacement of the mitral (MVR) or aortic valve (AVR). Anesthetic management was uniform in all patients and consisted of a high-dose fentanyl-pancuronium-oxygen regimen combined with benzodiazepines. Cardiopulmonary support was delivered using a pulsatile, low-flow, low-pressure perfusion regimen. During extracorporeal circulation flow was maintained at 1.8 $l/min/m^2$ during normothermia and at 1.1 $l/min/m^2$ during hypothermia. Body core temperature during cardiopulmonary bypass and during the early postoperative period was measured via a pulmonary artery catheter. Temperature during cardiopulmonary bypass decreased to 27.6 ± 0.7 °C. Mean total cardiopulmonary bypass time was 90 minutes (range: 34 - 150 minutes) and mean aortic cross-clamp time 63 minutes (range: 20 - 110 minutes). Weaning from cardiopulmonary support was started after rewarming the patients to normal body core temperatures. Postoperatively, all patients were admitted to the intensive care unit, and extubation of the trachea was performed within 14 hours.

Blood samples for determination of endotoxin serum concentration were collected before CPB (T1), immediately after CPB (T2), two hours (T3), and ten hours (T4) after CPB under sterile conditions. For determination of endotoxin blood was collected in vacuum collection tubes (KABI Diagnostica), separating plasma from the cellular elements by a gel after centrifugation at 600 g for 15 minutes. Measurements of endotoxin were performed in a kinetic, turbidimetric LAL (limulus amebocyte lysate)-assay, developed to take into account sample-related interferences with the LAL-endotoxin reaction by performing an individual internal standardisation in each sample. Diluted (1:5) and heated (10 min at 80°C) plasma samples were spiked with different endotoxin concentrations (1:10 dilutions, ranging from 5000 to 0.5 pg/ml), then LAL was added to each well. By this procedure, it can be avoided that external standard curves in water or in a pool of normal plasma have to be used. The increase in optical density was measured for 100 min at 30 sec intervals in a photometer (Thermomax, Molecular Devices) providing a constant temperature of 37 °C. The time of onset of the reaction was evaluated by the computer from the kinetics of each LAL-sample mixture. Based on these data, the endotoxin levels of the samples were determined using a mathematical model of a nonlinear curve fit, representing the internal standardisation. The endotoxin sensitivity of this procedure is 0.1 pg/ml (Reference endotoxin: NP3 (Novopyrexal, 100 pg NP3 = 1 EU EC-5; LAL: Associates of Cape Cod). The normal range of endotoxin levels in human plasma is < 1 pg/ml of endotoxin, representing the sensitivity for a diluted sample.

In 12 patients, determination of interleukin-6 in blood samples taken two hours after CPB (T3) was carried out using a commercially available ELISA method (R&D Systems, Inc. Minneapolis, USA). The sensitivity of this tests is 0.35 pg/ml.

For statistical analysis, not-detectable endotoxin levels were considered as 0.9 pg/ml because the detection limit of the assay was 1 pg/ml. Data are presented as mean ± SEM. Prebypass (T1) and postbypass (T2, T3, T4) endotoxin levels were compared using the one sample t-test. $^{*}P < 0.05$ was accepted as statistically significant. Regression analysis was performed according to the method of Pearson.

Results

Demographic data and the type of surgical procedure are shown in table 1. Prebypass control samples (T1) were negative for endotoxin in 13 patients and positive in 1 patient. Mean prebypass endotoxin concentration (T1) was 0.94 ± 0.03 pg/ml. Immediately after weaning from cardiopulmonary bypass (T2), endotoxemia was present in 10 patients (71%) in contrast to 4 patients (29%) without detectable serum endotoxin concentration. In the early postoperative period (T3, T4), 10 patients developed new endotoxemia. Postbypass endotoxin concentrations (T2) were significantly elevated to 2.84 ± 0.65 pg/ml. Endotoxin concentrations 2 hours (T3) and 10 hours (T4) after CPB were 3.44 ± 1.4 pg/ml and 3.05 ± 1.1 pg/ml (fig. 1). Two hours after the end of cardiopulmonary bypass interleukin-6 levels were increased to 180 ± 40 pg/ml.

The mean postoperative hyperthermic peak was 38.4 ± 0.1 °C, which showed no linear correlation to the endotoxin levels measured at any point of time (T2: $r = 0.01$; T3: $r = 0.15$; T4: $r = 0.13$). A regression plot of postoperative endotoxin levels (T4: 10 hours after CPB) and postoperative hyperthermic peaks is given in figure 2. Additionally, there was no linear correlation between postoperative IL-6 levels (T3) and postoperative hyperthermia (r = 0.3) (fig. 3).

No.	age/sex	operation
1	77/m	CABG (4X)
2	73/f	CABG(2x) +AVR
3	50/f	Redo-AVR
4	56/m	CABG (3x)
5	70/m	CABG (3x)
6	62/f	Redo-AVR
7	62/f	MVR
8	57/m	CABG (3x)
9	56/m	AVR
10	62/m	Redo-AVR
11	70/m	CABG (1x)
12	36/f	Redo-AVR
13	58/m	CABG (3x)
14	57/m	CABG (1x)

Table 1: Demographic data and type of surgery (Abbreviations see text)

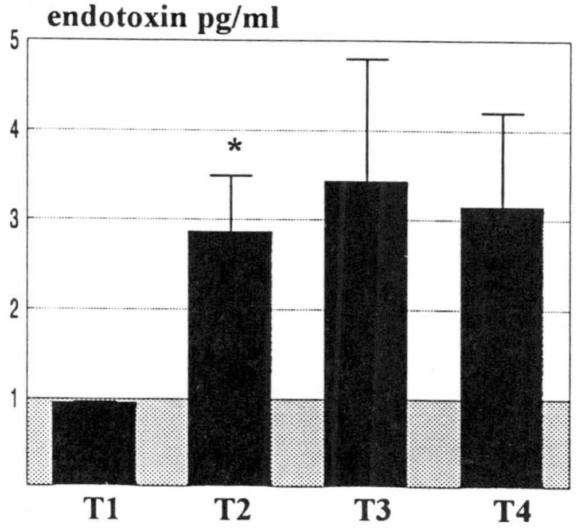

Fig. 1: Endotoxin concentration before (T1) immediately after (T2), two hours (T3) and ten hours (T4) after cardiopulmonary bypass *P<0.05

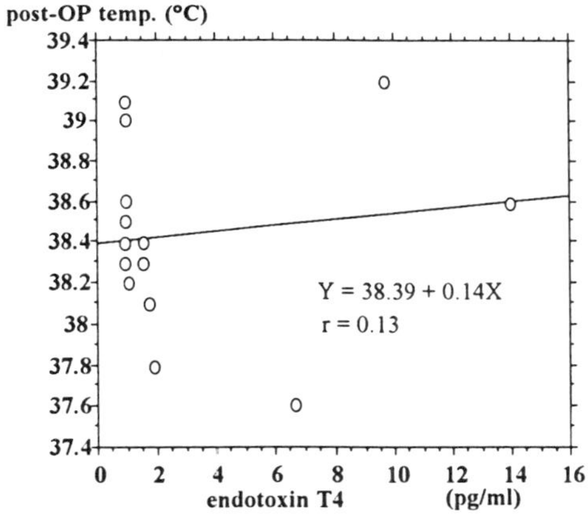

Fig. 2:Correlation between endotoxin (T4) and postoperative hyperthermic peak.

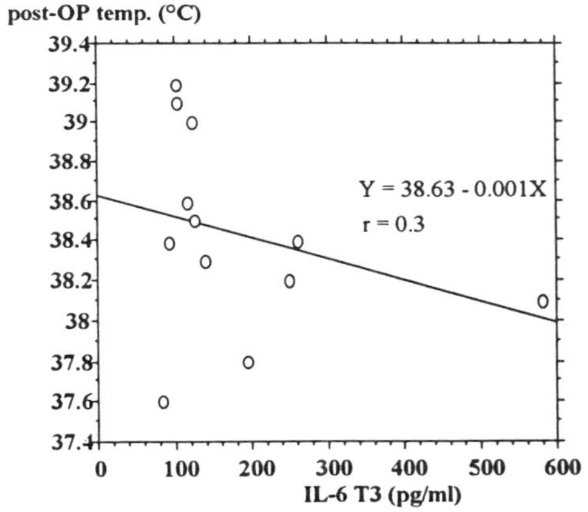

Fig. 3: Correlation between interleukin-6 concentration (T4) and post-operative hyperthermic peak.

Discussion

The postoperative course of body core temperature following cardiac surgery is characterized by an initial hypothermic to normothermic phase associated with shivering followed by hyperthermia. The maximal hyperthermic peak usually occurs 10 - 12 hours after surgery. This biphasic temperature course is similar to that seen in endotoxin-induced septic shock. Hyperthermia following cardiopulmonary bypass is believed to be part of a systemic inflammatory response syndrome characterized by an increase in extravascular water, organ dysfunction, vasoconstriction and coagulopathy (4). Various inflammatory mediators have been identified during and after cardiopulmonary bypass (4). In recent years, endotoxemia during and after cardiac surgery has been described (1,2,3). Endotoxin, a lipopolysaccaride of the cell wall of gram-negative bacteria, is a potent activator of inflammation and stimulates macrophages to release fever-inducing cytokines (5). Many of the biological effects of endotoxin are mediated by the local and systemic release of cytokines. Elevated levels of tumor necrosis factor (6) and interleukins 6 and 8 (7) have been demonstrated in relation to cardiopulmonary bypass. It has been hypothesized that this endotoxemia is caused by a translocation of endotoxin out of the intestines due to a hypoperfusion-induced damage of the mucosal barrier during cardiopulmonary bypass (3). Our results show that endotoxemia is present in 71% of the patients immediately after the end of cardiopulmonary bypass. A direct correlation between the level of circulating endotoxin and the maximal hyperthermic peak could not be demonstrated. Additionally, there was no correlation between the increased interleukin-6 concentration two hours after cardiopulmonary bypass and maximal postoperative hyperthermia. We conclude from these results that postoperative hyperthermia following cardiopulmonary bypass is the result of a multifactorial inflammatory process in which circulating endotoxin as activator of various fever-inducing cytokines may play a significant role. However, a direct correlation between postoperative endotoxin level and postoperative hyperthermia does not exist.

References

1. Nilsson, L, Kulander, L, Nystrom, SO, Eriksson, O (1990) : Endotoxins in cardiopulmonary bypass. J Thorac Cardiovasc Surg 100, 77-78.

2. Andersen, LW, Baek, L, Degn, H, Lehd, J, Krasnik, M, Rasmussen, JP (1987): Presence of circulating endotoxins during cardiac operations. J Thorac Cardiovasc Surg 93, 115-119.

3. Andersen, LW, Landow, L, Baek, L, Jansen, E, Baker, S (1993): Association between gastric intramucosal pH and splanchnic endotoxin, antibody to endotoxin, and tumor necrosis factor- concentrations in patients undergoing cardiolpulmonary bypass. Crit Care Med 21, 210-217.

4. Moat, NE, Shore, DF, Evans, TW (1993): Organ dysfunction and cardiopulmonary bypass: the role of complement and complement regulatory proteins. Eur J Cardio-thorac Surg 7, 563-573.

5. Michie, HR, Manogue, KR, Spriggs, DR (1988): Detection of circulating tumor necrosis factor after endotoxin administration. N Engl J Med 318, 1481-1486.

6. Jansen, NJ, Oeveren, W van, Brock, van den, Oudemans-van Straaten, HM, Stoutenbeck, CP, Joen, MC, Roozendaal, KJ, Eysman, L, Wildevuur, CR (1991): Inhibition by dexamethasone of the reperfusion phenomena in cardiopulmonary bypass. J Thorac Cardiovasc Surg 102, 515-525.

7. Finn, A, Naik, S, Klein, N, Leveninsky, RL, Srobel, S, Elliott, M (1993): Interleukin 8 release and neutrophil degranulation after paediatric cardiopulmonary bypass. J Thorac Cardiovasc Surg 105, 242-246.

Thermal Balance in Health and Disease
Advances in Pharmacological Sciences

SWEATING AND VASOCONSTRICTION THRESHOLDS DURING PROPOFOL SEDATION

K. Leslie, D. I. Sessler, A. R. Bjorksten,* M. Ozaki, M. Schroeder and S. Lin

Thermoregulation Research Laboratory, University of California, San Francisco, CA, 94143-0648. U.S.A., and *Department of Anesthesia, Royal Melbourne Hospital, Parkville, VIC, 3050. Australia.

Intraoperative hypothermia is common, and produces potentially serious complications. It results largely from anesthetic-induced inhibition of protective thermoregulatory repsonses (1,2). Typical surgical doses of volatile anesthetics (*e.g.*, isoflurane, enflurane) decrease the vasoconstriction threshold (core temperature triggering vasoconstriction) 2-4°C (3,4)and increase the sweating threshold (core temperature initiating sweating) ≈1°C (5,6).The result is an ≈4°C increase in the interthreshold range (core temperatures *not* triggering autonomic thermoregulatory responses).

The effects of intravenous anesthetics on thermoregulatory responses in humans are less well known. Fentanyl combined with nitrous oxide decreases the vasoconstriction threshold ≈3°C (7). Hynson *et al* demonstrated substantial impairment of thermoregulatory vasoconstriction during propofol/nitrous oxide anesthesia, with some volunteers demonstrating no thermoregulatory compensation at core temperatures as low as 33°C (8). However, several limitations of this study inspired us to re-evaluate the thermoregulatory effects of propofol.

Firstly, nitrous oxide itself has substantial effects on thermoregulatory thresholds (9).Secondly, core temperature was reduced by cooling sensate skin. Since skin temperature is an important thermoregulatory afferent signal, core-temperature thresholds generally cannot be determined from a cutaneous cooling protocol. Thirdly, the core temperature triggering sweating was not measured; this protocol, therefore, did not determine the inter-threshold range. And lastly, most volunteers were given only one dose of propofol, so a dose-response relationship was not established.

Accordingly, we tested the hypothesis that propofol increases the interthreshold range by raising the sweating threshold and decreasing the vasoconstriction threshold. We used a new model in which afferent thermal input from the legs was blocked by epidural anesthesia, and thermal manipulations were restricted to insensate lower-body skin.

Methods

We have so far studied three healthy male volunteers. None was obese, was taking medication, or was a serious athlete. The volunteers' height was 186 ± 3 cm (mean ± SD), weight 83 ± 4 kg, and age 31 ± 6 yr. Volunteers were studied on three randomly-ordered, consecutive days: 1) no propofol (control); 2) target propofol blood concentration 2 µg/ml; and, 3) target propofol blood concentration 4 µg/ml. Cooling was commenced at the same time each day to account for circadian temperature variations.

Protocol: An L2-3 epidural catheter and right internal jugular central venous catheter were inserted, using standard techniques. A vein in the non-dominant arm was cannulated for fluid and drug administration. Routine anesthetic parameters, including blood pressure, heart rate and oxygen saturation were monitored continuously.

Ambient temperature was maintained near 22°C. Upper-body (sensate) skin temperature was kept constant (35.5-36°C) by manipulating the output of a modified Bair Hugger® forced-air warmer (Augustine Medical, Inc., Eden Prarie, MN). The torso and arms were wrapped in plastic to prevent evaporative heat loss.

Induction of epidural anesthesia was preceded by one hour of cutaneous warming to minimize redistribution hypothermia (1). Vasoconstriction resulting from inadequate vascular volume was prevented by infusion of 15 ml/kg lactated Ringer's solution. Epidural anesthesia then was induced using ≈20 ml 2% 2-chloroprocaine (Chloroprocaine HCl, USP®, Abbott Laboratories, North Chicago, IL). A dermatomal level of sensory blockade near T10 was maintained using an infusion of 2% 2-chloroprocaine (≈20 ml/h). The study infusion then was commenced (on the control day, no propofol was administered). Propofol was infused using an infusion pump (Ohmeda, Inc., Liberty Corner, NJ) controlled by a computer programmed to target propofol blood concentrations of 2 and 4 µg/ml. Airway and ventilatory support were provided as necessary.

The insensate legs were wrapped in plastic and were warmed using a water mattress (Blanketrol II, Maxi-Therm blanket, Cincinnati Sub-Zero, Cincinnati, OH) set to 42°C, and a Bair Hugger® set at "high" (≈42°C). Warming was continued until the sweating threshold was identified. Core hypothermia then was induced by decreasing the temperature of the water mattress under the legs to 10°C and exposing the legs to convective and evaporative cooling. Cooling continued until the fingers vasoconstricted.

Measurements: Core temperature was measured at the tympanic membrane. The aural probe then was inserted by volunteers until they felt the thermocouple touch the tympanic membrane; appropriate placement was confirmed when volunteers easily detected a gentle rubbing of the attached wire. The probe was then securely taped in place, the aural canal occluded with cotton, and a gauze bandage positioned over the external ear. Tympanic membrane temperatures correlate well with distal esophageal temperatures during anesthesia (10). Area-weighted, upper-body skin temperature was computed from measurements at 7 sites and lower-body skin temperature was similarly derived from 8 sites, as previously described (6).

The rate of sweating was determined using a ventilated capsule situated on the chest, as previously described (6).The core temperature triggering a sweating rate of 40 $g \cdot m^{-2} \cdot h^{-1}$ was considered the threshold for sweating. Absolute right middle fingertip blood flow was quantified using venous-occlusion volume plethysmography at 5-min intervals, as previously described (11).The vasoconstriction threshold was defined as the core temperature triggering a sudden and sustained decrease of finger flow to <0.25 ml/min. Vascular tone in the legs was assessed using calf minus toe, skin-temperature gradients (3).Gradients <0°C identified vasodilation.

Blood was sampled from the central venous catheter at each threshold, and subsequently analysed for propofol content using high-pressure liquid chromatography.

Statistical Analysis: Linear regression was applied to sweating and vasoconstriction data from each volunteer. Results for the group are presented as means ± SD.

	SWEATING		VASOCONSTRICTION	
SUBJECT	Slope (°C·μg⁻¹·ml⁻¹)	r^2	Slope (°C·μg⁻¹·ml⁻¹)	r^2
1	-0.08	0.98	-0.45	0.87
2	-0.09	0.50	-1.16	1
3	-0.06	0.91	-0.68	0.99
Ave ± SD	-0.08 ± 0.01	0.80 ± 0.26	-0.76 ± 0.35	0.95 ± 0.07

Table: The slopes [change in threshold (°C) per change in blood propofol concentration (μg/ml)] for sweating and vasoconstriction during propofol sedation.

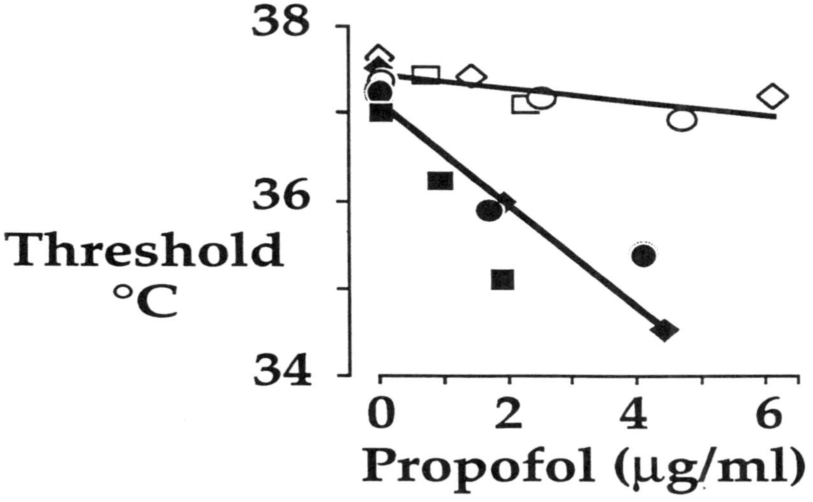

Fig. 1: Upper points (open markers) and line show the sweating threshold; lower points (closed markers) and line show vasoconstriction; each shape represents a single volunteer. Propofol, in sedation doses, thus markedly increases the interthreshold range, but does so without increasing the sweating threshold.

Results

The volunteers were mildly sedated when 2 µg/ml propofol blood concentration was targeted, and heavily sedated when the target was 4 µg/ml. Epidural-induced sensory block level at the sweating and vasoconstriction thresholds were similar (near T10). The legs remained vasodilated throughout epidural anesthesia in all volunteers. Upper-body skin temperatures were well controlled throughout each study, and comparable in the three volunteers.

The vasoconstriction threshold temperature decreased linearly with increasing propofol blood concentration, with a slope of -0.76 ± 0.35 $°C \cdot \mu g^{-1} \cdot ml^{-1}$. In contrast, the sweating threshold temperature was not affected by increasing propofol concentration. The interthreshold range was near 0°C during epidural analgesia alone, but increased linearly to ≈3°C when propofol blood concentration was 4 µg/ml (Table and Figure).

Discussion

Our results indicate that propofol, in sedation doses, substantially increases the interthreshold range. The range increases because propofol significantly decreased the vasoconstriction threshold. In contrast, the sweating threshold temperature remained nearly normal. This pattern differs distinctly from that produced by volatile anesthetics, which both decrease the vasoconstriction threshold (3,4) and increase the sweating threshold (5,6). Although the mechanism by which anesthetics impair thermoregulation remains to be determined, these data suggest that the specific thermoregulatory actions of propofol and the volatile anesthetics may differ.

Propofol is a potent inhibitor of thermoregulatory vasoconstriction, with sedation doses decreasing the threshold roughly 2°C. It is difficult to directly compare anesthetic potency of volatile and intravenous anesthetics. Comparison between our current and previous data (3,4) is further complicated because the studies of volatile agents used sensate skin for cooling and different criteria for vasoconstriction. Nonetheless, propofol appears to cause more inhibition than isoflurane (3) or enflurane (4).

Epidural anesthesia alone has significant effects on central thermoregulation (12).Studies in which the core was cooled by administration of cold intravenous fluids (with constant skin temperature), show that regional anesthesia increases the interthreshold range from normal values near 0.2°C (13) to ≈0.6-0.9°C (12).This increase may result because regional anesthesia markedly increases *apparent* (as opposed to actual) leg skin temperature.

In our study, where the insensate legs were used for cooling, the inter-threshold range was less than reported previously during regional anesthesia (12). This small interthreshold range may result because afferent thermal from the legs was not fully blocked (despite obliteration of consciously perceived leg thermal). Alternatively, it may simply reflect the small number of volunteers we have so far studied. Threshold differences induced by propofol administration can nontheless be attributed solely to the drug because epidural anesthesia was maintained on each study day. In conclusion, propofol significantly increases the interthreshold range, and does so by decreasing the vasoconstriction threshold. In contrast, the sweating threshold is not affected. This pattern of impairment differs significantly from that produced by volatile anesthetics.

Acknowledgments

These studies were supported by National Institutes of Health grants GM 39723 and GM 49670.

References

1. Sessler, DI, McGuire, J, Moayeri, A, Hynson, J (1991): Isoflurane-induced vasodilation minimally increases cutaneous heat loss. Anesthesiology 74, 226-232.

2. Belani, K, Sessler, DI, Sessler, AM, Schroeder, M, McGuire, J, Washington, D, Moayeri, A (1993): Leg heat content continues to decrease during the core temperature plateau in humans. Anesthesiology 78, 856-863.

3. Støen, R, Sessler, DI (1990): The thermoregulatory threshold is inversely proportional to isoflurane concentration. Anesthesiology 72, 822-827.

4. Washington, DE, Sessler, DI, McGuire, J, Hynson, J, Schroeder, M, Moayeri, A (1992): Painful stimulation minimally increases the thermoregulatory threshold for vasoconstriction during enflurane anesthesia in humans. Anesthesiology 77, 286-290.

5. Washington, D, Sessler, DI, Moayeri, A, Merrifield, B, Prager, M, McGuire, J, Belani, K, Hudson, S, Schroeder, M (1993): Thermoregulatory responses to hyperthermia during isoflurane anesthesia in humans. J Appl Physiol 74, 82-87.

6. Lopez, M, Ozaki, M, Sessler, DI, Valdes, M (1993): Physiological responses to hyperthermia during epidural anesthesia and combined epidural/enflurane anesthesia in women. Anesthesiology 78, 1046-1054.

7. Sessler, DI, Olofsson, CI, Rubinstein, EH (1988): The thermoregulatory threshold in humans during nitrous oxide-fentanyl anesthesia. Anesthesiology 69, 357-364.

8. Hynson, JM, Sessler, DI, Belani, K, Washington, D, McGuire, J, Merrifield, B, Schroeder, M, Moayeri, A, Crankshaw, D, Hudson, S (1992): Thermoregulatory vasoconstriction during propofol/nitrous oxide anesthesia in humans: Threshold and Sp_{O2}. Anesth Analg 75, 947-952.

9. Passias, TC, Mekjavic, IB, Eiken, O (1992): The effect of 30% nitrous oxide on thermoregulatory responses in humans during hypothermia. Anesthesiology 76, 550-559.

10. Cork, RC, Vaughan, RW, Humphrey, LS (1983): Precision and accuracy of intraoperative temperature monitoring. Anesth Analg 62, 211-214.

11. Rubinstein, EH, Sessler, DI (1990): Skin-surface temperature gradients correlate with finger-tip blood flow in humans. Anesthesiology 73, 541-545.

12. Kurz, A, Sessler, DI, Schroeder, M, Kurz, M (1993): Thermoregulatory response thresholds during spinal anesthesia. Anesth Analg 77, 721-726.

13. Lopez, M, Sessler, DI, Walter, K, Emerick, T, Ozaki, M Rate- and gender-dependence of the sweating, vasoconstriction, and shivering thresholds in humans. Anesthesiology, in press

Thermal Balance in Health and Disease
Advances in Pharmacological Sciences

EPIDURAL ANESTHESIA INCREASES APPARENT LEG TEMPERATURE AND DECREASES THE SHIVERING THRESHOLD

M. Ozaki, T. Emerick, D. I. Sessler, K. Walters and M. Schroeder

Thermoregulation Research Laboratory, University of California, San Francisco, California 94143-0648.

Hypothermia during regional anesthesia is as common, and nearly as severe, as that occurring during general anesthesia. Comparable thermal lability occurs in patients with spinal cord transections (1).Hypothermia results, in part, because lower core temperatures than usual are required to trigger shivering during major conduction anesthesia (2).Typically, the shivering threshold (core temperature triggering shivering) is reduced ≈0.6°C by regional anesthesia (3), but the etiology of this impairment remains unknown. Although small compared with the inhibition produced by general anesthetics, 0.6°C far exceeds the normal sweating-to-vasoconstriction interthreshold range of 0.2°C (4).

Induction of regional anesthesia usually is accompanied by a sensation of increased warmth in the affected body parts; patients, for example, often comment that their legs feel warmer after induction of epidural or spinal anesthesia. Some of this warming sensation certainly is a response to an actual increase in leg temperature, resulting from a core-to-peripheral redistribution of body heat. However, the ≈1°C actual increase in leg skin temperature may not fully explain the reported sensation of warmth. Furthermore, the sensation of relative warmth persists when surgical levels of anesthesia are established — and thus well after actual leg temperature can be directly sensed.

Conduction anesthesia blocks all thermal sensations; however, cold signals are disproportionately affected because at typical skin temperatures mostly cold receptors fire tonically (5). It thus seems likely that epidural and spinal anesthesia increase the leg temperature perceived by the thermoregulatory system. Because skin temperature contributes ≈5-20% to thermoregulatory control (6,7), increased *apparent* (as opposed to actual) leg temperature would produce a complimentary decrease in the core temperature triggering thermoregulatory shivering.

Accordingly, we tested the hypothesis that abnormal tolerance for hypothermia during epidural anesthesia coincides with an increase in apparent leg temperature. Apparent temperature was defined as the leg skin temperature required to induce a reduction in the shivering threshold comparable to that produced by epidural anesthesia.

Methods

We studied six young, healthy female volunteers. Four used oral contraceptives, and were studied during weeks one and two of their three-week-on/one-week-off contraceptive cycle; the remaining volunteers were studied during the first 10 days of the menstrual cycles. The volunteers' height was 164 ± 6 cm (mean ± SD), weight 56 ± 8 kg, and age 25 ± 2 yr.

Volunteers were studied on four randomly-ordered days: 1) leg skin temperature near 32°C (*cold*); 2) leg skin temperature near 36°C (*warm*); 3) leg skin temperature near 38°C (*hot*); 4) epidural anesthesia without leg warming (*epidural*). At each designated leg skin temperature, core hypothermia sufficient to evoke shivering was induced by central venous infusion of cold lactated Ringer's solution.

Protocol: Each subject was scheduled such that shivering occurred at nearly the same time on each of the four study days. Throughout the four study days, upper-body skin temperature was kept constant, near 35.7°C, by manipulating output of a Bair Hugger® forced-air warmer (Augustine Medical, Inc., Eden Prarie, MN). On the *cold* day, the legs were exposed to the ambient environment. On the other days, leg temperature was adjusted to target values using forced-air and circulating-water warmers.

Epidural anesthesia was induced using 15-20 ml 2% 2-chloroprocaine without epinephrine. Subsequently, a continuous infusion of 2% 2-chloroprocaine was administered at a rate of 13-18 ml/h to maintain a T10 sensory blockade level. In preliminary studies, we noticed that apparent temperatures were less in volunteers with constricted toes than those with dilated toes (although cold sensation was blocked in each case). Consequently, the epidural catheter position was adjusted if necessary and additional anesthetic administered to maintain toe vasodilation in each volunteer throughout anesthesia. Chloroprocaine was chosen as the epidural anesthetic because it is rapidly metabolized in plasma; central recirculation of this anesthetic *via* blood thus is unlikely to alter thermoregulatory responses.

Core hypothermia (at a rate of ≈0.8°C/h) was induced by central-venous infusion of lactated Ringer's solution cooled to ≈3°C. Cooling was begun after 1.5 h of leg exposure to ambient temperature on the *cold* day and 2.5 h of leg heating on the *warm* and *hot* days. Core cooling on the *epidural* day was initiated ≈1 h after induction of epidural anesthesia.

Monitoring: Core temperature was measured at the tympanic membrane. Visual inspection with an otoscope confirmed that the ear canal was free of wax in each volunteer. The aural probe then was inserted by volunteers until they felt the thermocouple touch the tympanic membrane; appropriate placement was confirmed when volunteers easily detected a gentle rubbing of the attached wire. The aural canal was occluded with cotton, the probe securely taped in place, and a gauze bandage positioned over the external ear. Tympanic membrane temperatures correlate well with distal esophageal temperatures in the perianesthetic period (8). Upper-body skin-surface temperatures were computed from measurements at 7 sites and leg skin-surface temperatures similarly were derived from 7 sites.

Calf minus toe, skin-surface temperature gradients were used as an index of leg perfusion (9). As in previous studies, we considered a gradient exceeding 4°C to indicate vasoconstriction, and a gradient less than 0°C to indicate vasodilation. Oxygen consumption was measured using a canopy-based metabolic monitor (Deltatrac™). A sudden and sustained increase in oxygen consumption (generally exceeding 150% of baseline values) identified shivering; the core temperature triggering shivering defined the threshold for this response.

Statistical Analysis: In each volunteer, linear regression was used to calculate the correlation between the shivering thresholds on the three non-epidural days and concurrent leg skin temperatures. The slope of this regression equation thus indicated the extent to which leg warming increased thermoregulatory tolerance for core hypothermia, and was expressed as a percentage leg skin contribution. The skin temperature which would have been required to produce the observed shivering threshold during epidural anesthesia then was interpolated from the regression; we designated this value as the *apparent* skin temperature.

The regression coefficients, apparent temperature, and differences between the apparent and the actual leg temperature during epidural anesthesia and those on the *cold* day were compared using unpaired, two-tailed *t* tests. Results are expressed as means ± SD; differences were considered statistically significant when $P < 0.01$.

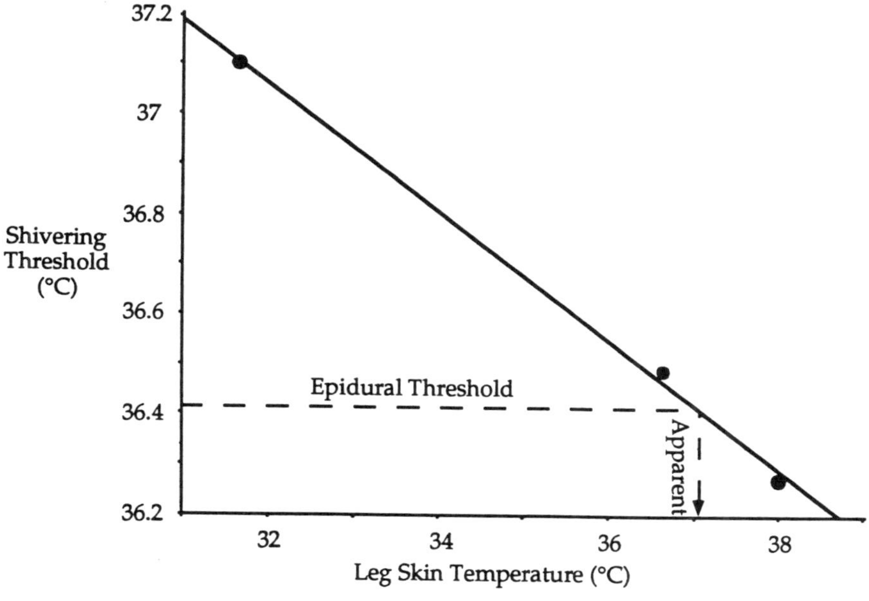

Fig. 1: Results in one volunteer. There was a good linear relationship (solid line) between the shivering thresholds and leg skin temperatures ($r^2 = 0.99$). The slope of the leg skin temperature *vs.* the shivering threshold regression indicates the extent to which leg skin temperature contributes to central thermoregulatory control of shivering. In this case, leg skin contributed 13% of the total thermal input. *Apparent* temperature during epidural anesthesia was interpolated from the regression of leg skin temperature against the shivering threshold (dashed lines), and was 37.3°C, which substantially exceeded leg temperatures on the cold day (typical clinical situation) and actual leg temperatures during epidural anesthesia.

Results

Upper-body skin temperatures were well controlled, with no statistically significant or clinically important differences among the volunteers, or on the different study days in individual volunteers.

The shivering threshold during epidural anesthesia was 0.7 ± 0.2°C less than that on the *cold* study day. There was a good linear relationship between the shivering threshold and leg skin temperature. The correlation coefficients (r^2) averaged 0.94 ± 0.06, and in all cases exceeded 0.85. The contribution of leg skin temperature to the shivering threshold was 15 ± 5% of the core contribution.

Apparent skin temperature significantly exceeded actual leg skin temperature both during epidural anesthesia (4.1 ± 1.0°C) and on the *cold* study day (5.8 ± 1.0°C). As an example, the results in one volunteer are shown in figure 1.

Discussion

Regional anesthesia may interfere with all components of thermoregulation: afferent sensing, central control, and efferent responses. Efferent thermoregulatory responses such as sweating, vasoconstriction, and shivering are clinically important responses which certainly are obliterated by regional anesthesia. However, there is considerable evidence that responses above the level of anesthetic-induced sympathetic and motor block also are abnormal. In 1972, Roe *et al.* observed that 1-3°C hypothermia was typical during spinal anesthesia, perceptively postulating that it resulted because the shivering threshold was ≈1°C below normal (2)and we have demonstrated similar abnormal tolerance to core hypothermia (3).

It is unlikely that anesthetics injected intrathecally or extra-durally diffuse rostrally in sufficient amounts to directly anesthetize higher regulatory centers. An alternative explanation is that epidurally administered anesthetics might indirectly impair central thermoregulation *via* absorption of local anesthetic into the blood stream and subsequent re-circulation to the central nervous system. However, plasma lidocaine concentrations similar to those observed during epidural anesthesia do not alter the vasoconstriction or shivering thresholds (10).Furthermore, chloroprocaine epidural anesthesia causes distinct thermoregulatory impairment, although this drug is rapidly metabolized in plasma. These data suggest that another mechanism mediates thermoregulatory impairment during regional anesthesia.

Both skin and core temperatures contribute to thermoregulation (7). Most warm receptors are nearly quiescent at skin temperatures ≤37°C (5). Consequently, at typical ambient temperatures, predominantly cold signals converge on the central thermoregulatory system. Regional anesthesia blocks all thermal information from the lower portion of the body, but tonic cold signals would be the major traffic disrupted. In this scenario, the central thermoregulatory system might interpret the absence of tonic cold input as relative warming, and therefore an increase in apparent leg temperature.

An increase in *apparent* leg temperature, combined with intact central thermoregulatory processing, would alter responses to core hypothermia in the same fashion as vigorous leg warming without anesthesia. Our data are consistent with this mechanism, and demonstrate that apparent leg skin temperature far exceed both actual leg temperature and leg temperature on the *cold* day (representing the typical clinical situation). Apparent leg tissue temperature similarly exceeded both actual tissue temperature during epidural anesthesia and on the *cold* day. The associated reduction in the shivering threshold on the *epidural vs. cold* study days was 0.7 ± 0.2°C, which is similar to values previously reported during epidural (11) and spinal (3) anesthesia.

We note, however, that quantifying the leg temperature required to produce thermoregulatory impairment comparable to that induced by regional anesthesia does not prove that a high apparent leg temperature is the *mechanism* of that impairment. While perhaps less likely, other potential mechanisms of impairment include direct inhibition of thermal integration at the level of the lumbar spinal cord. Integration does occur at that level, although most thermoregulatory processing occurs at higher levels, including the brainstem, midbrain, and hypothalamus.

Although a variety of thermoregulatory models have been proposed, the actual mechanisms by which thermal afferents initiate efferent responses remains virtually unknown. Consequently, it is unclear whether central control systems respond to integrated warm and cold afferent input, or whether warm and cold responses are independently processed from their respective afferent signals. A decreased shivering threshold during regional anesthesia would be expected in each case; however, warm responses such as sweating would be affected differently.

If both warm and cold regulatory responses are synchronously controlled by integrated afferent input, the sweating threshold would be comparably reduced during regional anesthesia. That is, a high apparent leg temperature would decrease regulatory tolerance for core hyperthermia. In contrast, the sweating threshold during epidural anesthesia likely would be increased by a system in which warm responses are largely controlled by warm afferent signals (rather than a combination of both warm and cold signals). This increase would result because anesthetic-induced inhibition of warm signals (even if they constitute a small fraction of the total thermal input) would *decrease* leg temperature apparent to the system integrating warm input and controlling sweating. We were unable to evaluate the sweating threshold in this protocol, but previous work suggests that the threshold may indeed be elevated by regional anesthesia (3).

A linear relation between skin and core temperatures triggering thermoregulation responses is consistent with numerous previous studies (6). For practical reasons, we evaluated shivering thresholds without anesthesia only at three different leg temperatures. Thus only three sets of data were included in our regression calculations. More data would increase the accuracy of the regression equations; nonetheless, the correlation coefficients were high and the relations clearly linear over the studied range.

Linear models usually include cutaneous contributions ranging from 5% (6) to ≈20% (7) of the core input. The exact cutaneous contribution, however, remains unclear because it has proven difficult to independently manipulate core and skin temperatures in humans. Certainly, the distinct contribution of leg skin temperature has not previously been established under steady-state conditions. Our data indicate that the shivering threshold is ≈15% determined by leg skin temperature. Since the legs represent roughly half the body surface area and the upper body is more sensitive to thermal stimuli, these data do not seem consistent with a cutaneous input less than 20%.

In summary, epidural anesthesia decreased the shivering threshold ≈0.7°C, a value far exceeding the normal interthreshold range of ≈0.2°C. Leg skin temperature contributed 15 ± 5% to central control of shivering. Apparent leg skin temperature was 37.8 ± 0.5°C, which significantly exceeded actual leg temperature during epidural anesthesia and on the "cold" day (the typical clinical condition). These data suggest that the shivering threshold is reduced during epidural anesthesia because anesthesia significantly increases apparent (as opposed to actual) leg skin temperature.

Acknowledgments

These studies were supported by Augustine Medical, Inc., and National Institutes of Health grants GM 39723 and GM 49670. Dr. Ozaki is the 1992-1993 Augustine Scholar.

References

1. Hales, JRS (1984): Thermal Physiology. In: Temperature regulatory set-point in paraplegics. Attia, M, Engel, P eds. New York: Raven Press, 79-82.

2. Roe, CF, Cohn, FL (1972): The causes of hypothermia during spinal anesthesia. Surg Gynecol Obstetr 135, 577-580.

3. Kurz, A, Sessler, DI, Schroeder, M, Kurz, M (1993): Thermoregulatory response thresholds during spinal anesthesia. Anesth Analg 77, 721-726.

4. Lopez, M, Sessler, DI, Walter, K, Emerick, T, Ozaki, M Rate- and gender-dependence of the sweating, vasoconstriction, and shivering thresholds in humans. Anesthesiology, in press

5. Pierau, F, Wurster, RD (1981): Primary afferent input from cutaneous thermoreceptors. Fed Proc 40, 2819-2824.

6. Wyss, CR, Brengelmann, GL, Johnson, JM, Rowell, LB, Silverstein, D (1975): Altered control of skin blood flow at high skin and core temperatures. J Appl Physiol 38, 839-845.

7. Benzinger, TH, Pratt, AW, Kitzinger, C (1961): The thermostatic control of human metabolic heat production. Proc Natl Acad Sci USA 47, 730-739.

8. Cork, RC, Vaughan, RW, Humphrey, LS (1983): Precision and accuracy of intraoperative temperature monitoring. Anesth Analg 62, 211-214.

9. Støen, R, Sessler, DI (1990): The thermoregulatory threshold is inversely proportional to isoflurane concentration. Anesthesiology 72, 822-827.

10. Glosten, B, Sessler, DI, Ostman, LG, Faure, EAM, Thisted, RA (1991): Intravenous lidocaine does not cause tremor or alter thermoregulation. Reg Anesth 16, 218-222.

11. Sessler, DI, Ponte, J (1990): Shivering during epidural anesthesia. Anesthesiology 72, 816-821.

Thermal Balance in Health and Disease
Advances in Pharmacological Sciences

HEAT BALANCE AND DISTRIBUTION DURING EPIDURAL ANESTHESIA

J. M. Hynson, D. I. Sessler, B. Glosten and J. McGuire

Thermoregulation Research Laboratory, University of California, San Francisco, California 94143-0648

Lumbar epidural anesthesia is commonly used for leg and lower abdominal surgical procedures (1). Local anesthetic agents injected into the epidural space produce anesthesia by blocking nerve conduction in the spinal cord and spinal nerve roots (1). Sympathetic neural blockade may increase lower body cutaneous and arterio-venous shunt blood flow; sensory blockade eliminates input from cutaneous thermal receptors; and motor blockade abolishes or reduces the effectiveness of shivering (2). Disruption of these mechanisms frequently leads to hypothermia (2-5). Core hypothermia often provokes intense thermoregulatory responses (vasoconstriction and shivering) in the unblocked upper body (2-4). We performed two studies to evaluate the etiology of the observed core hypothermia and changes in thermal balance during epidural anesthesia.

Thermal balance

The etiology of hypothermia during epidural anesthesia remains unknown. It is widely believed that increased cutaneous heat loss, due to cutaneous vasodilation, leads to negative heat balance. However, induction of general anesthesia increases cutaneous heat loss only slightly and hypothermia during general anesthesia is initially due to redistribution of body heat (6). We therefore tested the hypothesis that redistribution of body heat (from central to peripheral tissues) contributes substantially to hypothermia during epidural anesthesia.

Methods

We studied thermal balance in five healthy volunteers. After placement of a lumbar epidural catheter, each volunteer was exposed to an ambient temperature of 20°C for two hours. Epidural anesthesia was then induced by injection of 30-50 ml of 3% chloroprocaine. Additional anesthetic was given to achieve a thermal sensory block level to at least the fifth thoracic dermatome. Tympanic membrane temperature and skin temperature (area-weighted average of ten sites) were measured. Cutaneous heat loss was measured from 10 area-weighted sites using thermal flux transducers. Oxygen consumption, measured with a Deltatrac metabolic monitor, was converted to equivalent heat production.

Results

During the first hour of anesthesia, tympanic membrane temperature decreased 1.1 ± 0.3°C, and average skin temperature increased 0.9 ± 0.5°C. Cutaneous heat loss, increased 16 ± 6% (15 ± 5 W). However, upper-body shivering increased oxygen consumption and metabolic heat production increased even more (Figure 1).

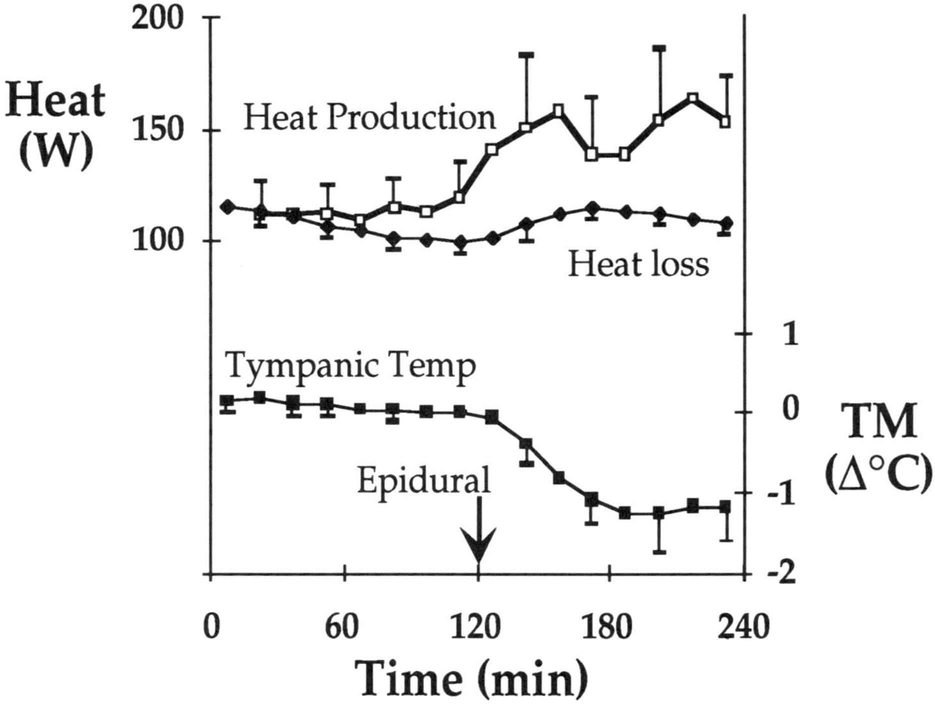

Fig. 1: Tympanic membrane temperature did not change significantly during the 120-min period before induction of epidural anesthesia (arrow), but decreased rapidly after. Mean skin temperature and cutaneous heat loss decreased slightly during the pre-induction period. Both increased following induction of anesthesia. Metabolic heat production increased more than cutaneous heat loss did; thermal balance, therefore, was positive. Tympanic membrane temperature decreased despite this positive thermal balance, indicating that core hypothermia results primarily from central-to-peripheral redistribution of body heat.

Average lower-body skin temperature increased in response to epidural anesthesia, while upper-body skin temperature did not (Figure 2). Cutaneous heat loss from the upper and lower body followed the same pattern, with all of the total increase induced by epidural anesthesia resulting from changes in the lower body.

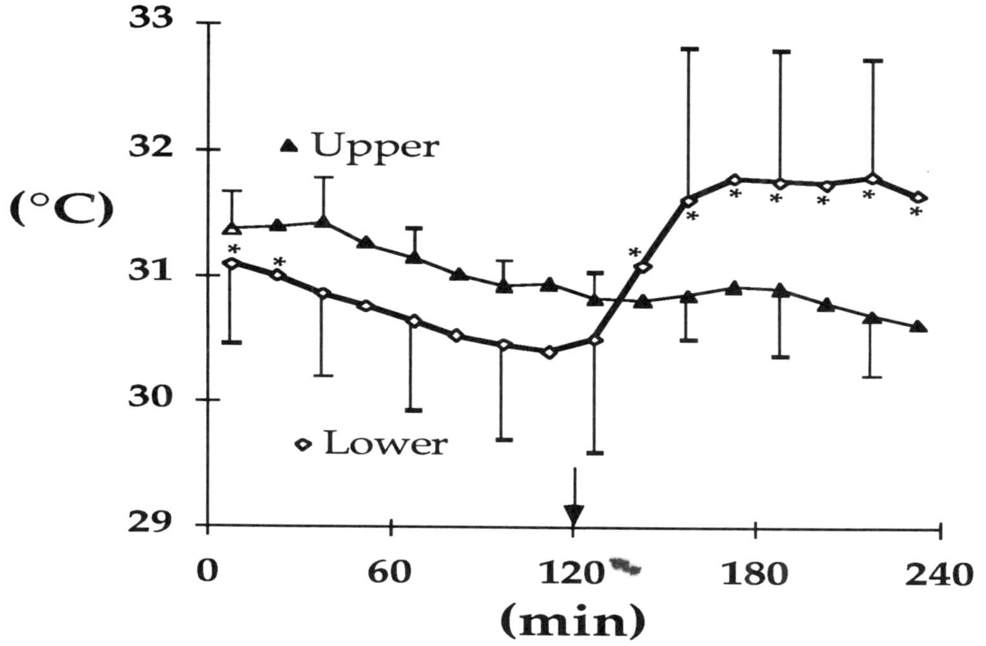

Fig. 2: Before induction of epidural anesthesia, average skin temperatures for the upper and lower body decreased slightly. Average skin temperature for the lower body increased significantly after induction of epidural anesthesia. In contrast, skin temperature of the upper body did not, reflecting the preserved sympathetic innervation of upper body skin. An asterisk (*) indicates values statistically different than those recorded during the 15-min period immediately before induction of epidural anesthesia.

Discussion

Leg tissue comprises a large portion of the peripheral tissue mass, and leg heat content decreases during thermoregulatory vasoconstriction (7). Induction of epidural sympathetic blockade increases leg blood flow and likely leads to transfer of heat into this tissue mass. During epidural anesthesia, increased heat production associated with shivering exceeded the increase in cutaneous heat loss. Heat balance therefore remained positive throughout anesthesia. The observed decrease in core temperature must thus result from redistribution of body heat from core to peripheral tissues.

Prevention of redistribution hypothermia

We have thus seen that, analogous to general anesthesia, hypothermia during epidural anesthesia is primarily due to redistribution of body heat. Redistribution hypothermia occurring during general anesthesia is difficult to treat (8), but can be prevented in two ways. One approach is to warm peripheral tissues before induction of anesthesia (9). When anesthesia is subsequently induced, there is less core-to-peripheral redistribution of heat because there is a smaller core-to-peripheral temperature gradient. The other approach is to increase peripheral heat content pharmacologically. Nifedipine, a potent vasodilating medication, decreases redistribution hypothermia when given 12 h before induction of general anesthesia (10). We thus tested the hypothesis that warming peripheral tissues before induction of epidural anesthesia would also minimize redistribution hypothermia.

Methods

Seven additional volunteers were each given two separate epidural anesthetics during a single study day. One anesthetic was preceded by a 2-h period of skin-surface warming (provided by a forced-air warmer set at 38°C). The other injection was preceded by 2 hours of exposure to an ambient temperature near 22°C. In either case, volunteers were exposed to an ambient temperature near 22°C after induction of anesthesia. Anesthesia was allowed to completely resolve before starting the second pre-induction treatment.

Results

Immediately before induction of epidural anesthesia, core temperatures were similar with and without pre-warming. Following induction of anesthesia, skin temperature and heat loss were greater when volunteers were pre-warmed; however, core temperature decreased less. The maximum decrease in core temperature during anesthesia after pre-warming was less than half of that measured after no pre-warming (0.3 ± 0.3°C *vs.* 0.7 ± 0.3°C) (Figure 3). Shivering was observed in four volunteers after pre-warming; however, all seven shivered during epidural anesthesia without pre-warming.

Discussion

Despite greater cutaneous heat loss, pre-warming largely prevented core hypothermia and effectively reduced the incidence of shivering. The effect of pre-warming on thermal balance was likely greater than that suggested by change in core temperature since more intense shivering in the unwarmed volunteers would have increased heat production and countered the decrease in core temperature.

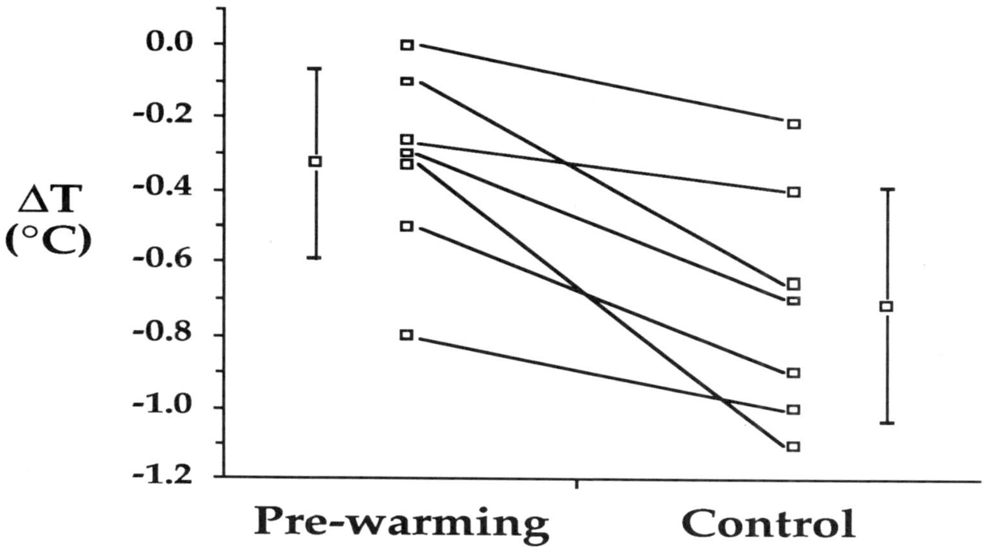

Fig. 3: The maximum decrease in tympanic membrane temperature in each volunteer after induction of epidural anesthesia was less when volunteers were pre-warmed.

Summary

These studies demonstrate that hypothermia during epidural anesthesia does not result from negative heat balance, *i.e.* heat loss to the environment in excess of metabolic heat production, but rather from redistribution of body heat from core to peripheral tissues. Cutaneous vasoconstriction in response to exposure in a cool environment effectively constrains heat to the core (7). Lumbar epidural anesthesia, by blocking sympathetic efferents to cutaneous arterioles, produces cutaneous vasodilation in the lower body (blocked area). Enhanced cutaneous blood flow increases skin temperature and cutaneous heat loss, but more importantly promotes transfer of heat from the thermal core to peripheral tissues. The result is core hypothermia. Warming peripheral tissues before induction of epidural anesthesia produces peripheral vasodilation and increases peripheral tissue heat storage, thus minimizing redistribution hypothermia.

References

1. Cousins, MJ, Bromage, PR (1988): Epidural neural blockade. In: Neural Blockade in Clinical Anesthesia and Management of Pain. 2nd Ed. Cousins, MJ, Bridenbaugh, P eds. Philadelphia: JB Lippincott Co, 253-254, 306-308.

2. Hynson, JM, Sessler, DI, Glosten, B, McGuire, J (1991): Thermal balance and tremor patterns during epidural anesthesia. Anesthesiology 74:680-690.

3. Glosten, B, Hynson, JM, Sessler, DI, McGuire, J. (1993): Preanesthetic skin-surface warming reduces redistribution hypothermia caused by epidural block. Anesth. Analg. 77:488-493.

4. Sessler, DI, Ponte, J (1990): Shivering during epidural anesthesia. Anesthesiology 72:816-821.

5. Frank, SM, Beattie, C, Christopherson, R, Norris, EJ, Rock P, Parker, S, Kimball, AW, Jr (1992): Epidural *versus* general anesthesia, ambient operating room termperature, and patient age as predictors of inadvertent hypothermia. Anesthesiology 77:252-257.

6. Sessler, DI, McGuire, J, Moayeri, A, Hynson, J (1991): Isoflurane-induced vasodilation minimally increases cutaneous heat loss. Anesthesiology 74:226-232.

7. Belani, K, Sessler, DI, Sessler, AM, Schroeder, M, McGuire, J, Merrifield, B, Washington, DE, Moayeri, A (1993): Leg heat content continues to decrease during the core temperature plateau in humans anesthetized with isoflurane. Anesthesiology 78:856-863.

8. Hynson, JM, Sessler, DI (1992): Intraoperative warming therapies: A comparison of three devices. J. Clin. Anesth. 4:194-199.

9. Hynson, JM, Sessler, DI, Moayeri, A, McGuire, J, Schroeder, M (1993): The effects of preinduction warming on temperature and blood pressure during propofol/nitrous oxide anesthesia. Anesthesiology 79:219-228.

10. Vassilieff, N, Rosencher, N, Sessler, DI, Conseiller, C, Lienhart, A: Nifedipine and intraoperative core body temperature in humans. Anesthesiology, in press.

Thermal Balance in Health and Disease
Advances in Pharmacological Sciences

EPIDURAL AND SPINAL ANESTHESIA ALTER THERMOREGULATORY RESPONSE THRESHOLDS

M. Schroeder, M. Ozaki, A. Kurz, D. I. Sessler, R. Lenhardt, A. Moayeri, K. M. Noyes, E. Rotheneder and M. Kurz

Thermoregulation Research Laboratory, University of California, San Francisco, California 94143-0648, and the Department of Anesthesia and Intensive Care, University of Vienna.

The core temperature triggering thermoregulatory action defines the *threshold* for that response. Elevated core temperatures trigger sweating and active pre-capillary vasodilation whereas progressively lower temperatures trigger arterio-venous shunt vasoconstriction and shivering. Temperatures between the sweating and vasoconstriction thresholds, which do *not* trigger thermoregulatory responses, identify the *interthreshold range* (1).

In 1972, Roe and Cohn reported that hypothermia was common during spinal anesthesia, and perceptively postulated that the hypothermia resulted from a failure to initiate thermoregulatory shivering (2). The shivering threshold in their patients averaged 1°C less than normal values. However, vasoconstriction (above the level of the block), remained intact during spinal anesthesia and was triggered at a *normal* temperature. Such divergence between the shivering and vasoconstriction thresholds is an unusual pattern of thermoregulatory impairment and cannot be explained by a simple alteration in afferent thermal input.

Sweating thresholds have not been measured during epidural or spinal anesthesia. Consequently, the interthreshold range and pattern of thermoregulatory impairment during major conduction anesthesia remain unknown. However, the sweating threshold appears to be elevated in patients with spinal cord transections (3). Similarly, these patients apparently permit wider core temperature fluctuations than individuals with intact spinal core function (4).

Epidural anesthesia prevents surgical pain primarily by decreasing conduction in spinal nerve roots. In contrast, anesthetic injected into the subarachnoid space directly impairs spinal cord function. Spinal anesthesia also tends to provide a "denser" block than epidural anesthesia. Furthermore, peripheral thermal afferent signals are initially integrated in the spinal cord (5).Consequently, spinal and epidural anesthesia may influence thermoregulatory processing differently.

To investigate the mechanisms of thermoregulatory impairment during regional anesthesia, we first tested the hypothesis that spinal anesthesia comparably reduces the shivering and vasoconstriction thresholds in volunteers. We then tested the hypothesis that the core-temperature thresholds triggering thermoregulatory sweating, vasoconstriction, and shivering are similar in volunteers given epidural and spinal anesthesia. And finally we compared shivering thresholds under clinical conditions in patients randomly assigned to epidural or spinal anesthesia.

Methods

Protocols For Thresholds During Spinal Anesthesia in Volunteers: We studied seven healthy female volunteers during the first 10 days of their menstrual cycles (6). The volunteers were minimally clothed and reclined on their backs in a room having an ambient temperature of 23.1 ± 1.1°C. Each volunteer participated on two study days. On one we evaluated thermoregulatory responses to hypothermia during spinal anesthesia; on the other day, responses were evaluated without anesthesia. Order of the study days was randomly assigned. On the spinal anesthesia day, volunteers were given 1000 ml of warmed lactated Ringer's solution before induction of anesthesia (4 ml of 0.5% bupivacaine). Arterial blood pressure remained within 15% of control values. On the non-anesthetic study day, the protocol was similar except spinal anesthesia was not administered. Throughout each study day, upper-body skin temperature was maintained near 35°C by insulating the skin with an appropriate number of cotton blankets. The lower body of each volunteer was initially warmed using forced-air (Bair Hugger®) and circulating-water heating until they were vasodilated in the fingers. Core hypothermia then was induced by central venous administration of lactated Ringer's solution cooled to ≈4°C. Additionally, we turned off the circulating water-mattress and uncovered the legs.

Protocol For Thresholds During Epidural and Spinal Anesthesia in Volunteers: We studied six young, healthy men. They each participated on three study days: *epidural anesthesia*, *spinal anesthesia,* and *no anesthesia.* On each day, the volunteers were warmed *via* the skin surface until sweating was detected, and then cooled by central venous administration of cold lactated Ringer's solution until vasoconstriction and shivering were triggered. On one of the study days, lumbar epidural anesthesia was then induced using 15-20 ml 2% 2-chloroprocaine without epinephrine. Subsequently, chloroprocaine was infused at a rate of 13-18 ml/h to maintain a sensory block level near T10. On the spinal anesthesia day, volunteers were given 1000 ml of warmed lactated Ringer's solution iv. Anesthesia then was induced by subarachnoid administration of 4 ml of 0.5% isobaric bupivacaine. On the *no anesthesia* day, the protocol was similar except anesthesia was not administered. Active warming was applied only to upper body; the lower body was covered by cotton blankets. Heating was continued until the sweating threshold was identified. Mean skin temperature at the sweating threshold subsequently was kept constant for the duration of the study. Core hypothermia then was induced by central venous administration of cold lactated Ringer's solution.

Protocol For Shivering Thresholds in Surgical Patients: We studied 21 patients undergoing elective knee and calf surgery. Upon arrival to the operating suite, 10 ml/kg of unwarmed iv fluid was administered. The patients were randomly assigned to receive epidural or spinal anesthesia. Epidural anesthesia was induced and maintained with 25 ml of 0.5% bupivacaine. Spinal anesthesia consisted of a single intrathecal bolus of 4 ml of 0.5% bupivacaine. The patients were covered with a single layer of surgical draping; no other warming measures were taken until they shivered.

Measurements For All Three Studies: Core temperature was measured at the tympanic membrane. The aural probe then was inserted by volunteers until they felt the thermocouple touch the tympanic membrane; appropriate placement was confirmed when volunteers easily detected a gentle rubbing of the attached wire. The probe was then securely taped in place, the aural canal occluded with cotton, and a gauze bandage positioned over the external ear. Tympanic membrane temperatures correlate well with distal esophageal temperatures during anesthesia (7). Area-weighted, upper-body skin temperature was computed from measurements at 7 sites and lower-body skin temperature was similarly derived from 8 sites.The rate of sweating was determined using a ventilated capsule situated on the chest, as previously described (8). The core temperature triggering a sweating rate of 40 $g \cdot m^{-2} \cdot h^{-1}$ was considered the threshold for sweating. Absolute right middle fingertip blood flow was quantified using venous-occlusion volume plethysmography at 5-min intervals, as previously described (9).The vasoconstriction threshold was defined as the core temperature triggering a sudden and sustained decrease of finger flow, usually to <0.1 ml/min. Shivering was evaluated by oxygen consumption, which was measured using a Deltatrac™. The tympanic membrane temperature triggering a marked (*e.g.*, 30%) and sustained increase in oxygen consumption was considered the threshold for shivering. Ranges are defined by the difference in respective thresholds. All data are presented as means ± SD; $P < 0.01$ identified statistically significant differences.

RESULTS

Thresholds During Spinal Anesthesia in Volunteers: Spinal anesthesia usually produced a T9-T10 sensory block level. There were no statistically significant or clinically important differences in upper body temperatures at any of the thresholds. Lower body temperatures were comparable during vasoconstriction and shivering with, and without, spinal anesthesia.

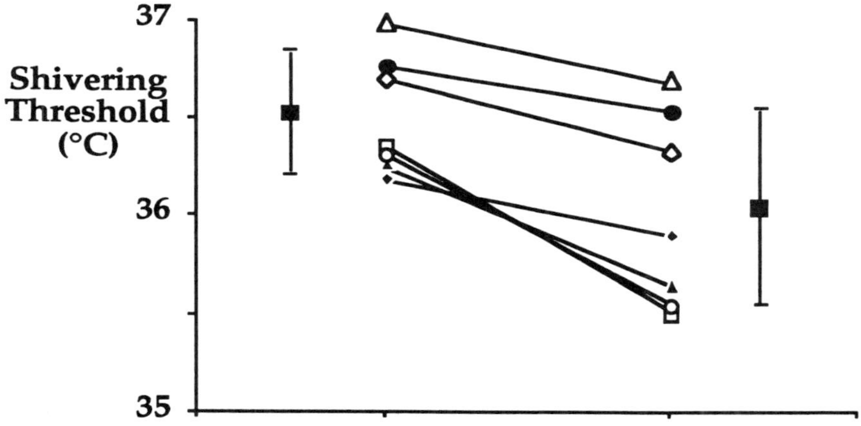

Fig. 1: The shivering thresholds in each individual with, and without, spinal anesthesia. The symbols correspond to the same volunteers shown in Figure 1. The threshold was significantly lower during spinal anesthesia, by 0.5 ± 0.2°C.

Spinal anesthesia significantly decreased the thresholds for vasoconstriction and shivering, and the decrease in each was ≈0.5° C (Table 1, Figure 1). Consequently, the vasoconstriction-to-shivering range was not altered by anesthesia.

	No Spinal	Spinal	Difference
Vasoconstriction (°C)	37.0 ± 0.2	36.6 ± 0.4*	0.4 ± 0.3
Shivering (°C)	36.5 ± 0.3	36.0 ± 0.5*	0.5 ± 0.2
Constriction-to-Shivering Range (°C)	0.5 ± 0.2	0.6 ± 0.4	-0.1 ± 0.2

Table 1: Spinal anesthesia significantly decreased the thresholds (°C) for vasoconstriction and shivering, and the decreased in each was ≈0.5°C. Consequently, the vasoconstriction-to-shivering range also decreased ≈0.5°C with, and without, anesthesia. Asterisks (*) indicate statistically significant differences between the two study days.

Thresholds During Epidural and Spinal Anesthesia in Volunteers: Both epidural and spinal anesthesia produced T8-T12 dermatomal blocks. There were no statistically significant differences in upper body or lower body skin temperatures at any of the thresholds (Table 2).

		No Anesthesia	Epidural	Spinal
Upper Body	**Sweating**	36.5 ± 0.6	36.7 ± 0.3	36.9 ± 0.4
	Vasoconstriction	36.6 ± 0.4	36.6 ± 0.2	36.6 ± 0.3
	Shivering	36.2 ± 0.5	36.1 ± 0.4	35.8 ± 0.6
Lower Body	**Sweating**	35.6 ± 0.7	36.3 ± 0.4	36.4 ± 0.8
	Vasoconstriction	36.0 ± 0.3	36.2 ± 0.5	36.2 ± 0.6
	Shivering	35.4 ± 0.4	35.8 ± 0.6	35.8 ± 0.8

Table 2: Average upper body and lower body, skin-surface temperatures (°C) at the sweating, vasoconstriction, and shivering thresholds were comparable on each study day.

The vasoconstriction threshold was unchanged by epidural and spinal anesthesia. Regional anesthesia slightly increased the sweating threshold and slightly decreased the shivering threshold Consequently, the sweating-to-vasoconstriction (interthreshold) range, the vasoconstriction-to-shivering range, and the sweating-to-shivering range all increased significantly compared to the nc anesthesia day. The interthreshold ranges were 0.6 ± 0.2 and 0.6 ± 0.1°C, respectively, in the volunteers during epidural and spinal anesthesia; in contrast, the range was only 0.3 ± 0.1°C without anesthesia. The thresholds and ranges were similar during epidural and spinal anesthesia (Table 3).

	No Anesthesia	Epidural	Spinal
Sweating	36.9 ± 0.2	37.1 ± 0.3	37.4 ± 0.3*
Vasoconstriction	36.6 ± 0.2	36.5 ± 0.2	36.7 ± 0.3
Shivering	35.7 ± 0.4	35.4 ± 0.5	35.5 ± 0.5
Sweating-to-Constriction	0.3 ± 0.1	0.6 ± 0.2*	0.6 ± 0.1*
Vasoconstriction-to-Shivering	0.9 ± 0.4	1.2 ± 0.5*	1.2 ± 0.4*
Sweating-to-Shivering	1.2 ± 0.4	1.7 ± 0.5*	1.9 ± 0.4*

Table 3: Regional anesthesia significantly increased all ranges. Values are expressed in °C. Asterisks (*) indicate statistically significant differences *vs.* no anesthesia.

Shivering Thresholds in Surgical Patients: Morphometric characteristics, ambient temperature, and block level (T8-T12) were comparable in the two groups. Initial tympanic membrane temperatures were similar with each type of anesthesia. The shivering thresholds were virtually identical in patients given epidural (36.4 ± 0.6°C) and spinal (36.3 ± 0.4°C) anesthesia.

Discussion

The mechanism by which regional anesthesia decreases cold-response thresholds remains unclear, but may result from an apparent (as opposed to actual) increase in lower body temperature. Specifically, at typical leg skin temperatures (≈33°C), cold-receptor activity pre-dominates (10). Thus, when regional anesthesia blocks all thermal input, the regulatory system may largely perceive loss of tonic cold signals and interpret the change as relative leg warming. In this scenario, reduction in the vasoconstriction and shivering thresholds by regional anesthesia will depend critically on the lower body skin temperature during the trial and be obliterated by a sufficiently high skin temperature. To the extent that regional anesthesia induces an apparent increase in leg skin temperature, the thermoregulatory system thus would tolerate more core hypothermia than usual. The clinical expression of such tolerance would be unusually low vasoconstriction and shivering thresholds when leg skin is at a typical (cool) temperature.

The data from our first volunteer study confirm Roe and Cohn's critical observation that the shivering threshold is significantly reduced during spinal anesthesia (2).The slightly greater reduction they observed (≈1°C *vs.* ≈0.5°C) likely reflects variability in the study populations or protocols (they evaluated older patients who, perhaps, were sedated). Comparable reductions in the vasoconstriction and shivering thresholds during spinal anesthesia would be expected on the basis of an increased apparent leg temperature. It is likely that the preserved vasoconstriction threshold previously reported (2) resulted from anesthetic-induced reduction in vascular volume.

Consistent with an increase in apparent leg temperature, vasoconstriction and shivering thresholds were not significantly decreased at mean lower body skin temperatures near 36°C in our second set of volunteers — a value far exceeding that typical in patients who are not actively warmed (11) and that in our first volunteer study (≈33°C). It remains unclear whether central control systems respond to integrated warm and cold afferent input, or whether warm and cold responses are independently processed from their respective afferent signals. Decreased vasoconstriction and shivering thresholds would be expected in each case during regional anesthesia with unwarmed legs; similarly, little change in the cold-response threshold would be expected when the legs are sufficiently warmed. However, warm responses such as sweating would be affected differently: the threshold would be reduced if central control systems respond to integrated warm and cold afferent input, but increased if warm and cold responses are independently processed. This increase would result because anesthetic-induced inhibition of warm signals (even if they constitute a small fraction of the total thermal input) would *decrease* leg temperature apparent to the system integrating warm input and controlling sweating.

The observed increases in the sweating thresholds during regional anesthesia thus are most consistent with independent thermoregulatory processing of warm and cold afferent signals. The magnitude of the increase in the sweating threshold presumably is directly related to leg skin temperature, as is the decrease in vasoconstriction and shivering thresholds. The net result is, therefore, that regional anesthesia likely expands the interthreshold range at any leg temperature. This pattern of thermoregulatory impairment is similar to that produced by isoflurane and enflurane anesthesia (1), although its mechanism presumably differs and its magnitude is far less.

The sweating, vasoconstriction, and shivering thresholds were comparable in volunteers during epidural and spinal anesthesia. Similarly, the shivering thresholds were virtually identical during each type of anesthesia in patients. These data suggest that the two types of regional anesthesia comparably alter thermoregulatory processing and also are consistent with our theory that impairment results largely from blocked afferent temperature signals.

Acknowledgments

These studies were supported by Augustine Medical, Inc., and National Institutes of Health grants GM 39723 and GM 49670.

References

1. Lopez, M, Sessler, DI, Walter, K, Emerick, T, Ozaki, M Rate- and gender-dependence of the sweating, vasoconstriction, and shivering thresholds in humans. Anesthesiology, in press

2. Roe, CF, Cohn, FL (1972): The causes of hypothermia during spinal anesthesia. Surg Gynecol Obstetr 135, 577-580.

3. Tam, H, Darling, RC, Cheh, H, Downey, JA (1978): The dead zone of thermoregulation in normal and paraplegic man. Can J Physiol Pharmacol 56, 976-983.

4. Hales, JRS (1984): Thermal Physiology. In: Temperature regulatory set-point in paraplegics. Attia, M, Engel, P eds. New York: Raven Press, 79-82.

5. Simon, E (1974): Temperature regulation: The spinal cord as a site of extrahypothalamic thermoregulatory functions. Rev Physiol Biochem Pharmacol 71, 1-76.

6. Kurz, A, Sessler, DI, Schroeder, M, Kurz, M (1993): Thermoregulatory response thresholds during spinal anesthesia. Anesth Analg 77, 721-726.

7. Cork, RC, Vaughan, RW, Humphrey, LS (1983): Precision and accuracy of intraoperative temperature monitoring. Anesth Analg 62, 211-214.

8. Lopez, M, Ozaki, M, Sessler, DI, Valdes, M (1993): Physiological responses to hyperthermia during epidural anesthesia and combined epidural/enflurane anesthesia in women. Anesthesiology 78, 1046-1054.

9. Rubinstein, EH, Sessler, DI (1990): Skin-surface temperature gradients correlate with fingertip blood flow in humans. Anesthesiology 73, 541-545.

10. Pierau, F, Wurster, RD (1981): Primary afferent input from cutaneous thermoreceptors. Fed Proc 40, 2819-2824.

11. Glosten, B, Sessler, DI, Faure, EAM, Støen, R, Thisted, RA, Karl, L (1992): Central temperature changes are not perceived during epidural anesthesia. Anesthesiology 77, 10-16.

Thermal Balance in Health and Disease
Advances in Pharmacological Sciences

CIRCADIAN RHYTHMS OF BODY TEMPERATURE IN PATIENTS WITH ANOREXIA NERVOSA, HYPOTHERMIA AND HYPOPHOSPHATEMIA

S. Yoshiue, T. Kanamori, H. Sekine, A. Maeda, M. Ishida, J. Ogata[1] and T. Kato

1st Department of Internal Medicine, Saint Marianna University School of Medicine, 2-16-1, Sugao Miyamae-ku, Kawasaki, Kanagawaken, 216 and [1]Kyouritsu Women's University, Faculty of Home Economics, 2-2-1, Hitotsubashi, Chiyoda-ku, Tokyo, 1011, Japan

Anorexia nervosa is a prolonged illness principally affecting young girls after puberty. The clinical manifestations include self-induced, severe weight loss, amenorrhea and psychopathology. The etiology has been studied extensively (1, 2) but remains unknown. Central body temperature in patients with anorexia nervosa may be reduced by up to 1°C. Patients with marked weight loss show a fall in plasma levels of LH, FSH and estrogens reflecting hypothalamo-pituitary dysfunction. This has led to the suggestion that the hypothermia in anorexia nervosa may reflect a disorder of the hypothalamic thermoregulatory centers (3). Measurements of the thermal responses to exercise demonstrated that patients with anorexia nervosa can regulate body temperature in a thermoneutral environment (4). The onset of vasodilation following immersion of an arm in a hot water bath is delayed in these patients (5) indicating a shift in the threshold temperature for vasodilation (6). Our own studies confirmed the hormonal dysfunction although circadian temperature rhythms were normal (7). Itoh (8) has suggested that hypothalamic dysfunction may be a dominant factor in the pathophysiology of anorexia nervosa. In case reports of two severely affected patients the hypothermia was accompanied by respiratory distress, hyperventilation and hypophosphatemia (9-11). The present study was undertaken to investigate the relationship of the hypothermia and hypophosphatemia in this rare syndrome.

Methods

The core temperature (Tc) circadian rhythm was determined as described previously (12) in 10 female hospital patients with a diagnosis of anorexia nervosa (6) or bulimia nervosa (4). The biochemical data (total serum protein, serum calcium and phosphorus) were obtained from 2,207 controls (642 females, 1,565 males), 21 females with anorexia nervosa and 20 with bulimia nervosa.

Results

The average clinical course of the anorexia nervosa patients was 2.3 y; 3 patients recovered whereas the condition of the remainder was unchanged. Of those with bulimia nervosa the period of treatment averaged 5.5 y with one recovery, of the other 3 patients their condition was unchanged or deteriorated. In the anorexia nervosa patients the values for serum protein, calcium and phosphorus were 6.3±0.6 g/dl, 8.4±0.5 mg/dl and 2.8±0.6 mg/dl, respectively. Corresponding values in patients with bulimia nervosa were 7.1±0.4 g/dl, 9.6±0.4 mg/dl and 4.1±0.3 mg/dl. The circadian rhythm of Tc and hand temperature in the 10 patients is illustrated in Fig. 1. The mean levels in Tc in the patients with anorexia nervosa were below the normal range; other patterns were essentially normal.

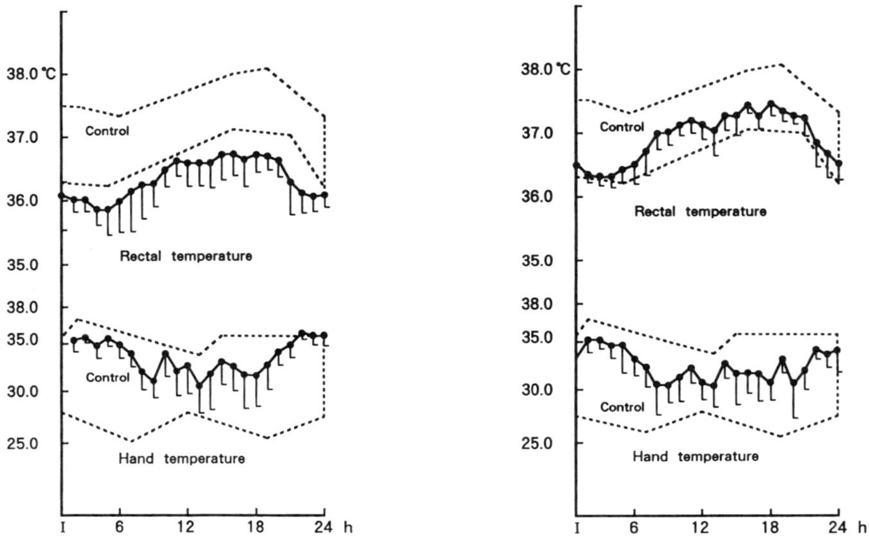

Fig. 1: Circadian rhythm of Tc in anorexia nervosa (left) and bulimia nervosa (right).

A representative case of anorexia nervosa was a 20 y old woman who, at the age of 16 y, was 158 cm tall and weighed 50 kg. She began to diet, her body weight fell to 26 kg and she became amenorrhagic. She was admitted to hospital with acute respiratory distress and unable to walk. Moderate hypophosphatemia (2 mg/dl) was present. Following treatment by total parenteral nutrition serum phosphate fell to 0.5 mg/dl. Circadian Tc rhythms on three subsequent admissions to hospital were abnormal (Fig. 2) and hand temperatures were dissociated. Serum phosphate levels were low on all occasions.

Discussion

There do not appear to be any previous reports of anorexic patients exhibiting hypothermia, abnormal circadian temperature rhythms and hypophosphatemia. During the past 33 years two patients have died who manifest moderate hypophosphatemia and a low axillary temperature. One severely undernourished girl with moderate hypophosphatemia and a low body temperature recovered when treated by total parenteral nutrition.

The occurrence of hypophosphatemia and abnormal circadian Tc rhythms in anorexia nervosa could result from multiple organ failure. The possible series of changes leading to the development of hypothermia as a result of abnormal rhythms is depicted in Fig. 3.

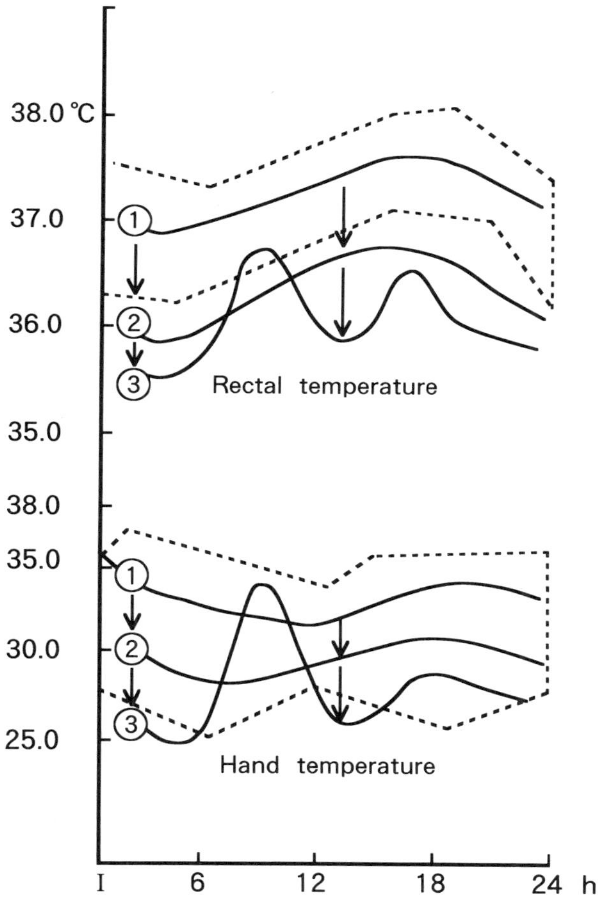

Fig. 2: Circadian rhythms in a representative patient.

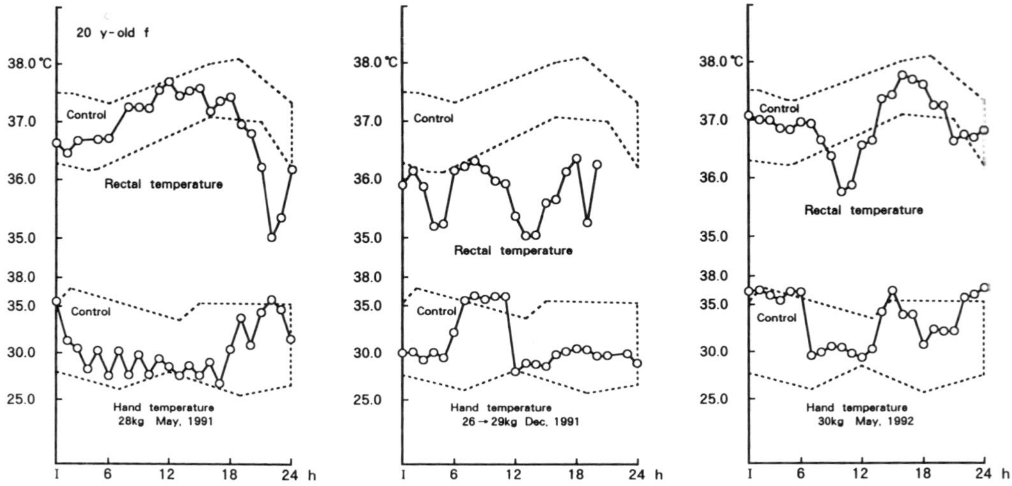

Fig. 3: Changes in the circadian rhythms of Tc with the onset of hypothermia.

Acknowledgements

The authors are grateful to Dr. M. Iriki of the Department of Physiology, Yamanashi Medical College for assistance with the interpretation of the data.

References

1. Garfinkel, PE, Kaplan, AS (1986): Anorexia nervosa: Diagnostic conceptualizations. In: Handbook of Eating Disorders. Physiology, Psychology, and Treatment of Obesity, Anorexia, and Bulimia. Brownell, KD, Foreyt, JP eds. New York: Basic Books, 266-282.

2. Demitrack, MA, Szostak, C, Weingartner, H (1992): Cognitive dysfunction in eating dis orders: A clinical psychobiological perspective. In: Cognitive Neuropsychology in Clinical Practice. Margolin, DI eds. New York, Oxford: Oxford University Press, 96-127.

3. Mecklenburg, RS, Loriaux, DL, Thompson, RH, Andersen, AE, Lipsett, MB (1974) Hypothalamic dysfunction in patients with anorexia nervosa, Medicine 53, 147-159.

4. Davis, CTM, Fohlin, L, Thoren, C (1977): Temperature regulation in anorexia nervosa. J Physiol 268, 8.

5. Wakeling, A, Russell, GFM (1970): Disturbances in the regulation of body temperature in anorexia nervosa. Psychol Med 1, 30-39.

6. Luck, P, Wakeling, A (1980): Altered thresholds for thermoregulatory sweating and vasodilatation in anorexia nervosa. Br Med J 281, 906-908.

7. Yamada, K, Takeuchi, H, Yazumi, T, Ito, H, Yoshizawa, H, Yoshiue, S, Ishida, M (1986): Study of body temperature in anorexia nervosa. The Autonomic Nervous System 23, 1016.

8. Ito, H (1990): Simultaneous measurement of temperatures at different sites in clinical entities displaying hypothermia. The Autonomic Nervous System 27, 281-291.

9. Knochel, JP (1977): The pathophysiology and clinical characteristics of severe hypophosphatemia. Arch Intern Med 137, 203-220.

10. Gustavsson, CG, Eriksson, L (1989): Acute respiratory failure in anorexia nervosa with hypophosphatemia. J Intern Med 225, 63-64.

11. Levy, LA (1980): Severe hypophosphatemia as a complication of the treatment of hypothermia. Arch Intern Med 140, 128-129.

12. Yoshiue, S, Yoshizawa, H, Itoh, H, Yazumi, T, Nakamura, M, Sekine, H, Takahashi, T, Kanamori, H, Suzuki, N, Ihashi, N, Sakaki, R, Ishida, M, Ogata, J (1992): Circadian body temperature in patients with psychogenic prolonged hyperthermia. In: Thermoregulation: The Pathophysiological Basis of Clinical Disorders. Lomax, P, Schönbaum, E eds. Basel: Karger, 157-166.

Thermal Balance in Health and Disease
Advances in Pharmacological Sciences

CIRCADIAN RHYTHMS OF HOT FLASHES AND BODY TEMPERATURE IN MENOPAUSAL WOMEN

R. Freedman[1], D. Norton, S.Woodward, G. Cornélissen and F. Halberg

Behavioral Medicine Laboratory, Wayne State University and Chronobiology Laboratory, University of Minnesota [1]C. S. Mott Center, 275 E. Hancock, Detroit, MI 48201 USA

Hot flashes (HFs) are the most common symptom of the climacteric and affect the vast majority of menopausal women. HFs are subjectively experienced as sensations of internal heat in the chest, neck, and face, and are accompanied by profuse sweating in these areas. Substantial cutaneous vasodilation occurs, particularly in the extremities, as well as modest tachycardia (1). Core body temperature measured from the esophagus and tympanum generally declines following HFs (2,3).

Circadian variations in thermoregulatory responses including core and peripheral temperatures, heat conductance, and sweating have been repeatedly demonstrated in human studies (4,5). These variations suggest that the thermoregulatory effector responses of HFs might also demonstrate temporal patterns. A previous study (6) found a significant 24-hour rhythm of self-reported HFs, but no physiological measurements were performed. We therefore performed ambulatory skin conductance monitoring to objectively demarcate HFs and recorded core body temperature with radiotelemetry pills in menopausal women with and without HFs.

Pilot Study

Seventeen menopausal women reporting frequent HFs received 24-hour ambulatory monitoring of sternal skin conductance level (SCL) to detect HFs using the methods described below (1). 199 HFs were recorded. A cosine function was fitted to the time distribution of HFs using a least-squares method. This procedure demonstrated a significant ($p<.01$) circadian rhythm having an acrophase at about 1800h.

Methods

Ten menopausal women (54.2 ± 3.4 years old) reporting frequent HFs and five menopausal women (52.8 ± 2.2 years old) reporting no history of HFs were recruited using newspaper ads for research volunteers. All had been amenorrheic for at least one year. The symptomatic women weighed an average of 161.5 ± 33.4 lbs whereas the asymptomatic women weighed 160.1 ± 26.7 lbs. Three symptomatic women and two asymptomatic women were African-American; the rest were Caucasian. All gave written informed consent according to procedures approved by our Institutional Review Board.

Women were recorded for one 24-hour period starting at about 1000h. Skin conductance level was measured with 2 Ag/AgCl electrodes, 1.5 cm in diameter, filled with 0.05M KCl paste and attached over the sternum. They were connected to a 0.5V constant voltage circuit and one DC channel of an Oxford Medilog 9000 recorder (Oxford, UK). The Medilog tapes were replayed at 20X real time with an Oxford Medilog playback unit, an analog-to-digital converter, and a computer. The computer was programmed to define a hot flash as a skin conductance increase ≥ 2 μmho in 30 seconds. We have shown previously that this sternal skin conductance response has a high correspondence with self-reported HFs and does not occur in premenopausal women during ambulatory monitoring (1). Using these methods, hot flashes were detected in all the symptomatic women (range 1 to 22 per 24 hours). No hot flashes were detected in the asymptomatic women.

Core body temperature was recorded every 30 sec. using an ingested radiotelemetry pill ("CorTemp"). The data were stored in memory by the Cortemp recorder and subsequently loaded into an IBM PC computer for further analysis.

Core temperatures were graphically displayed for individual recordings and manually edited to eliminate artifacts. Approximately 90 min of data at the beginning of each recording were eliminated to allow egress of the telemetry pill from the stomach. The core temperature data were averaged in one-hour blocks and analyzed using the cosinor method. Each data series was analyzed by a single cosinor involving the least-squares fit of a 24-hour cosine curve to the data (7). The following rhythm parameters are thus obtained: the MESOR, a rhythm-adjusted mean; the circadian amplitude, a measure of the extent of predictable change within a day; and the circadian acrophase, a measure of timing of overall high values recurring each day. The residual sum of squares is used for hypothesis testing, i.e., for testing the zero-amplitude (no-rhythm) assumption and to obtain 95% confidence intervals for each parameter. The circadian characteristics for each woman in each group were further summarized by population-mean cosinor, a procedure examining the extent of clustering of circadian amplitude-acrophase pairs within a group (8). Rejection of the zero-amplitude (no-rhythm) assumption with this procedure lends itself to the generalization of results to the population. Parameter tests (9) were used to compare the circadian patterns of core body temperature between women with and without hot flashes and also to compare the timing of the circadian rhythm in core body temperature with that in the incidence of hot flashes, for those women who exhibited these symptoms.

Results

A circadian rhythm can be validated with statistical significance for temperature data from each woman, Table 1. A least-squares spectrum indicates that the circadian rhythm is invariably the most prominent component. Table 2 summarizes the population results. A circadian rhythm in core body temperature can be validated for menopausal women with hot flashes ($P=0.0001$). For women without hot flashes, a circadian rhythm is of borderline statistical significance ($P=0.058$). A comparison of the circadian characteristics between these two groups indicates a difference of borderline statistical significance for the MESOR ($P=0.066$) and for the circadian acrophase as compared to women without hot flashes. These results are illustrated in Figure 1, obtained by stacking the data from all women in 24 hourly classes over an idealized 24-hour day. The one-way analysis of variance rejects the hypothesis of equality of class means (F=10.136; P<0.001 for women with hot flashes; F=3.658; P<0.001 for women without hot flashes). The circadian pattern is readily apparent.

Table 1: Circadian rhythm characteristics of core body temperature in menopausal women with and without hot flashes*

ID	PR	P	MESOR ± SE (°C)	Amplitude ± SE	Acrophase (95% CI)	
			Women with hot flashes			
503	85	<0.001	36.78 ± 0.05	0.72 ± 0.07	-249°	(-238°, -261°)
515	48	0.001	36.86 ± 0.03	0.21 ± 0.05	-254°	(-226°, -282°)
516	66	<0.001	36.54 ± 0.05	0.41 ± 0.06	-221°	(-202°, -239°)
518	89	<0.001	36.71 ± 0.04	0.76 ± 0.06	-220°	(-211°, -230°)
519	80	<0.001	36.91 ± 0.03	0.40 ± 0.04	-258°	(-245°, -272°)
520	61	<0.001	36.55 ± 0.05	0.36 ± 0.06	-218°	(-197°, -240°)
521	39	0.006	36.49 ± 0.04	0.21 ± 0.06	-219°	(-185°, -254°)
522	62	<0.001	36.49 ± 0.02	0.19 ± 0.03	-236°	(-215°, -257°)
525	82	<0.001	36.67 ± 0.04	0.44 ± 0.06	-199°	(-183°, -216°)
527	61	<0.001	36.66 ± 0.05	0.37 ± 0.07	-223°	(-201°,-244°)
			Women without hot flashes			
602	81	<0.001	36.88 ± 0.03	0.36 ± 0.04	-277°	(-264°, -289°)
603	89	<0.001	36.79 ± 0.04	0.54 ± 0.05	-253°	(-240°, -268°)
604	88	<0.001	36.63 ± 0.03	0.42 ± 0.04	-246°	(-235°, -257°)
608	68	<0.001	36.86 ± 0.04	0.37 ± 0.06	-211°	(-192°, -230°)
609	48	<0.001	36.96 ± 0.04	0.23 ± 0.05	-320°	(-292°, -348°)

Table 2: Comparison of circadian rhythm characteristics of core body temperature between menopausal women with and without hot flashes*

Hot Flashes? (N)	PR	P	MESOR ± SE (C°)	Amplitude ± SE	Acrophase (90%CI)	
Yes (10)	67.3	0.001	36.67 ± 0.05	0.39 ± 0.06	-229°	(-217°, -242°)
No (5)	74.8	0.058	36.82 ± 0.07	0.33 ± 0.10	-225°	(-229°, -291°)
Group Comparison:			P=0.066	0.587	P=0.214	0.079

*PR=percent rhythm (proportion of variability accounted for by fitted model), P=P-value from test of zero-amplitude (no-rhythm) assumption; acrophase in (negative) degrees, with 360° ≅ 24 hours; 0°=local midnight; SE=standard error; 95% CI=95% confidence interval; 90% CI=90% confidence interval.

Overall, the women without hot flashes have higher temperatures than women with hot flashes. Figure 1 also shows that the rise in core body temperature after awakening is much slower for the women without hot flashes than for women with hot flashes, a feature that may contribute to the difference in acrophase.

The incidence of hot flashes varied from 1 a day to 22 a day in the 10 menopausal women studied. This short record notwithstanding, a circadian rhythm in the incidence of hot flashes can be demonstrated with statistical significance (P=<0.05) in 6 out of the 10 women; in another case, the circadian rhythm is of borderline statistical significance, Table 3. A population summary indicates the overall statistical significance of the circadian pattern in the incidence of hot flashes (P=0.020).

Table 3: Circadian rhythm characteristics of the incidence of hot flashes in menopausal women*

ID	PR	P	MESOR ± SE	Amplitude ± SE	Acrophase (95%CI)
503	50	<0.001	0.501 ± 0.109	0.705 ± 0.154	-265°(-238°, -292°)
515	37	0.008	0.417 ± 0.122	0.601 ± 0.172	-290°(-254°, -327°)
516	36	0.009	0.917 ± 0.142	0.687 ± 0.201	-322°(-284°, -359°)
518	54	<0.001	0.834 ± 0.126	0.881 ± 0.178	-321°(-297°, -346°)
519	19	0.113	0.209 ± 0.127	0.394 ± 0.179	-218°(0, 0)
520	4	0.663	0.709 ± 0.131	0.169 ± 0.185	-313°(0, 0)
521	9	0.385	0.043 ± 0.042	0.083 ± 0.059	- 83°(0, 0)
522	28	0.032	0.292 ± 0.113	0.456 ± 0.160	-210°(-163°, -256°)
525	26	0.040	0.667 ± 0.103	0.401 ± 0.146	-163°(-113°, -212°)
527	24	0.057	0.459 ± 0.110	0.397 ± 0.155	-260°(0, 0)
Population Summary:	28.7	0.020	0.505 ± 0.100	0.305(0.094,0.516)	-276°(-215°, -311°)

*PR=percent rhythm (proportion of variability accounted for by fitted model), P=P-value from test of zero-amplitude (no-rhythm) assumption; acrophase in (negative) degrees, with 360° ≅ 24 hours; 0° = local midnight; SE=standard error; 95% CI=95% confidence interval.

A polar plot summarizing the circadian rhythm in core body temperature and the incidence of hot flashes is provided in Figure 2. In this display, the directed line (vector) represents by its length the amplitude and by its direction the acrophase of the circadian rhythm of each variable. The acrophase is shown vs. a circular scale of 360° equated to 24 hours with midnight shown on top as 0°. The ellipse around the tip of each vector represents the 95% confidence region for the joint estimate of the circadian amplitude assumption. It can also be seen from Figure 2 that the circadian acrophase of the incidence of hot flashes lags behind the peak in core body temperature. In view of the different units in which core body temperature and the incidence of hot flashes are expressed, a parameter test cannot be applied directly for the comparison of acrophases. In order to proceed with a test of acrophases, results were first normalized. The normalization used entailed equating all MESORs to 1 and rescaling the amplitudes in such a way as to achieve an acrophase-unweighted average of 1 for each group. Parameter tests applied to these normalized circadian parameters indicate that the difference in circadian acrophase is of borderline statistical significance (0.056). A comparison of the circadian amplitude and acrophase considered jointly reveals a P-value of 0.050. Examination of the results further indicates that the average acrophase-weighted-amplitude is 0.95 for core body temperature and 0.64 for the incidence of hot flashes, a result pointing to the larger dispersion of acrophases for the latter variable as compared to the former. A

FIGURE 1

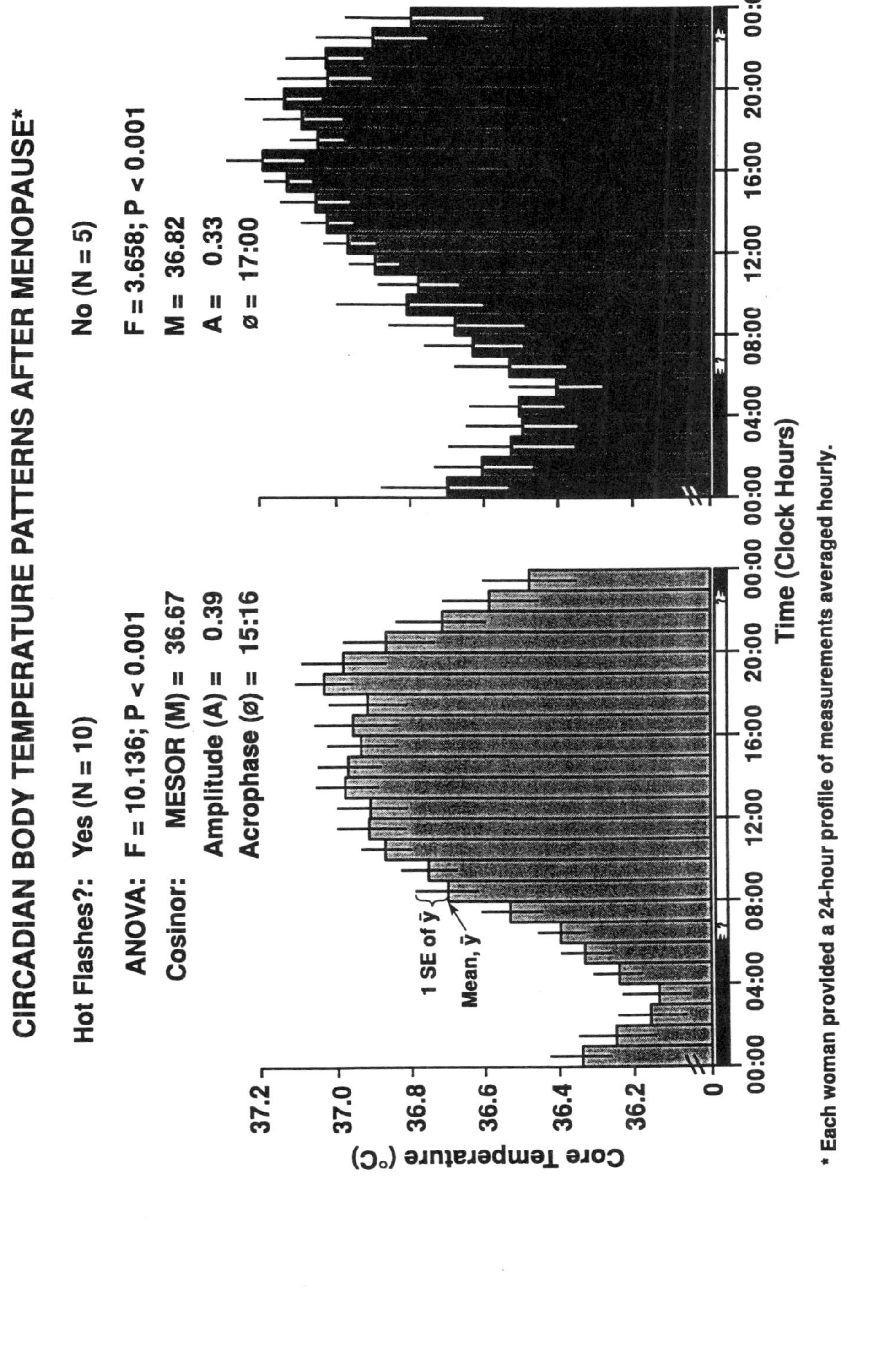
CIRCADIAN BODY TEMPERATURE PATTERNS AFTER MENOPAUSE*
Hot Flashes?: Yes (N = 10)
ANOVA: F = 10.136; P < 0.001
Cosinor: MESOR (M) = 36.67
Amplitude (A) = 0.39
Acrophase (ø) = 15:16
No (N = 5)
F = 3.658; P < 0.001
M = 36.82
A = 0.33
ø = 17:00
1 SE of ȳ
Mean, ȳ
Core Temperature (°C)
37.2
37.0
36.8
36.6
36.4
36.2
0
00:00
04:00
08:00
12:00
16:00
20:00
00:00
Time (Clock Hours)
* Each woman provided a 24-hour profile of measurements averaged hourly.

FIGURE 2

CIRCADIAN RHYTHM OF CORE BODY TEMPERATURE (A) AND INCIDENCE OF HOT FLASHES (B) IN POSTMENOPAUSAL WOMEN

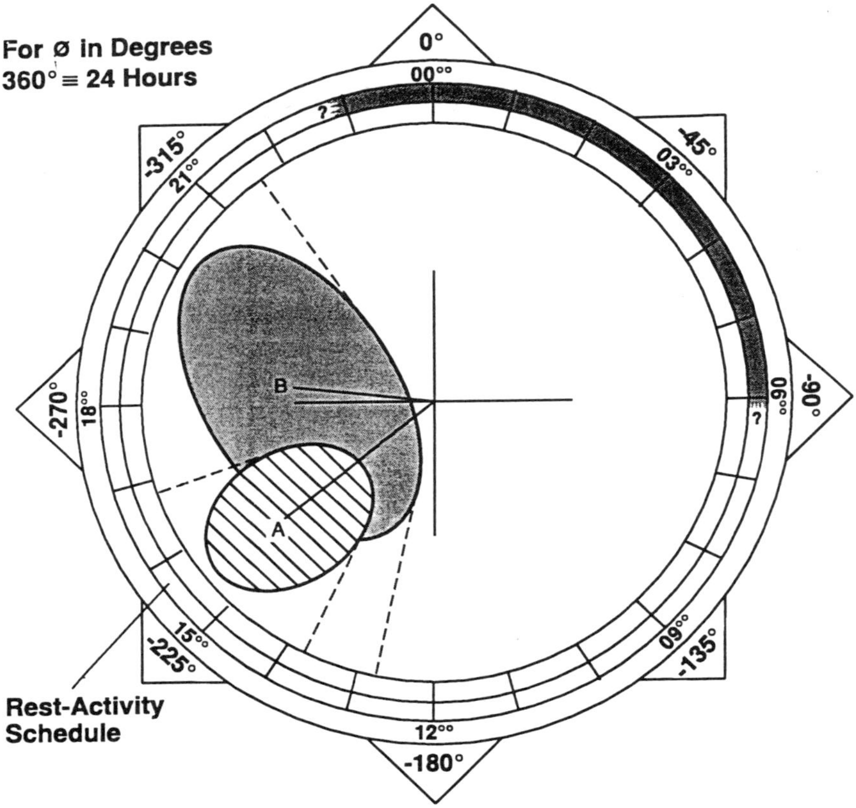

Population-Mean Cosinor

Key	P	N	PR	MESOR ± SE	Amplitude*	Acrophase (ø)*
A	0.001	10	67	36.67 0.05	0.39 (0.19 0.58)	-229° (-206 -252)
B	0.020	10	29	0.50 0.09	0.30 (0.05 0.65)	-276° (-189 -323)

P = Probability of hypothesis: Amplitude = 0; N = Number of series used by the mean cosinor technique to find ellipse.

* Conservative 95% confidence limits (parentheses) derived from cosinor ellipse.

linear regression of the circadian acrophase as a function of the MESOR of the incidence of hot flashes suggests that a larger incidence of hot flashes is associated with a later circadian acrophase (P=0.016). An association is also found between the circadian amplitude of hot flashes and that of core body temperature (P=0.040).

An individual comparison of circadian acrophases indicates that in six cases the acrophase of core body temperature leads and in four cases it lags behind the acrophase of the incidence of hot flashes. Accordingly, a paired t-test of the difference in circadian acrophase between these two variables is not statistically significant. This result qualifies the difference in amplitude-acrophase pair of the normalized circadian parameters of the two variables reported above. When two groups are formed, depending on whether the acrophase of core body temperature leads or lags behind that of the incidence of hot flashes, a difference in the MESOR of the incidence of hot flashes is of borderline statistical significance (P=0.055). This result is in keeping with that of the linear regression between the MESOR and the circadian acrophase of hot flashes reported above.

Discussion

The results obtained herein warrant further study on larger groups of menopausal women. Caution is indicated in the interpretation of our results in view of the fact that when the ten menopausal women exhibiting hot flashes are arbitrarily subdivided into two subgroups of 5 women, a statistically significant difference in MESOR (P=0.036) is found, pointing to inter-individual differences in the circadian pattern of core body temperature. Such a difference may perhaps underlie a difference in the incidence of hot flashes, albeit this hypothesis cannot be tested with the sample available.

In the present investigation we found significant temporal patterning of HFs in a pilot study and in subsequent work where HFs were identified physiologically. Using cosinor analysis, a statistically significant circadian rhythm with an acrophase at 1800h was fitted to the distribution of HFs, although it explained only 29% of the variance. It is likely that other rhythms with other frequencies also perturb these data. A previous study (6) found two groups of subjects having distinct temporal HF patterns: one group with an acrophase around 1800h and a second group with a peak between 0400 and 1000h. We did not identify the second group, but it is possible that this finding would emerge in a larger group of subjects. In addition, the previous study used only self-reported HFs, which are subject to errors of recall, particularly during sleep.

The circadian rhythm of core body temperature of the flashers may be phase-advanced relative to that of the non-flashers and to the rhythm of HFs. We have previously demonstrated increased sympathetic activation measured by plasma MHPG (the main metabolite of brain norepinephrine) and by the HF response to adrenergic drugs in symptomatic compared to asymptomatic women (10,11). Research by others has demonstrated a circadian variation of plasma MHPG whose acrophase (1500h) is similar to the acrophase of body temperature of our symptomatic group (1524h). The central noradrenergic nervous system is intimately involved in thermoregulation and we suspect that the MHPG rhythm may be driving the rhythm of HFs. We have not yet measured MHPG over 24-hour spans in our subjects, however, and this hypothesis must be tentative at present. The MESOR for core temperature of the symptomatic women appeared to be lower than that of the asymptomatic group (P=0.06). Previous studies have shown that considerable heat is lost during HFs, through sweating and cutaneous vasodilation, and that core temperature decreases following HFs (1-3). We, therefore, interpret the reduction in core temperature in the symptomatic women to heat lost during HFs.

In conclusion, a circadian rhythm has been demonstrated for the core body temperature of menopausal women with or without hot flashes, as well as for the incidence of hot flashes. Women with hot flashes tend to have a lower core body temperature as compared to those without hot flashes. There is also a slight tendency for the acrophase of the incidence of hot flashes to precede the time of peak core body temperature, albeit this difference may reflect more a larger dispersion in individual acrophases than a true acrophase difference.

References

1. Freedman, RR (1989): Laboratory and ambulatory monitoring of menopausal hot flashes. Psychophysiology 26(5), 573-579.

2. Tataryn, IV,, Lomax, P, Bajorek, JG, Chesarek, W, Meldrum, DR, Judd, HL (1980): Postmenopausal hot flushes: A disorder of thermoregulation. Maturitas 2, 101-107.

3. Kronenberg, F, Cote, LJ, Linkie, DM, Dyrenfurth, I, Downey, JA (1984): Menopausal hot flashes: thermoregulatory, cardiovascular, and circulating catecholamine and LH changes. Maturitas 6, 31-43.

4. Smith, RE (1969): Circadian variations in human thermoregulatory responses. J Appl Physiol 26(5), 554-560.

5. Aschoff, J, Heise, A (1972): Thermal conductance in man: Its dependence on time of day and on ambient temperature, In: Advances in Climatic Physiology. Itoh, S, Ogata, K, Yoshimura, H eds. New York:Springer, 334-348.

6. Albright, DL, Voda, AM, Smolensky, MH, Hsi, B, Decker, M (1989): Circadian rhythms in hot flashes in natural and surgically-induced menopause. Chronobiol Int 6(3), 279-284.

7. Halberg, F (1969): Chronobiology. Ann Rev Physiol 31, 675-725.

8. Halberg, F, Tong, YL, Johnson, EA (1967): Circadian system phase: an aspect of temporal morphology; procedures and illustrative examples. Prov International Congress of Anatomists. In: The Cellular Aspects of Biorhythms, Symposium on Biorhythms. Berlin: Springer-Verlag, 20-48.

9. Bingham, C, Arbogast, B, Cornélissen, GG, Lee, JK, Halberg, F (1982): Inferential statistical methods for estimating and comparing cosinor parameters. Chronobiologia 9, 397-439.

10. Freedman, RR, Woodward, S (1992): Elevated α_2-adrenergic responsiveness in menopausal hot flushes: Pharmacologic and biochemical studies. In Thermoregulation: The Pathophysiological Basis of Clinical Disorders. Lomax, P, Schînbaum, E, eds. Basal:Karger, 6-10.

11. Freedman, RR, Woodward, S, Sabharwal, SC (1990): α_2-adrenergic mechanism in menopausal hot flushes. Obstet Gynecol 76(4), 573-578.

12. DeMet, EM, Halaris, AE, Gwirtsman, HE, Reno, RM (1985): Diurnal rhythm of 3-methoxy-4-hydroxyphenylglycol (MHPG): Relationship between plasma and urinary levels. Life Sci 37, 1731-1741.

Thermal Balance in Health and Disease
Advances in Pharmacological Sciences

CARDIOVASCULAR AND ADRENERGIC MANIFESTATIONS OF COLD STRESS

S. M. Frank, S. N. Raja, L. A. Fleisher, C. Beattie, M. S. Higgins and M. J. Breslow

Department of Anesthesiology/Critical Care Medicine, Carnegie 442, Johns Hopkins Hospital, 600 N. Wolfe St., Baltimore, Md. 21287, USA

Introduction

Ischemic cardiac morbidity is the most common cause of death in the United States in patients undergoing surgery as well as in the non-perioperative setting.(1). Coronary artery disease is ultimately the predisposing factor in patients who suffer cardiac death, but usually a triggering event disrupts the delicate balance between myocardial oxygen supply and demand and leads to ischemia and infarction. Some of the factors that have been shown to induce myocardial ischemia include tachycardia, hypertension, hypotension, thrombosis, and cold stress. Bainton et al. (2) described a two-fold increase in the death rate from ischemic cardiac events in the winter relative to the summer. This study was a simple epidemiologic description of the relationship between time of year, ambient outdoor temperature, and mortality. These results have been confirmed by other investigators as well.(3) Neither study, however, has determined, or even proposed a mechanism to explain the higher death rate.

We have recently described an increased incidence of early postoperative myocardial ischemia in patients undergoing peripheral vascular surgery who developed inadvertent hypothermia (<35 °C) during surgery.(4) Although age was a confounding variable, the association between temperature and ischemia was independent of age by subgroup and multivariate analyses. Like the Bainton study, we did not identify a mechanism to explain the hypothermia/ischemia relationship. Evidence for a physiologic mechanism is needed to confirm this relationship. It is possible that by maintaining normothermia in the perioperative period, the incidence of ischemic cardiac morbidity may be reduced in the 30-70% of surgical patients who develop inadvertent hypothermia. The classic model of cold stress that has been used to elicit myocardial ischemia is the cold pressor test (CPT). This test was described over 50 years ago(5) and involves immersion of the hand into a cold ice bath for a period of one to three minutes. Earlier studies described an increased blood pressure and heart rate immediately upon immersion. Subsequently, both epinephrine and norepinephrine have been shown to increase during CPT.(6-8) In addition, the test has been shown to induce myocardial ischemia in patients with coronary artery disease, presumably from vasoconstriction. Coronary vascular resistance increases significantly during CPT, and this effect is prevented by the α-adrenergic antagonist, phentolamine.(9) Although the CPT is not a model of

whole body cold stress, these findings provide insight into the potential mechanism for cold-induced ischemic cardiac events.
The data presented here provide an overview of the studies we have conducted to examine the cardiovascular manifestations of cold stress and the adrenergic response. These studies were conducted in patients undergoing surgery, as well as in volunteers without surgical disease.

Methods and Results

Study #1. Perioperative Ischemia Randomized Anesthesia Trial (PIRAT)

Methods: 100 patients undergoing lower extremity revascularization for arterial insufficiency were enrolled in a randomized trial comparing epidural and general anesthesia. The primary investigational outcome was Holter monitor documented myocardial ischemia, determined by a cardiologist who was blinded to the clinical course. Anesthetic and hemodynamic management were controlled by strict prospectively designed protocols. Ischemia was defined as one or more mm of ST segment depression lasting more than one minute. The incidence of ischemia was compared in the epidural and general groups. Ischemia occurring in the first 24 hours following surgery was also analyzed according to oral sublingual temperature recorded upon admission to the intensive care unit immediately following surgery. The incidence of ischemia was compared in the hypothermic (<35°C) and normothermic (≥35°C) patients. Subgroup and multivariate analyses were used to adjust for potential confounding variables (anesthetic and age).

Results: Of the 100 patients enrolled in the study, 33 were admitted with a sublingual temperature < 35°C. The hypothermic and normothermic groups were compared demographically. The groups were similar with regard to history of myocardial infarction, angina, coronary bypass surgery, and preoperative ß-adrenergic blocker therapy. Mean age was greater in the hypothermic (70 ± 2 years) than in the normothermic (62 ± 1 years) group (p=0.001). Table 1 shows the incidence of myocardial ischemia in occurring in the first 24 postoperative hours in the hypothermic and normothermic patients with subgroup analysis by age and type of anesthesia. The incidence of myocardial ischemia in the hypothermic patients was 2-3 times greater than the incidence in the normothermic patients. The greater incidence of myocardial ischemia in the hypothermic patients was evident in the younger and older subgroups as well as in the epidural and general anesthetic subgroups. To determine the independent predictors of myocardial ischemia, the following variables were entered into a multiple logistic regression model: temperature, age, gender, diabetes, intraoperative blood loss, history of angina or myocardial infarction, preoperative ß-adrenergic blocker therapy, history of smoking, hypertension, or coronary artery bypass surgery, postoperative hematocrit, shivering, and anesthetic technique. Temperature upon admission to the ICU was the only variable associated with myocardial ischemia (p=0.02) in this model.

Table 1. Incidence of Myocardial Ischemia by Hypothermia, Age, and Type of Anesthesia

	Patient Age		Type of Anesthesia		Totals
	> 65 years	≤ 65 years	Epidural	General	
Hypothermic (< 35° C)	10/24 (42)	2/9 (22)	4/10 (40)	8/23 (35)	12/33 (36) *
Normothermic (≥ 35 °C)	5/28 (18)	4/39 (10)	5/36 (14)	4/31 (13)	9/67 (13)
Totals	15/52 (29)$	6/48 (13)	9/46 (20)	12/54 (22)	21/100 (21)

* p < 0.01 versus Normothermic, $ p = 0.05 versus < 65 years. Pecentages are given in parentheses.

Study #2. Perioperative Body Temperature Randomized Trial

Methods: 49 surgical patients were enrolled in a randomized trial comparing routine care with skin-surface warming during the intraoperative period and the first two postoperative hours. Inclusion criteria (all of which needed to be met) were 1) age > 60 years, 2) scheduled thoracic, abdominal, or lower extremity vascular surgery, and 3) either documented coronary artery disease (CAD) or ≥ 2 risk factors for CAD by criteria used in previous studies.(10) Routine care was defined as one layer of surgical drapes during surgery, and one or two warmed cotton blankets in the postoperative period. Skin-surface warming was accomplished with a forced-air warmer (Warm Touch™, Mallinckrodt Medical Inc., St. Louis, MO). Core temperature (tympanic), and forearm minus fingertip skin-surface temperature gradient were measured (an index of vasoconstriction). This gradient has been previously shown to correlate with blood flow.(11) Plasma levels of epinephrine and norepinephrine were measured by HPLC. The routine care and the skin-surface warming groups were compared by ANOVA with repeated measures.

Results: The routine care (n=23) and the skin-surface warming (n=26) groups were similar with regard to age, sex, diabetes, preoperative ß-adrenergic blockers, type of anesthesia or surgery, requirement for postoperative mechanical ventilation, and type and quality of postoperative analgesia as indicated by pain scores. Temperature and catecholamine data are shown in Figure 1. Mean core temperature upon admission to the ICU was lower in the routine care (35.3±0.1°C) than in the skin-surface warming (36.6±0.1°C) group (p=0.0001). Core temperature was lower, and forearm fingertip gradient was greater in the routine care group during the first 3 postoperative hours. In the routine care group, mean epinephrine level was increased above the preoperative baseline at one hour postoperatively. Epinephrine level did not change from baseline in the skin-surface warming group. Mean norepinephrine level was elevated above baseline at ICU admit, at one hour and at three hours postoperatively. Norepinephrine was unchanged from baseline in the skin-surface warming group.

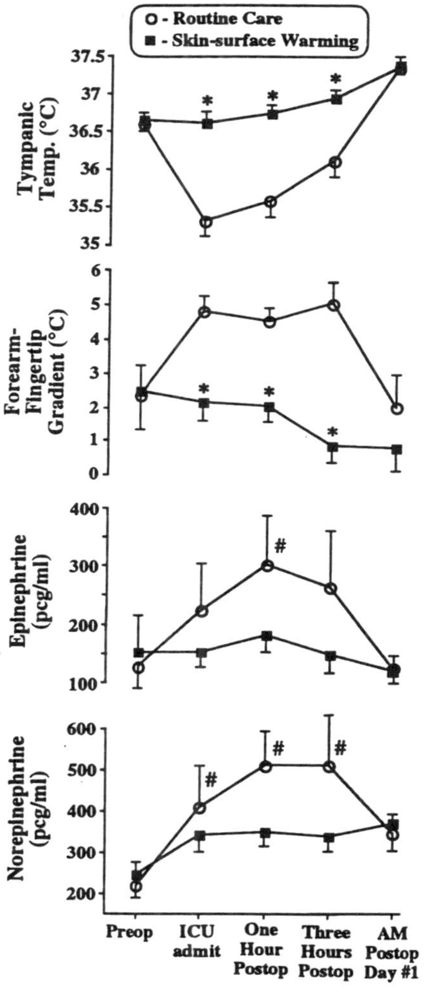

Fig. 1: Core temperature, skin-surface temperature gradient, epinephrine, and norepinephrine levels are shown in surgical patients randomized to routine care or skin-surface warming groups. In the early postoperative period, the routine care group had a lower mean core temperature, a greater degree of peripheral vasoconstriction (as indicated by forearm fingertip gradient), and significantly elevated epinephrine and norepinephrine levels. * - $p < 0.01$ vs. routine care, # - $p < 0.05$ vs. preop.

Study #3. Cold Intravenous Infusion Study

Methods: Seven volunteers (1 female, 6 males) were enrolled in a study to examine the adrenergic response to acutely induced core hypothermia. Thirty ml/kg of 4°C normal saline was infused over a 30 minute period through an intravenous catheter in the right arm. Core temperature (tympanic) and forearm minus fingertip skin-surface temperature gradient (left arm) were measured. Blood samples were drawn at five times during the study for determination of plasma levels of epinephrine and norepinephrine. Data were analyzed by ANOVA with repeated measures.

Results: All seven volunteers (ages 18-33) completed the study and were included in the analysis. Core temperature, forearm fingertip skin-surface temperature gradient, epinephrine and norepinephrine plasma levels are shown in Figure 2. The 30 minute cold infusion is indicated on the X-axis. Mean core temperature fell to 35.6±0.6 °C and forearm fingertip gradient increased rapidly to 6.4±1.6 °C indicating vasoconstriction. Mean norepinephrine level increased four-fold by the end of the infusion, then returned toward baseline. Mean plasma epinephrine level did not change significantly.

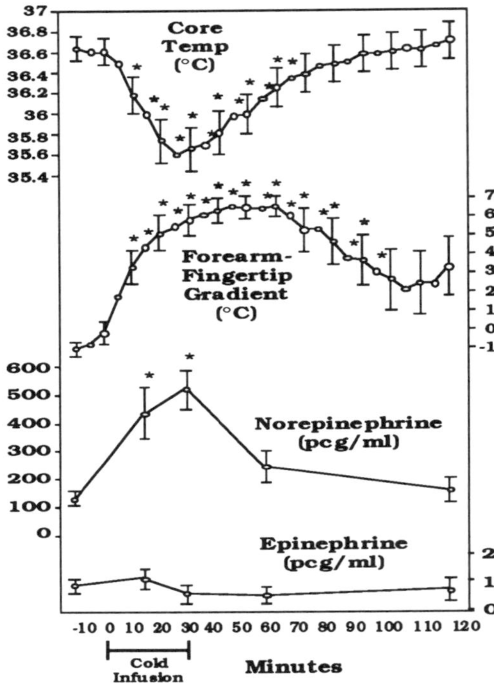

Fig. 2: Core temperature, skin-surface temperature gradient, plasma norepinephrine, and epinephrine levels are shown before, during and after intravenous infusion of 30 ml/kg cold saline (4°C). The infusion was delivered over a 30 minute period as indicated on the X-axis. Core hypothermia was associated with significant vasoconstriction, and a fourfold increase in norepinephrine. Epinephrine was unchanged. * - $p < 0.01$

Study #4. Cold Pressor Test and the Effects of α-adrenergic Blockade

Methods: Six male volunteers were enrolled to determine the effects of α-adrenergic blockade on peripheral vasoconstriction during cold pressor test (CPT). CPT was performed using a 2 minute immersion of the right hand in a cold water bath (5-7°C). The test was performed twice; before and immediately after the administration of phentolamine (1 mg/kg) given intravenously over a 10 minute period. Vasoconstriction in the left hand was assessed by two methods: laser Doppler blood flow (Perimed-PF4, Piscataway, NJ) and forearm minus fingertip skin-surface temperature gradient. Laser Doppler data were analyzed as the maximum percent change from baseline during the two minute CPT, with baseline defined as the average blood flow (laser Doppler units) over the three minute period prior to the CPT. Skin-surface gradient data were analyzed as the change in temperature gradient (Δ °C) over the two minute duration of the CPT. The pre- and post- phentolamine data were compared using a paired t-test.

Results: All six volunteers completed the study and were included in the analysis. Significant vasoconstriction occurred during CPT before α-adrenergic blockade, but was significantly reduced after α blockade. A typical laser Doppler response to CPT before and after phentolamine is shown in Figure 3.

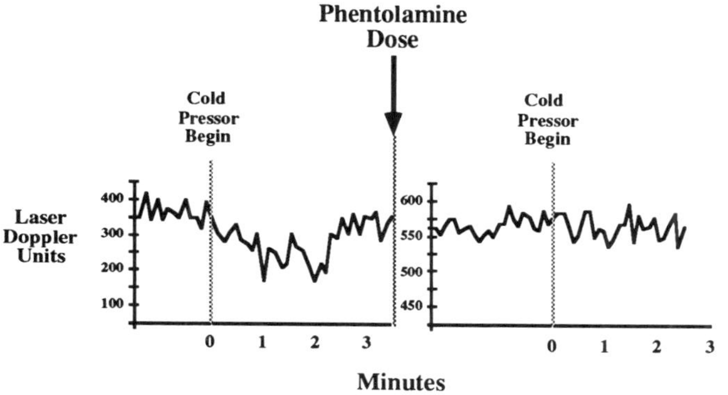

Fig. 3: Laser Doppler blood flow in the left index finger is shown during cold pressor test (CPT) with the right hand immersed in an ice bath (5-7°C). A two minute immersion was performed before and after the administration of intravenous phentolamine (1 mg/kg). The data from this volunteer illustrate the typical vasoconstriction that occurs during CPT and the attenuation of this response by α-adrenergic blockade.

Both techniques that were used to measure the vascular response (forearm fingertip temperature gradient and laser Doppler blood flow) indicated a greater degree of vasoconstriction before phentolamine, relative to the response after phentolamine. These data are shown in Figure 4.

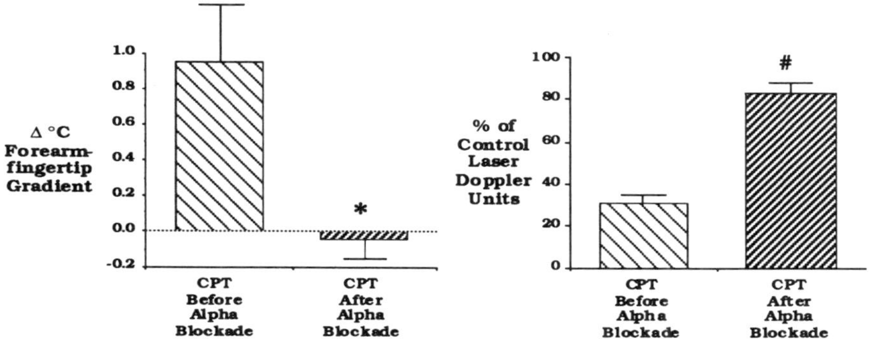

Fig. 4: Peripheral vasoconstriction in the left hand was assessed during CPT (right hand immersion) before and after the administration of phentolamine (1 mg/kg). Vasoconstriction as measured by both forearm fingertip skin-surface temperature gradient and laser blood flow was significantly attenuated by α-adrenergic blockade. * - $p<0.05$, # - $p<0.001$.

Conclusion

Cold induced myocardial ischemia occurs in both the ambulatory and perioperative settings, but the mechanisms responsible for this phenomenon are not well understood. All three models of cold stress presented here elicited a significant adrenergic response. The first model involved a whole body cold exposure (surgical patients), the second primarily a core exposure (cold intravenous fluid), and the third a peripheral exposure (cold pressor test). All three were associated with peripheral vasoconstriction which appears to be norepinephrine mediated. The peripheral vasoconstriction induced by the cold pressor test was blocked by an α-adrenergic antagonist. Based on previous studies that have demonstrated the vasoconstrictor effect of norepinephrine in the coronary vasculature,(12) it is conceivable that there is concomitant α-mediated vasoconstriction in the peripheral and coronary vessels during cold stress. Further studies will be necessary to determine the clinical significance of these findings, and to assess the effects of sympatholytic drugs for the treatment or prevention of cold-induced cardiac morbidity.

Acknowledgements

The authors gratefully acknowledge assistance from Susan Kelly R.N. and Kenneth Kubos Ph.D as well as the nursing staff in the Acute Care Unit and the Surgical Intensive Care Unit.
Supported in part by the National Institutes of Health grants GM38177, NS-26363, OPD-GCRC 5MO1RR00722, and Mallinckrodt Medical Inc.

References

1. Mangano DT: (1990) Perioperative cardiac morbidity. Anesthesiology 72, 153-184.

2. Bainton D, Moore F, Sweetnam P: (1977) Temperature and deaths from ischemic heart disease. Br J Prev Soc Med 31, 49-53.

3. Rose G: (1966) Cold weather and ischaemic heart disease. Brit J Prev Soc Med 20, 97-100.

4. Frank SM, Beattie C, Christopherson R, Norris EJ, Perler BA, Williams GM, Gottleib SO: (1993) Unintentional hypothermia is associated with postoperative myocardial ischemia. Anesthesiology 78, 468-476.

5. Hines EA, Brown GE : (1936) The cold pressor test for measuring the reactibility of the blood pressure: Data concerning 571 normal and hypertensive subjects. Am Heart J 11, 1-9.

6. Cummings MF, Steele PM, Mahar LJM, Frewin DB, Russell WJ: (1983) The role of adrenal medullary catecholamine release in the response to a cold pressor test. Cardiovasc Res 17, 189-191.

7. Cuddy RP, Smulyan H, Keighley JF, Markason CR, Eich RH: (1966) Hemodynamic and catecholamine changes during a standard cold pressor test. Am Heart J 71, 446-454.

8. McLean JK, Sathasivam P, MacNaughton K, Graham TE: (1992) Cardiovascular and norepinephrine responses of men and women to two cold pressor tests. Can J Physiol Pharmacol 70, 36-42.

9. Mudge GH, Grossman W, Mills RM, Lesch M, Braunwald E: (1976) Reflex increase in coronary vascular resistance in patients with ischemic heart disease. N Engl J Med 295, 1333-1337.

10. Mangano DT, Browner WS, Hollenberg M, London MJ, Tubau JF, Tateo IM: (1990) Association of perioperative myocardial ischemia with cardiac morbidity and mortality in men undergoing noncardiac surgery. N Engl J Med 323, 1781-1788.

11. Rubinstein EH, Sessler DI: (1990) Skin-surface temperature gradients correlate with fingertip blood flow in humans. Anesthesiology 73, 541-545.

12. Vatner SF, Higgins CB, Braunwald E: (1974) Effects of norepinephrine on coronary circulation and left ventricular dynamics in the conscious dog. Circ Res 34, 812-823.

ADRENERGIC MODULATION OF THE SURGE IN TAIL SKIN TEMPERATURE INDUCED BY LHRH AND 2 DEOXYGLUCOSE

M. J. Katovich, G. Strubbe and B. E. Wright.

Department of Pharmacodynamics, College of Pharmacy, Box 100487 JHMHC, University of Florida, Gainesville, FL, 32610 (USA)

We previously observed (1-5) that administration of an opioid antagonist to morphine-dependent rats results in a prompt rise in tail skin temperature (TST) and a subsequent fall in colonic temperature (Tc). Additionally, an elevated heart rate and hypersecretion of luteinizing hormone (LH) are observed during these temperature changes in this animal model (2). These physiological changes are similar in magnitude, duration and temporal association with those exhibited in women undergoing a flushing episode. These flushing episodes are more pronounced in ovariectomized rats (4), and are attenuated with administration of estrogen (4,5) or α-adrenergic agents (1,3). These same therapeutic agents have been used effectively to alleviate the flushes observed in women (6-8). From both our animal studies and clinical trials it appears that some degree of adrenergic activation may be responsible for the flushing episodes.

In 1980 Lomax et al (9) first demonstrated that central administration of luteinizing hormone releasing hormone (LHRH) can elevate skin temperature in the rat. We have similarly demonstrated that an elevation in tail skin temperature can be produced in naive rats following administration of either an LHRH agonist (10) or 2 deoxyglucose (2DG) (11). These 'flush' responses are similar to those observed during withdrawal in the morphine dependent rat. The purpose of the current study was to evaluate if the elevations in skin temperature associated with administration of either LHRH or 2DG were also attenuated with the α-adrenergic agents which have been previously effective in our morphine-dependant animal model and in clinical trials. Positive results would suggest that this morphine dependency may not be necessary for the development of an animal model for the menopausal hot flush.

Methods

Female Charles Rivers CD rats, initially weighing 175-225 g were housed in groups of 2 in hanging stainless steel cages in a room maintained at 25±1°C and illuminated from 05:00 to 19:00 hr. Food and water were provided ad libitum. All animals were ovariectomized 2-3 weeks prior to the temperature studies.

For central drug administration, standard stereotaxic techniques were used to implant cannula guides into the lateral cerebral ventricles (icv) one week prior to the temperature experiments. On the day of the experiment, animals were lightly restrained in wire mesh cages. Colonic temperature (Tc) was measured with a copper-constantan thermocouple inserted 6 cm beyond the anus and taped to the base of the tail. Tail skin temperature (TST) was measured with a second copper constantan thermocouple taped to the dorsal region of the tail. Temperatures for all 12 animals (6 control and 6 experimental for each study) were measured simultaneously with a data acquisition and control system (CYBORG) interfaced to an Apple IIe computer. Temperatures were recorded at 5 min intervals. All rats were allowed a one hour adjustment period after they were placed in the restraining cages to obtain baseline temperatures. In the first study the effects of central administration of an LHRH agonist ([Des-Gly^{10},im-Benzyl-D-His^{6}, Pro^{9}]-LHRH, ethyl amide, Bachem California, Torrance, CA) was evaluated. In these studies central administration of either phentolamine (50 μg/5μl, icv, Sigma Chemical Co., St. Louis Mo.) or clonidine (10μg/5μl, icv, Sigma) was administered thirty min prior to administration of LHRH (10 or 20 μg/5μl, icv). In subsequent studies the same adrenergic agents were administered 30 min prior to 2-deoxy-glucose (2DG, 750 mg/kg, ip, Sigma). In each study control animals were administered vehicle alone. Temperatures were monitored at 5 min intervals until TST values were restored to baseline conditions (1 or 2 hours). In an final group of studies, systemic injection of DSP-4 [N-(2-chloroethyl)-N-ethyl-2-bromobenzylamine hydrochloride] was administered (50 mg/kg, ip, Research Biochemicals International, Natick, MA.) 5 days prior to the first temperature study. The same animals were also studied 6 days later (11 days after administration of DSP-4). In these separate DSP-4 studies all animals received either LHRH (20 μg/5μl, icv) or 2DG (750 mg/kg, ip) following the 60 min equilibration period. All temperature studies were performed between 9 and 12 A. M. in a room maintained at 26±1°C.

Data were summarized by calculating the mean maximal changes in temperature irrespective of time and the integrated area under the temperature curves produced by each rat for each of the groups and analyzed by Student's t-test with significance set at 95% confidence limits. For the DSP-4 studies data were analyzed by an ANOVA and the significance between groups determined by Fisher's LSD post hoc test.

Results

In the first study (Fig. 1) central administration of LHRH resulted in a significant rise in TST in the vehicle-treated rats which was completely abolished with central administration of clonidine (10 μg/5μl, icv). The maximal rise in TST is reached within 15-20 min of LHRH administration. Minimal effects of either drug were observed in colonic temperature. Table 1 summarizes the results of central clonidine and phentolamine on the temperature response to central administration of LHRH (10 and 20 μg/5μl, icv). Basal TST and Tc were similar between the groups for all of the studies. There was no significant difference in the temperature responses observed with the two doses of LHRH. The maximal rise in TST (the 'flush'-response) in the control groups ranged from 2 to 5°C. Pretreatment with centrally administered phentolamine or clonidine essentially abolished the rise in TST. These results are similarly manifested when the area under the temperature curve are evaluated. The negative numbers for some of the areas indicate that during the 60 min following administration of LHRH the mean changes in TST were reduced below baseline levels. Results for the effects of LHRH and the pretreatment with α-adrenergic agents on core temperature are not as consistent. Administration of LHRH either had no effect or slightly reduced Tc over the 60 min period. Only with the low dose of LHRH did clonidine pretreatment alter this effect (data not shown).

Figure 2 summarizes the results of administration of 2DG to naive rats. Control animals responded to 2DG administration with an elevation of TST and a corresponding fall in Tc. Pretreatment with clonidine (10 µg/5µl, icv) significantly reduced the rise in TST without having an appreciable effect on the fall in Tc. In both 2DG studies (Table 2) basal TST and Tc were similar for both the control and treated groups. Only clonidine treatment significantly reduced the changes in TST induced by 2DG; phentolamine pretreatment was ineffective. It is interesting to note that the peak temperature response for the 2DG studies occurs much later than the response to LHRH and the fall in Tc is of greater magnitude with 2DG.

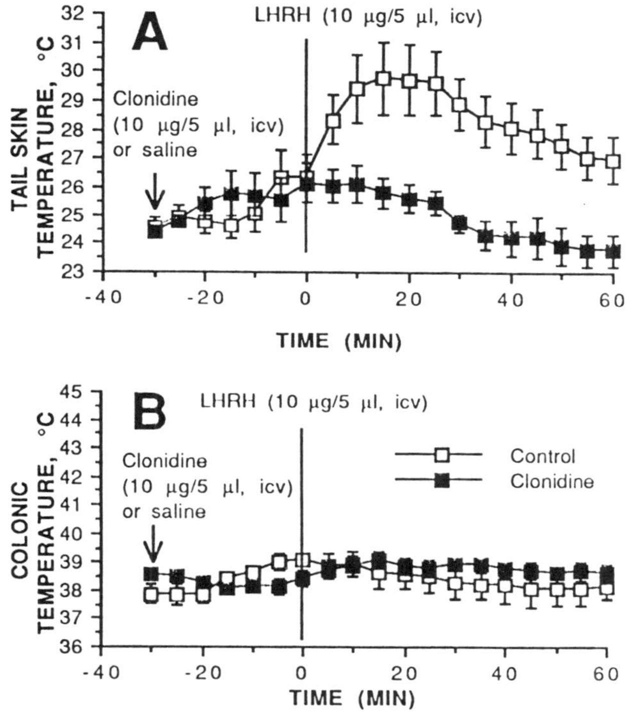

Fig. 1: Effect of systemic administration of LHRH (10µg/5µl icv) at time zero on mean tail skin temperature (A) and colonic temperature (B). Animals were pretreated with clonidine (10µg/5µl icv, filled squares), or saline (5µl icv, open squares) 30 min prior to LHRH. Each point represents the mean of 5 or 6 animals and the vertical bar represents one standard error of the mean.

Results from the DSP-4 studies reveal significant differences between the LHRH and 2DG temperature responses. As in the previous studies, basal TST and Tc were similar between experimental groups. In the LHRH studies the maximal increases in TST was 3.4±0.7°C, 5 days following DSP-4 treatment and a similar 2.7±1.0°C, 11 days after DSP-4 treatment. Both these values were significantly lower than that of the control group (5.4±0.3°C). Areas under the 60 min TST curves likewise were significantly different at 76.9±31.2, 95.2±34.5 and 175.8±20.6 °C-min for the 3 groups, respectively. For the 2DG studies significance was only observed in the 5 day DSP-4 group. The rise in maximal TST was 3.8±0.8°C in controls which was similar to the 3.9±1.0°C in the 11 day DSP-4 group. Both values were significantly higher than the 0.6±0.2°C observed in the 5 day DSP-4 group. Likewise the area under the two hour TST curve were similar between the 11 day (161.7±68.2°C-min) and control (165.4±48.5°C-min) group which was significantly higher than the -4.6±12.1°C-min observed in the 5 day DSP-4 group.

Table 1: The effects of central phentolamine or clonidine on the TST response to central administration of LHRH in female rats.

Experimental Group	Body Weight (gms)	Basal TST (°C)	Basal Tc (°C)	Max rise in TST† (°C)	Area under 1 hr TST curve (°C-min)
LHRH (10μg/5μl icv)					
control	249±6+	26.3±0.8	39.1±0.2	3.8±1.3	127±57.8
clonidine (10μg/5μl icv)	258±6	26.2±0.7	38.4±0.3	0.3±0.2*	-73.2±25.8*
control	256±6	25.8±0.2	38.3±0.4	3.8±0.9	97.2±24.1
phentolamine(50μg/5μl icv)	262±6	26.4±0.6	37.5±0.4	0.9±0.7*	-10.5±36.7*
LHRH (20μg/5μl icv)					
control	266±7	24.9±0.8	38.3±0.2	1.9±4	119.4±46.7
clonidine (10μg/5μl)	272±4	24.6±0.4	37.6±0.5	0.5±.2*	11.4±11.7*
control	264±5	23.4±0.1	37.7±0.2	5.1±1.2	162.9±55.7
phentolamine (50μg/5μl)	261±9	24.2±0.6	37.4±0.4	1.4±1.0*	-35.1±34.6*

+ data expresses as mean ± one standard error of the mean
* significantly different from control (p<0.05) † maximal rise in TST is irrespective of time

Tab. 2: The effects of central phentolamine (50μg/5μl, icv) and clonidine (10μg/5μl) on the TST response to peripheral administration of 2 deoxyglucose (2DG, 750mg/kg,ip) in female rats.

Experimental Group	Body Weight (gms)	Basal TST (°C)	Basal Tc (°C)	Max rise in TST† (°C)	Area under 2 hr TST
Saline + 2DG	273±5+	25.0±0.4	38.6±0.3	3.7±0.9	171.3±53.8
Clonidine + 2DG	267±10	24.8±0.4	38.1±0.2	1.0±0.5*	-10.0±33.4*
Saline + 2DG	291±9	25.0±0.3	38.4±0.2	4.4±1.0	142.7±46.3
Phentolamine + 2DG	288±14	25.8±0.5	38.9±0.2	3.4±1.1	65.1±37.0

+ data expresses as mean ± one standard error of the mean
* significantly different from control (p<0.05) † maximal rise in TST is irrespective of time

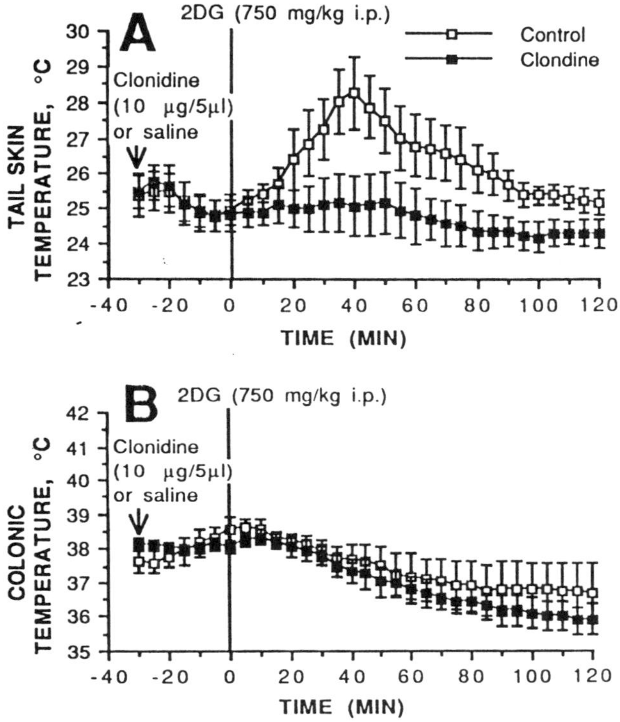

Fig. 2: Effect of systemic administration of 2DG (750mg/kg, ip) at time zero on mean tail skin temperature (A) and colonic temperature (B). Animals were pretreated with clonidine (10µg/5µl icv, filled squares), or saline (5µl icv, open squares) 30 min prior to 2DG. Each point represents the mean of 5 or 6 animals and the vertical bar represents one standard error of the mean.

Discussion

We have previously reported that the morphine-dependent rat may be a useful model in which to study the mechanisms involved in the menopausal hot flush. In these dependent animals administration of naloxone or spontaneous withdrawal from morphine (2) produce thermal, cardiac and endocrine manifestations which are similar in magnitude, duration and temporal association to those manifested in women undergoing a hot flush. We have also demonstrated in this model that the thermal response is more sensitive to the absence of estrogens (4) and can be abolished by central or peripheral administration of clonidine, ST-91 [(2-(2,6-diethylphenyl-imino)-2-imidazolidine), Boehringer Ingelheim Laboratories] and phentolamine (1,3). Likewise we have demonstrated that LHRH antagonists (10) and hyperglycemia (12) can each significantly reduce the surge in TST induced during morphine withdrawal. The current study has demonstrated that the LHRH flush response can be abolished by treatment with either phentolamine or clonidine, suggesting an adrenergic component to the flush response. It therefore appears that morphine dependency is not necessary to produce the flush response. Collectively, these results suggest some mechanistic relationship between LHRH, glucose and the adrenergic system in the flush response.

Lomax et al (9) was the first to report that central administration of LHRH results in a marked rise in skin temperature in the rat. We also have demonstrated that central administration of LHRH results in an elevated tail skin temperature. This response is more sensitive in morphine-dependent animals (10) suggesting that opiates play a role in modulating the thermal responses. We previously demonstrated that the thermal response to LHRH is centrally mediated and not associated with secretion of LH (10). Others also have demonstrated that LH itself does not appear to be the trigger for the hot flush (13). We suggested that LHRH's mechanism of action may be as a neuromodulator to some thermosensitive neurons; activation of these pathways could result in a flush-like response.

The thermosensitive neurons of the hypothalamus are known to be gluco-, osmo-, and steroid-sensitive, which suggests that temperature-sensitive neurons have multiple interactions with various regulatory systems. We have demonstrated that cellular hypoglycemia, induced by 2DG (11) or administration of insulin (12) does result in a flush-like response. Similar reports have been observed in man (14,15). This hypoglycemic condition results in the activation of the sympathetic nervous system which acts as a counter-regulatory mechanism intended to maintain glucose homeostasis (16). While the magnitude of the TST surge is similar following administration of LHRH and 2DG, the profile of the responses differ slightly. The 2DG response is slower in onset and longer in duration. This difference likely reflects the time needed to transport the glucose analogue across the blood brain barrier to induce intracellular glucose deprivation. There is also a more marked decrease in the Tc with 2DG administration. Although clonidine attenuated the TST response, phentolamine was without effect. These two adrenergic agents act at different sites; this may account for the difference in the observed responses. Phentolamine is an α_1 and α_2 adrenergic antagonist (17), whereas clonidine, a partial agonist, acts presynaptically via an α_2-adrenergic mediated mechanism to inhibit sympathetic activity (17). By inference, since central α_2 receptor activation antagonizes the temperature response, blockade by phentolamine of α_1- (but not α_2-) adrenergic receptors likely located postsynaptically dampens the TST response to central administration of LHRH.

Destruction of peripheral noradrenergic neuronal activity with DSP-4 (5 days study) similarly attenuated the TST response to both LHRH and 2DG administration. However, upon peripheral noradrenergic recovery, while central noradrenergic neuronal activity is still compromised with the DSP-4 neurotoxin (18), the LHRH induced TST surge is still attenuated while the 2DG induced surge is unaffected. This suggests that the TST response to peripheral administration of 2DG is due more to peripheral than to central adrenergic pathways. Collectively, the results suggest that adrenergic neuronal involvement is necessary for the rise in TST following administration of either LHRH or 2DG. They also suggest that peripheral noradrenergic neuronal activity may play a greater role in the 2DG-induced temperature surge than in the LHRH-induced surge. Thus, LHRH and hypoglycemia may not follow a common pathway to produce a hot flush response. Furthermore, the LHRH-induced flush response may be a better animal model by which to study the flush response in naive animals.

Acknowledgements

The authors would like to acknowledge Ms. Linda Panchou for preparation of the manuscript and Phylis Raynor and Kim Hanley for technical assistance. This work was supported by NI grant HD-18133.

References

1. Katovich, M.J., Pitman, D.L. and Barney, C.C. (1990):Mechanisms mediating the thermal response to morphine withdrawal in rats. Proc. Soc. Exp. Biol. Med. 193, 129-135.

2. Simpkins J.W., Katovich, M.J., and Song, I-C. (1983):Similarities between morphine withdrawal in the rat and the menopausal hot flush. Life Sci 32, 1957-1966.

3. Katovich, M.J., Simpkins, J.W. and Barney, C.C. (1987): α-Adrenergic mediation of the tailskin temperature response to naloxone in morphine-dependent rats. Brain Res. 426, 55-61.

4. Katovich, M.J., and Simpkins, J.W. Further evidence for an animal model for the flush (1989). In: Thermoregulation: Research and Clinical Applications. Lomax, P. and Schönbaum, E., Eds., Karger, Basel, 1215-1330.

5. Katovich, M.J., and O'Meara, J. (1987):Effect of chronic estrogen on the skin temperature response to naloxone in morphine-dependent rats. Can. J. Physiol. Pharmacol. 65, 563-567.

6. Tulandi, T., Samarthji, L., and Kinch, R.A. (1983):Effect of intravenous clonidine on menopausal flushing and luteinizing hormone secretion. Brit. J. Obstet. Gynec. 990, 854-857.

7. Freedman, R.R., Woodward, S. and Sabharwal, S.C. (1990):α-adrenergic mechanism in menopausal hot flushes. Obstet. Gynecol. 76, 573-578.

8. Lomax, P., Schönbaum, E. (1993):Postmenopausal hot flushes and their management. Pharmac. Ther. 57, 347-358.

9. Lomax, P., Bajorek, J.G., Chesarek, W. and Tataryn, I.V. (1980):Thermoregulatory effects of luteinizing hormone releasing hormone in the rat. Thermoregulatory Mechanisms and their therapeutic Implications, In: Cox, B., Lomax, P., Milton, A. S. and Schönbaum, E. Eds., Karger, Basel, 208-211.

10. Katovich, M.J., Simpkins, J.W., and O'Meara, J. (1989):Effects of acute central LHRH administration on the skin temperature response in morphine dependent rats. Brain Res. 494, 85-94.

11. Simpkins, J.W., Andreadis, D.K., Millard, W.J. and Katovich, M.J. (1990):The effects of cellular glucoprivation on skin temperature regulation in the rat. Life Sci. 47, 107-115.

12. Simpkins, J.W., Katovich, M.J. and Millard, W.J. (1990):Glucose modulation of skin temperature responses during morphine withdrawal in the rat. Psychopharmacol. 102, 213-220.

13. Meldrum, D.R., Erlick, Y., Lu, J.K.H. and Judd, H.L. (1982):Objectively recorded hot flushes in patients with pituitary insufficiency. J. Clin. Endocrinol. Metab. 50, 685-689.

14. Santiago, J.V., White, N.H., Skor, D.A., Levandoski, L.A., Bier, D.M. and Cryer, P.E. (1984): Defective glucose counterregulatory, limits intensive therapy of diabetes mellitus. Am. J. Physiol. 247, E215-E220.

15. Bussell, K.L., Murphy, F.Y., Nicks, S.A. and Vesely, D.L. (1987):Facial flushing secondary to hypoglycemia. J. Med. 18, 123-132.

16. Cryer, P.E. and Gerich, J.E. (1983):Relevance of glucose counterregulatory systems to patients with diabetes: critical roles of glucagon and epinephrine. Diabetes Care 6, 95-99.

17. Blaschke, T.F. and Melmon, K.L. (1980):Antihypertensive agents and the drug therapy of hypertension. In: The Pharmacological Basis of Therapeutics, Gilman, A.G., Goodman, L.S. and Gilman, A. Eds, MacMillan, N.Y., 793-810.

18. Jonsson, G., Hallman, H. Ponzio, F. and Ross, R. (1981):DSP-4 [N-(2-chlorethyl)-N-ethyl-2-bromobenzylamine)-a useful tool for central and peripheral noradrenaline neurons. Europ. J. Pharmacol. 72, 173-188.

Section V:

FEVER AND ANTIPYRESIS

Thermal Balance in Health and Disease
Advances in Pharmacological Sciences

FEVER AND ANTIPYRESIS

M. J. Kluger

Institute for Basic and Applied Medical Research - The Lovelace Institutes
2425 Ridgecrest Dr., SE   Albuquerque, New Mexico 87108

Fever is one of the most common host defense responses of an infected organism. In my 1991 review on fever (1) I presented a table that contained examples of fevers in 26 species of ectothermic animals from lizards to leeches. Since that time fevers have been demonstrated in a lizard belonging to the Teiidae family (*Callopistes maculatus*), both in response to injection of bacteria (2), and to emotional stress (3), and in the marine toad, *Bufo marinus* in response to injection of LPS (4).

It is clear from dozens of studies that the rise in body temperature associated with fever is a highly regulated process, involving numerous sophisticated physiological and behavioral responses. As for many regulatory processes there appear to be multiple factors that influence the regulated body temperature during fever -- some raise the thermoregulatory set-point and others prevent the set-point from rising too high. The former are called "endogenous pyrogens," and the latter "endogenous antipyretics" or "endogenous cryogens." In this review, I will briefly discuss evidence for the roles of cytokines and hormones in modulating these thermoregulatory changes during infection. In addition, I will present some of our most recent data on the phenomenon of "stress-induced hyperthermia," which in many ways has the characteristics of true fevers (i.e. an elevated thermoregulatory set-point).

There are many substances that have been touted as endogenous pyrogens, and this list continues to grow. For example, Shapiro et al. have recently reported that ciliary neurotrophic factor (CNTF) is an endogenous pyrogen (5). In this well-designed and executed study, it was shown that iv injection of CNTF into rabbits produces a dose-dependent rapid-onset fever. These fevers were blocked by pretreatment of the rabbits with indomethacin, a potent antipyretic drug. But is CNTF an endogenous pyrogen or is it simply a pyrogen, which when injected at pharmacologic doses causes fever? In my review article (1) I argued that before concluding that a substance truly was in the natural pathway between contact with some pathogen and the elevation in the thermoregulatory set-point, it was important for that substance to obey a series of rules that are in some ways analogous to Koch's postulates for determining whether a putative microorganism was responsible for a particular disease. These criteria are briefly discussed below:

Criterion #1. Application of the putative endogenous pyrogen (EP) at the hypothetical site of action results in a rise in body temperature. For example, if the anterior hypothalamus is the theoretical site of action of EP, then injection of the putative EP at this locus should induce fever.

Criterion #2. When an exogenous pyrogen induces a fever, (a) the released putative EP has some quantitative relationship to the magnitude of the rise in temperature, and (b) the putative EP is released in amounts similar to those that cause a similar rise in temperature when the putative endogenous pyrogen is given at the hypothetical site of action.

Fevers vary in magnitude and duration. Suppose a putative EP is secreted from fixed macrophages in the liver into the circulation. If so, when an animal is injected with an exogenous pyrogen (e.g. LPS), the EP should be released from the liver into the circulation roughly in proportion to the resultant fever. Moreover, when a putative EP is infused into an experimental animal to achieve the plasma concentration observed during fever, the rise in body temperature should be virtually indistinguishable from that following injection of the exogenous pyrogen.

Criterion #2, however, is not essential because it assumes that the administration of an exogenous pyrogen does not influence (e.g. up-regulate) the receptor for a putative EP. For example, were LPS either directly or indirectly to increase the number of receptors for an EP, then the plasma concentration of that EP might vary in an "all-or-none" mode, rather than in proportion to the fever. Furthermore, this criterion assumes that during the course of the fever other factors (e.g. endogenous cyrogens) are not playing an influential role. Nevertheless, it is probable that whatever turns out to be the "circulating EP" will fulfill this criterion to some extent.

Criterion #3. The rise in body temperature induced by the putative EP should be associated with appropriate thermoregulatory effector responses, which are indicative of an elevation in thermoregulatory set-point (e.g. induction of peripheral vasoconstriction, shivering, appropriate behaviors).

As described above, fever is not simply a rise in body temperature, but rather a condition in which the thermoregulatory set-point is elevated. As such, the febrile organism will use a variety of thermoregulatory behaviors to elevate body temperature. These effector mechanisms will vary depending on the environmental temperature.

Criterion #4. Substances that block the *production* of the putative EP should prevent the rise in body temperature.

If drug X blocks the production of a putative EP, the development of fever should be prevented. If drug X does block the production of a putative EP, but does not prevent fever, or only attenuates fever, then other pyrogens must also be involved in fever.

Criterion #5. Substances that block the *action* of the putative EP should prevent the rise in body temperature.

As in criterion #4, if drug X blocks the action of a putative EP, the result should be complete prevention of fever. If drug X does block the action of a putative EP, but does not prevent fever, or only attenuates fever, then other pyrogens must also be involved in influencing the fever.

A similar set of criteria was developed for demonstrating whether a substance was an endogenous antipyretic or cryogen (EC). In many respects, some of the most convincing evidence supporting the notion that a specific substance is an endogenous pyrogen or cryogen has come from studies using specific drugs or antibodies to block the *in vivo* actions of putative EPs or ECs.

Interleukin-1 and Fever

Since its discovery (i.e. cloning) in 1984, interleukin-1 (IL-1) has been touted by virtually all investigators working in the area of fever research as being an important circulating mediator of fever. Surprisingly, there is little evidence that the circulating levels of IL-1 actually rise during most fevers. Most investigators have been unable to show any correlation between fever and circulating IL-1 using bioassays or immunoassays. Recently, Derijk and Berkenbosch (6) have shown some dose-dependent rises in circulating IL-1β using an immunoassay, but the specific relationship between this measured immunoreactive IL-1β and fever is not yet clear. To my knowledge, there is no evidence that IL-1α circulates in the blood during fevers. There are data, however, that shows that IL-1β (but not IL-1α) is an endogenous pyrogen based on using neutralizing antibodies. Studies from Rothwell's group (7), and from my laboratory (8) have shown that pretreatment of rodents with antiserum to IL-1β attenuated LPS-induced fevers. Klir et al. have recently shown that microinjection of anti-IL-1β directly into the anterior hypothalamus of the rat caused a 58% reduction in fevers caused by ip injection of LPS (9). Since there is little evidence that IL-1β crosses from the circulation into the cns at the level of the anterior hypothalamus, these data support the hypothesis that peripheral injection of LPS causes a rise in IL-1β within the cns, and it is probably not necessary for IL-1β to be a circulating EP.

Tumor Necrosis Factor α and Fever

Tumor necrosis factor α (TNF) is pyrogenic when injected in sufficiently large doses into experimental animals or humans. But, is TNF an EP? What happens when the biological actions of TNF is blocked using antibodies or binding proteins to TNF? Studies in rabbits in which monoclonal antibodies to TNF were injected have shown that this resulted in reduced fevers (10,11). Unfortunately, in these studies no data were presented indicating whether the antibody neutralized the TNF bioactivity. Did this antibody neutralize the biological activity of the TNF, or actually increase the biological activity by preventing the clearance of the TNF from the circulation? When a polyclonal antibody to TNF was injected into rabbits, which was shown to neutralize TNF bioactivity, this resulted in *larger* elevations in body temperature (12). Long et al. (13,14) have also shown that treatment of rats with neutralizing antibody to TNF led to larger LPS-induced fevers, and Smith and Kluger (15) have shown that treatment with antiserum to TNF actually prevents the fall in body temperature observed in rats implanted with an MCA tumor. In fact, systemic injections of low doses of TNF into rats actually has been shown to attenuate fevers in rats (16). Dr. John Klir and co-workers in my laboratory have repeated these studies and have confirmed that sub-pyrogenic doses of TNF injected ip will actually reduce LPS-induced fevers in rats. Injection of TNF directly into the cns via intracerebroventricular or intrahypothalamic injections had no antipyretic action, suggesting to us that the cryogenic action of TNF resides outside the cns. In addition, it was found that ip injection of a soluble TNF receptor (Immunex Corp.), which binds to and blocks the biological activity of TNF, led to larger LPS-induced fevers in both rats and mice (Drs. John Klir and Wieslaw Kozak, ongoing studies). Could TNF actually be an EC or endogenous antipyretic? When the TNF is effectively neutralized, or low (subpyrogenic doses) of TNF are injected, fevers are enhanced. But when an antibody is injected that may not neutralize the TNF, perhaps this may result in the TNF actually remaining biologically active in the circulation for a longer period of time, and this be the explanation for lowered fevers.

Kwiatowski et al. (17), in a very exciting study, have shown that treatment of human patients with a monoclonal antibody to TNF led to reduced fevers in children suffering from cerebral malaria. Based on these data, they concluded that TNF is an EP in humans with malaria. However, the circulating TNF levels in the patients treated with the antibody were actually higher (as measured by immunoassay) than in the patients injected with vehicle. The authors of this study argued that the antibody inhibited TNF "by retaining it in the circulation and reducing its availability to tissue receptors." This is a reasonable explanation, although probably not more reasonable than using the interpretation made above -- namely that the antibody used in this study actually increased TNF's activity in the circulation, and it was the enhanced TNF activity that resulted in the antipyresis. One way to test this hypothesis would be to inject subpyrogenic doses of TNF into human patients with malaria to see if this would produce antipyresis or enhance fever.

Interleukin-6 and Fever

There have been many studies in the past several years that have shown that injection of IL-6 causes fevers in experimental animals. Some have suggested that IL-6 cannot be an important pyrogen because such large doses of IL-6 are needed to induce fever. What was not acknowledged was that the plasma and cns concentrations of biologically-active IL-6 rise up to several thousand-

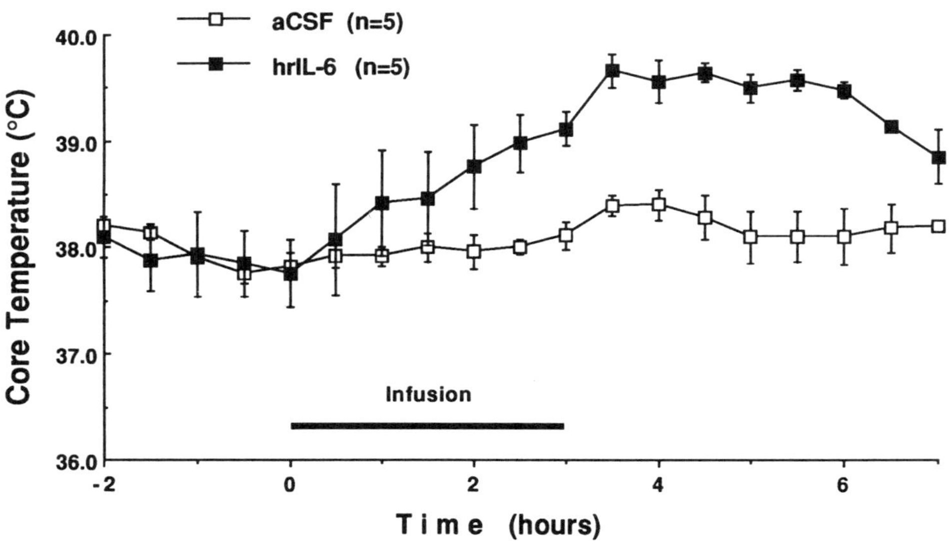

Fig. 1: Core temperature of rats during intrahypothalamic infusion of human recombinant (hr) IL-6 (15.6 U in 0.13 µl/min) or artificial cerebrospinal fluid (aCSF: 0.13 µl/min)(from reference 20).

fold in response to injections of relatively modest amounts of LPS (18). Roth et al. (19) and Klir et al. (20) have shown that both the circulating and the hypothalamic concentrations of IL-6 rise during fevers in both guinea pigs and rats. When IL-6 is infused into the hypothalamus of rats at a dose that simulates that seen during LPS-induced fevers, this resulted in fevers (Figure 1). These data, which fulfill criterion #2 described above, support the hypothesis that IL-6 is indeed an EP. There are several *in vitro* and *in vivo* data supporting the hypothesis that the rise in IL-6 is at least partially due to the release of IL-1β. Klir et al. (9) have shown that microinjection of a neutralizing antibody to IL-1β into the anterior hypothalamus not only attenuates fevers caused by ip injection of LPS, but also blocks the rise in hypothalamic IL-6. Since the injection of the antibody to IL-1β had no significant effect on circulating concentrations of IL-6, these data support the hypothesis that the IL-1β-induced rise in hypothalamic IL-6 is the result of locally-produced IL-6.

Endogenous Antipyretics

As noted above, there is evidence that TNF may be an endogenous antipyretic, which exerts its actions outside the cns. In addition to TNF, there are many hormones, which rise during infections that have antipyretic action. The most notable of these are alpha melanocyte stimulating hormone (α-MSH) and arginine vasopressin (AVP). Space does not permit an extensive review of the antipyretic actions of these two peptides, and the interested reader should see the excellent reviews by Lipton and Catania (21) and Kasting (22).

Within the past couple of years it has become apparent that glucocorticoids are endogenous cryogens (23,24). First it was shown that adrenalectomy resulted in larger LPS-induced fevers in rats (23). But, adrenalectomy results in the loss of more than simply glucocorticoids. However, when glucocorticoids were given back to adrenalectomized rats, it was found that only when the plasma concentration of corticosterone reached levels found during periods of stress (ca. 250 ng/ml) did this result in the attenuation of LPS-induced fever (24). These data support the hypothesis that in the intact rat, the effects of glucocorticoids are not permissive, and that elevated levels of corticosterone are needed to suppress fever. In this study it was also shown that the administration of RU38486, the potent progesterone and glucocorticoid antagonist, led to larger fevers in intact rats. These data provided an independent means (using a pharmacologic approach) to test the hypothesis that glucocorticoids are involved in modulating fevers. Where is the site of glucocorticoid negative-feedback? We have recently found that the anterior hypothalamus is the probable site of glucocorticoid negative-feedback (24) (Figure 2).

Psychological Stress-Induced Fever

When a person (25) or an experimental animal is exposed to psychological stress, this often results in a rapid rise in body temperature, which is not associated with any rise in physical activity (Figure 3). This rise in body temperature in response to stressful situations has been called "Lampenfieber" in German, which literally translates into "lamp fever." The closest English-equivalent is probably "stage-fright." Lampenfieber is associated with the chills, teeth-chattering, and brief waves of cold and shivering -- all characteristics of the rising phase of fever.

There are considerable data in animal models supporting the hypothesis that exposure of rats to psychological stress (exposure to an "open field" or simply switching the cages of rats) causes a rise in body temperature that is compatible with this being a rise in thermoregulatory set-point (i.e. a fever). For example, antipyretic drugs attenuate this rise in body temperature (27,28). The elevation in body temperature occurs in a cool microclimate and is associated with appropriate physiological responses such as skin vasoconstriction (29,30). The rise in body temperature in stressed rats is also modulated by TNF, such that pretreatment of rats with antiserum to TNF leads to larger rises in body temperature (31). This is similar to the findings discussed above for LPS-induced fever. Exposure of rats to an open field also results in a rise in the circulating concentrations of IL-6 (32).

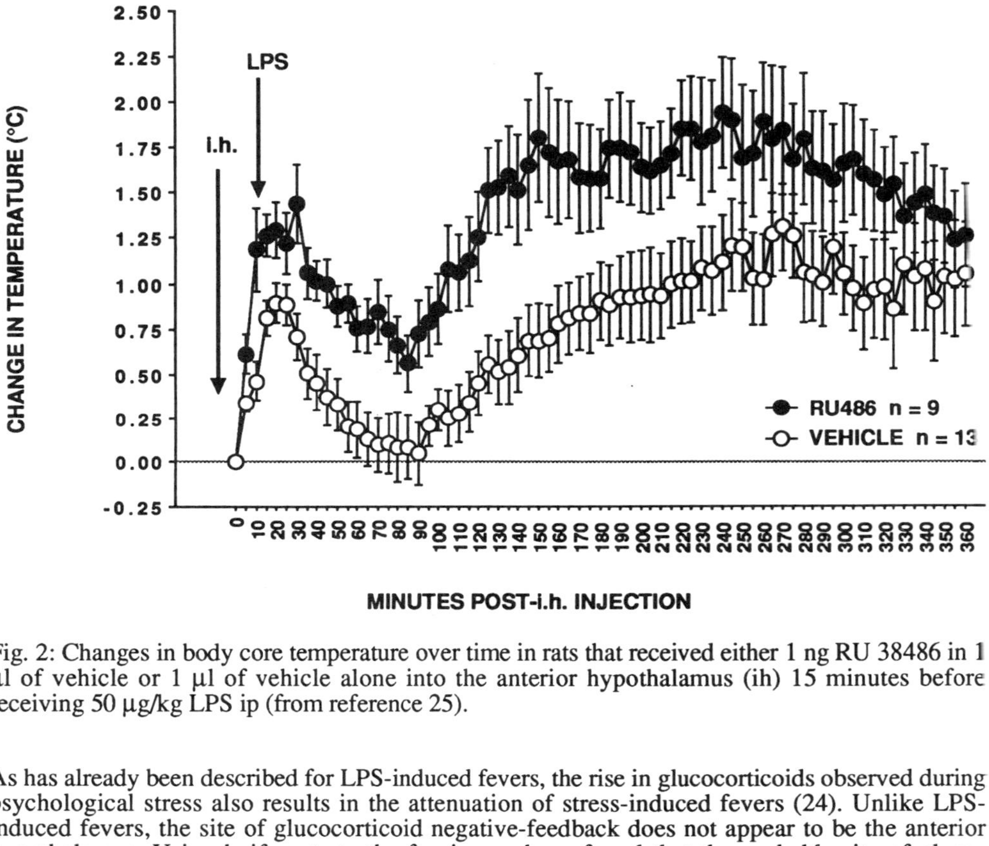

Fig. 2: Changes in body core temperature over time in rats that received either 1 ng RU 38486 in 1 µl of vehicle or 1 µl of vehicle alone into the anterior hypothalamus (ih) 15 minutes before receiving 50 µg/kg LPS ip (from reference 25).

As has already been described for LPS-induced fevers, the rise in glucocorticoids observed during psychological stress also results in the attenuation of stress-induced fevers (24). Unlike LPS-induced fevers, the site of glucocorticoid negative-feedback does not appear to be the anterior hypothalamus. Using knife cuts to the fornix, we have found that the probable site of glucocorticoid feedback during stress-induced fevers is the hippocampus (Morrow et al., manuscript in preparation). Another difference between LPS-induced fevers and psychological stress-induced fevers it that we were unable to attenuate psychological fevers by pretreating rats with neutralizing antiserum to IL-1β. These data indicate to us that precise cytokine pathways (and cns sites of action) involved in these two types of fevers are different.

Van Gool et al. (33) have shown that the administration of epinephrine to rats resulted in a rise in the circulating concentration of IL-6, and this response was blocked by the administration of propranolol. Dr. Jin Fang Liao et al. in my laboratory have recently confirmed these data using an isolated-perfused rat liver (IPRL) preparation. Infusion of 1 µm/ml of epinephrine into the IPRL resulted in significant increases in the concentration of IL-6 measured in the fluid leaving this liver preparation. It is not yet clear whether catecholamines also cause the release of IL-6 in the anterior

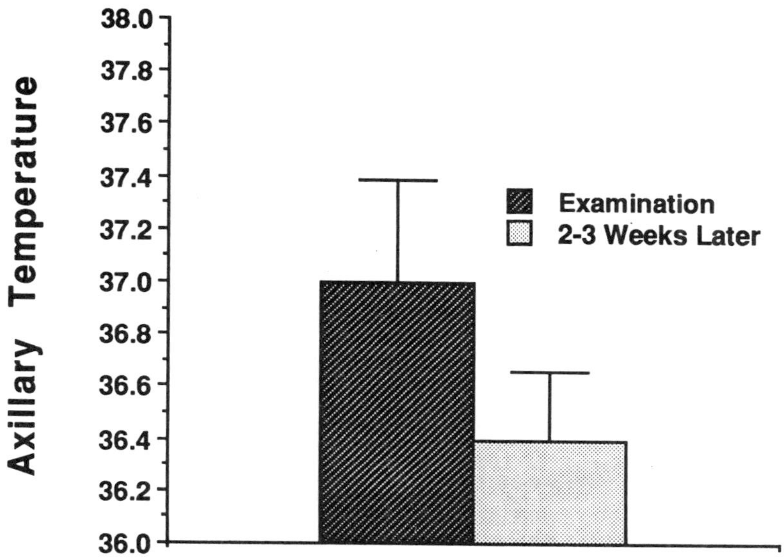

Fig. 3: Average body temperatures (± SD) of 22 residents who were taking their yearly examinations in psychiatry and again two to three weeks later at the same time of day (adapted from reference #26).

hypothalamus, the putative site of pyrogenic action of this cytokine during LPS-induced fevers. In preliminary studies, we have found that the concentration of IL-6 rises not only in the anterior hypothalamus, but also in the dentate gyrus following an ip injection of LPS. We have not yet investigated whether this area of the cns also shows a rise in IL-6 during psychological stress.

A comparison of the similarities between LPS- and psychological stress-induced rises in body temperature is shown in Table 1.

Table 1. Similarities Between LPS- and Stress-Induced Rises in Body Temperature

- attenuated by cyclooxygenase inhibitors (i.e. sodium salicylate, indomethacin)
- occur in cool ambient temperatures
- modulated by antiserum to TNF (rises in body temperature are increased)
- plasma concentration of IL-6 rises (although not as much during psychological stress)
- modulated by glucocorticoids

Summary

Fevers are regulated rises in body temperature. There are numerous cytokines and hormones that are considered putative endogenous pyrogens or endogenous antipyretics or cryogens. The demonstration that a substance (e.g. cytokine X) causes a fever (or anapyrexia) when injected into an animal is not sufficient evidence to conclude that that substance is an endogenous pyrogen (or endogenous cryogen). Evidence was presented that supports the hypothesis that IL-1β is involved in LPS-induced fever, and the site of action of this cytokine is the anterior hypothalamus. IL-1β probably is inducing fever via the release of the cytokine IL-6 within this same area of the cns. Because of limitations in space, virtually nothing was said regarding the roles of prostaglandins or the OVLT in the development of fever. These are exciting areas of investigation that deserve reviews of their own. It is clear that most, but not all, fevers are mediated via prostaglandins. Stitt et al. (34) has speculated that the OVLT produces PGE_2 and the primary site of action of this prostaglandin is not the preoptic-anterior hypothalamus, as generally assumed, but rather near or in the OVLT. Rotondo et al. (35) have speculated, based on correlation studies, that it is the circulating level of PGE_2 that is important during fever (rather than brain level). These data were based on collection of venous blood draining the ear of rabbit. Since most prostaglandin is thought to be cleared by the lungs, what is more important is the arterial concentrations of PGE, and Morimoto et al. (36) have recently shown that the *arterial* concentration of PGE_2 in febrile rabbits does not rise during fever. Thus, it is probably brain (perhaps OVLT), rather than peripheral PGE_2 that is important in fever. I believe that there are sufficient data to support the renaming of "stress-induced *hyperthermia*" as "stress-induced *fever*." The precise pathways between contact with a pathogen or exposure to a psychologically-stressful situation are not identical. However, there is considerable evidence indicating that stress-induced rises in body temperature are regulated, thus fulfilling the necessary criterion for this phenomenon to be called a fever.

References

1. Kluger, M.J. (1991): Fever: role of pyrogens and cryogens. Physiological Reviews 71:93-127.

2. Hallman, G.M., Ortega, C.E., Towner, M.C., and Muchlinski, A.E. (1990): Effect of bacterial pyrogen on three lizard species. Comp. Biochem. Physiol. 96A:383-386.

3. Cabanac, M., and Gosselin, F. (1993): Emotional fever in the lizard *Callopistes maculatus*. FASEB J. 7:A18.

4. Sherman, E., Baldwin, L., Fernandez, G., and Deurell, E. (1991): Fever and thermal tolerance in the toad *Bufo marinus*. J. Therm. Biol. 16:297-301.

5. Shapiro, L., Zhang, X.-X., Rupp, R.G., Wolff, S.M., and Dinarello, C.A. (1993): Ciliary neurotrophic factor is an endogenous pyrogen. Proc. Natl. Acad. Sci. USA 90:8614-8618.

6. Derijk, R. and Berkenbosch, F. (1992): Development and application of a radioimmunoassay to detect interleukin-1 in rat peripheral circulation. Am. J. Physiol. E:1092-E1098.

7. Rothwell, N. L, Busbridge, N.J., Humphray, H., and Hissey, P. (1989): Central actions of interleukin-1ß on fever and thermogenesis. In: The Physiological and Pathological Effects of Cytokines, C.A. Dinarello, M.J. Kluger, M.C.Powanda, and J.J. Oppenheim, editors. New York: Wiley-Liss, 307-311.

8. Long, N.C., Otterness, I., Kunkel, S.L., Vander, A.J., and Kluger, M.J. (1990): The roles of interleukin-1ß and tumor necrosis factor in lipopolysaccharide-fever in rats. Am. J. Physiol, 259:R724-R728.

9. Klir, J.J., McClellan, J.L., and Kluger, M.J. (1994): Interleukin-1β causes the increase in anterior hypothalamic interleukin-6 during LPS-induced fever in the rat. Am. J. Physiol., In Press.

10. Nagai, M., Saigusa, T., Shimada, Y., Inagawa, H., Oshima, H. and Iriki, M. (1988): Antibody to tumor necrosis factor (TNF) reduces endotoxin fever. Experientia 44:606-607.

11. Kawaski, H., Moriyama, M., Ohtani, Y., Naitoh, Y., Naitoh, M. Tanaka, A., and Nariuchi, H. (1989): Analysis of endotoxin fever in rabbits by using a monoclonal antibody to tumor necrosis factor (cachectin). Infection and Immunity 57:3131-3135.

12. Mathison, J.C., Wolfson, E., and Ulevitch, R.J. (1988): Participation of tumor necrosis factor in the mediation of gram negative bacterial lipopolysaccharide-induced injury in rabbits. J. Clin. Invest. 81:1925-1937.

13. Long, N.C., Kunkel, S.L., Vander, A.J., and Kluger, M.J. (1990): Antiserum against TNF enhances LPS fever in the rat. Am. J. Physiol., 258:R332-R337.

14. Long, N.C., Otterness, I., Kunkel, S.L., Vander, A.J., and Kluger, M.J. (1990): The roles of interleukin-1ß and tumor necrosis factor in lipopolysaccharide fever in rats. Am. J. Physiol., 259:R724-R728.

15. Smith, B.K., and Kluger, M.J. Anti-TNFα antibodies normalized body temperature and enhanced food intake in tumor-bearing rats. (1993): Am. J. Physiol. 265:R615-R619.

16. Long, N.C., Morimoto, A., Nakamori, T., and Murakami, N. (1992): Systemic injection of TNF-α attenuates fever due to IL-1β and LPS in rats. Am. J. Physiol. 263:R987-R991.

17. Kwiatkowski, D., Molyneux, M.E., Stephens, S., Curtis, N., Klein, N., Pointaire, P., Smit, M., Allan, R., Brewster, D.R., Grau, G.E., and Greenwood, B.M. (1993): Anti-TNF therapy inhibits fever in cerebral malaria. Quart. J. Medicine 86:91-98.

18. LeMay, L. G. , Vander, A.J., and Kluger, M.J. (1990): The role of IL-6 in fever in the rat. Am. J. Physiol., 258:R798-R803.

19. Roth, J., Conn, C.A., Kluger, M.J., and Zeisberger, E. (1993): Kinetics of systemic and intrahypothalamic interleukin-6 and tumor necrosis factor during endotoxin fever in the guinea pig. Am. J. Physiol., 265:R653-R658.

20. Klir, J.J., Roth, J., Szelenyi, Z., McClellan, J.L., and Kluger, M.J. (1993): Role of hypothalamic interleukin-6 and tumor necrosis factor α in LPS-fever in rat. Am. J. Physiol., 265:R512-R517.

21. Lipton, J.M., and Catania, A. Pyrogenic and inflammatory actions of cytokines and their modulation by neuropeptides: techniques and interpretations. (1993): In, Neurobiology of Cytokines, Part B, edited by E.B. De Souza, Methods in Neurosciences, Volume 17, San Diego: Academic Press, 61-77.

22. Kasting, N.W. Criteria for establishing a physioloigcal role for brain peptides. (1989): A case in point: the role of vasopressin in thermoregulation during fever and antipyresis. Brain Res. Rev. 14:143-153.

23. Coehlo, M.M., Souza, G.E.P., and Pela, I.R. Endotoxin-induced fever is modulated by endogenous glucocorticoids in rats. (1992): Am. J. Physiol. 263:R:423-R427.

24. Morrow, L.E., McClellan, J.L., Conn, C.A., and Kluger, M.J. (1993): Glucocorticoids alter fever and IL-6 responses to psychological stress and to lipopolysaccharide. Am.J. Physiol., 264:R1010-1016.

25. McClellan, J.L., Klir, J.J., Morrow, L.E., and Kluger, M.J. (1994): The central effects of glucocorticoid receptor antagonist RU38486 on lipopolysaccharide and stress-induced fever. Am. J. Physiol., Submitted for Publication.

26. Marazziti, D., Di Muro, A. and Castrogiovanni, P. (1992): Psychological stress and body temperature changes in humans. Physiology and Behavior. 52:393-395.

27. Briese, E., and Cabanac, M. (1980): Emotional fever and salicylate. In: Contributions to thermal physiology, edited by Szelenyi, Z, and Szekely, M. Oxford: Pergamon Press, 161-163.

28. Kluger, M. J., O'Reilly, B., Shope, T. R., and Vander, A. J. (1987): Further evidence that stress hyperthermia is a fever. Physiol. and Behavior 39:763-766.

29. Long, N.C., Vander, A.J., and Kluger, M.J. (1990): Stress-induced rise of body temperature in rats is the same in warm and cool environments. Physiol. Behavior 47:773-775.

30. Briese, E., and Cabanac, M. (1991): Stress hyperthermia: physiological arguments that it is a fever. Physiol. Behavior 49:1153-1157.

31. Long, N.C., Kunkel, S.L., Vander, A.J., and Kluger, M.J. (1990): Antiserum against tumor necrosis factor increases stress hyperthermia in rats. Am. J. Physiol., 258:R591-R595.

32. LeMay, L.G., Vander, A.J., and Kluger, M.J. (1990): The effects of psychological stress on plasma interleukin-6 activity in rats. Physiology and Behavior, 47:957-961.

33. van Gool, J., van Hendrina, V., Helle, M., and Aarden, L.A. (1990): The relation among stress, adrenalin, interleukin 6 and acute phase proteins in the rat. Clin. Immunol. Immunpathology 57:200-210.

34. Stitt, J.T., Shimada, S.G., and Bernheim, H.A. (1984): Microinjection of zymosan and lipopolysaccharide into the organum vasculosum laminae terminalis of rats enhances their febrile responsiveness to endogenous pyrogen. In: Thermal Physiology, edited by J.R.S. Hales, NY, Raven Press, p. 555-558.

35. Rotondo, D.H., Abul, H.T., Milton, A.S., and Davidson, J. (1988): Pyrogenic immunomodulators increase the level of prostaglandin E_2 in the blood simultaneously with the onset of fever. Eur. J. Pharmacol. 154:145-152.

36. Morimoto, A., Morimoto, K., Watanabe, T., Sakata, Y., and Murakami, N. (1992): Does an increase in prostaglandin E_2 in the blood circulation contribute to a febrile response in rabbits? Brain Res. Bull. 29:189-192.

Thermal Balance in Health and Disease
Advances in Pharmacological Sciences

SELECTIVE BRAIN COOLING DURING FEVER ?

G. Kuhnen

Physiologisches Institut, Justus-Liebig-Universität, Aulweg 129, D-35392 Giessen (Germany)

Selective brain cooling (SBC) is a mechanism, by which many mammalian species can cool their brains below the temperature of the arterial blood (Tblood). In these species, arterial blood destined for the brain passes the carotid rete where it is cooled by venous blood draining the nasal mucosa (1). SBC occurs during normothermia and hyperthermia, increases with carotid blood temperature, is in principle independent of panting, and causes a dissociation of brain temperature (Tbrain) and Tblood (2). SBC is controlled by the venous return from the nasal mucosa which can be directed either via the angularis oculi vein to cause SBC, or via the facial vein to flow directly back to the trunk without SBC (3). SBC is regarded as a mechanism to protect the brain against heat damage. Fever is a condition during which Tblood rises significantly and the thermoregulatory thresholds for heat loss are elevated. If the protection against heat damage is the main function of SBC, it should be fully active during fever.

Previous studies showed controversial results. While sheep seem to have normal SBC during fever (4), SBC in oxen appeared to be reduced (5), and in rabbits SBC changed during fever (6). This study in goats was designed to determine SBC during fever and to compare these results with data measured before in non-febrile goats (2).

Methods

The experiments were performed in 6 female goats (32-53 kg, 2-4 years). Four animals, which had been previously used in other experiments, were prepared with carotid loops and an arterial venous shunt (7). In these goats Tblood was measured intravasally in the carotid artery and the aortic arch while Tbrain was measured in the hypothalamus. In 3 other animals Tblood was measured via a guide tube in the thyroid branch of the carotid artery. Tbrain was again measured in a guide tube in the hypothalamus. Fever was induced by i.v. injection of bacterial LPS (*Salmonella typhosa*, 0.5 µg/kg). Between two LPS-injections was an interval of about 14 days. During the experiments metabolic rate (MR) and respiratory evaporative heat loss (REHL) were determined. Apart from Tblood and Tbrain, the surface temperatures of the ear (Tear) and of the nasal mucosa (Tnm) were measured. SBC was calculated as Tblood minus Tbrain. The slope of SBC was determined as the change of SBC per °C change of Tblood and reflects the dissociation of Tbrain from Tblood. Ambient temperature and relative humidity were 20 °C and 55 %, respectively.

Results

Figure 1 shows the time course of a single experiment. About 10 minutes after the injection of LPS (*Salmonella typhosa*, 0.5 µg/kg) Tblood and Tbrain started to increase. Although there was SBC at the beginning of the experiment (Tblood was higher than Tbrain) SBC diminished in spite of increasing Tblood. While reaching the first temperature peak a small degree of SBC occurred again. During the following decrease of Tblood SBC was enhanced and Tbrain was clearly lower than Tblood. But during the second increase of Tblood SBC diminished and was then activated again during decreasing Tblood. The single temperature phases during fever were accompanied by different thermoregulatory responses. The first temperature increase was accompanied by a decrease in blood flow to the head and through the nasal mucosa, and an increase of MR. The following decrease of Tblood correlated with an increase of REHL. The second increase of Tblood was caused by a high level of vasoconstriction and by a small increase of MR. During the second phase skin blood flow increased as indicated by Tear. This single experiment is typical in so far as it shows clear differences in relation to SBC between different phases of fever. During the phases of increasing Tblood SBC was nearly zero while during the phases with decreasing Tblood, SBC was clearly developed. The absolute degree of SBC was different in single experiments.

The mean fever response of 16 experiments in 6 goats could be described as follows: 13 ± 4 min after the injection of LPS Tblood increased to a first peak at 40.0 ± 0.3 °C (time 34 ± 11 min). The second peak was reached after 230 ± 26 min with a mean Tblood of 40.6 ± 0.5 °C.

In figure 2 the relationship between Tbrain and Tblood during single phases of fever is shown. The mean Thypo is plotted as a function of Tblood for 16 experiments in 6 goats. Generally all phases showed a small degree of SBC which increased during the phases with decreasing temperatures. There was a large difference in the degree of SBC occurring in febrile and in non-febrile goats. For instance, at a Tblood of 40.5 °C SBC was about 3 to 8 times higher in non-febrile goats compared to febrile goats. In febrile goats it is surprising that Tbrain increased nearly in parallel with Tblood (Fig. 2) because in non-febrile goats SBC was enhanced with increasing Tblood and caused a clear dissociation between Tbrain and Tblood (slope 0.8). The slope of SBC during fever was nearly zero (slopes: first rise of Tblood 0.09, first fall of Tblood 0.04, second rise of Tblood 0.12, second fall of Tblood 0.40).

In a single experiment the thresholds of SBC were determined by increasing and decreasing carotid blood temperature around the threshold temperature of SBC. This was done by extracorporeal heat exchangers acting on the carotid blood. The threshold for SBC was taken at that carotid blood temperature at which carotid blood and brain temperatures became equal. The fever response was represented by trunk temperature measured as blood temperature in the aortic arch. The results shown in figure 3 demonstrate that the threshold for SBC changed in synchrony with the phases of the fever response. The threshold followed the course of trunk temperature. However, the threshold seemed to decrease more quickly than Tblood as indicated by the higher SBC during the phases with decreasing Tblood.

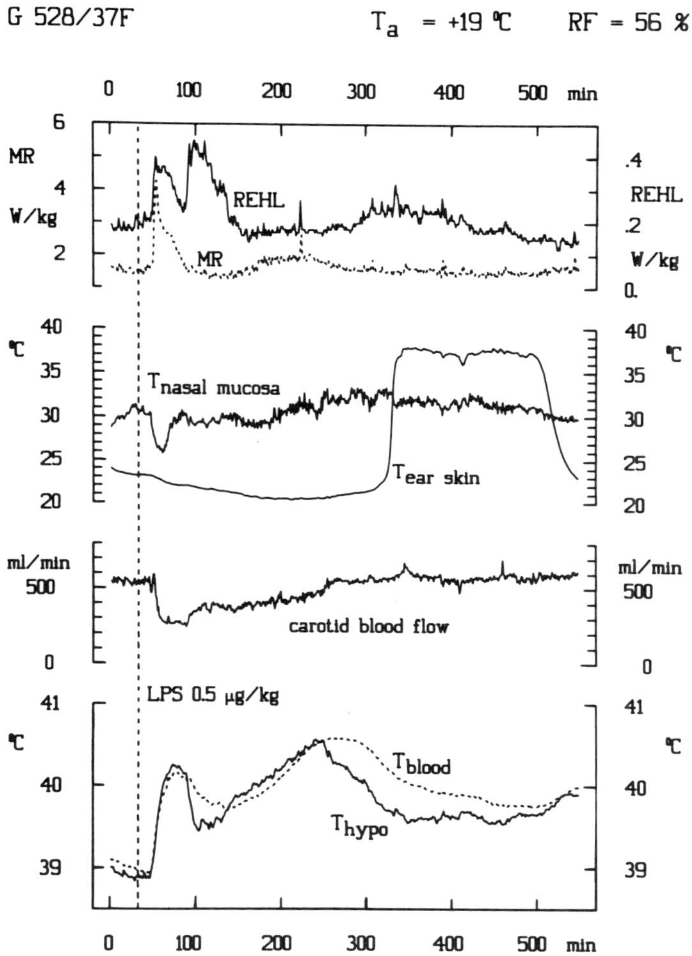

Fig. 1: Fever induced by i.v. injection of LPS (*Salmonella typhosa*, 0.5 µg/kg) in a single experiment. REHL: respiratory evaporative heat loss. MR: metabolic rate. $T_{nasal\ mucosa}$, $T_{ear\ skin}$: surface temperatures of the nasal mucosa and the ear skin, respectively. T_{blood}: blood temperature measured in the carotid artery. T_{hypo}: hypothalamic temperature representing brain temperature. T_{hypo} lower than T_{blood} indicates selective brain cooling.

Discussion

SBC occurred during fever as shown before in sheep and ox. SBC in febrile goats was generally in the range of 0.1 to 0.4 °C. However, in febrile sheep SBC of 0.5 to 1.0 °C was determined (4),

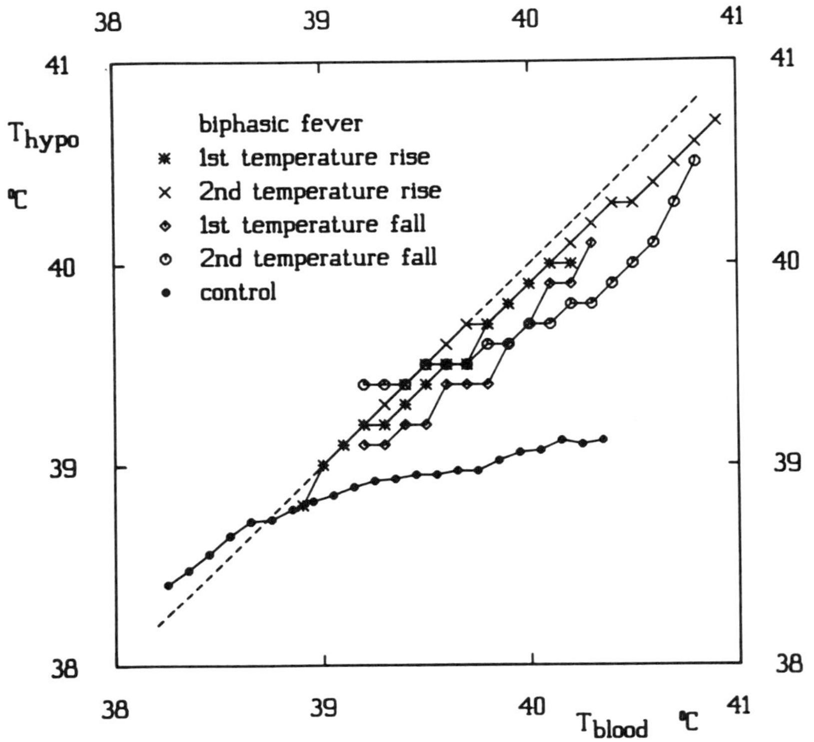

Fig. 2: Mean hypothalamic temperatures (T_{hypo}) are plotted vs. blood temperatures (T_{blood}) during different phases of fever (defined as sections with nearly linear changes of T_{blood}). Grand mean from 6 goats and 16 experiments. Fever was induced by LPS-injection (0.5 µg/kg, *Salmonella typhosa*). The control values of non-febrile goats are from Kuhnen & Jessen (2).

which is much larger than in goats. The reference temperature used in sheep for calculating SBC was rectal temperature. Therefore, it could be speculated that the degree of SBC was overestimated because rectal temperature does not correctly represent Tblood (8). In febrile goats the degree of SBC was significantly reduced in comparison to non-febrile goats, for instance the mean SBC in febrile goats at a Tblood of 40.5 °C was only one fourth compared to non-febrile goats. The degree of SBC depends on the phase of the fever response. SBC was larger during phases with decreasing Tblood compared to phases with increasing Tblood. Similar effects can be seen if one compares the SBC of febrile (5) and non-febril oxen (9). Although the possibility of SBC in rabbits is controversial (10, 11), the course of Thypo in relation to core temperature in rabbits presented by Werner & Graener (6) was very similar to what was found in goats. In rabbits Thypo was lower than core temperature mainly during phases with decreasing temperatures.

In non-febrile goats SBC increased significantly with Tblood (0.8 °C SBC per °C increase of Tblood) and produced a dissociation of Tbrain from Tblood, which may function as a protection against heat damage. However, SBC in febrile goats showed no correlation with Tblood. It must

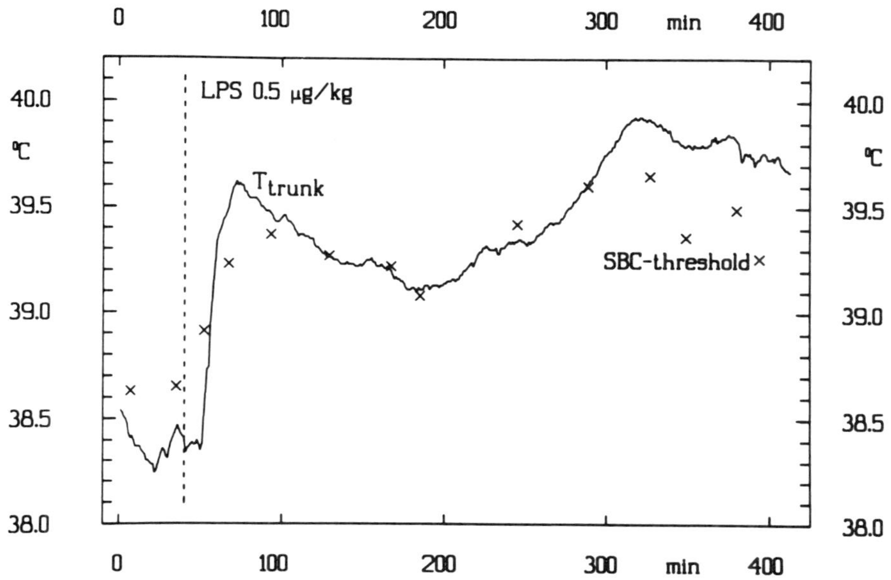

Fig. 3: Determination of threshold temperatures of selective brain cooling (SBC-threshold x) during fever in a single experiment. Trunk temperature (T_{trunk}) was measured in the aortic arch while the temperature of the blood supply of the head (blood of the carotid artery) was manipulated by extracorporeal heat exchangers to determine the threshold of SBC.

be taken into account that the correlation between SBC and Tblood was determined at Tblood above the threshold in non-febrile goats and around the threshold in febrile goats. The increase of the threshold temperature for SBC during fever was shown in a single experiment. Although the changes of Tblood used in this experiment were too small for a correct calculation of the slope of SBC, it appeared smaller as in non-febrile goats. These effects must reduce SBC during fever. The reduction of SBC by fever was no artificial effect of fever induced by LPS, because this effect was also seen in an unrestrained goat during fever of unkown origin (12).

It is concluded that fever inhibits SBC by elevating the threshold and reducing the slope of SBC. Thus, the capability of SBC to act as a protective mechanism against heat damage is reduced during fever.

Summary

16 experiments in 6 goats have shown that during LPS-induced fever selective brain cooling (SBC) was significantly reduced in comparison to non-febrile controls. The degree of SBC depended on the phase of fever. SBC was larger during phases with decreasing blood temperatures compared to phases with increasing blood temperatures.

Acknowledgements

This study was supported by DFG Ku 807/1-2.

References

1. Baker, MA (1982): Brain cooling in endotherms in heat and exercise. Annu Rev Physiol 44, 85-96.
2. Kuhnen, G, Jessen, C (1991): Threshold and slope of selective brain cooling. Pfluegers Arch 418, 176-183.
3. Johnsen, HK, Blix, AS, Mercer, JB, Bolz, KD (1987): Selective cooling of the brain in reindeer. Am J Physiol 253, R848-R853.
4. Laburn, HP, Mitchell, D, Mitchell, G, Saffy, K (1988): Effects of tracheostomy on brain and body temperatures in hyperthermic sheep. J Physiol (Lond) 406, 331-344.
5. Chesy, G, Kadziela, W, Lachowski, A, Caputa, M, Kozak, W, Wieczór, H (1980): Pyrogen fever and selective brain cooling in the ox. In: Contribution to thermal physiology. Szelényi, Z, Székely, M eds. Oxford London New York: Pergamon Press, 197-199.
6. Werner, J, Graener, R (1984): Dynamic properties of pyrogen-induced fever in the rabbit. In: Thermal Physiology. Hales, JRS ed. New York: Raven Press, 557-563.
7. Kuhnen, G, Jessen, C (1992): Effects of selective brain cooling on mechanisms of respiratory heat loss. Pfluegers Arch 421, 204-208.
8. Bligh, J (1957): The relationship between the temperature in the rectum and the blood in the bicarotid trunk of the calf during exposure to the heat stress. J Physiol (Lond) 136, 393-403.
9. Chesy, G, Caputa, M, Kadziela, W, Kozak, W, Lachowski, A (1983): The influence of ambient temperature on brain homeothermia in the ox (*Bos taurus*). J therm Biol 8, 259-263.
10. Kluger, MJ, D'Alecy, LG (1975): Brain temperature during reversible upper respiratory bypass. J Appl Physiol 38, 268-271.
11. Baker, MA (1980): Anatomical and physiological adaptations of panting mammals to heat and exercise. In: Environmental physiology: aging, heat and altitude. Horvath, SM, Yousuf, MK eds. Amsterdam: Elsevier North Holland, 121-146.
12. Jessen, C (1994): Blood and brain temperatures of an unrestrained goat during an 11-month period. In: Integrative and cellular aspects of autonomic functions: temperature and osmoregulation. Pleschka, K, Gerstberger, R eds. Paris: John Libbey Eurotext, 329-338

Thermal Balance in Health and Disease
Advances in Pharmacological Sciences

CYTOKINE LEVELS DURING THE EARLY PHASE OF THE ENDOTOXIN FEVER

L. Janský[1], S. Vybíral[1], D. Pospíšilová[2], J. Roth[3], J. Dornand[4], E. Zeisberger[3] and J. Kamínková[2]

[1]Department of Comparative Physiology, Faculty of Science, Charles University, 128 00 Prague 2, Vinicná 7, [2]Department of Immunology, 2nd Faculty of Medicine, Charles University, Prague (Czech Republic),[3]Physiologisches Institut, Klinikum der JustusLiebig-Universität, Giessen (Germany) and [4]INSERM U65, Université de Montpellier II, Montpellier (France)

Fever induced by lipopolysaccharide (LPS) causes release of endogenous pyrogens into circulation. Since cytokines have a capacity to induce fever, when injected peripherally or centrally, the view prevails that they act as endogenous pyrogens (8). It is presumed that cytokines acting directly or indirectly on the hypothalamus induce fever by elevating the thermoregulatory set point.

Several authors have described striking increases in cytokine concentrations or activity in blood (1, 2, 4, 9, 10, 11, 12) and in brain (3, 5, 6, 7, 11, 12) during the late phase of the fever. No data are available so far about cytokine production during the early phase of the fever, when the seting of the thermostat to the higher level occurs. Therefore, IL1, TNFα and IL6 concentrations in blood and in cerebrospinal fluid of guinea pigs were estimated during the first 45-60 min following LPS injections.

Methods

Experiments were performed on guinea pigs chronically equipped either with catheters in the left carotid artery or with push-pull cannulae stereotaxically implanted into the rostral hypothalamus.

Febrile responses to i.m. injections of bacterial LPS (E. coli 20 μg/kg) were monitored by measurement of the abdominal and skin temperatures. Sampling of blood plasma was performed at the time immediately before LPS injection and 15, 30 and 45 min after LPS injection. Push-pull samples were collected during these time periods: -15 - 0 min, 0 - 15 min, 15 - 30 min, 30 45 min and 45 - 60 min.

Determinations of cytokines were done by bioassays. IL_6 determinations were based on measurement of a dose dependent growth of the B9 hybridoma cell line. Determinations of IL_1 were based on measurements of a dose dependent growth of the CTTL cell line, stimulated by IL_2, which was generated from the NOB-1 cell line (subclone EL 461). Determinations of TNFα were based on the cytotoxic effect of this cytokine on the cell line L929 subclone. The results were expressed in pg/ml.

Results

Administration of the LPS resulted in dramatic increases of plasmatic TNFα, IL_1, and IL_6 concentrations. TNFα concentrations increased by several orders, while concentrations of IL_6 increased by about 10 fold, only. The start of the IL_6 increase seems to be delayed behind that of TNFα. No steady state levels of cytokines were reached, indicating that their concentrations might further increase in the later phase of the fever (Fig. 1).

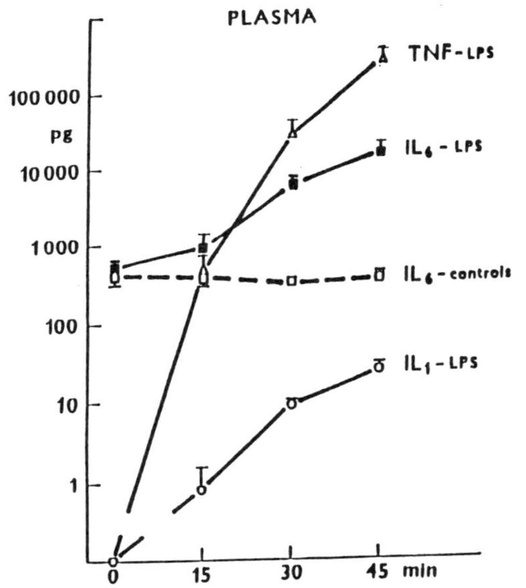

Fig. 1: Concentrations of cytokines (IL_1, TNFα, IL_6) in plasma of control (interrupted line) and LPS treated (solid lines) guinea pigs. No TNFα and IL_1 was detectable in the blood of control animals. Means +S.E.

In cerebrospinal fluid administration of LPS resulted in small, but significant increases of TNFα and IL_6 concentrations (by about 300 - 400%). The concentrations of IL_6 did not increased until 15 min after injection and reached a steady level 30 min after injection, already. The onset of the increased concentration of TNFα was delayed behind that of IL_6. 60 min after LPS injection concentrations of TNFα and IL_6 increased by about 300-400%. No IL_1 was observed in cerebrospinal fluid after LPS (Fig. 2, 3).

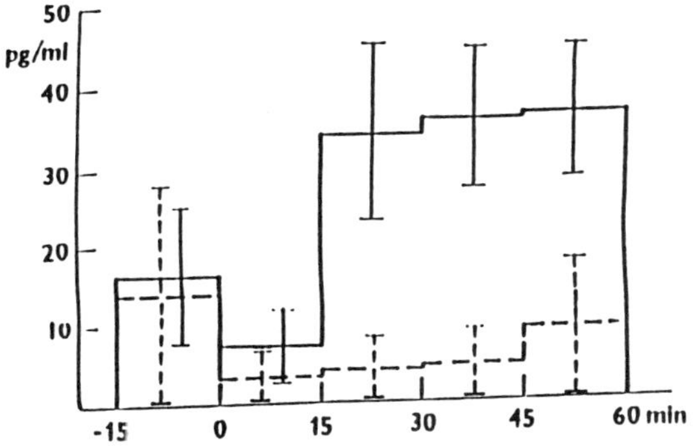

Fig. 2: Concentrations of IL6 in cerebrospinal fluid of control (interrupted line) and LPS treated (solid lines) guinea pigs. Means and ± S.E.

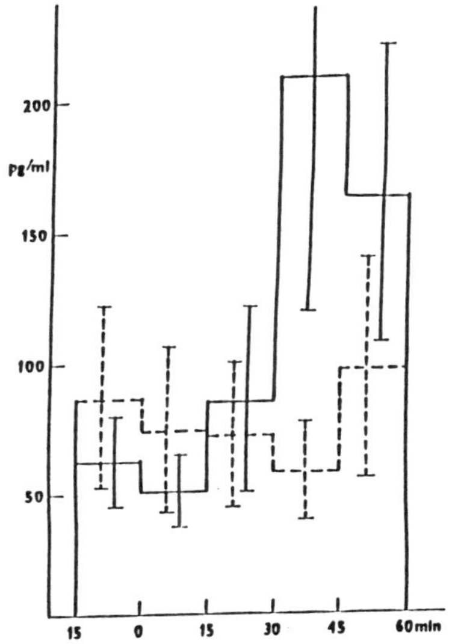

Fig 3: Concentrations of TNFα in cerebrospinal fluid of control (interrupted line) and LPS treated (solid line guinea pigs. Means and ± S.E.

Discussion

Cytokine levels in plasma of control and LPS treated animals and men were measured by several authors during the late phase of the fever (1, 2, 4, 9, 10, 11, 12).

In accordance with data from the literature our measurements performed in the early phase of the fever (upto 45 min after LPS injections) also document a sharp increase in the concentration of cytokines. TNFα levels increase faster than the levels of IL_6 and IL_1.

No evidence from the literature is available about cytokine levels in cerebrospinal fluid in the early phase of the fever when the reseting of the body thermostat occurs. In our experiments concentrations of TNFα and IL_6 increased by about three to four times after LPS. A new observation is that the IL_6 concentrations increased earlier than that of TNFα and reached a steady state level 30 min after LPS injection.

Since the cytokines were detected in the brain at the time of the onset of the fever (within the first 30 min after LPS injection) it is tempting to conclude that cytokines in the brain are involved in reseting of the body thermostat to the higher level during fever.

Evidence from the literature indicates that the increased levels of cytokines usually persist for several hours, when the body temperature returns to the normal level, already. The physiological role of these increased concentrations of cytokines during the late phase of the fever remains unclear.

References

1. Byerley, LO, Alcock, NW, Starness, Jr, HF (1992): Sepsis-induced cascade of cytokine mRNA expression: correlation with metabolic changes. Am J Physiol 261, E728-E735.

2. Cannon, JG, Tompkins, RG, Gelfand, JA, Michie, HR, Stanford, GG, van der Meer, HWM, Endres, S, Lonnemann, G, Corsetti, J, Chrenow, B, Wilmore, DW, Wolff, SM, Burke, JF, Dinarello, CA (1990): Circulating interleukin-1 and tumor necrosis factor in septic shock and experimental endotoxin fever. J Infect Dis 161, 79-84.

3. Coceani, F, Lees, J, Dinarello, CA (1988): Occurrence of interleukin-1 in cerebrospinal fluid of the conscious cat. Brain Res 446, 245-250.

4. Foster, KD, Conn, CA, Kluger, MJ (1992): Fever, tumor necrosis factor, and interleukin-6 in young mature, and aged Fischer 344 rats. Am J Physiol 262, R211-R215.

5. Frei, K, Leist, TP, Meager, A, Gallo, P, Leppert, D, Zinkernagel, RM, Fontana, A (1988): Production of B cell stimulatory factor-2 and interferon in the central nervous system during viral meningitis and encephalitis. J Exp Med 168, 449-453.

6. Gijbels, K, van Damme, J, Proost, P, Put, W, Carton, H, Billiau, A (1990): Interleukin 6 production in the central nervous system during experimental autoimmune encephalomyelitis. Eur J Immunol 20, 233-235.

7. Klir, JJ, Roth, J, Szelenyi, Z, McClellan, JL, Kluger, MJ (1993): Role of hypothalamic interleukin-6 and tumor necrosis factor-α in LPS fever in rat. Am J Physiol 265, R512-R517.

8. Kluger, MJ (1991): Fever: role of pyrogens and cryogens. Physiol Rev 71, 93-127.

9. Lemay, DR, Lemay, LG, Kluger, MJ, D´Alecy, LG (1990): Plasma profiles of IL-6 and TNF with fever-inducing doses of lipopolysaccharide in dogs. Am J Physiol 259, R126-R132.

10. Lemay, LG, Otterness, IG, Vander, AJ, Kluger, MJ (1990): In vivo evidence that the rise in plasma IL 6 following injection of a fever-inducing dose of LPS is mediated by IL 1β. Cytokine 2, 199-204.

11. Lemay, LG, Vander, AJ, Kluger, MJ (1990): Role of interleukin 6 in fever in rats. Am J Physiol 258, R798-803.

12. Roth, J, Conn, CA, Kluger, MJ, Zeisberger, E (1993): Kinetic of systemic and intrahypothalamic IL-6 and tumor necrosis factor during endotoxin fever in guinea pigs. Am J Physiol 265, R653-R658.

Thermal Balance in Health and Disease
Advances in Pharmacological Sciences

GLUCOSE HOMEOSTASIS IN FEVERS INDUCED BY TWO DOSES OF BACTERIAL ENDOTOXIN (LPS), INTERLEUKIN-1β (IL-1β) OR TUMOR NECROSIS FACTOR-α (TNFα) ADMINISTERED 24 HOURS APART

A. Höche and W. Riedel

Max-Planck-Institut für physiologische und klinische Forschung, W.G. Kerckhoff-Institut, Parkstraße 1, and Department of Anaesthesia and Intensive Care Medicine, Kerckhoff-Klinik, Benekestraße 2-8, D-61231 Bad Nauheim, Germany.

Glucose homeostasis is maintained by a fine balance between the anabolic hormone insulin and counterregulatory hormones glucagon, cortisol, catecholamines and growth hormone. This balance is frequently lost during trauma and infection. Depending on stage and severity of inciting injury, hyper- or hypoglycemia is observed. Upon stimulation by bacterial lipopolysaccharide (LPS), the cytokines IL-1β and TNFα, produced by phagocytic cells, play a critical regulatory role in the host response to trauma and infection. They have been found to affect protein, lipid and carbohydrate metabolism in peripheral tissues, while at the central level they are believed to act as the endogenous pyrogens. Most animals and man react to repeated doses of LPS with a diminished fever, the phenomenon defined as early pyrogenic tolerance. Its underlying mechanism seems to reside in the periphery, with a reduced capability of cytokine synthesis in reticulo-endothelial cells, since central sensitivity to LPS or IL-1β with respect to producing fever remains unchanged in the tolerant state (5). Application of IL-1β and TNFα to rabbits has revealed that carbohydrate metabolism is altered by diverting glucose away from aerobic oxidation towards enhanced anaerobic metabolism, with lactate serving as the precursor for gluconeogenesis (4). Combined with elevated levels of insulin, these events define the peripheral resistance to insulin frequently observed in endotoxemia. The present study investigates a possible alteration of the insulin and glucagon secretion profile in the early pyrogenic state. Additionally, the sensitivity of pancreatic β cells to glucose is tested with bolus glucose injections in the afebrile, febrile and tolerant state.

Methods

A total of 33 female Chinchilla rabbits (Charles River, Kisslegg, Germany), wheighing from 3.5 to 4.5 kg, sitting unrestrained in a temperature-controlled rabbit box kept at 28°C, were studied. Core temperature was measured with a thermistor inserted 10 cm deep into the rectum. A central ear artery was cannulated, under local anesthesia, for recording blood pressure and heart rate, but also for blood sampling. Continuous or bolus injections of drugs were performed via an ear vein, punctured under local anesthesia. The experiments were started around 0900 h, when core temperature and cardio-respiratory parameters had been stable for at least 45 min. Pyrogenic tolerance was studied in 13 rabbits by i.v. injection of two bolus doses, exactly 24 h apart, of first 1 μg/kg LPS, then, 4 weeks later, 1 μg/kg recombinant human (rh) IL-1β, and then, again 4 weeks later, 10 μg/kg rh TNFα. Rh IL-1β was a gift from Dr. Y. Hirai of Otsuka Pharmaceutical (Tokushima, Japan), rh TNFα was purchased from Bachem (Heidelberg, FRG). Body temperatures and blood samples were taken every 15 min during 3 h observation periods on each day. Plasma glucagon and insulin were estimated by radioimmunoassay, cortisol by ELISA (all Serono, Freiburg, FRG) and glucose enzymatically. To test the secretory capacity of pancreatic α and β cells in the cytokine-stimulated state, 800 mg glucose per kg body weight were i.v. administered, within 5 min, to 10 rabbits, either 30 min following a continuous infusion of 200 ng/kg/min IL-1β, or 15 min following a bolus injection of 10 g/kg TNFα, repeating the procedure exactly 24 h later. To estimate the sensitivity of pancreatic cell to glucose in early pyrogenic tolerance, 800 mg glucose per kg body weight were given, within 5 min, to 10 rabbits first in the afebrile state, or 90 min after the first or second dose of LPS. Sensitivity changes of the evoked insulin responses were calculated by linear estimations of the increase of insulin versus blood glucose within the period of 90 min following glucose application.

Statistics. All data are reported as means ± SE. The statistical significance of changes from control of different variables was estimated by paired t-test using the modified t-statistic calculated by Bonferroni's method. Slopes of curves were determined by the method of least squares. Differences between the slopes were tested using Student's paired t-test, with a $P < 0.05$ considered to be a significant difference.

Results

As shown by the data in Table 1, a transformation of a biphasic fever pattern into a short-lasting monophasic one is observed after a second dose of LPS given 24 h later. In contrast, injection of IL-1β evoked a similar core temperature rise on both days, whereas the second dose of TNFα produced a higher fever. Neither LPS nor the two cytokines affected blood pressure on either day. LPS induced with a latency of 60 min hyperglycemia on day one. 24 hours later, the rabbits were hypoglycemic and the second dose of LPS was without any effect on blood glucose. These reactions seem to correlate with a rise of glucagon on day one, and a slight decrease of glucagon on day two. Insulin showed a biphasic rise on day one, with maxima around 30 and 120 min after LPS, whereas only a single peak occurred 30 min after LPS on day two. The effects of IL-1β on blood glucose levels, and on insulin and glucagon secretion were similar on both days. There was only a short-lasting increase of insulin and glucagon 15 min following IL-1β, followed by a rapid development of hyperglycemia, which closely correlated with a sustained increase of heart rate from 186 ± 5 to 312 ± 10 b/min. The first dose of TNFα caused a short-lasting increase of insulin, without affecting blood glucose and glucagon levels, while the second dose 24 h later was followed by a prolonged increase of insulin, again without alterations of glucagon and blood glucose levels. Cortisol secretion rose immediately after the first dose of LPS, while 24 h later, at still elevated cortisol levels, the second dose of LPS induced a nearly fourfold higher secretion of cortisol. The effects of IL-1β on cortisol secretion were likewise exaggerated by the second dose, while TNFα seemed to have no additional effect on cortisol secretion, taking into account a similar rise of cortisol in saline-treated rabbits between 0900 and 1200 h.

The secretory capacity of pancreatic α and β cells following exposure to IL-1β and TNFα was tested in 10 other rabbits using an i.v. bolus injection of 800 mg/kg glucose. This caused in the control and cytokine-treated state a nearly identical increase of plasma glucose from 135 ± 3 to 435 ± 15 mg/dl within 15 min. Plasma glucose levels then linearly decayed, reaching control levels 60 min after the injection. As shown in Fig. 1, the experimentally induced hyperglycemia caused within 30 min in the afebrile state a rise of insulin from 6.6 ± 0.2 to 47.1 ± 11.3 μU/ml on day one, and from 6.8 ± 0.2 to 39.6 ± 10.9 μU/ml on day two. During a continuous infusion of IL-β insulin rose after 30 min from 7.8 ± 0.2 to 99.1 ± 18.2 μU/ml on day one, and from

	min	Core Temperature (°C)	Glucose (mg/dl)	Insulin (μU/ml)	Glucagon (pg/ml)	Cortisol (ng/ml)
LPS						
Day 1	0	39.6 ± 0.1	134 ± 3	10.5 ± 1.3	262 ± 20	1.30 ± 0.44
	30	39.9 ± 0.1*	138 ± 2	20.8 ± 4.6*	281 ± 22	2.48 ± 0.61*
	60	40.3 ± 0.1*	147 ± 6*	12.1 ± 1.9	401 ± 25*	5.55 ± 1.17*
	90	40.4 ± 0.1*	222 ± 9*	10.2 ± 2.4	340 ± 19*	8.86 ± 1.41*
	120	40.5 ± 0.2*	243 ± 15*	23.1 ± 6.7*	313 ± 18*	10.99 ± 1.57*
	180	40.8 ± 0.2*	238 ± 13*	23.4 ± 7.9*	365 ± 15*	14.29 ± 2.14*
Day 2	0	40.4 ± 0.1$	117 ± 2$	13.5 ± 2.4	300 ± 17	16.45 ± 2.87$
	30	40.8 ± 0.1*	114 ± 2	25.9 ± 6.5*	270 ± 15	37.03 ± 4.86*
	60	41.1 ± 0.1*	117 ± 2	22.7 ± 2.0*	258 ± 14*	33.95 ± 4.20*
	90	41.1 ± 0.1*	116 ± 2	19.9 ± 1.6*	282 ± 13	30.53 ± 4.58*
	120	40.8 ± 0.2	116 ± 3	8.8 ± 1.2	225 ± 13*	39.40 ± 4.39*
	180	40.5 ± 0.2*	112 ± 5	10.2 ± 1.7	226 ± 12*	53.60 ± 6.45*
IL-1β						
Day 1	0	39.6 ± 0.1	140 ± 2	15.7 ± 3.4	248 ± 21	0.24 ± 0.09
	30	39.9 ± 0.1*	134 ± 3	16.9 ± 3.4	348 ± 24*	1.28 ± 0.31
	60	40.1 ± 0.2*	161 ± 7*	21.1 ± 3.5*	249 ± 20	3.94 ± 0.81*
	90	40.3 ± 0.1*	148 ± 4	15.4 ± 2.3	238 ± 17	5.08 ± 0.89*
	120	40.4 ± 0.1*	155 ± 4*	17.7 ± 1.7	245 ± 14	6.30 ± 0.92*
	180	40.9 ± 0.2*	152 ± 5*	24.7 ± 6.0*	251 ± 12	8.18 ± 0.83*
Day 2	0	39.3 ± 0.1	131 ± 5	24.5 ± 4.8$	220 ± 19	0.30 ± 0.16
	30	39.8 ± 0.1*	160 ± 4*	30.3 ± 4.6*	267 ± 18*	6.26 ± 0.24*
	60	40.1 ± 0.2*	160 ± 10*	26.8 ± 6.0	229 ± 18	11.10 ± 1.27*
	90	40.3 ± 0.2*	159 ± 10*	30.2 ± 7.0*	247 ± 14	12.00 ± 1.31*
	120	40.4 ± 0.2*	161 ± 7*	27.3 ± 4.8	202 ± 15	13.72 ± 1.36*
	180	40.8 ± 0.2*	164 ± 6*	25.6 ± 5.5	174 ± 16*	15.56 ± 1.42*
TNFα						
Day 1	0	39.7 ± 0.1	140 ± 3	20.5 ± 4.5	255 ± 5	0.30 ± 0.08
	30	39.9 ± 0.1*	141 ± 2	50.2 ± 14.8*	254 ± 12	2.07 ± 0.20*
	60	40.1 ± 0.1*	144 ± 2	26.5 ± 6.7	247 ± 5	1.90 ± 0.30*
	90	40.1 ± 0.1*	143 ± 2	24.1 ± 6.8	253 ± 10	1.83 ± 0.42*
	120	40.0 ± 0.1*	146 ± 2	16.2 ± 2.1	250 ± 12	1.70 ± 0.30*
	180	39.9 ± 0.1*	146 ± 2	22.4 ± 4.6	256 ± 3	0.50 ± 0.11
Day 2	0	39.6 ± 0.1	142 ± 2	24.4 ± 4.8	270 ± 13	1.33 ± 0.15
	30	40.0 ± 0.1*	147 ± 3	71.0 ± 23.1*	272 ± 8	4.28 ± 0.44*
	60	40.4 ± 0.1*	147 ± 3	44.2 ± 12.1*	243 ± 2	4.03 ± 0.38*
	90	40.3 ± 0.1*	147 ± 3	35.7 ± 10.1*	272 ± 7	3.30 ± 0.39*
	120	40.1 ± 0.2*	147 ± 3	34.5 ± 10.0*	248 ± 5	4.73 ± 0.17*
	180	40.4 ± 0.1*	149 ± 3	28.9 ± 6.9	236 ± 8	4.40 ± 0.61*

Tab. 1: Effects of two injections of LPS, IL- 1β and TNFα 24 hours apart. * Significant difference from control values of same day, $ significant difference from control values of day one.

7.0 ± 0.2 to 117.1 ± 19.8 µU/ml on day two. 15 min following application of TNFα, insulin increased within 30 min from 14.5 ± 0.8 to 212.5 ± 58.4 µU/ml on day one, and from 19.2 ± 1.3 to 341.6 ± 56.6 µU/ml on day two, with significantly elevated levels of 38.6 ± 7.4 µU/ml 120 min after glucose injection.

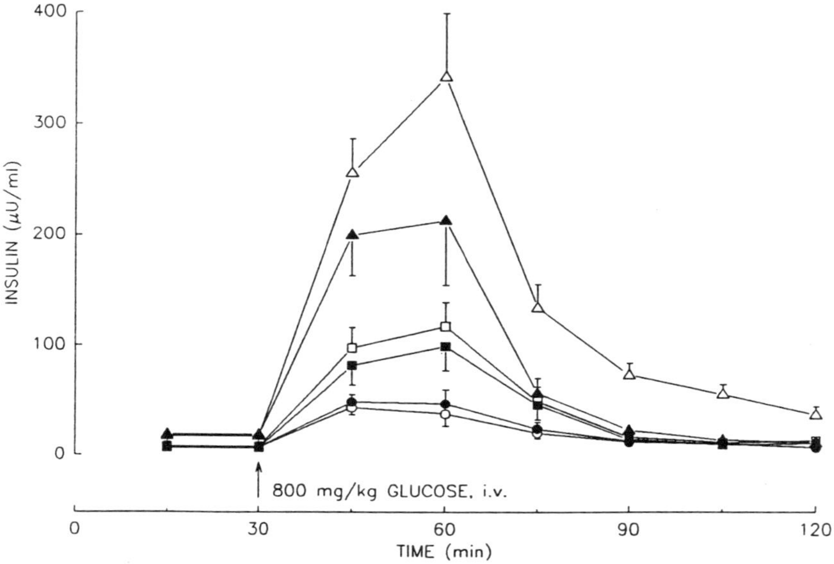

Fig. 1. Effects of 800 mg/kg glucose i.v. on insulin secretion in 10 rabbits after injection of saline (○, ●), 1 µg/kg IL-1β (□, ■) or 10 µg/kg TNFα (Δ, ▲). Closed symbols: 1st dose, open symbols: 2nd dose given 24 h later.

Glucagon decreased within 15 min following the application of glucose from 245 ± 12 to 223 ± 12 pg/ml in the control group, and from 238 ± 10 to 216 pg/ml in the TNFα-treated group. In the IL-1β-treated group glucagon fell from 235 ± 16 to 184 ± 8 pg/ml within 30 min on day one, and from 220 ± 13 to 166 ± 10 pg/ml on day two.

Fig. 2 shows the changes of sensitivity of the pancreatic β-cells to glucose in the afebrile, febrile and pyrogenic tolerant state. The hyperglycemia elicited by bolus injections of 800 mg/kg glucose was of similar magnitude and duration as reported for the preceeding group. In the afebrile state insulin was found to rise with a relationship of 38.5 ± 13.5 µU/ml insulin per 100 mg plasma glucose. The relationship was significantly reduced in LPS-fever on day one to 16.2 ± 8.8 µU/ml insulin, while in the tolerant state the sensitivity of the insulin producing cells to glucose was significantly elevated, with 83.5 ± 24.5 µU/ml per 100 mg increase of plasma glucose.

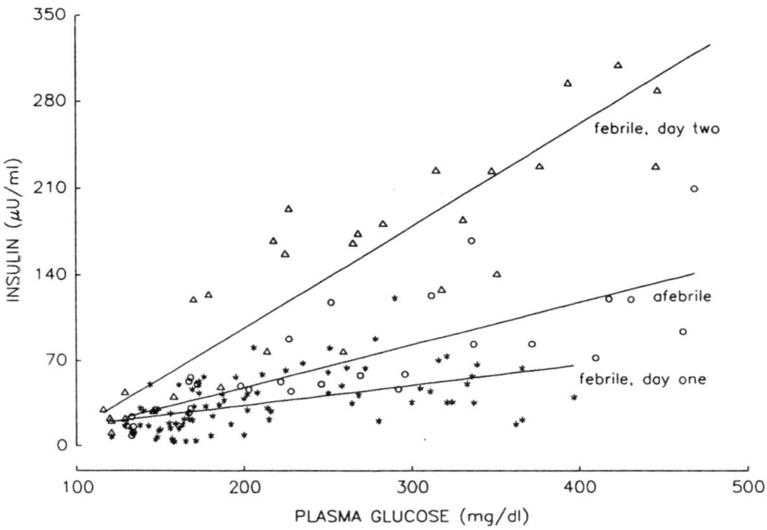

Fig. 2. Relationship between insulin secretion and blood glucose levels in the afebrile (o), acutely febrile (∗) and pyrogenic tolerant state (Δ).

Discussion

The present results indicate that hyperglucagonemia and an increased visceral sympathetic activity are essential in maintaining glucose production in endotoxemic rabbits. The hyperglycemia induced by the first dose of LPS coincides with the pattern of increased visceral sympathetic activity (3), while the tolerant state is characterized by attenuated sympathetic activation (1), which together with lowered glucagon levels may lead to hypoglycemia. Among the cytokines believed to act as putative mediators of fever, IL-1 has been shown to increase visceral, especially adrenal sympathetic activity (2). The present study shows that IL-1β increases heart rate and transiently glucagon secretion, in contrast to TNFα, which alters neither heart rate nor glucagon nor plasma glucose levels, but stimulates insulin secretion. This could indicate that in the pyrogenic tolerant state, characterized by low glucagon and high insulin levels, the capacity of macrophages to synthesize IL-1 is lowered, but is elevated for TNFα. The effect of a single injection of TNFα on insulin production seems to be a long-lasting one and could thus explain the higher insulin levels following a second injection of TNFα as well as the augmented sensitivity of the pancreatic β cell to glucose after a second dose of TNFα, or in the early pyrogenic state. Elevated cortisol levels seem to be without effect on glucose production in pyrogenic tolerance.

References

1. Jones, SB, Westfall, MV, Sayeed, MM (1988): Plasma catecholamines during *E. coli* bacteremia in conscious rats. Am J Physiol 254, R470-R477.

2. Niijima, A, Hori, T, Aou, S, Oomura, Y (1991): The effects of interleukin-1 on the activity of adrenal, splenic and renal sympathetic nerves in the rat. J Auton Nerv Syst 36, 183-192.

3. Riedel, W, Kozawa, E, Iriki, M (1982): Renal and cutaneous vasomotor and respiratory rate adjustments to peripheral cold and warm stimuli and to bacterial endotoxin in conscious rabbits. J Auton Nerv Syst 5, 177-194.

4. Tredget, EE, Yu, YM, Zhong, S, Burini, R, Okusawa, S, Gelfand, JA, Dinarello, CA, Young, VR, Burke, JF (1988): Role of interleukin 1 and tumor necrosis factor on energy metabolism in rabbits. Am J Physiol 255, E760-E768.

5. Yamashiro, O, Morimoto, A, Sakata, Y, Watanabe, T, Murakami, N (1993): Febrile and metabolic tolerance to endotoxin and human recombinant interleukin-1 in rabbits. Am J Physiol 264, R1180-R1185.

Thermal Balance in Health and Disease
Advances in Pharmacological Sciences

ENDOTOXAEMIA IS NORMALLY A LIMITING FACTOR IN HEAT TOLERANCE

J. R. S. Hales[1] and M. Nagai[2]

[1]School of Physiology and Pharmacology University of New South Wales Kensington (Sydney) 2033, Australia. [2]Department of Physiology, Medical University of Yamanashi, Yamanashi, 409-38, Japan.

Introduction

During heat stress, whether at rest in hot environments or during exercise, the thermoregulatory requirements for increased blood flow (BF) and volume in skin to promote heat loss are commonly in balance with the cardiovascular requirement for an adequate arterial pressure to maintain the BF in vital tissues; this balance is achieved to a significant extent by a redistribution of blood away from the gastrointestinal tract (see 1). When heat stress becomes severe, the high levels of skin BF are not maintained, but decline supposedly due to cardiovascular demands conflicting with thermoregulatory requirements (see 2). However, sweating or panting also decline, i.e., thermolytic mechanisms overall are not maintained at their maximal levels.

With the recognition that many features of heat stroke resemble endotoxin shock (disseminated intravascular coagulation, petecheal haemorrhages, diarrhoea, hyperkalaemia, and commonly hypotension and acidosis) and that low gastrointestinal BF can lead to abnormally high movement of endotoxins from the gastrointestinal tract into the blood stream (3), it has been proposed that endotoxin could act during severe heat stress, to inhibit thermolytic mechanisms and/or enhance heat production as in development of fever. Thus, heat tolerance is enhanced by (a) development of endotoxin tolerance (4), or by (b) reducing the gastrointestinal contents before heat exposure (5), or by (c) pretreatment with oral antibiotics (6). Two questions remained, first, to further test the theory that endotoxin effects normally contribute to the rate of rise in core temperature during heat exposure, the effects of pretreatment with indomethacin were determined. Secondly, it was questionable whether an additional endotoxin load would have any effect during severe heat stress, i.e., since core temperature is already raised by the external heat load, the matching of core temperature to the raised hypothalamic set-point might be achieved without any additional actions by the endotoxins; the present study has therefore also determined the effects on core temperature and thermolytic mechanisms, of a dose of lipopolysaccharide (LPS) given during severe heat stress.

Methods

Ten conscious, adult New Zealand White rabbits, of either sex and weihing 1.9 - 3.8 kg, were each exposed twice to an environment of 35°C dry bulb (d.b.) with 40% relative humidity (r.h.). The first exposure was taken as the "control" response and with the second, either (a) indomethacin (s.c., 20 mg kg^{-1}) was given approx 40 min before heat exposure, or (b) LPS (from *E.coli*, i.v. 150 or 300 ng kg^{-1}) was administered when rectal temperature (T_{re}) had risen 0.5°C. T_{re} was measured using a thermocouple inserted 5cm and respiratory frequency was recorded as torso movements via a pneumograph and pressure transducer. Indices of changes in skin BF were obtained as ear skin surface temperature (T_{sk}) and blood cell flux using a laser-Doppler (Moor Inst., MBF3D). No statistical analyses were performed because of the small, variable number of observations made to date.

Table 1: Mean (±S.E.) values for various parameters in rabbits exposed to 35°C d.b. with 40% r.h. normally (CONTROL) and after indomethacin (INDO) or lipopolysaccharide (LPS) treatment. T_{re}= rectal temperature; T_{re}= rate of rise in ΔT_{re}; Time = time taken for T_{re} to reach 41°C; f= respiratory frequency when T_{re} =41°C; T_{sk}= ear skin temp when T_{re}=41°C; ΔSkBF= maximum magnitude of blood flux increase in ear skin.

Parameter	CONTROL	INDO	LPS
Pre- heat T_{re} (°C)	38.74 ± 0.06	39.05 ± 0.09	38.69 ± 0.11
ΔT_{re} (°C h^{-1})	1.05 ± 0.07	0.78 ± 0.14	1.36 ± 0.10
Time (min)	124 ± 7	143 ± 7	95 ± 6
f (breaths min^{-1})	394 ± 36	436 ± 25	194 ± 63
T_{sk} (°C)	39.15 ± 0.16	38.81 ± 0.20	38.08 ± 0.57
ΔSkBF	1.6 ± 0.4	2.2 ± 0.6	------

Results

Heat exposure was discontinued in control experiments when T_{re} reached 41 - 41.5°C so as to not jeopardise the well-being of the animal. The second heat exposure was imposed 3-6 days later and was discontinued when T_{re} reached 41.6-42°C. All animals survived.

Fig. 1 illustrates the different body temperature responses obtained. That is, T_{re} rose at a lower rate with indomethacin and at a higher rate with LPS, compared to the same animal when untreated. Table 1 contains detailed data which show that from a normal T_{re} which was not significantly different between the groups, indomethacin caused T_{re} to rise at a lower, and LPS at a higher rate (1.05 vs. 0.78 vs. 1.36°C per h). Respiratory frequency was marginally higher with indomethacin and lower with LPS. Skin temperature did not differ significantly between the groups. The laser-Doppler index of skin BF showed a greater increase in indomethacin treated compared with control conditions, but has not yet been recorded after LPS treatment.

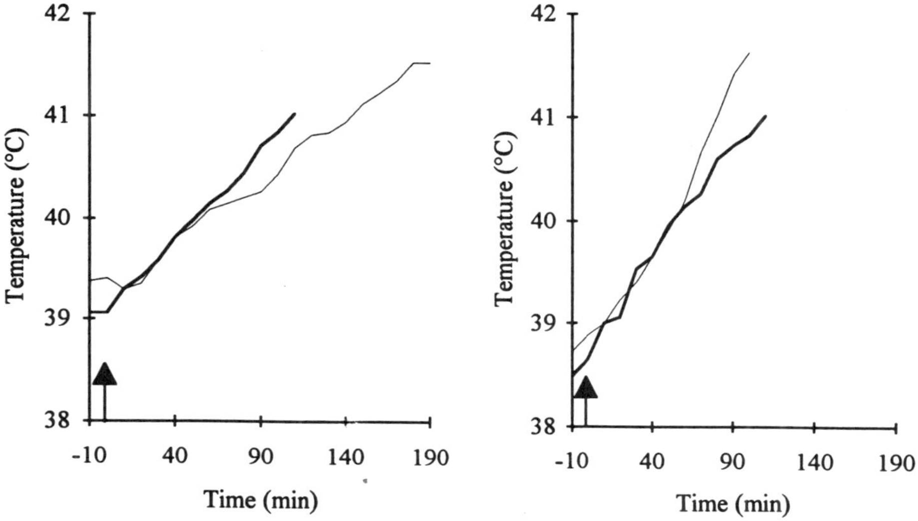

Fig. 1: Rises in rectal temperatures of two rabbits exposed to 35°C d.b. with 40% r.h., normally (bold lines) and after treatment (fine lines) with indomethacin (left) or LPS (right). Arrows show the start of heat exposure.

Discussion

In environments approximating thermoneutrality, indomethacin alone has no effect on body temperature (7), however, the normal rise in core temperature following administration of endotoxin can be prevented by prior treatment with indomethacin due to its blockade of prostaglandin production (8). Therefore, the present lower rates of T_{re} rise in animals treated with indomethacin show that these animals were normally being subjected to increased endotoxin load when exposed to heat. The source of such endotoxin must be the gastrointestinal tract, since circulating endotoxin levels increase during severe heat stress and this can be prevented by prior treatment with oral antibiotics (6). As discussed in detail elsewhere (9), gram-negative bacteria, from which LPS is derived, are normally abundant in the gastrointestinal tract. Normally there is a 'gut barrier', such that the relatively small quantities of LPS entering the blood stream are readily detoxified by the reticuloendothelial system. The barrier is diminished by low gastrointestinal BF, a condition known to prevail during heat stress, so that quantities of LPS entering the blood stream exceed the detoxifying capacity of the reticuloendothelial system; it is also possible that the system itself has a diminished capacity in hyperthermia and associated conditions. The increased rate of rise in T_{re} following LPS treatment is also an important observation. That is, in environments approximating thermoneutrality, endotoxins raise the hypothalamic set-point temperature and therefore core temperature is raised by reducing heat loss and/or increasing heat production (10). In a severely hot environment, because core temperature is already elevated it was feasible that endotoxin, whether from the gut or administered, would have no effect. Thus, in the particular rabbits used and under our particular conditions, the heat-induced increases in endotoxin levels were still insufficient to evoke maximal effects.

Skin temperature did not differ significantly, however, this is not entirely unexpected because of the poor relationship between high T_{sk} and BF in hot environments. Laser-Doppler indices of ear skin BF measured in two heat exposed animals indicated a greater magnitude of increase after indomethacin than under control conditions. Therefore the observed changes in thermolytic mechanisms appear small but are in the correct direction, ie., indomethacin had no apparent effect on panting but caused increases in skin BF, whereas LPS depressed panting and we are as yet unable to comment on skin BF changes. It may be that alterations in panting heat loss (not reflected in frequency) or in metabolic heat production also contributed to the final temperature changes.

In *conclusion*, these observations are consistent with the hypothesis that endotoxins from gastrointestinal bacteria normally contribute to the limitation of tolerance of severe heat stress by acting, as in the normal development of fever, to raise body temperature by reducing heat loss and/or increasing heat production. That is, the apparent failing of thermolytic mechanisms during advanced stages of severe heat stress leading to heat stroke appears to be due not only to conflicts between cardiovascular and thermoregulatory requirements, but also to development of endotoxaemia leading to conflicting thermoregulatory requirements.

References

1. Rowell, LB (1983): Cardiovascular adjustments to thermal stress. In: Handbook of Physiology Sec. 3, The Cardiovascular System Vol. III, Part 2. Shepherd, JT, Abboud, FM eds. Bethesda: Amer. Physiol. Soc., 967 - 1023.

2. Hales, JRS (1987): Proposed mechanisms underlying heat stroke. In Heat Stress: Physical Exertion and Environment. Hales, JRS, Richards, DAB eds. Amsterdam: Elsevier, 85 - 102.

3 Gaffin, SL, Gathiram, P, Wells, MT, Brock-Utne, JG (1986): Effect of corticosteroid prophylaxis on lipoplysaccharide levels associated with intestinal ischemia in cats. Crit Care Med 14, 889 - 891.

4. Du Bose, DA, Basamania, K, Maglione, L, Rowlands, J (1983): Role of bacterial endotoxins of intestinal origin in rat heat stress mortality. J Appl Physiol 54, 31 - 36.

5. Bynum, G, Brown, J, DuBose, D, Marsili, M, Leav, I, Pistole, TG, Hamlet, M, LeMaire M Caleb, B (1979): Increased survival in experimental dog heat stroke after reduction of gut flora. Aviat Space Environ Med 50, 816 - 819.

6. Butkow, N, Mitchell, D, Laburn, H, Kendi, E (1984): Heat stroke and endotoxaemia in rabbits. In: Thermal Physiology. Hales, JRS ed. New York: Raven Press, 511 - 514.

7. Itami,T, Ema, M, Kanoh, S (1986): Antipyretic mechanism of indomethacin in rabbits. J Pharmacobiodyn 9, 271-275.

8. Fitzpatrick, FA, Wynalda, MA (1976): In vivo suppression of postaglandin biosynthesis by non-steroidal anti-inflammatory agents. Prostaglandins 12, 1037 - 1051.

9. Hales, JRS, Hubbard, RW, Gaffin, SL (1994): Limitation of heat tolerance. In Handbook of Physiology - Adaptation to the Environment. Fregly, MJ, Blatteis, CM eds. Am Physiol Soc/Oxford Univ. Press: New York, (in press).

10. Hammel, HT, (1965): Neurons and temperature regulation. In: Physiological Controls and Regulations. Yamamoto, WS, Brobeck, JR, eds. Philadelphia: Saunders, 71 - 97.

Thermal Balance in Health and Disease
Advances in Pharmacological Sciences

THE EFFECT OF EXO- AND ENDOPYROGENS ON SEROTONIN METABOLISM IN THE NUCLEUS RAPHE DORSALIS OF THE RAT

H. W. Clement[1], C. Ruwe[1], D. Gemsa[2] and W. Wesemann[1]

[1]Institute of Physiological Chemistry, Dept. of Neurochemistry, Hans-Meerwein-Str., and [2]Institute of Immunology, Philipps-University, D-35033 Marburg, Germany

Introduction

The effects of immune stimulants and cytokines on sleep and body temperature have recently been summarized (1). However, the exact mechanism how and where cytokines may control sleep regulation and temperature in the central nervous system is still not well understood. The serotoninergic transmitter system of the nucleus raphe dorsalis (NRD) has been shown to be involved in sleep and temperature regulation. Although serotonin (5-HT) release was found to be high in most brain areas during wakefulness, the NRD differed in that a high release of 5-HT could be observed during sleep (2). The aim of our study was to examine whether 5-HT metabolism, as measured by the extracellular serotonin metabolite 5-hydroxyindoleacetic acid (5-HIAA) in the NRD, could be altered by immune stimulants and cytokines. In addition, the effect of normal light-dark cycle, immobilization and the application of drugs on body temperature and 5-HT metabolism in the NRD was included for comparison with the effect of cytokines.

Materials and Methods

Animals

Male Lewis rats (Zentralinstitut für Versuchstierzucht, Hannover, Germany), weighing 300 ± 50 g, were used for all experiments. They were single-housed under a 12:12 h light:dark cycle (light from 07.00 to 19.00 h) at 23°C and constant humidity (50%) for at least 6 weeks prior to experimentation.

Voltammetric recordings: Light-dark cycle and immobilization experiments

The levels of extracellular 5-HIAA concentrations in the NRD were followed with electrochemically pretreated carbon multifibre electrodes (50 µm thick and 500 µm long). The carbon fibre working electrode was implanted 2 weeks prior to the experiments under chloral hydrate anaesthesia (500 mg/kg i.p.) using a David Kopf stereotaxic instrument (Bilany, Düsseldorf, Germany). The stereotaxic coordinates of the NRD were used according to the atlas of Paxinos and Watson (3): A=5,2 mm, L=0 mm, V=5,7 mm; 45° angle from the perpendicular. The position of the electrode was controlled histologically at the end of the experiment.A tungsten wire auxiliary electrode and an Ag/AgCl reference electrode were fixed together with the carbon fibre electrode and an electrical connector with acrylic dental cement on the skull.Differential pulse voltammetric recordings were performed using a modified Polarorecord E 506 (Metrohm, Filderstadt, Germany). Potential sweep: 0 to 500 mV with 10 mV/s scan rate, measuring pulse: amplitude: 50 mV, duration 60 ms. The voltammetric peak at 280 mV was pharmacologically characterized as 5-HIAA. For testing the effect of immobilization, measurements started 40 min prior to the experiment while the rat was still in its cage. Rats were immobilized by fixing the feet in holes of a board with sticking plaster. A gauze bandage was wrapped two times around the animal's body. The fixation procedure was carried out within 3 min. Immobilization was applied for a total of 10 min starting with the fixation procedure. Measurements were continued for 3 h after the beginning of immobilization. In order to follow the 24 h light-dark rhythm, measurements started in the light phase and were stopped after 48 h.

Measurements of drug effects

In short term experiments the voltammetric recordings were performed in the anaesthetized rat using carbon monofibre electrodes (12 µm thick and 500 µm long). The rats were pretreated with allopurinol (20 mg/kg i.p.) 1 h before anaesthesia to inhibit uric acid production. The carbon fibre electrode was stereotaxically implanted with the above mentioned coordinates. Auxiliary (stainless steel) and reference electrode (Ag/AgCl) were placed on the skull. Voltammetric recordings were performed every 5 min, started 40 min prior to substance application, and were continued for 3 h after drug application.

Temperature recordings

To study the action of immune stimulants, cytokines, 8-hydroxy-2-(di-n-propylamino)-tetraline (8-OHDPAT), and ethanol, temperature recordings were performed in parallel with voltammetric recordings. Measurements were performed with a rectal thermoelement when interleukin-1ß (Il-1ß), tumor necrosis factor-alpha (TNF-α) or muramyldipeptide (MDP) were injected and with a telemetric temperature transmitter (Elvax Mini Mitter Co, Sunriver, USA) when lipopolysaccharide (LPS), 8-OHDPAT, or ethanol were injected. The implantation of the transmitter in the peritoneal cavity was performed 2 weeks prior to the experiment as descibed elsewhere (4)

Chemicals

MDP and LPS were obtained from Sigma, Deisenhofen, Germany, 8-OHDPAT from RBI Biotrend, Köln, Germany, TNF-α was a generous gift of BASF/Knoll AG, Ludwigshafen, Germany, and IL-1ß was donated by Dr. F. Seiler, Behringwerke, Marburg, Germany.

Results

In vivo voltammetry with carbon fibre electrodes was used to study the 5-HT metabolism in the NRD of the rat. By this method, extracellular levels of 5-HIAA could be monitored, paralleling in most cases extracellular 5-HT concentrations although exceptions have been reported (4).As shown in Figure 1, the intraperitoneal application of LPS led to a 100 % increase of extracellular 5-HIAA levels in the NRD when compared to control rats. This LPS effect on central 5-HT metabolism was accompanied by a 0.4 °C rise of the body temperature.

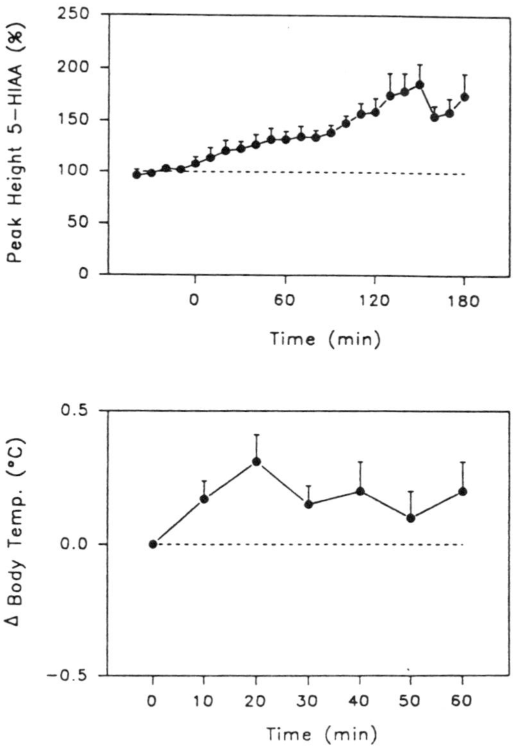

Fig. 1: Effect of intraperitoneally injected LPS (10 µg/kg) on extracellular 5-HIAA levels in the NRD (top) and on body temperature (bottom). Results are means ± S.E.M. of data obtained from 5 animals per group. Top: Results are expressed in percentage of the mean of 40 min preexposure measurements. Bottom: Results are expressed as differences to the respective control group.

The endogenous pyrogen IL-1ß which is thought to mediate partially the effects of the exogenous pyrogen LPS was also able to evoke a 5-HIAA increase in the NRD to the same extent as LPS. However, the effect on body temperature was slower but more pronounced when compared to LPS (Fig. 2).

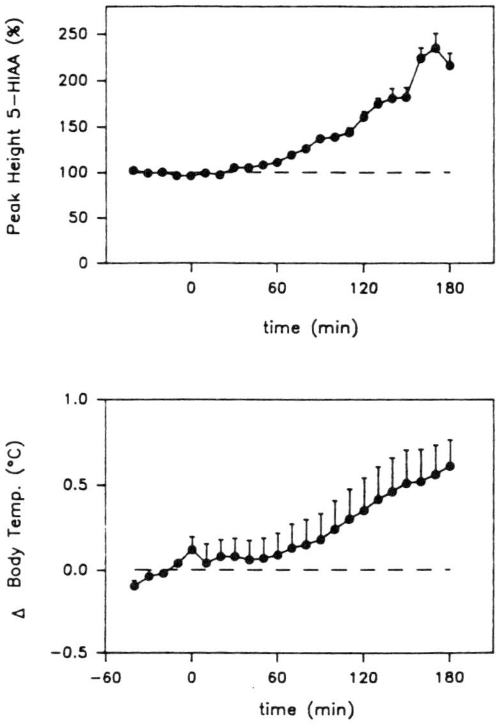

Fig. 2: Effect of intravenously injected IL-1ß (10^5 U) on extracellular 5-HIAA levels in the NRD (top) and on body temperature (bottom). Results are means ± S.E.M. of data obtained from 5 animals per group. Top: Results are expressed in percentage of the mean of 40 min preexposure measurements. Bottom: Results are expressed as differences to the respective control group.

For the sake of brevity, the other data are summarized in Table 1 and will be shortly addressed in the following text. Intravenous injection of TNF-α (10 µg/kg) displayed the strongest enhancement on extracellular 5-HIAA levels and increased the body temperature by 1.5 °C. Intraperitoneal application of MDP enhanced 5-HIAA concentrations and rectal temperature similar to LPS. During the light phase, extracellular 5-HIAA levels were at a minimum whereas a 5-HIAA peak was noted 3-4 h after light was turned off. For body temperature, a parallel change has been reported in the literature (5). Immobilization for 10 min led to an increase of extracellular 5-HIAA in the NRD by about 150 %, paralleling the previously described augmentation of body temperature (6). Lowering the body temperature by 8-OHDPAT (30 µg/kg) or ethanol (1 g/kg) was strongly associated with a 5-HIAA decrease by 50 % or 60 %, respectively.

Table 1: Effects of immune stimulants, cytokines, temperature-decreasing agents and environmental changes on body temperature and extracellular 5-HIAA levels in the NRD of the rat.

	Body Temperature	extracellular 5-HIAA in the NRD
LPS	↑	↑
MDP	↑	↑
IL-1	↑	↑
TNF-alpha	↑	↑
8-OHDPAT	↓	↓
Ethanol	↓	↓
Immobilization 10 min.	↑	↑
Light	↓	↓
Dark	↑	↑

Discussion

Our data clearly demonstrate that the immune stimulants LPS or MDP and the cytokines IL-1ß or TNF-α not only increased body temperature but also stimulated serotoninergic neurotransmission which indicates that the NRD is involved in both sleep and temperature regulation. This concept is further supported by the findings that an experimentally decreased body temperature by application of ethanol or 8-OHDPAT was closely associated with a decline of the 5-HT metabolite 5-HIAA. It has previously been reported that a decreased 5-HT release after 8-OHDPAT was paralleled by sleep reduction while an increase of extracellular 5-HIAA during immobilization appeared to be associated with sleep rebound (8, 9). Along this line, our data on immobilization confirm the previous results (9).

The parallel changes of body temperature and extracellular 5-HIAA levels induced by exogenous pyrogens such as LPS or MDP and endogenous pyrogens such as Il-1ß or TNF-α do not necessarily imply a direct effect of these compounds on the serotoninergic system. It appears to be more likely that a cascade of intermediate compounds is additionally involved which could be separately blocked, leaving one chain of events, i. e. fever induction intact but blocking the other, i. e. somnolence. Support for this notion has already been forwarded by experiments showing that acetaminophen may block the pyrogenic effect of MDP and anisomycin the pyrogenicity of IL-1ß but not their somnogenic effects (1).

The rather complicated network of interactions is exemplified by the presently known effects of IL-1 in the brain: inducibility by TNF-α (10), stimulation of CRF and ACTH (1) which is known to inhibit sleep (9) whereas the ACTH metabolite CLIP is known to promote sleep (9). In addition, IL-1 may affect the noradrenergic system (11) as well as the release of growth hormone releasing factor and insulin (1). Our results reported here may add further information by showing that exogenous and endogenous pyrogens may strongly stimulate the serotoninergic system of the NRD.

References

1. Krueger, JM, Opp, MR, Kapás, L, Kimura-Takeuchi, M, Toth, L (1993): Muramyl Peptides and Interkeukin-1 in Sleep Regulation. In: Endocrine and Nutritional Control of Basic Biological Functions. Lehnert, H, Murison, R, Weiner, H, Hellhammer, D, Beyer, J eds. Seattle: Hogrefe & Huber, 347-354.

2. Cespuglio, R, Sarda, N, Gharib, A, Chastrette, N, Houdouin, F, Rampin, C, Jouvet, M (1990): Voltammetric detection of the release of 5-hydroxyindole compounds throughout the sleep-waking cycle of the rat. Exp Brain Res, 80, 121-128.

3. Paxinos, G, Watson, C (1982): The rat brain in stereotaxic coordinates. New York: Academic Press.

4. Clement, HW, Schäfer, F, Ruwe, C, Gemsa, D, Wesemann, W (1993): Stress-induced changes of extracellular 5-hydroxyindoleacetic acid concentrations followed in the nucleus raphe dorsalis and the frontal cortex of the rat. Brain Res 614, 117-124.

5 Alföldi, P, Rubicsek, G, Cserni, G, Obál, F (1990): Brain and core temperatures and peripheral vasomotion during sleep and wakefulness at various ambient temperatures in the rat. Eur J Physiol 417, 336-341.

6. Jacob, JJ, Suaudeau, C, Vidal, C (1983): Temperature changes induced by non painful stress in the rat. In: Environment, Drugs and Thermoregulation. Lomax, P, Schönbaum, E eds. Basel: Karger, 121-124.

7. Morgan, WW, Rudeen, PK, Pfeil, KA (1975): Effect of immobilization stress on serotonin content and turnover in regions of the rat brain. Life Sci 17, 143-150.

8. Hjorth, S, Magnusson, T (1988): The 5-HT_{1A} receptor agonist, 8-OHDPAT, preferentially activates cell body 5-HT autoreceptors in rat brain in vivo. Naunyn-Schmiedebergs Arch Pharmacol 338, 463-471.

9. Houdouin, F, Cespuglio, R, Gharib, A, Sarda, N, Jouvet, M (1991): Detection of the release of 5-hydroxyindole compounds in the hypothalamus and the n. raphe dorsalis throughout the sleep-waking cycle and during stressful situations in the rat: a polygraphic and voltammetric approach. Exp Brain Res 85, 153-162.

10. Zeisberger, E, Roth, J (1993): Neurobiological concepts of fever generation and suppression. Neuropsychobiol 28, 106-109.

11. Dunn, AJ (1993): Infection as a stressor: a cytokine-mediated activation of the hypothalamo-pituitary-adrenal axis? Ciba-Found Symp 172, 226-239.

EFFECTS OF LPS AND INHIBITORS OF EICOSANOID SYNTHESIS ON SHORT TERM FEVER RESPONSE AND PRODUCTION OF HYPOTHALAMIC PGE_2

J. Kaplanski, L. Abramovich and V. Fraifeld

Department of Clinical Pharmacology, Ben Gurion University of the Negev, Beer Sheva 84120, Israel

Bacterial endotoxin (Lipopolysaccharide, LPS) is a pyrogen that causes fever in a variety of animals. LPS can both act directly on the thermoregulatory center in the hypothalamus or indirectly by inducing endogenous pyrogen synthesis. Endogenous pyrogen also act on the hypothalamus to stimulate synthesis of eicosanoids to initiate fever (1). Prostaglandin E_2 (PGE_2) is considered to be the major prostanoid that can modify the function of the thermosensetive neurons in the hypothalamus from normal to febrile state (2,3). This state of changes in the body temperature is a result of various internal changes that lead to shift in the hypothalamic "set-point". In this situation, all body systems are geared to adjust and maintain body temperature with the new "set-point" as reference. In previous studies we have shown that dexamethasone (DEXA) and indomethacin (INDO) inhibitors, of eicosanoid synhesis, reduced the hyperthermic fever response, induced by LPS and inhibit the synthesis of PGE_2 in the hypothalamus after 24 hours (4,5). The purpose of the present investigation was to complete the picture and examine the possible relationship between prostaglandins and body temperature after short time response to LPS and inhibitors of eicosanoid synthesis.

Methods

Male Sprague-Dawley rats 14-18 weeks old weighing 350-450 g were maintained in a room kept at 22 ± 1°C and relative humidity of 35-55% at least 2 weeks before commencing the experiments. The rats were housed on wood shavings, 2 per cage with water and food ad libitum.

Rats were injected (ip) either with lipopolysaccharide (LPS, E. coli, 100 mg/rat, 250mg/rat) or vehicle (control, 0.5 ml). In the first experiments 2 hr before LPS administration, animals were injected with INDO (50 mg/kg ip) in $NaHCO_3$ 2% (1,0ml/rat). In the second experiment rats were injected with DEXA (5 mg/kg ip) 48 hr, 24 hr and 2 hr before LPS administration.

Rectal temperature was measured before and during each treatment with a thermistor probe (Anriterm HL 600, Anritsu Meter Co, LTD, Japan) inserted 5 cm into the rectum.

Following the final rectal temperature measurement 2hr after the LPS and 4 hr after INDO and DEXA administration, the rats were decapitated and the anterior hypothalami, weighing 35 mg, were excised. One hypothalamus was incubated per vial (6 vials per group) for 3h in Krebs-Henselite buffer, containing glucose (0.2%) in an environment of O_2 (95%) and CO_2 (5%) at 37°C. The buffer was replaced every 60 minutes. The buffer samples that were recovered were frozen, and at a later time assayed for PGE_2 by radioimmunoassay (6). Statistical evaluation was carried out using Student's t-test (two-tailed) and factorial analysis.

Results

A biphasic pattern of rectal temperature was found after LPS administration (Fig.1). An initial drop in the rectal temperature after 2 hr (-1.5°C), 3 hr (-1.0°C), 4 hr (-0.8°C), after 6 hrs was followed by an increase in rectal temperature which peaked after 24 hr (+0.9°C).

DEXA administration to LPS-treated rats prevented the drop in the rectal temperature at each point measured in the hypothermic phase (Fig.1).

On the other hand, INDO did not prevent the initial decrease in rectal temperature induced by LPS treatment (Fig. 2).

As shown in Fig.3 LPS at 2 hr caused a considerable, more than 2 fold increase, ($p < 0.05$) in hypothalamic PGE_2 production. Both DEXA and INDO prevented the LPS-induced increase in hypothalamic PGE_2 synthesis ($p < 0.05$).

Discussion

Lipopolysaccharide effects on body temperature pattern depend on many factors (7). In a previous study (8), as in this study, at an ambient temperature of 22°C in adult male rats LPS induced a biphasic pattern of rectal temperature. This agrees with some reports (9) and in varience with others (10). In the present study hypothalamic PGE_2 synthesis rose 2 hrs after treatment with LPS and remained high until 24 hr post injection. This rise did not correlate with rectal temperature changes. At 2 hr there is a drop in rectal temperature and at 24 hr there is an increase in rectal temperature. INDO and DEXA in the 24 hr LPS-treated rats revert rectal temperature and hypothalamic PGE_2 (4,5). In this investigation INDO and DEXA reduced the elevation in hypothalamic PGE_2 production, induced by LPS. The effect of these inhibitors of eicosanoid synthesis on rectal temperature is different: INDO did not prevent, but DEXA did prevent the initial phase of LPS-induced decrease in rectal temperature. DEXA inhibits both generation of prostanoids and leukotrienes via inhibition of phospholipase A_2 (11). INDO interferes only with the production of prostanoids. It might be that in the hypothermic phase prostanoids (PGE_2, PGD_2, etc.) (12) play a minor role and leukotrienes or some other factor involving steroids interference play a major role in this phenomenon.

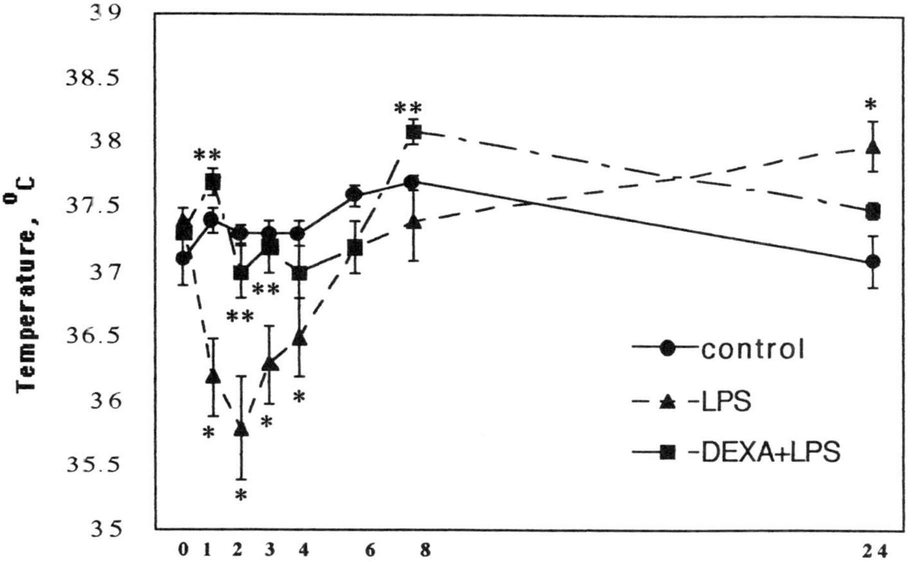

Fig. 1: Effect of DEXA on rectal temperature of LPS-treated rats. * $p < 0.05$ vs control, ** $p < 0.05$ vs control.

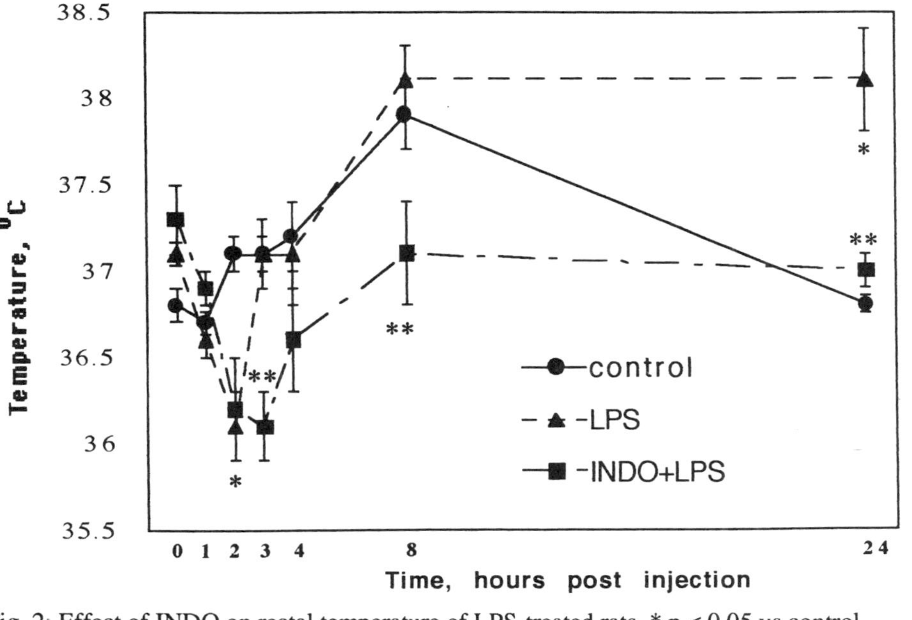

Fig. 2: Effect of INDO on rectal temperature of LPS-treated rats. * $p < 0.05$ vs control, ** $p < 0.05$ vs LPS.

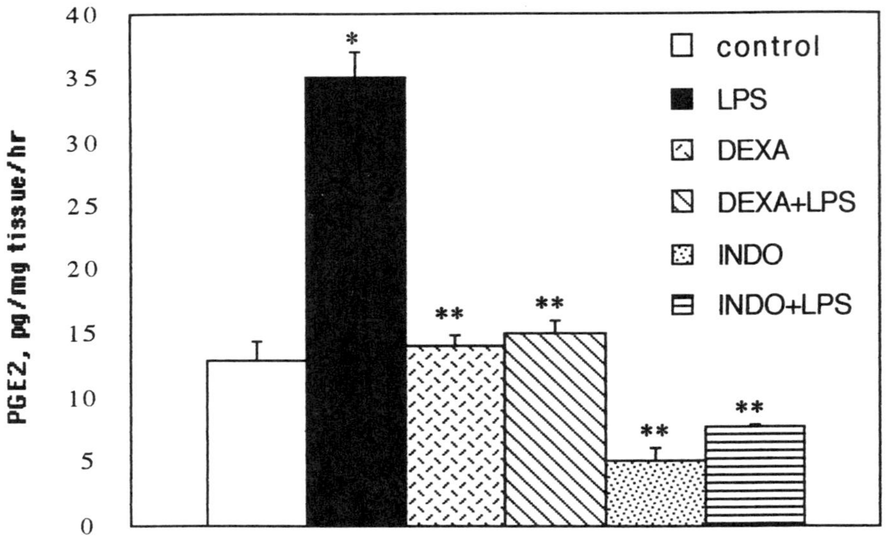

Fig. 3: Effect of DEXA and INDO on hypothalamic PGE_2 production 2 hrs post LPS administration in rats. * $p < 0.05$ vs control, ** $p < 0.05$ vs LPS.

References

1. Dinarello, CS, Cannon, J and Wolff, SM (1988): New concepts on the pathogenesis of fever. Rev Inf Dis 10, 168-189.

2. Coceani, F (1991): Prostaglandins and fever. Facts and controversies. In: Fever: Basic Mechanisms and Managment. Mackowiak, P ed. New Jork: Raven Press, 59-70.

3. Rothwell, NJ (1992): Eicosanoids, thermogenesis and thermoregulation. Prostaglandins, Leukotrienes and Essential Fatty Acids 46, 1-7.

4. Kaplanski, J, Hadas, H and Sod-Moriah, UA (1989): Effect of Dexamethasone on fever and thermoregulation in acutely heat-exposed rats. In: Thermal Physiology. Mercer, JB ed, Elsevier Science Publishers B.V, 407-412

5. Kaplanski, J, Hadas, H and Sod-Moriah, UA (1989): Effect of Indomethacin on hyperthermia induced by lipopolysaccharides or high ambient temperature in rats. In: Proceedings of the 7th International Symposium Pharmacology of Thermoregulation, Odense, Denmark. Schonbaum, L ed, Thermoregulation: Researh and Clinical Applications,176-178.

6. Danon, A, Leibson, J and Assouline, J (1993): Effect of aspirin, indomethacin, flufenamic acid and paracetamol on prostaglandin output from rat stomach and renal papilla in-vitro and ex-vivo. J Pharm Pharmacol 35, 576-579.

7. Ford, DM and Klugman, KP (1980): Body mass and sex as determining factors in the development of fever in rats. J Physiol Lond 304, 125-128.

8. Hadas, H, Sod-Moriah, UA and Kaplanski, J (1989): The effect of lipopolysaccharide on fever and thermoregulation in rats. In: Thermal Physiology. Mercer, JB ed, Elsevier Science Publishers B.V, 407-412.

9. Morimoto, A, Ono, T, Watanabe, T and Murakami, N (1986): Fever in rats during normal and dehydrated conditions. J Appl Physiol 61, 2060-2066.

10. Feldberg, W and Saxena, PN (1975): Prostaglandins endotoxin and lipid A on body temperature in rats. J Physiol Lond 304, 43-50.

11. Milton, AS, Abul, HT, Davidson, J and Rotondo, D (1989): Antipyretic mechanism of action of dexamethasone. In: Proceedings of 7th International Symposium Pharmacology of Thermoregulation, Odense, Denmark. Schonbaum, L ed, Thermoregulation: Research and Clinical Application, 74-77.

12. Veno, R, Narumiya, S, Ogorochi, T, Nakayama, T, Ishikawa, Y and Hayaishi, O (1982): Role of prostaglandin D_2 in the hypothermia of rats caused by bacterial lipopolysaccharide. Proc Natl Acad Sci 79, 6093-6097.

Thermal Balance in Health and Disease
Advances in Pharmacological Sciences

PROLONGED "SLIGHT" FEVER IN RATS INDUCED BY ADMINISTRATION OF FREUND'S COMPLETE ADJUVANT: EFFECTS OF INDOMETHACIN AND *NO*-RELATED DRUGS

I. N. Semenenya, A. V. Gourine and V. N. Gourine

Institute of Physiology, Academy of Sciences of Belarus, Skorina Street 28, Minsk 220725, Republic of Belarus, CIS

The states characterized by lasting increases in body temperature (within 37.1-38.0°C, measured in armpit) and called "slight" fever are widespread in clinical practice (1,2). Experimental simulation of "slight" fever would enable us to investigate the mechanisms of persistence of this state and to propose adequate pharmacological corrections.

Febrile responses may be initiated by a number of agents including endotoxins, viruses as well as by hypersensitivity and immune responses (1,3). It was shown previously that administration of a potent immunostimulant Freund's Complete Adjuvant (FCA) to experimental animals led to the development of local delayed-type hypersensitivity. In conditions of prolonged antigenic stimulation there was a migration of large numbers of leukocytes (mainly, lymphocytes and neutrophils) from blood to the "inflammatory" site (I.N. Semenenya, A.V. Gourine and V.N. Gourine, unpublished data). These activated cells are known to release a variety of messengers that confer immunity (some of them with pyrogenic activity). Therefore, we suggested that the development of the local delayed-type hypersensitivity response under the influence of FCA might be accompanied by a prolonged rise in body temperature. To test this hypothesis, we investigated the effects of Freund's Complete Adjuvant on body temperature of rats at various ambient temperatures.

Methods

Adult male rats weighing 200-250 g were used in all experiments. The animals were housed four per cage and had free access to food and water. The experiments were carried out in a temperature controlled room at 15°C, 22°C, or 30°C. Rectal temperature was measured with a thermistor probe inserted 3 cm into the rectum. FCA injections as well as daily temperature measurments were performed between 09:00-10:00h. Freund's Complete Adjuvant containing 0.1% killed and dried *Micobacterium butyricum* (CALBIOCHEM, cat. 344289), indomethacin (SIGMA Chemical, product no. I7378), the nitric oxide donor - molsidomine (N-(ethoxycarbonyl)-3-(4-morpholinyl)-

sydnoneimine, SIGMA Chemical,product no. M2901), the nitric oxide-synthase inhibitor N-nitro-L-arginine (SIGMA Chemical, product no. N5501) were used.

FCA of different dosage was injected into the hind paws of experimental animals. A possible involvement of prostaglandin and nitric oxide (NO) synthesis in the action of FCA was studied by determining the effects of intraperitoneal (i.p.) injections of indomethacin (10.0mg/kg), molsidomine (5.0 mg/kg), and N-nitro-L-arginine (25.0 mg/kg) on rectal temperature in FCA-treated animals.

Data were expressed as means ± SE. Comparisons between the experimental and control groups were analyzed for statistical significance using Student's t-test. The significance level was ensured at $P<0.05$.

Results and discussion

At 15°C, administration of FCA induced a dose-related significant rise in body temperature, which was observed in the course of 8-9 days. The results are presented in Table 1.

Tab. 1: Time course of the changes in rectal temperature in FCA treated rats (Ta=15°C)

		Change in rectal temperature °C Mean ± SE					
FCA (mL)	n	Days 1	2	3	4	8	10
Control	24	0.05±0.04	0.25±0.06	0.11±0.03	0.05±0.04	0.65±0.04	0.35±0.04
0.05	6	0.49±0.26	0.63±0.29	0.61±0.12*	0.52±0.09*	0.51±0.32	0.30±0.14
0.10	6	0.85±0.25*	0.71±0.21	0.79±0.14*	0.95±0.21*	1.47±0.28*	0.74±0.20
0.15	24	1.25±0.02*	1.06±0.03*	0.94±0.04*	0.67±0.04*	1.28±0.05*	0.49±0.11
0.20	6	0.73±0.19*	1.15±0.19*	0.92±0.25*	1.07±0.18*	0.61±0.35	0.55±0.49
0.30	6	0.84±0.25*	0.93±0.14*	0.85±0.26*	0.68±0.32	0.35±0.18	0.21±0.21

*Significantly different from controls (P<0.05).

In a neutral or warm environment (Ta=20-30°C) the rise in body temperature induced by application of FCA was observed only in the first two days (data not shown). The results show that the development of local delayed-type hypersensitivity in rats under the influence of FCA is accompanied by prolonged rise in body temperature. This "slight" fever may originate from the pyrogenic action of cytokines released by activated lymphocytes and neutrophils which are accumulated in essential quantities at the site of immunostimulant application.

It is generally believed that endogenous pyrogens cause fever by raising the thermoregulatory "set point" via the release of prostaglandins of the E series (PGEs) in the hypothalamus (4). Like PGEs, the diffusible stable free radical gas nitric oxide (NO) is a potent smooth muscle relaxant in various vascular beds (5,6). Like PGEs (7), NO has been shown to function as neural messenger or neuromodulator in the central nervous system (5,6,8). Furthermore, recent evidence favours a role for NO as an effector molecule in immunological reactions (5,6).

The effects of indomethacin (an inhibitor of prostaglandin synthesis), molsidomine (the nitric oxide donor), and N-nitro-L-arginine (an inhibitor of NO-synthase) on "slight" fever induced by administration of FCA are presented in Tables II and III. As shown, in FCA treated animals i.p. injection of indomethacin (10 mg/kg) led to a sharp decrease in rectal temperature (Table II). The suppression of body temperature by indomethacin suggests that prostaglandin synthesis is responsible for the body temperature rise during FCA-induced "slight" fever.

Tab. 2: Effects of systemically administered indomethacin (10.0 mg/kg) or molsidomine (5.0 mg/kg) on rectal temperature in FCA treated rats (Ta=15°C)

	n	Dose,ml	Tbo,°C	Drug	Tbo 90min,°C
1. Control	16		36.71 ± 0.08		36.82 ± 0.09
2. FCA	8	0.15	37.64 ± 0.15^{1}	Saline	37.85 ± 0.11^{1}
3. FCA	8	0.15	37.70 ± 0.11^{1}	INDO	$36.91 \pm 0.13^{*}$
4. FCA	8	0.15	37.73 ± 0.13^{1}	M	$37.09 \pm 0.12^{*}$

Abbreviations and symbols: Tbo,°C: mean (±SE) rectal temperature 72h after injection of FCA (2,3,4). Tbo 90min,°C: mean (±SE) rectal temperature 90 min after i.p. injections of drugs (3,4) or control saline (2). INDO: indomethacin. M: molsidomine. ^{1}Significantly different ($P<0.05$) from controls(1). *Significantly different ($P<0.05$) from 2.

Table 2 also shows that i.p. injection of molsidomine (5.0mg|kg) induced a Tbo fall in FCA treated rats. There is evidence that cytokines stimulate the expression of the inducible NO-synthase in endothelial and smooth muscle cells as well as in macrophages, neutrophils and lymphocytes (5,6). Therefore, we suggest that stimulation of the inducible NO-synthase by cytokines released by activated leukocytes and generation of nitric oxide can modulate the magnitude of FCA-induced "slight" fever.

Tab. 3: Effect of systemically administered N-nitro-L-arginine (25.0 mg\kg) on rectal temperature in FCA treated rats (Ta=15°C)

	n	Tbo,°C	Tbo 24h,°C	Tbo 48h,°C
1.Control	16	37.11±0.08	36.93±0.11	36.70±0.08
2.FCA	8	37.02±0.12	37.80±0.11[1]	37.48±0.10[1]
3.FCA+L-NA	8	36.89±0.09	38.45±0.13[1*]	38.00±0.12[1*]

Abbreviations and symbols: Tbo,°C: mean (±SE) basal rectal temperature. Tbo 24h(48h),°C: mean (±SE) rectal temperature 24h(48h) after injection of FCA (2,3). L-NA: N-nitro-L-arginine. [1]Significantly different ($P<0.05$) from controls (1). *Significantly different ($P<0.05$) from 2.

Acute i.p. injection of a NO-synthase inhibitor, N-nitro-L-arginine (25.0 mg/kg) 2.5h after the FCA application significantly enhanced "slight" fever in the first two days (Table 3). There is also evidence that N-nitro-L-arginine when administered intraperitoneally to rats, resulted in marked, irreversible *ex vivo* inhibition of brain NO-synthase. Furthermore, following 90% inhibition of enzyme activity with N-nitro-L-arginine, brain NO-synthase activity did not return to normal for at least 5 days (9). We speculate that nitric oxide is involved in the central mechanisms of thermoregulation during "slight" fever as one of the possible endogenous antipyretics.

Conclusion

The temperature response after single injection of Freund's Complete Adjuvant to the hind paws of experimental rats enables us to consider this state of the organism as a model of prolonged "slight" fever which is affected by indomethacin and NO-related drugs.

Acknowledgements

This research was supported by ASB grant B1-120 to Valery N. Gourine.

References

1. Atkins, E. and Bodel, Ph. (1979): Clinical fever: its history, manifestations and pathogenesis. Fed Proc 38, 57-63.

2. Kimura, M., Tano, Y., Tomizawa, S. and Hirano, Y. (1987): Clinical study on slight fever. Kawasaki Med J 13, 133-137.

3. Iriki, M. (1988): Fever and fever syndrome - current problems. Jap J Physiol 38, 233-250.

4. Stitt, J.T. (1986): Prostaglandin E as the neural mediator of the febrile response. Yale J Biol Med 59, 137-149.

5. Moncada, S., Palmer, R.M.J. and Higgs, E.A. (1991): Nitric oxide: physiology, pathophysiology, and pharmacology. Pharmacol Rev 43, 109-142.

6. Nathan, C. (1992): Nitric oxide as a secretory product of mammalian cells. FASEB J 6, 3051-3064.

7. Gilbert, J.C., Davison, D.V. and Wyllie, M.G. (1978): Studies of the physiological roles of prostaglandins in the central nervous system. Neuropharmacology 17, 417-431.

8. Snyder, S.H. (1992): Nitric oxide and neurons. Curr Opin Neurobiol 2, 323-327.

9. Dwyer, M.A., Bredt, D.S. and Snyder, S.H. (1991): Nitric oxide synthase: Irreversible inhibition by L-N-nitro-arginine in brain *in vitro* and *in vivo*. Biochem Biophys Res Commun 176, 1136-1141.

Thermal Balance in Health and Disease
Advances in Pharmacological Sciences

VENTRAL SEPTAL AVP RELEASE IN THE FEBRILE HYPERTENSIVE AND NORMOTENSIVE RAT

M. L. Earle[1], T. Horn[1], N. Kasting[2], R. Landgraf[3] and Q. J. Pittman[1]

[1]Neuroscience Research Group, University of Calgary, 3330 Hospital Drive N.W., Calgary, Alberta, Canada. T2N 4N1. [2]Department of Physiology, University of British Columbia, Vancouver, B.C. Canada. [3]Department of Psychiatry, Max Planck Institute, Munich, Germany.

Introduction

The neuropeptide arginine vasopressin (AVP) is thought to act within the brain, specifically the ventral septal area (VSA) as a central neurotransmitter involved in febrolysis (2, 5, 10). A role for endogenous AVP as a pressor agent involved in central cardiovascular regulation has also been suggested (9, 10, 11). Previous studies from our laboratory have suggested that, in the one kidney, one clip (1K1C) Goldblatt hypertensive rat, there is an inhibition of the febrile response to prostaglandin E_1 (PGE_1). The method used to induce hypertension produced a very dramatic increase in systolic blood pressure during the initial four days following clipping of the renal artery. The aim of the present study was firstly, to determine if this rapid rise in blood pressure was necessary to produce an inhibition of the febrile response, and secondly, to determine if there is any change in the release of AVP into the VSA, determined using the technique of push-pull perfusion. We produced a more slowly developing hypertension by removing one kidney and clipping the renal artery leading to the other kidney at the same time, rather than four days earlier as used in the study by Fyda et al., (3).

Methods

Male Sprague-Dawley rats (Charles River Inc., Quebec, Canada) weighing 100-140g (at the time of surgery) were housed 3-4 per cage, and maintained at 20°C with access to laboratory rat chow and water *ad libitum*. Systolic blood pressure recordings were taken repeatedly over several weeks in awake rats using the tail cuff plethysmography system (IITC Inc. New Jersey, U.S.A. (1)). Following baseline tail cuff recordings of systolic blood pressure, rats were anaesthetized (sodium pentobarbital, 50 mg/kg, i.p.) and under aseptic conditions, the left renal artery, vein and nerve were isolated and ligated, and the left kidney was removed taking care not to damage the adrenal glands. The right kidney was exteriorized and the renal artery isolated by removing the overlying fascia and a small aluminum clip was placed over the renal artery so as to occlude approximately

80% of blood flow. The kidney was replaced in the abdomen and the incisions closed. The controls for these experiments were "sham-operated" rats in which the left kidney was removed but the right kidney was left intact. Rats were then placed in a Kopf stereotaxic frame and a lateral cerebral ventricular (icv) cannula and a unilateral guide cannula were implanted.

The rats were allowed to recover and topical bacitracin (Baciguent, Upjohn) was applied to the wounds and several drops of acetaminophen (giving a final concentration of approximately 0.2 mg/ml) were added to the drinking water for 24 hours. At either day 4 or 18 following surgery, systolic blood pressures were recorded as described above and the rats were anaesthetized with urethane (1.6 g/kg). Once anaesthetized, the rats were placed onto heating pads with a rectal probe to monitor body temperature, until a stable body temperature was achieved (approximately 30 mins to 1 hour). At this time, push-pull perfusion of the VSA was performed. Using artificial cerebrospinal fluid (aCSF) which was freshly prepared each day with the peptidase inhibitor bacitracin (0.025g) added to the perfusion fluid. The push-pull perfusion cannula was lowered through the guide cannula so that the tip rested in the VSA. The initial 15 mins perfusion fluid was discarded and the samples for analysis were collected over a 30 min period (20 μl/min, resulting volume 600 μl per sample). Prostaglandin E_2 (PGE_2) was injected icv, by gravity flow via a stainless steel injection needle (27 gauge) attached to PE-20 tubing. The perfusates were stored immediately at -20°C for radioimmunoassay for AVP content. Three sample periods were used, the first was a 30 min collection period prior to the injection of PGE_2, the second was a sample collected for the 30 min period following the injection of pyrogen and the final sample was the 30 min sample collected following this. The perfusates were analyzed using a specific radioimmunoassay for AVP (4). Prior to and throughout the duration of push-pull perfusion the animal's rectal temperature was monitored and recorded every 10 mins.

At the conclusion of the study, perfusion sites were identified histologically. The maximum changes in temperatures were analyzed using a Student t-test, as were the systolic blood pressures. Changes in AVP content were analyzed statistically using the analysis of variance (ANOVA) followed by a Scheffe post-hoc test where appropriate. Results were considered significant if $p<0.05$.

Results

Systolic blood pressures at day 4 were not significantly different from the control presurgery levels, although by day 18 the 1K1C rats had significantly increased systolic blood pressures, these increases were significantly greater than in the sham operated rats (table 1), indicating that there was a slowly developing hypertension over the 18 day period. At neither day 4 nor day 18 following surgery was any inhibition of the febrile response to PGE_2 observed in either the sham-operated or 1K1C rats. In addition to these animals, a further group of animals were used, however, the peptidase inhibitor bacitracin was not added to the perfusate, and consequently no peptide was detected with the radioimmunoassay. However, the fever data collected from these animals were pooled, giving a much larger final sample size for the fever data (day 4, n=25 1K1C, n=14 sham-operated rats; day 18, n=18 1K1C, n=12 sham-operated rats). Even with this much larger sample size (figure 1) there was no inhibition of the febrile response observed in the 1K1C rats compared to their sham-operated control rats. Consistent to this there was no observable increase in AVP content of the perfusates (table 2), other than that observed at day 18 (sham-operated rats, sample 2). Although of statistical significance this latter observation was well within previously reported baseline levels of AVP in perfusates of the VSA (6) and is thus probably not of biological significance.

	Sham-operated	± S.E.M.	1K1C	± S.E.M.
Baseline	95.8 (n=4)	6	96.8 (n=10)	4.7
Day 4	100.9	7.8	102.7	3.4
Baseline	96.4 (n=3)	1.3	101.6 (n=8)	4.5
Day 18	137.3	13.1	179	7.8

Tab. 1: Systolic blood pressures prior to (baseline) and at day 4 or day 18 following 1K1C or sham-operated surgery.

	Sample 1	Sample 2	Sample 3
Day 4: Sham (n=4)	1.3 ± 0.3	1.4 ± 0.2	1.5 ± 0.6
Day 4: 1K1C (n=10)	1.8 ± 0.2	1.5 ± 0.3	1.7 ± 0.5
Day 18: Sham (n=3)	1.1 ± 0.2	1.6 ± 0.1	1.2 ± 0.2
Day 18: 1K1C (n=8)	1.2 ± 0.2	1.0 ± 0.1	1.0 ± 0.1

Tab. 2: AVP content (pg/ml) of perfusate of the VSA in the three samples collected. PGE_2 (icv) was given by gravity flow prior to the collection of sample 2. All samples were collected for a 30 min period and analyzed using a specific radioimmunoassay for AVP content.

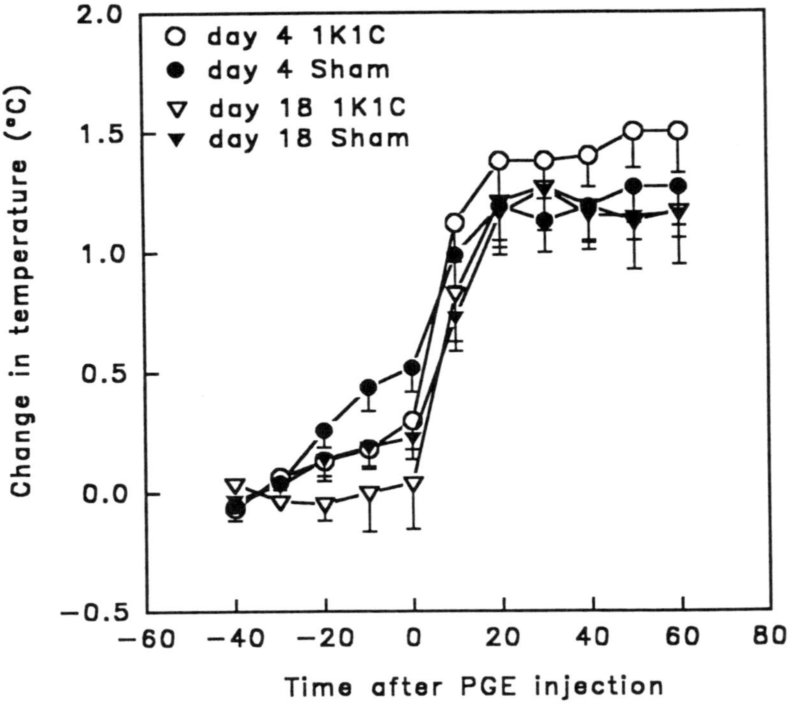

Fig. 1: Graph showing change in temperature (°C) in the 1K1C rats at days 4 (open circle) and days 18 (open triangle) and sham operated rats at days 4 (closed circle) and days 18 (closed triangle). PGE_2 injection (icv, gravity flow) was given at time zero (indicated by the arrow). There were no significant differences in febrile responses between these groups.

Discussion

The present study suggests that, using a slower developing hypertension, there is no inhibition of the febrile response to PGE_2, given icv, at either days 4 or 18. Consistent to this there was no elevation of AVP detected in the push-pull perfusates from the VSA.

The present study, using a slightly different protocol than that used in the previous study of Fyda et al., (3), suggests that the inhibition of the febrile response observed in the initial studies is not the result of an elevation of blood pressure *per se*, but is more likely the result of the rapid increase in blood pressure that was evident in the initial study. Fyda et al., (2) undertook the 1K1C surgery by removing one kidney four days prior to clipping the remaining kidney. In their hands this protocol resulted in a rapid rise in blood pressure from approximately 115 mmHg which rose to 170 mmHg after 5 days, a rise of 55 mmHg. This rise in blood pressure was at least nine times greater than that observed in the present study (approximately 97 mmHg to 103 mmHg) and is the most likely explanation for the inhibition of the febrile response these researchers observed. These data would suggest that there is no direct link between the 1K1C hypertensive process and the febrile response to PGE_2, and that the initial observations of Fyda et al., (3) should be interpreted with caution.

Although the initial studies were performed in conscious animals, the present study was undertaken in the urethane anaesthetized rat, since push-pull perfusion is difficult (although not impossible) in the conscious animal. Cardiovascular reflexes are well maintained under urethane anaesthesia compared to those observed when other anaesthetics are used (7) and the pyrogenic and thermoregulatory responses have been well described (8). Baseline AVP release into the VSA has also been well studied (6), and is consistent with the present study. Landgraf et al., (6) described an increase in AVP detected in the perfusates of the VSA during PGE fever, and attributed the release of AVP to the defervescence observed. AVP was thus suggested to act as a "thermal brake", controlling the height of the febrile response, and aiding in the return of body temperature to it's "set point". The present study suggests that AVP release does not occur during the onset of the febrile response, however, perfusates were not collected during the defervescence phase, and indeed there may have been altered AVP release during this time period.

Thus this study demonstrates that in 1K1C hypertensive and normotensive rats, using a more slowly developing hypertension, there is no inhibition of the febrile response to PGE_2, given icv. In addition, the present study shows no increase in AVP detected in push-pull perfusates of the VSA, during the rising and sustained phases of PGE_2-induced fever (60 mins following icv injection). We conclude that a more rapid increase in blood pressure is required to induce the inhibition of the febrile response observed by Fyda et al., (3).

References

1. Bunag, RD, Butterfield, J (1982): Tail-cuff blood pressure measurements without external preheating in awake rats. Hypertension, 4, 898-903.

2. Cooper, KE, Kasting, NW, Lederis, K, Veale, WL (1979): Evidence supporting a role for endogenous vasopressin in natural suppression of fever in the sheep. J Physiol (Lond) 295, 33-45.

3. Fyda, D, Veale, W, Pittman, Q (1993): Suppression by central AVP of prostaglandin E1 hyperthermia in one-kidney, one-clip Goldblatt hypertensive rats. Am J Physiol 264, R143-R148. 4. Kasting, NW, Carr, DB, Martin, JB, Blume, H, Bergland, R (1983): Changes in cerebrospinal fluid and plasma vasopressin in the febrile sheep. Can J Physiol Pharmacol 61, 427-431.

5. Kasting, NW, Cooper, KE, Veale, WL (1979): Antipyresis following perfusion of brain sites with vasopressin. Experientia 35, 208-209.

6. Landgraf, R, Malkinson, TJ, Veale, WL, Lederis, K, Pittman, QJ (1990): Vasopressin and oxytocin in rat brain in response to prostaglandin fever. Am J Physiol 259, R1056-R1062.

7. Maggi, CA, Meli, A (1986): Suitability of urethane anaesthesia for physiopharmacological investigations in various systems. Part 2: cardiovascular systems. Experientia 42, 292-297.

8. Malkinson, TJ, Cooper, KE, Veale, WL (1988): Physiological changes during thermoregulation and fever in urethan-anaesthetized rats. Am J Physiol 255, R73-R81.

9. Matsuguchi, FM, Sharabi, FM, Gordon, FJ, Johnson, AK, Schmid, PG (1982): Blood pressure and heart rate responses to microinjection of vasopressin into the nucleus tractus solitarius region of the rat. Neuropharmacology 12, 687-693.

10. Pittman QJ, Landgraf R: Vasopressin in thermoregulation and blood pressure control; in Jard S, Jamison R (eds): Vasopressin. John Libbey Eurotext Ltd, 1991, pp 177-184.

11. Rohmeiss, P, Becker, H, Dietrich, R, Luft, F, Unger, T (1986): Vasopressin: mechanism of central cardiovascular action in conscious rats. J Cardiovasc Pharmacol 8, 689-696.

Thermal Balance in Health and Disease
Advances in Pharmacological Sciences

HYPEROSMOTIC STIMULATION OF THE HYPOTHALAMIC PARAVENTRICULAR AND SUPRAOPTIC NUCLEI ALTERS BODY TEMPERATURE IN THE NORMOTHERMIC AND FEBRILE RAT

T. Horn[1], T. J. Malkinson[1], Q. J. Pittman[1] and R. Landgraf[2]

[1]Neuroscience Research Group and Department of Medical Physiology, University of Calgary, 3330 Hospital Drive N.W., Calgary, Alberta, T2N 4N1, Canada.
[2]Max Planck Institute of Psychiatry, Clinical Institute, Munich, Germany.

Introduction

There is considerable evidence that the nuclei supraopticus (SON) and paraventricularis (PVN) contribute to endocrine and autonomic regulation, but little is known about the involvement of these nuclei in temperature regulation and fever. However, pharmacological stimulation of these areas results in an enhanced sympathetically mediated thermogenesis in the brown adipose tissue. (1, 2). PVN-lesioned rats develop a lower fever than normal animals in response to central application of prostaglandine E_2 (PGE_2) or intraperitoneal (i.p.) E. coli endotoxin but responses to heat or cold were unaltered (3). In addition to having mechanisms to elevate body temperature, the body also has mechanisms to lower it, in particular the peptide vasopressin (AVP) appears to be important in antipyresis. It was shown that electrical stimulation of either the guinea pig SON (4) or the rabbit PVN (5) attenuates PGE_1-fever. It is now a well accepted principle that AVP, released into the ventral septal area (VSA) during fever, is an endogenous and centrally acting antipyretic substance (6). Recently a mean to stimulate the PVN and SON was established utilizing direct stimulation of hypothalamic nuclei with hypertonic saline via microdialysis. Hyperosmotic NaCl delivered via microdialysis into the SON or PVN resulted in increased AVP levels, both, into extracellular fluid of these nuclei (7) and into the blood (8) as well as increased extracellular levels of amino acids in the perfusion area (9). Both, behavioral and autonomic changes accompany these alterations in transmitter release. In this study we utilized the model of direct hypertonic NaCl-stimulation via microdialysis within hypothalamic nuclei PVN and SON to test whether the stimulus has an effect on temperature regulation. We tested:

1) the effect of hypertonic stimulation utilizing microdialysis with 1M NaCl-aCSF within the SON or PVN on the body temperature in normothermic and febrile animals.

2) the involvement of V1-receptors in the VSA during PGE_1-fever in the mechanisms of the antipyretic effect due to hypertonic stimulation within the SON.

3) the effect of PGE_1-fever on the elevated and basal intra-SON release of AVP.

Methods

U-shaped or concentric probes (7) were stereotaxically implanted with their tips in or adjacent to the SON or PVN of urethane-anaesthetized (1.6g/kg) Sprague-Dawley rats. All animals received an additional guide cannula for intracerebroventricular (icv.) injections and were placed on regulated heating pads to stabilize their body temperature for a period of 120 min before any treatment. During that initial period, the dialysis probes were perfused with isotonic aCSF.

The animals with SON-implanted probes were divided into four groups. One group was perfused continuously with aCSF and received an icv. injection of prostaglandine E_1 (PGE_1 ,120ng/ 5µl, n=5). The second group (n=7) was treated with the same dose of PGE_1 icv., but the microdialysis perfusion medium was replaced for a 30-min period with 1M NaCL-aCSF, beginning 20 min before the icv injection. The third group (n=5) was perfused in the same manner with 1M NaCL-aCSF but received an icv. injection of 5 µl vehicle (saline). A control group (n=5) was perfused continuously with aCSF and was icv. injected with 5 µl saline. PVN-implanted animals underwent a similar protocol: one group (n=5) was dialysed with isotonic aCSF and received an icv injection of 120 ng PGE_1 in 5 µl saline. The temperature response of this group was compared with a second group (n=5) which was dialysed with 1M NaCL-aCSF for a 30-min interval starting 20 min before the icv. injection. In other experiments, SON-implanted rats with icv. cannulae received bilaterally implanted guide cannulae for microinjections into the VSA. These animals were divided into three groups. One (control, n=8) was dialysed in the SON with isotonic aCSF throughout the experiment. Bilateral saline injections of 1 µl into the VSA were given 15 min prior to the PGE_1-application (120 ng). In the second and third group the isotonic dialysis medium was replaced for a period of 30 min with 1M-NaCL aCSF. PGE_1 was injected at 10 min after the start of the hypertonic SON stimulation. The second group received bilateral injections into the VSA of 400 pmol V_1-receptor antagonist $d(CH_2)_5Tyr(Me)AVP$ dissolved in 1 µl saline over 1min and the third group 1 µl saline injections into the VSA.

To study a possible intranuclear release of AVP during fever, dialysates were collected in 30-min intervals. Therefore the SON of febrile animals (120 ng PGE_1 icv.) was dialysed with either isotonic or 1M NaCl-aCSF starting 10 min before the PGE_1-application. Then both groups were dialysed with isotonic aCSF for two sample periods.At the end of the experiment (i.e. 90 min later) each animal was again dialyzed with 1 M NaCl-aCSF. The dialysates were frozen at -20°C for later analysis for AVP by radioimmunoassay. To study the effect of exogenously applied AVP into the SON, another subset of animals (n=6) was dialysed with an AVP-containing aCSF (100ng AVP/100 µl aCSF) for a period of 30 min followed by a 30-min pulse of 1M NaCl-aCSF 90 min later.

The temperature responses were recorded in 10-min intervals. Data were statistically analyzed using one way analysis or Kruskal-Wallis analysis of variance followed by post-hoc tests as appropriate and the results are expressed as mean ± SEM. The placement of the probe was histologically determined in serial frontal cryotome sections treated using standard staining procedures.

Results

Controls showed a stabile body temperature during the recording period, indicating, that neither the icv injection nor the microdialysis in the SON with isotonic NaCl-aCSF affects the normal body temperature (Fig.1). The delivery of 120ng PGE_1 into the ventricle resulted in the known febrile

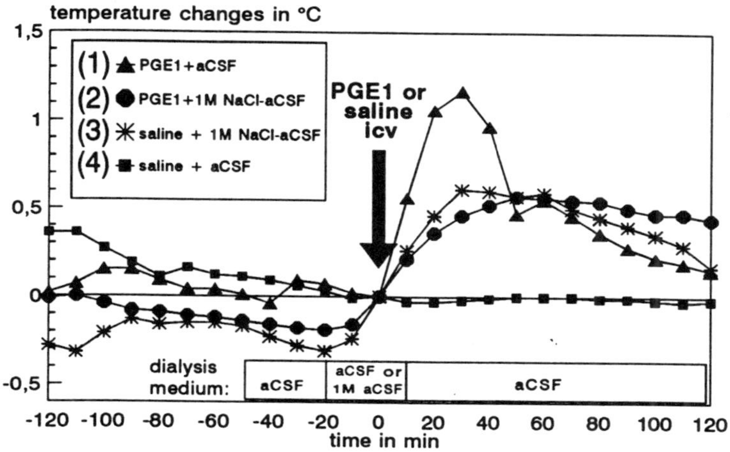

Fig. 1: Changes in body (colonic) temperature in response to (1) 120 ng PGE_1 / 5μl icv. during simultaneous dialysis of the SON with isotonic aCSF (PGE_1+aCSF, n=5), (2) PGE_1 icv. during hypertonic stimulation of the SON (PGE_1+ 1M NaCl-aCSF, n=7), (3) saline icv. during hypertonic stimulation of the SON (saline+1M NaCl-aCSF, n=5) or (4) saline icv. during dialysis of the SON with isotonic aCSF (saline +aCSF, controls, n=5). The arrow and the bar at the bottom indicate the time course of the treatments. Significant differences ($p<0.05$): 1 vs. 2,3 (20-40 min), (2) vs (4) (20-120 min), (1), (3) vs (4) (20-70 min).

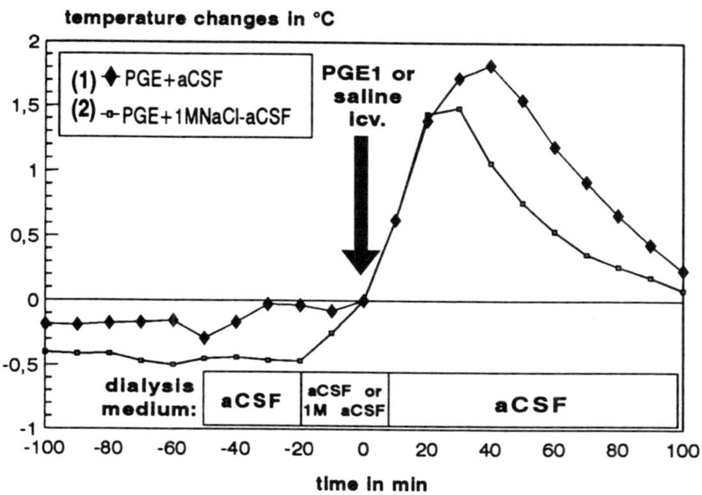

Fig. 2: Changes in body (colonic) temperature in response to (1) 120 ng PGE_1/5 μl icv. during simultaneous dialysis of the PVN with isotonic aCSF (PGE_1+ aCSF, n=5) and (2) PGE_1 icv. during hyperosmotic stimulation of the PVN with 1M NaCl-aCSF (PGE_1+1M NaCl-aCSF, n=5). The arrow and the bar at the bottom indicate the time course of the treatments. Significant differences ($p<0.05$): 1 vs. 2 (40-60 min).

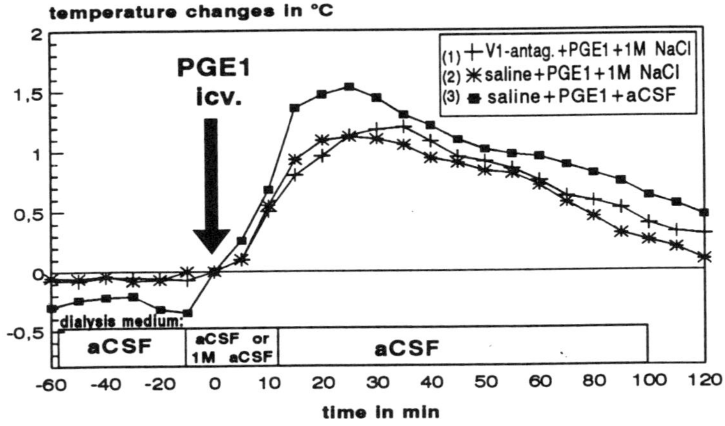

Fig. 3: Changes in body (colonic) temperature in response to (1) 120 ng PGE_1/ 5 μl saline, dialysis of the SON with 1M NaCl-aCSF and a bilateral 400 pmol V1-antagonist application / 1μl into the VSA 15 min prior to the injection (V1-antagonist+PGE_1+1M NaCl aCSF, n=7) or (2) 120 ng PGE_1/ 5 μl saline, dialysis of the SON with 1M NaCl-aCSF and a bilateral saline application into the VSA 15 min prior to the injection (saline + PGE_1 + 1M NaCl-aCSF, n=6) or (3) 120 ng PGE_1/ 5 μl saline, dialysis of the SON with 1M NaCl-aCSF and a bilateral saline application into the VSA 15 min prior to the injection (saline + PGE_1+ aCSF, n=8). Statistical significance ($p<0.05$): (3) vs. (1), (2) (15-30 min) No significant difference was observed when (1) was compared with (2).

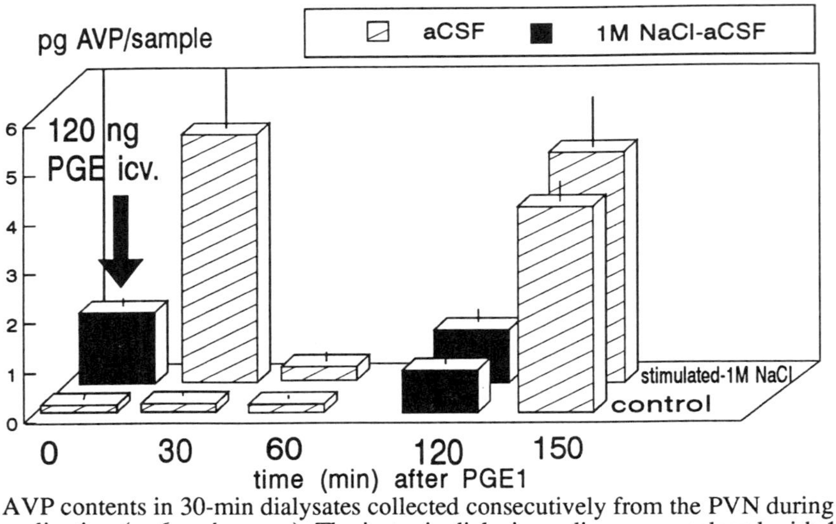

Fig. 4: AVP contents in 30-min dialysates collected consecutively from the PVN during and after PGE_1 application (n=6 each group). The isotonic dialysis medium was replaced with 1M NaCl-aCSF in the stimulated 1M NaCl-group during the PGE_1-application. In both groups, the isotonic dialysis medium was replaced by 1M NaCl-aCSF again at 120 min. Besides the expected changes of intranuclear AVP-release in response to the delivery of 1M NaCl-aCSF there was no significant effect of the PGE_1 treatment on the basal (control) or evoked (simulated-1M NaCl) AVP-release.

response with a maximum of 1.17 ± 0.37 °C ($p<0.01$) above baseline. When such febrile animals were dialysed in the SON with 1M NaCl-aCSF, the peak of the fever response to PGE_1 was significantly reduced by approx 50% ($p<0.05$). However the delivery of 1M NaCl-aCSF *per se* resulted in an significant increase in body temperature of 0.6 ± 0.15°C ($p<0.05$) in nonfebrile animals. In animals which were dialysed with isotonic aCSF via probes located within the PVN the icv. application of 120ng PGE_1 resulted in a typical temperature change of 1.82±0.26°C (Fig. 2). When such animals received a 30-min pulse of 1M NaCl-aCSF, the temperature rise peaked at 1.48±0.18°C. A significantly reduced fever response was observed from 40 to 60 min after the PGE_1-application ($p<0.05$). Pretreatment by bilaterally infused V1-antagonist into the VSA failed to abolish the antipyretic effect of 1M NaCl-aCSF in the SON (Fig. 3). Perfusion of AVP in the SON did not alter basal body temperature. The analysis of the AVP content in the dialysis samples of the SON revealed that during PGE_1-fever, the intranuclear AVP-release remained at basal levels. The evoked AVP-release to 1M NaCl-aCSF (at 0 min and 30 min) during fever also did not differ from that to a second 30-min perfusion with 1 M Nacl-aCSF after the fever response (at 120 and 150 min, respectively, Fig. 4).

Discussion

Our results clearly indicate that the stimulation of the SON and adjacent areas with hypertonic NaCl-aCSF results in an elevation of body temperature in the urethane-anaesthetized rat. While this procedure causes intranuclear release of AVP (7, 8) and many amino acids (9) we have been unable to activate the temperature changes by applying AVP directly into the SON. As applying glutamate to the SON (2) or PVN (1, 10) activates brown adipose tissue, locally released glutamate could be responsible for the elevation in temperature. The neuronal pathways responding to these changes are not known; the SON is not thought to have extensive CNS projections (11), making it possible that activation of neurons of the lateral hypothalamic area adjacent to the SON may be responsible for these effects. It is interesting that perfusion of the nearby posterior hypothalamic area with high Na^+ has previously been shown to increase body temperature (12).

Despite the slight hyperthermic effect of the osmotic stimulation, we observed an antipyretic effect of the direct NaCl-hypertonic stimulus when applied to the SON or PVN of PGE_1-febrile animals. Since it is known that this stimulation causes AVP- release in the septal area (13), an antipyretic effect of such released peptide acting on V1-receptors in the VSA seemed to be possible. However, in our study, the application of V1-antagonist into the septal area during hyperosmotic stimulation of the SON of febrile animals failed to abolish the antipyretic effect. Thus, the present results indicate an involvement of SON and PVN (or adjacent tissue) in thermoregulation, but there is no evidence of the involvement of AVP in the body temperature response to direct osmotic stimulation of the SON. At this point further investigations are necessary to clarify the role of the intranuclear dynamics of neurotransmitters in thermoregulation. The mechanisms underlying the effects of hypertonic stimulation of the SON and PVN on parameters of the autonomic regulation remain to be determined

Acknowledgements

Supported by a NATO Collaborative Research Grant, Volkswagen Foundation and the Medical Research Council of Canada. TH is a DAAD and Government of Canada Awardee, and QJP is an Alberta Heritage Foundation for Medical Research Scientist.

References

1. Amir, S (1990) Stimulation of the paraventricular nucleus with glutamate activates interscapular brown adipose tissue thermogenesis in rats. Brain Res. 508, 152-155.

2. Amir, S, De Blasio, S (1991) Activation of brown adipose tissue thermogenesis by chemical stimulation of the hypothalamic supraoptic nucleus. Brain Res. 563, 349-352.

3. Horn, T, Wilkinson, M, Landgraf, R, Pittman, QJ (1994) Reduced febrile responses to central and peripheral pyrogens following lesions of the hypothalamic paraventricular nucleus. Am. J. Physiol. (in press).

4. Zeisberger, E (1989) Antipyretic action of vasopressin in the ventral septal area of the guinea pig's brain. In: Thermoregulation: Research and Clinical Applications. Lomax, P, Schönbaum, E eds. Basel: Karger, 65-68.

5. Ruwe, WD (1992) Electrical stimulation of the paraventricular nucleus attenuates pyrogen fever in the rabbit. Brain Res. 588: 181-190.

6. Kasting, NW (1989) Criteria for establishing a physiological role for brain peptides. A case in point: the role of vasopressin in thermoregulation during fever and antipyresis. Brain Res. Rev. 14, 143-153.

7. Landgraf, R, Ludwig, M (1991) Vasopressin release within the supraoptic and paraventricular nuclei of the rat brain: osmotic stimulation via microdialysis. Brain Res. 558, 191-196.

8. Neumann, I, Ludwig, M, Engelmann, M, Pittman, QJ, Landgraf, R, (1994) Simultaneous microdiaysis in blood and brain: Oxytocin and vasopressin release in response to central and peripheral osmotic stimulation in the rat. Neuroendocrinology 58, 637-645.

9. Horn, T, Bauce, L, Landgraf, R, Pittman, QJ (1992) Microdialysis with hyperosmotic NaCl causes central release of amino acids and dopamine. Soc. Neurosci. Abstr 18 (2) 421.10.

10. Martin, DS, Haywood, JR (1992) Sympathetic nervous system activation by glutamate injections into the paraventricular nucleus. Brain Res. 577, 261-267.

11. Swanson, LW, Sawchenko, PE (1983). Hypothalamic integration: organisation of the paraventricular and supraoptic nuclei. Ann. Rev. Neurosci. 6, 269-324.

12. Meyers, RD, Simpson, CW, Higgins, D, Nattermann, RA, Rice, JC, Redgrave, P, Metcalf, G (1977) Hypothalamic Na^+ and Ca^{++} Ions and temperature set-point: New Mechanisms of action of a central or peripheral thermal challenge and intrhypothalamic 5-HT, NE, PGE_1 and pyrogen. Brain Res. Bull. 1, 301-327.

13. Landgraf, R, Ludwig, M., Engelmann, M (1993) Central release of vasopressin and its role in behavioral performance. In: Vasopressin. Gross, P, Richter, D, Robertson, GL eds, London: Libbey, 607.

Thermal Balance in Health and Disease
Advances in Pharmacological Sciences

EFFECTS OF IMMOBILIZATION ON FEBRILE, NEUROENDOCRINE AND IMMUNOLOGICAL RESPONSES IN GUINEA PIGS

E. Zeisberger and J. Roth

Physiologisches Institut, Klinikum der JLU Giessen, Aulweg 129, 35392 Giessen, Germany

Fever is induced by actions of mediators of the immune system on the central nervous system (1). A febrile rise in body temperature is beneficial because it accelerates the activation of the immune defence. However, both, an extremely elevated body temperature, as well as an activated immune system, are potentially dangerous, and must be restricted to limited time periods. Hypothalamic neuroendocrine cells activating the hypothalamic-pituitary-adrenal (HPA) axis play an important role in the suppression of the immune system. It seems that these cells activate also a number of antipyretic mechanisms that control the febrile rise in body temperature and prevent its increase to lethal values. The activation of the HPA-system can thus be regarded as a negative feedback mechanism in the complex neuroendocrine reflex aiming at reestablishing of the normal state in the immune and thermoregulatory systems (2). It has long been known that the HPA-system can be activated also by a number of non-immune external or internal stimuli (3,4). These stimuli at distinct levels of intensity are called stressors and the neuroendocrine reaction evoked by their action, stress. Due to the competition of immune and non-immune stimuli in the activation of the HPA-system, the febrile response to immune signals can be limited or suppressed by different stressors. Immobilization stress proved to be a very effective inhibitor of fever in guinea pigs (5). Since the effects of immobilization on neuroendocrine and immunological responses in the guinea pig and their time courses are not known, we describe, in the present paper, three experimental series elucidating some of these effects.

Methods

Animals. The experiments were performed in several groups of guinea pigs. The animals were equipped either with intraarterial catheters for repeated, stress-free, blood sampling or stereotactically implanted guide cannulae, through which push-pull perfusion cannulae could be introduced into the limbic brain septum (for details cf. 6). *Fever induction and measurement of body temperature*. Endotoxin fever was induced by intramuscular injection of bacterial lipopolysaccharide (LPS from E. coli, 0111:B4, 20 µg/kg, Sigma Co., St. Louis, U.S.A.). Abdominal temperature was measured by use of battery-operated bio-telemetry transmitters (VM-FH-Discs, Mini-Mitter Co., Sun River, OR, U.S.A.) implanted intraperitoneally. Output (frequency in Hz) was monitored by a mounted antenna placed under each animal's cage (RA 1000 radioreceivers, Mini-Mitter Co., Sunriver, OR, U.S.A.) and multiplexed by means of a BCM 100 consolidation matrix to an IBM personal computer system. A Dataquest IV acquisition system (Data Sciences Inc., St. Paul, MN, U.S.A.) was used for auto-matic control of data collection and analysis. In the second series, colonic temperature was measured every 30 min starting 2 h before pyrogen application in unrestrained animals. For this purpose a thin plastic coated thermocouple was repeatedly reinserted 6 cm beyond the anus for 2 min at each time point of temperature measurement. Before the experiments the animals were accustomed to this procedure which did not cause excitement or any apparent changes in their body temperature. *Samples.* Blood plasma samples or hypothalamic push-pull perfusates were collected 1 h before and 1 h, or 3 h after the injection of LPS or solvent, from outside the animal cage, without exciting or restricting the animals. This was very important, because stress suppresses the febrile responses.

Assays. Cytokines: The activities of the two cytokines, tumor necrosis factor (TNF) and interleukin-6 (IL-6) in blood plasma were detected by the following methods. Determination of TNF was based on the cytotoxic effect of TNF on the mouse fibrosarcoma cell line WEHI 164 subclone 13. The percentage of viable cells incubated either with different concentrations of TNF standard or with biological samples was measured by use of the dimethylthiazol-diphenyltetrazolium bromide (MTT) colorimetric assay. Determination of IL-6 was based on the dose-dependent growth stimulation of IL-6 on the B 9 hybridoma cell line. After incubation with different concentrations of IL-6 standard or with biological samples, the MTT assay was used to measure the number of living cells (for details and literature c.f. 6). *Hormones:* Cortisol in plasma was measured by use of a commercially available radioimmuno-assay (Biermann GmbH-Diagnostica, Bad Nauheim, F.R.G.). Arginine vasopressin (AVP) in plasma and push-pull perfusate was measured using a sensitive radioimmunoassay in the laboratory of the Max-Planck-Institut für Physiologische und Klinische Forschung in Bad Nauheim (courtesy of Prof. Dr.E.Simon), c.f. (7) for details.

Evaluation: Values are expressed as means ± standard error. A one-factor analysis of variance (ANOVA) of data was used to test whether there were differences in temperature response to LPS injection between experimental and control immobilized and free moving animals followed by the Scheffe post-hoc test. The differences in fever indices, concentrations of hormones and activities of cytokines between immobilized and free moving animals were compared by Student´s t-tests.

Results

In freely moving animals, the body temperature remained stable in controls injected with saline, whereas pyrogen-injected animals developed a biphasic febrile increase of body temperature characterized in different series by fever indices (temperature increase integrated over 6 h) from 4° to 7° C • h. In immobilized animals injected with pyrogen, the febrile increase in body temperature was completely prevented. On the contrary the immobilized animals developed a hypothermia resulting in negative fever indices varying in different series from -2° to -7° C • h. In the first series depicted in Fig.1, the febrile changes in body temperature are compared in freely moving and immobilized animal groups (n = 6). The blood samples were taken 1 h before and 1 h as well as 3 h after the injection of pyrogen or solvent, and the cortisol levels in blood plasma were determined. The cortisol levels in plasma remained stable (below 190 ng/ml) in controls injected with saline, whereas in pyrogen-injected, freely moving, animals they increased from 204 to 719 and 1 289 ng/ml 1 h and 3 h after the injection. The immobilized animals had 3 times higher cortisol levels in plasma before the injection of pyrogen than freely moving animals. The levels further increased after the pyrogen injection and reached the maximal levels after just 1 h which were attained by freely moving animals after 3 h. Also the levels of AVP were elevated in immobilized animals before the injection of

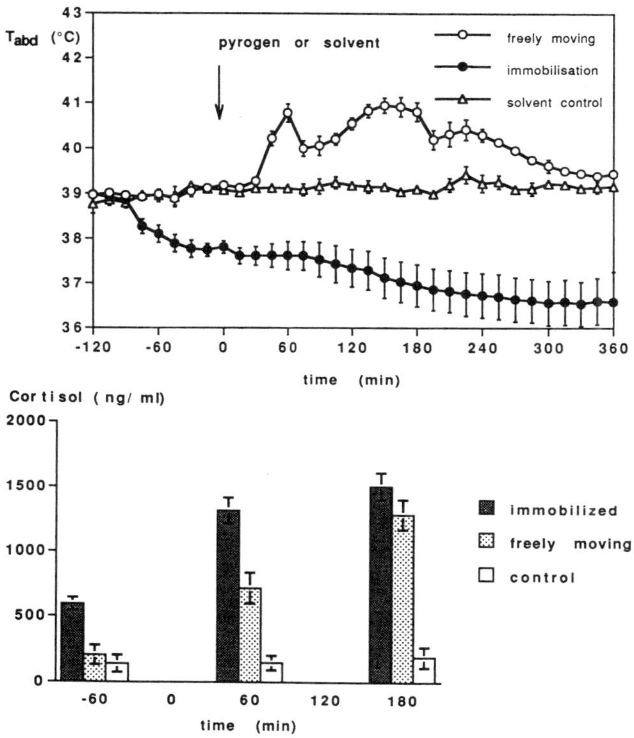

Fig. 1: Above: Comparison of changes in abdominal temperature registered by means of radiotelemetry in groups of freely moving (open symbols) and immobilized (filled symbols) guinea pigs (n = 6), injected at time 0 by pyrogen or solvent. Below: Cortisol levels in blood plasma samples taken 1 h before and 1 h or 3 h after the injection of pyrogen or solvent (n = 6 at each time point under all 3 experimental conditions).

pyrogen, both in plasma and push-pull perfusate from the limbic brain septum, as indicated by data from the second series depicted in Fig. 2. In this series the response to bacterial pyrogen was investigated alternatively in the same group of animals (n = 20) in immobilized or freely moving states. In some of these animals (n= 6) equipped either with permanently implanted arterial catheters or intraseptal guide cannulae for push-pull perfusion, samples of blood plasma or of perfusate from the ventrolateral septum were obtained 1 h before the injection of pyrogen and analyzed for AVP content. From the increased cortisol and AVP levels it can be derived that the immobilization stress per se activated the HPA axis. To investigate whether this activation was sufficient to suppress the immune system, in the third experimental series blood samples were taken 1 h before as well as 3 h after the injection of bacterial pyrogen in a group of guinea pigs (n = 8) either in immobilized or in freely moving states and analyzed for activities of two cytokines, TNF and IL-6. The results of this series are depicted in Fig. 3. From the whole cytokine cascade two cytokines were picked up, having their maximal activities either in the early phase (TNF) or in the later phase (IL-6) of the febrile response. In comparison to pyrogen-injected freely moving animals (white columns), in immobilized animals the release of the both cytokines was depressed to about one third (shaded columns).

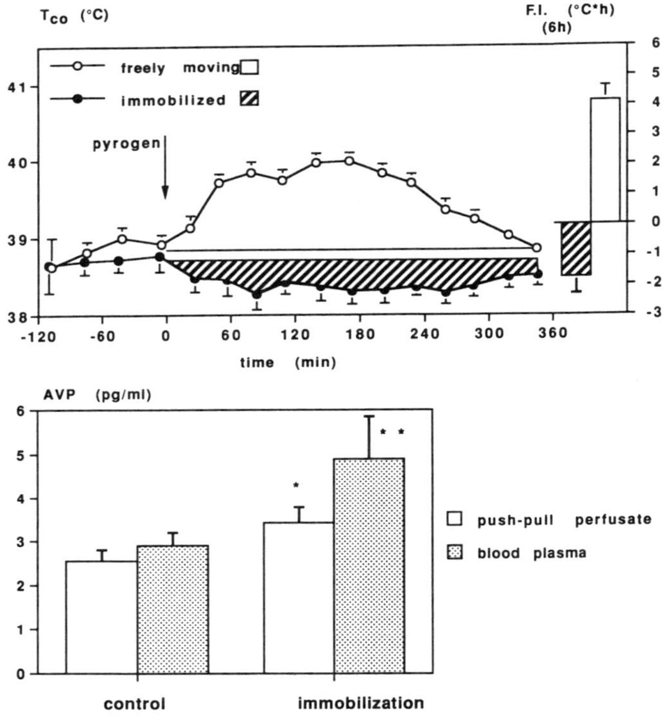

Fig. 2: Above: Changes in colon temperature measured repeatedly in the same group of animals investigated either in immobilized or freely moving state before and after the injection of bacterial pyrogen. F.I. = fever index, an integrated temperature change in 6 h after the injection of pyrogen. Below: AVP-levels in blood plasma (shaded columns) and in push-pull perfusate from the brain septum (white columns) in freely moving (control) and immobilized animals 1 h before the injection of pyrogen.

Discussion

The results of all three series clearly demonstrate the suppression of fever by immobilization and confirm thus the previous reports (5). The febrile rise of body temperature was completely abolished, and in many cases rather a hypothermia developed during the immobilization. This may be caused by the method we used. Immobilization in a stretched position prevents the animals from curling up in the cold and thus from reducing the heat loss by means of behavior. The different extent of hypothermic reactions may be ascribed to a different extent of stretching used by several experimentators and to differences in the animals´experiences with the procedure (some animals being immobilized for the first time, in others the immobilization was repeated). It is known that the expression of behavioral and neurochemical responses evoked by stressors depend on recent

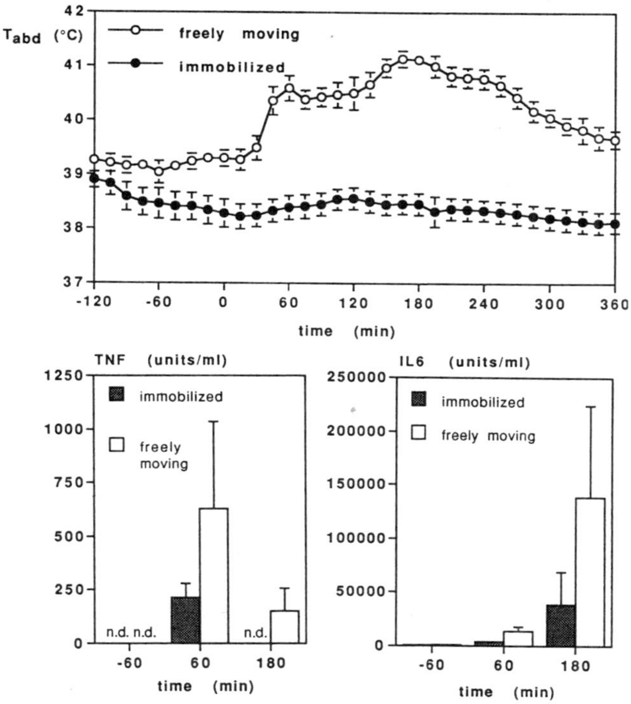

Fig. 3: Above: Changes in abdominal temperature before and after the injection of pyrogen given at time 0. The temperature was registered by means of radiotelemetry in a group of guinea pigs (n = 8) investigated either in immobilized or freely moving states. Below: Cytokine activities in blood plasma samples taken 1 h before, as well as 1 h or 3 h after the injection of pyrogen (n.d. = non-detectable, n = 4 at each time point under both experimental conditions).

individual experience, a phenomenon described as habituation or tolerance (8). Also genetic factors determine the expression of behavioral and neurochemical responses evoked by stressors. There are striking differences between the responses of different species or strains of animals to the same stressor. Whereas the guinea pig keeps quiet during the immobilization and reduces the motor activity, the rat visibly increases the motor activity, becomes aggressive and develops a hyperthermia (9, 10). In spite of the absence of visible signs of excitement, guinea pigs respond to immobilization by an activation of the HPA axis. This can be evidenced by 3-fold increased levels of the stress hormone cortisol in blood plasma in comparison to freely moving animals. Cortisol and melanocyte-stimulating hormones released by the activation of the HPA axis down-regulate the immune response to pyrogen as evidenced by the depressed release of cytokines. Due to a negative feedback action of cortisol the hypothalamic neurons produce less corticotropin releasing hormone (CRH), but more AVP, as evidenced by immunocytochemical studies (2), and documented by increased AVP levels in plasma and push-pull perfusate from the brain. The immune stress evoked by injected pyrogen stimulates the HPA axis already during the febrile rise in body temperature. The maximal levels of released cortisol are reached 3 h after the injection of pyrogen in freely moving animals at the time when the heat dissipation initiates the fever decline. Immobilized animals respond to pyrogen injection in spite of down-regulated immune signals by maximal activation of the HPA axis reached already 1 h thereafter, probably due to sensitization to the immune stimuli by the non-immune stressors. Due to a negative feedback action of cortisol the hypothalamic neurons start earlier to produce AVP instead of CRH . Also central antipyretic systems are activated sooner. Thus in immobilized animals the depressed immune signals are not able to induce the febrile rise of body temperature because they are weak and are counteracted by activated antipyretic systems from the very beginning.

Acknowledgements

This study was supported by the DFG, project Ze 183 4/1. Excellent technical assistance of Birgit Störr is gratefully acknowledged.

References

1. Kluger, MJ (1991): Fever: Role of pyrogens and cryogens. Physiol Rev 71, 93-127.
2. Zeisberger, E, Roth, J, Kluger, MJ (1994): Interactions between the immune system and the hypothalamic neuroendocrine system during fever and endogenous antipyresis. In: Integrative and Cellular Aspects of Autonomic Functions. Pleschka, K, Gerstberger, E eds. London, Paris: John Libbey Eurotext, 181-190.
3. Harbuz, MS, Lightman, SL (1992): Review: Stress and the hypothalamo-pituitary-adrenal axis: Acute, chronic and immunological activation. J Neuroendocrin 134, 327-339.
4. Lilly, MP, Gann, DS (1992): The hypothalamic-pituitary-adrenal-immune axis: A critical assessment. Arch Surg 127, 1463-1474.
5. Zeisberger, E (1990): The role of septal peptides in thermoregulation and fever. In: Thermoreception and Temperature Regulation. Bligh, J, Voigt, K eds. Berlin, Heidelberg: Springer, 273-283.
6. Roth, J, Conn, CA, Kluger, MJ, Zeisberger, E (1993): Kinetics of systemic and hypothalamic IL-6 and tumor necrosis factor during endotoxin fever in guinea pigs. Am J Physiol 265, R653-R658.
7. Roth, J, Schulze, K, Simon, E, Zeisberger, E (1992): Alteration of endotoxin fever and release of arginine-vasopressin by dehydration in the guinea pig. Neuroendocrinology 56, 680-686.
8. Zeisberger, E, Roth, J (1994): Central regulation of adaptive responses to heat and cold. In: APS Handbook of Physiology, Section 4, "Adaptation to the Environment". Fregly, MJ, Blatteis, CM eds. Bethesda: Oxford University Press, in press.
9. Long, NC, Morimoto, A, Nakamori, T, Yamashiro, O, Murakami, N (1991): Intraperitoneal injections of prostaglandin E_2 attenuate hyperthermia induced by restraint or interleukin-1 in rats. J Physiol (London) 444, 363-373.
10. Long, NC, Morimoto, A, Nakamori, T, Murakami, N (1991): The effect of physical restraint on IL-1β- and LPS-induced fever. Physiol Behav 50, 625-628.

Thermal Balance in Health and Disease
Advances in Pharmacological Sciences

MODIFICATIONS OF FEBRILE TEMPERATURE REGULATION BY ACCLIMATION

U. Beckmann and J. Werner

Institut für Physiologie, Ruhr-Universität Bochum, MA 4/59, D-44780 Bochum, Germany

Fever is known to be rather an adjustment than a disturbance of the thermoregulatory system. The febrile body temperature elevation is caused by a concurrent shift of the control characteristics of all thermoregulatory effector mechanisms. Depending on ambient conditions, heat production and heat conservation mechanisms are activated and heat loss mechanisms are reduced (1). This process is usually explained by the concept of an adjustment of the control system to a shifted set point (2,3). Acclimation, too, is a process which influences the regulatory mechanisms of the temperature control system. In general, thermo-adaptive modifications enable the organism to cope with a given environmental stress more efficiently. These modifications may consist, on the one hand, of morphological changes, which primarily modify the controlled system, such as changes in thermal insulation of the body shell. On the other hand, functional adaptive effects can alter effector-capacities or bring about adjustments of the control characteristics like shifts of threshold body temperatures for the activation of thermoregulatory effectors (4,5). Adaptive alterations of control mechanisms have been found on the level of thermoreception (6) and of afferent pathways (7,8). So the question arises which influences thermo-adaptive changes may have on the action of regulatory mechanisms during fever and on the extent of febrile temperature elevation. This question was investigated by studies on pyrogen-induced fever in acclimated rabbits.

Methods

Experiments were carried out with several groups of New Zealand White rabbits which were acclimated to different ambient temperatures (T_a) :

(NA)	thermoneutral adaptation	(8 weeks at T_a=22°C, n=13)
(CA)	cold adaptation	(8 weeks at T_a= 2°C, n=20)
(WA)	warm adaptation	(8 weeks at T_a=30°C, n=20).

The progress of the adaptation process was studied once a week under resting conditions at an ambient temperature corresponding to that of the adaptation procedure. The fever course after injection of exogenous pyrogen (LPS, 0.5 µg/kg i.v.) was investigated both before and after adaptation in a climatic chamber at ambient temperatures of 2°C, 22°C and 30°C. Half of the cold and warm adapted groups were used for control experiments with an injection of isotonic saline (n=10, each).

The following regulatory variables were continously recorded: rectal temperature (T_{rect}), mean skin temperature (T_S), and ear skin temperature ($\bar{T}_{S\ ear}$), metabolic heat production (M), respiratory evaporative heat loss (REHL), heart frequency (HR), respiratory frequency (F_{resp}), tidal volume (V_T), EMG, and ear blood flow (BFl).

Results

Acclimation Effects

During the 8 week acclimation course cold adaptation evoked a considerable improvement of fur insulation as could be seen from an increase of hair mass within a standard area by about 30 % and a significant decrease of fur conductance. Mean skin temperature rose by about 1.5°C, whereas rectal temperature remained unchanged. Metabolic heat production at T_a=2°C was decreased by 10 %, i.e. the same core temperature was obtained at a lower energy cost. As could be concluded from the afebrile control experiments at different ambient temperatures, this decrease of metabolic rate could primarily be attributed to a downward shift of the characteristic of the passive system, due to the improvement of body shell insulation as a morphological effect of cold adaptation. Under afebrile conditions, there were no distinct functional changes of the effector characteristic of heat production. Heat loss by means of respiratory evaporation and ear blood flow was reduced after cold adaptation at all ambient temperatures. The decrease of REHL was due to a shift of the effector characteristic of respiratory frequency to a higher body temperature, i.e. an increase of panting threshold.

During the course of warm acclimation, there were no changes in fur density , mean skin temperature, or thermal conductance of the fur, which would indicate a change of body shell insulation. Warm adaptation resulted in a significant decrease of rectal temperature and mean body temperature of about 0.4°C. At T_a=30°C, a slight decrease of ear blood flow and ear convective heat loss could be observed, and respiratory frequency was reduced with unchanged tidal volume, resulting in a significant decrease of REHL. The results of control experiments at different T_a after warm adaptation showed a downward shift of the thresholds for elevations of both ear blood flow and respiratory evaporative heat loss. Nevertheless, the maximum values of heat defence mechanisms in the warm environment were reduced by warm adaptation, because the decrease of body temperature was even greater than the deviations of effector characteristics. So, an enhanced efficiency of total heat loss had to be concluded. Metabolic heat production, which showed no differences at T_a=30°C after warm adaptation, was significantly reduced in the afebrile control at thermoneutral and low ambient temperatures. This decrease of metabolic rate was due to a distinct shift of shivering threshold to a lower body temperature.

Influences on the Fever Process

Intravenous injection of LPS caused a biphasic rise of rectal temperature, which lasted for approximately 6 hours (see Fig.1). At T_a=2°C, in the cold acclimated rabbits the local fever

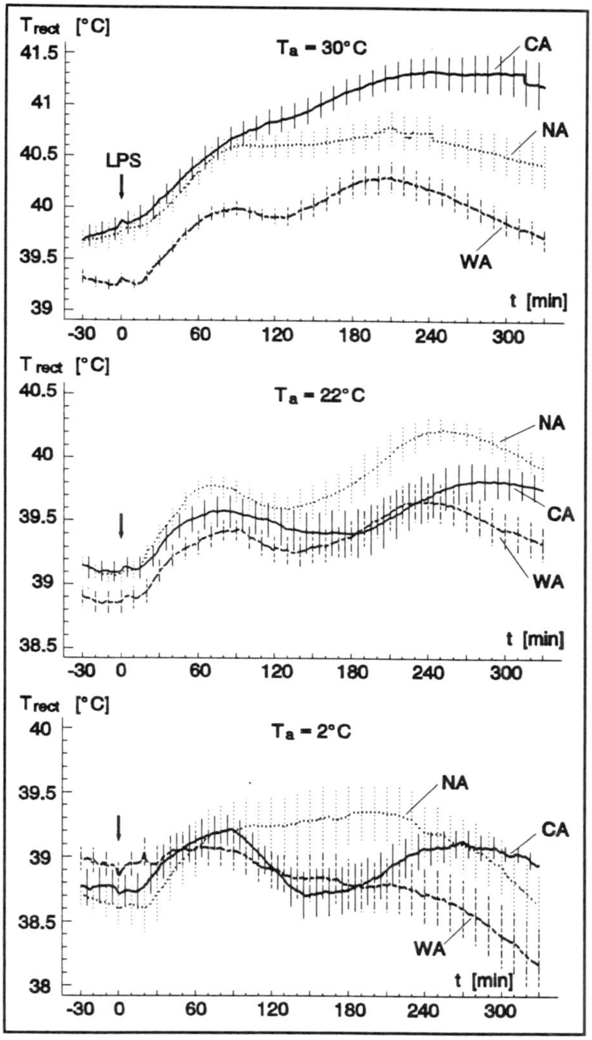

Fig. 1: Time course of rectal temperature in LPS-fever; mean ± S.E. of non-acclimated (NA), cold-(CA), and warm acclimated (WA) rabbits

minimum was essentially lowered, and the second fever peak was significantly lower and delayed as compared to the non-acclimated animals. In thermoneutral environment, both maxima of the biphasic core temperature course were decreased after cold adaptation, and the local minimum was prolonged. So, unlike in afebrile conditions, cold adaptation caused a decrease of rectal temperature during fever. At high ambient temperature, however, cold adaptation brought about a large overshoot in the rise of T_{rect} and an elevated regulatory load-error, leading to severe hyperthermia in some animals. The total extent of temperature elevation in fever, as expressed by the integrative fever index for a time period of 300 minutes (FI_{300}), was significantly decreased after cold adaptation in a thermoneutral and cold environment, but increased at T_a=30°C. The morphological modifications promoted body heat accumulation, an effect which was obviously maladaptive in the warm environment.In the warm acclimated rabbits at T_a=30°C, as in the afebrile control, rectal temperature was lowered by about 0.5°C also during fever. After both temperature maxima of the fever course, there was an accelerated temperature decline, and the regula-tory load-error was minimized, indicating the improved efficiency of heat loss mechanisms. In the thermo-neutral environment, too, core temperature in fever was significantly decreased after warm adaptation, and the fever duration was reduced. In the cold, rectal temperatures of the warm acclimated rabbits fell far below their starting values, leading to hypothermia in some animals, even during fever. After warm adaptation, the rabbits were no longer able to keep their core temperatures at a constant level in the cold environment or to carry out a febrile temperature rise. For all ambient conditions applied, the fever index FI_{300} after warm acclimation was significantly lower than before. In the cold it even had negative values.

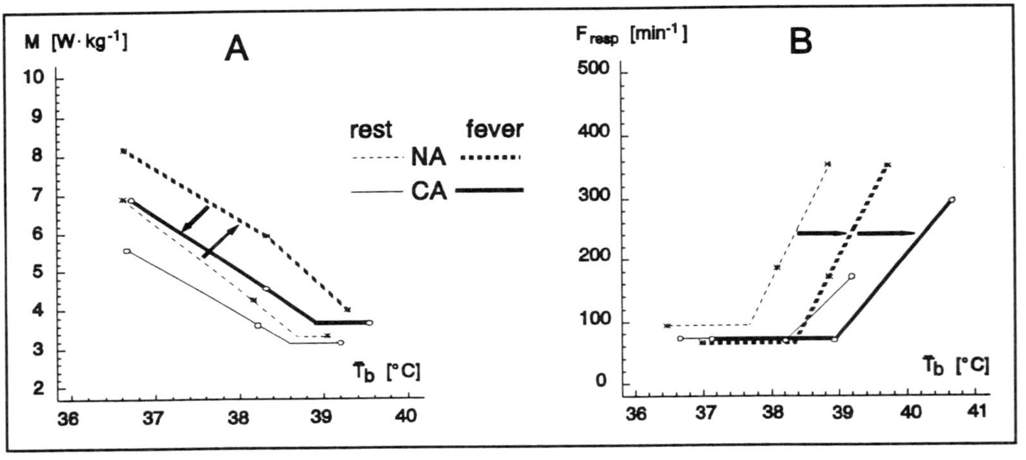

Fig. 2: Effector characteristics of metabolic rate (A) and respiratory frequency (B) of cold-(CA) and non-acclimated (NA) rabbits at rest and in fever; mean values at different T_a

The diminutions of the febrile temperature rises observed in the cold and thermoneutral environment after cold acclimation could be attributed mainly to changes in metabolic heat production, which is the dominant febrile effector at lower ambient temperatures. In the cold, there was a strong decrease of metabolic rate after the first fever maximum, even below the level of afebrile values before injection of pyrogens, while at T_a=22°C the local minimum of M was prolonged. At both ambient temperatures, the rise of heat production to the second maximum was delayed in time and initiated at a lower body temperature. These effects are in accordance with the observed behaviour of rectal temperature after cold adaptation. As can be seen in Fig.2A, these modifications in metabolic rate can be described by a shift of the effector characteristic to a lower mean body temperature ($\bar{T}_b$) as a functional effect of cold adaptation, which could be found only in febrile conditions. The diagram shows the M/$\bar{T}_b$-relationship under resting conditions before injection of pyrogens (thin lines) and at the time of the first fever maximum of metabolic rate (thick lines) in the experiments at different ambient temperatures. The pyrogen-induced upward shift of the effector characteristic in non-acclimated animals (change from the thin broken to the thick broken line) was nearly compensated by an opposite shift of the effector characteristic during fever caused by cold acclimation (thick unbroken line).

The observed elevated body temperature in fever at high ambient temperature after cold adaptation was mainly due to the increment of body shell insulation. But also the reductions of heat loss mechanisms developed during cold acclimation supported the strong rise in temperature. Fig.2B shows the effector characteristics of respiratory frequency before injection of pyrogens (thin lines) and in the final decline phase of fever (thick lines). The elevation of panting threshold caused by fever (change from the thin to the thick broken line) was enhanced by the additional threshold shift caused by cold adaptation (change to the thick unbroken line). So, during cold acclimation, heat loss mechanisms adapted in a way which, in climatic conditions inverse to the acclimation conditions, was detrimental to the efficiency of control. The start of panting in the decline phase of fever at T_a=30°C was shifted to a T_{rect} elevated by 0.5°C, but was not changed in time, showing that the time course of effector control in the fever process was unaffected by adaptation. This also indicates that the regulatory mechanisms during fever did not in this case react on the adaptive modifications which had caused an enhanced regulatory load-error.

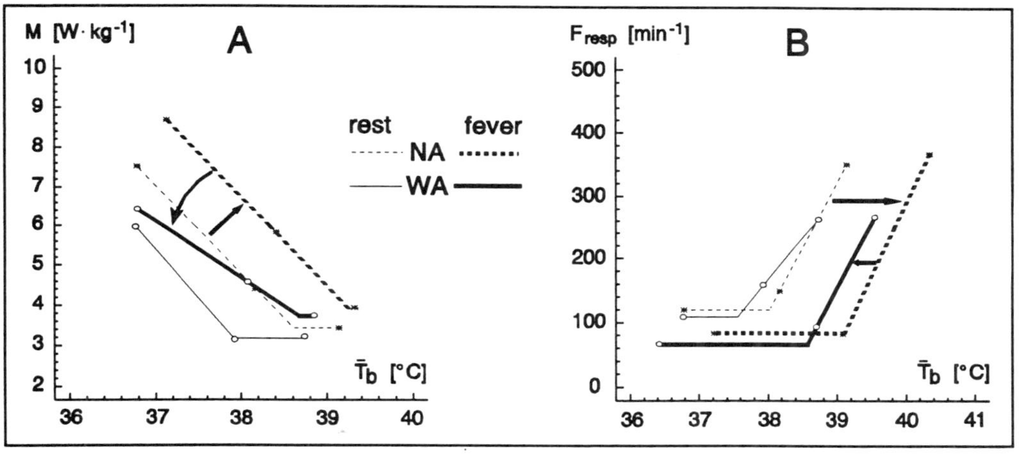

Fig. 3: Effector characteristics of metabolic rate (A) and respiratory frequency (B) of warm-(WA) and non-acclimated (NA) rabbits at rest and in fever; mean values at different T_a

Warm acclimation caused a decrease of core temperature in afebrile and in febrile conditions by a similar amount. This could be attributed to a concurrent downward shift of the effector characteristics of all autonomic effector mechanisms. The acceleration of temperature declines and the reduction of fever duration at T_a=22°C were mainly due to the decrease of the threshold for an elevation of metabolic heat production (see Fig. 3A). This shift of shivering threshold also caused a significant delay in the start of shivering, as measured by the EMG, during the onset-phase of fever. In the cold environment, the decrease of metabolic rate in fever after warm adaptation was even stronger. This effect is described by a reduced gain of the effector characteristic. As in the afebrile state before LPS-injection, also in the decline phase of fever (see thick lines in Fig. 3B) the threshold for an elevated respiratory frequency was decreased. But, because of the much lower $\bar{T}_b$ in the warm acclimated animals, F_{resp} was lower during panting at high ambient temperature, causing a reduction of respiratory water loss of about 50 %. The same typical modifications apply to the effector characteristic of ear blood flow after warm adaptation, with the exception that the basal level of ear blood flow during vasoconstriction in the cold environment was elevated by warm adaptation.

Conclusions

As far as cold acclimation is concerned, it can be concluded that in thermoneutral and cold environments, where metabolic heat production is the major fever effector, the morphological adaptive changes do not lead to an enhancement of febrile temperature elevation. The influence of functional controller adjustments of cold adaptation, concerning the effector control of metabolic heat production, is predominant and causes not only a reduction of the metabolic costs of fever, but also an attenuation of the fever-induced elevation of body temperature. Whereas in the heat, cold acclimation favours the development of a fatally high increase of temperature, so for the fever process in a warm environment, cold adaptation turns out to be maladaptive.Warm adaptation brings about a decrease of body temperature in afebrile as well as in febrile conditions and is suited to lower the thermoregulatory strain of fever. This effect is due to a concurrent shift of the effector characteristics for metabolic heat production, for panting, and for peripheral vasodilatation to a lower body temperature and an obviously enhanced efficiency of heat loss mechanisms, which reduces the regulatory load-error during the dynamic fever process at high T_a. In the cold, however, warm acclimation may encounter hypothermic problems, even during fever. Thermoregulatory mechanisms during fever make use of effector control in the way it has been adjusted by acclimation. The results show only little evidence that control mechanisms in fever were reacting on regulatory load-errors caused by the thermo-adaptive changes. This is most distinct in environmental conditions inverse to the conditions of acclimation. So, the fever process seems to superimpose on the acclimation process more or less independently.

Supported by Deutsche Forschungsgemeinschaft (We 919/1)

References

1. Graener, R, Werner, J (1986): Dynamics of endotoxin fever in the rabbit. J Appl Physiol 60, 1504-1510

2. Stitt, J (1993): Central regulation of body temperature. In: Exercise, heat, and thermoregulation. Gisolfi, CV, Lamb, DR, Nadel, ER eds. Dubuque: Brown & Benchmark, 1-39

3. Werner, J (1988): Functional mechanisms of temperature regulation, adaptation, and fever: Complementary system theoretical and experimental evidence. Pharmac Ther 37, 1-23

4. Brück, K, Zeisberger, E (1987): Adaptive changes in thermoregulation and their neuropharmacological basis. Pharmac Ther 35, 163-215

5. Werner, J (1990): Models of cold and warm adaptation. In: Thermoreception and temperature regulation. Bligh, J, Voigt, K eds. Heidelberg: Springer-Verlag, 224-228

6. Hensel, H, Schäfer, K (1982): Static and dynamic activity of cold receptors in cats after long-term exposure to various temperatures. Pflügers Arch 392, 291-294

7. Brück, K, Hinckel, P (1982): Thermoafferent systems and their adaptive modifications. Pharmac Ther 17, 357-381

8. Werner, J, Schingnitz, G, Hensel,H (1981): Influence of cold adaptation on the activity of thermoresponsive neurons in thalamus and midbrain of the rat. Pflügers Arch 391, 327-330

Thermal Balance in Health and Disease
Advances in Pharmacological Sciences

V_2-RECEPTOR-MEDIATED HYPERPYRETIC EFFECT OF ARGININE VASOPRESSIN IN THE VENTRAL SEPTAL AREA OF THE RABBIT BRAIN

A. A. Romanovsky, L. Janský and V. N. Gourine

Institute of Physiology, Belarusian Academy of Sciences, Minsk 200725, Belarus

The antipyretic action of arginine vasopressin (AVP) has been documented in various species [for ref. see (1)] and is thought to be mediated via V_1-like receptors in the ventral septal area (VSA) of the brain (2). In rabbits, perfusion of the VSA with AVP was reported to cause an obvious antipyresis (3, 4). In the same species, however, the microinjection of a comparable amount of AVP into the dorsal septum enhanced fever (5). It is not clear whether this inconsistency reflects the structure-specificity or the technique-dependence of the thermal effect of the peptide. To clarify this, we investigated the thermal effect of AVP administered into the rabbit VSA, *i.e.*, the same structure as in the studies of Naylor *et al.* (3) and Malkinson *et al.* (4), but by the method of microinjection, *i.e.*, the same technique as in the work of Bernardini *et al.* (5). Preliminary results of this study have been reported (6).

Methods

Seventy-three experiments were performed in female New Zealand albino rabbits weighing 2.5-3.1 kg. The animals were anesthetized by intravenous (i.v.) injection of sodium pentobarbital (40 mg/kg), and a stainless steel cannula (o.d. 1.0 mm, i.d. 0.5 mm) was implanted into the VSA (3.0-4.5 mm anterior to the bregma, 0.7-1.0 mm lateral to the sagittal suture, 12.0-12.5 mm deep from the skull surface); the coordinates were based on the results of test implantations and the atlas by Tsvetkova (7). The animals were allowed a week's recovery before experimentation. During the experiments, rectal temperature (T_r) was measured for 1 h before and 5 h after i.v. and/or intraseptal (i.s.) injections of the following drugs: lipopolysaccharide (LPS) *S. typhi*, prostaglandin (PG) E_2, AVP, V_1-antagonist $d(CH_2)_5Tyr(Me)AVP$ (MAVP), V_2-agonist 1-desamino-8-D-AVP (DDAVP), and pyrogen-free saline (PFS). The decision to induce PG-hyperthermia by administering the drug i.s. was based on data (8) showing a high sensitivity of the VSA to PGE_1. The drugs were delivered i.s. in a volume of 1-5 μl (in the case of MAVP, 10 μl) over a 2-min period. The experiments were performed at 18-22°C. After the animals were sacrificed and their brains removed and fixed, the localization of cannula tips was verified histologically (Fig. 1).

Results

The basal value of T_r was 39.0 ± 0.1°C (average for all experiments). The i.s. administration of AVP (5 μg) resulted in an increase and prolongation of the T_r response to LPS (200 ng/kg, i.v.) and to PGE_2 (500 ng, i.s.) (Figs. 2A and 2B, respectively). AVP *per se* caused a slight, gradual rise in T_r, indistinguishable, however, from that induced by the i.s. injection of PFS (5 μl) (Fig. 2C). The V_2-agonist DDAVP (5 μg, i.s.), which has a smaller molecular weight than the parent peptide, was more effective than AVP (5 μg, i.s.) in enhancing the T_r response to the threshold dose of PGE_2 (250 ng, i.s.) (Fig. 3A). The V_1-antagonist MAVP, when administered i.s. at a dose of 10 μg, *i.e.*, 20 times higher than the effective dose in the rat VSA (2), did not attenuate either PGE_2 (μg, i.s.)-hyperthermia or LPS (1 μg/kg, i.v.)-fever (Figs. 3B and 3C, respectively). MAVP (10 μg, i.s.) also did not prevent the enhancing effect of AVP (5 μg, i.s.) on the T_r response to PGE_2 (500 ng, i.s.) (Fig. 3D), but rather, *vice versa*, significantly ($p < 0.05$) increased this effect during the defervescence phase of PGE_2-hyperthermia (not shown in the figure). Thermal responses to the i.s. injection of DDAVP (5 μg) or MAVP (10 μg) alone (not illustrated) did not differ from those to the i.s. administration of AVP or PFS.

Discussion

Our data show that, when injected into the rabbit VSA, AVP enhanced and prolonged both LPS fever and (to a lesser extent) PGE_2-hyperthermia. To the best of our knowledge, this is the first report of the hyperpyretic action of AVP in this area of the brain, which is generally considered to

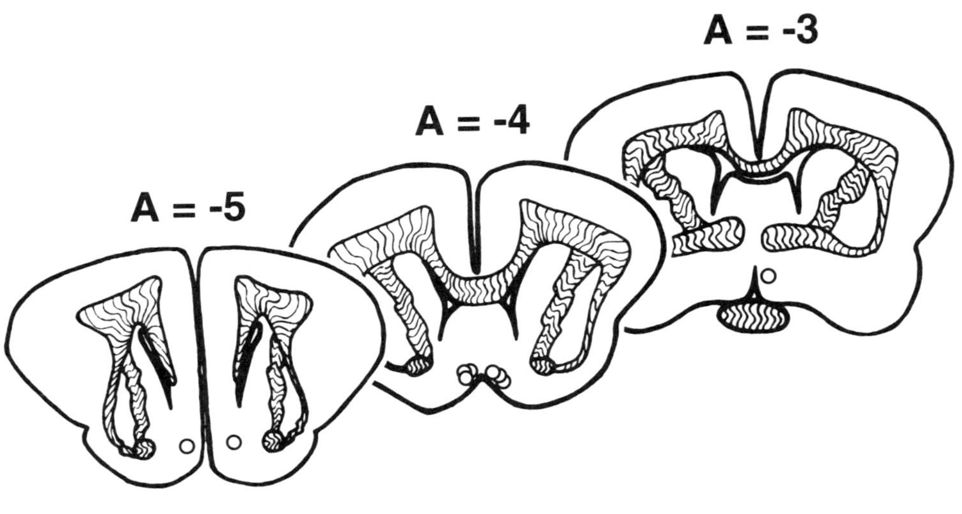

Fig. 1. Microinjection loci (nine in total) are demonstrated in the rostal, central, and caudal portions of VSA. The coordinate A (mm), according to the atlas of Tsvetkova (7), shows the distance from bregma in the rostral (-) direction.

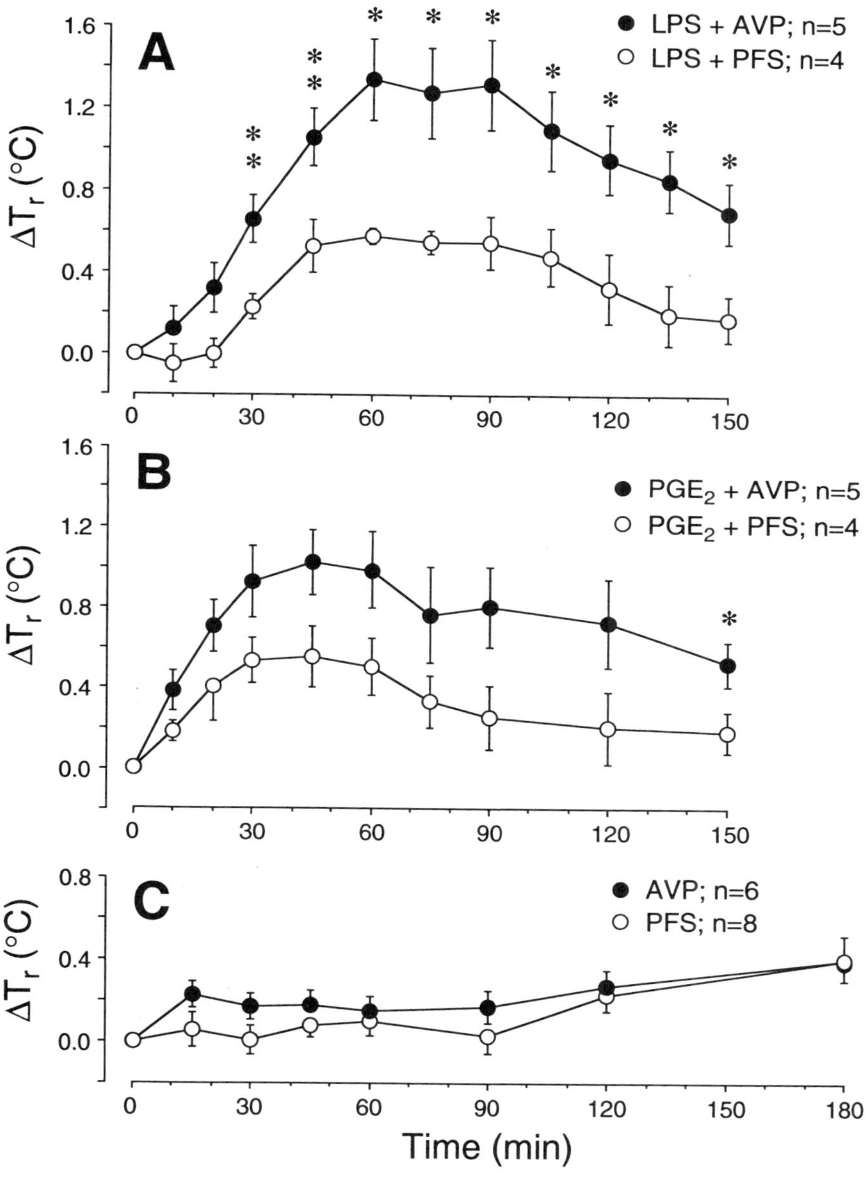

Fig. 2. Changes in rectal temperature (ΔT_r) induced by the i.s. injection of AVP (5 μg) or PFS: **A** — 5 min after the i.v. injection of LPS (200 ng/kg) or PFS; **B** — 5 min after the i.s. injection of PGE_2 (500 ng) or PFS; and **C** — under afebrile conditions (no pretreatment). $^*p < 0.05$; $^{**}p < 0.01$.

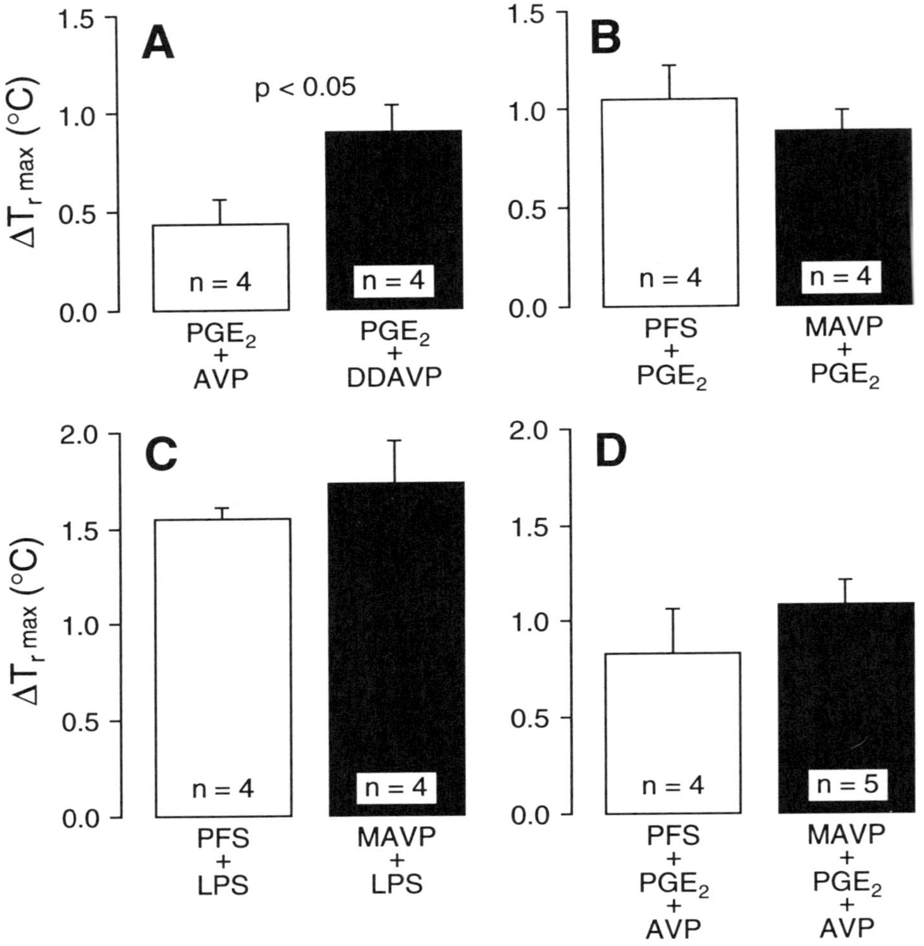

Fig. 3. Maximal changes in rectal temperature ($\Delta T_r max$) induced by: **A** — the i.s. injection of DDAVP (5 μg) or AVP (5 μg) 5 min after the i.s. injection of PGE_2 (250 ng); **B** — the i.s. injection of MAVP (10 μg) or PFS 15 min before the i.s. injection of PGE_2 (1 μg); **C** — the i.s. injection of MAVP (10 μg) or PFS 15 min before the i.v. injection of LPS (1 μg/kg); and **D** — the i.s. injection of MAVP (10 μg) or PFS at 15 min before the two i.s. injections: PGE_2 (500 ng) and, 5 min later, AVP (5 μg).

be responsible for the initiation of antipyresis. Since DDAVP, a selective V_2-agonist, exhibited higher hyperpyretic potency than the parent peptide and since MAVP, a selective V_1-antagonist, did not attenuate the hyperpyretic effect of AVP, our data would suggest that the fever-enhancing effect of the peptide was mediated by receptors similar to the peripheral V_2-type. The presence of

V_2-receptors in the hypothalamus has been shown (9), and their role in mediating the thermal effect of AVP in fever (although antipyretic rather than hyperpyretic) has been proposed (10).

There is some evidence of the site-specificity of the central thermoregulatory action of AVP. This peptide has been recently shown to exert different effects on thermosensitive neurons in slice preparations of the caudal *vs.* rostral part of the preoptic/anterior hypothalamus (PO/AH) (11). In the present study, we have noticed that, under afebrile conditions, the i.s. administration of AVP into the caudal portion of the VSA (on the border of the PO/AH) usually resulted in higher T_r rises than its administration into the rostral VSA. Nevertheless, contradictory effects of the i.s. AVP on fever [antipyresis (3, 4) *vs.* hyperpyresis (5)] are likely to be due to the difference in the method of drug administration, rather than site-specificity. Indeed, we were not able to find, under the conditions of both LPS-fever and PGE_2-hyperthermia, any consistent difference between the hyperpyretic effects of AVP injected into the central portion of the VSA, its caudal part, and its very rostral part. Moreover, AVP microinjected into the dorsal septum has been reported to produce hyperpyresis as well (5). On the other hand, when the neuropeptide was administered into the VSA by another technique - perfusion - it caused obvious antipyresis (3, 4).

Since low and high rates of delivery of the drug by perfusion *vs.* injection are analogous to low and high doses, this discrepancy may reflect the dose-dependence of the effect. The common ability of many biologically active substances to exert physiologically opposite effects in low and high doses by acting through different receptors often leads to confusion in interpretation of the results (12). On the other hand, the microinjection-induced brain tissue injury and the consequent, nonspecific release of PGs (13) may contribute to the difference between the thermal responses to perfusion and injection. Although neither AVP itself nor any of its analogues modified the thermal response to the i.s. injection *per se* (responses to AVP, DDAVP, MAVP, and PFS were practically the same), it cannot be *a priori* excluded that, under the conditions of fever or PGE_2-hyperthermia, when intrabrain concentrations of both PGs and AVP are changed, the injection-associated local release of PGs, which develops in any case, may influence AVP-PGE interactions in a different way and cause different resultant changes in body temperature.

In conclusion, the present data show that administration of AVP in the rabbit VSA by injection, in contrast to perfusion, results in a paradoxical, fever-enhancing effect mediated through V_2-like receptors. We consider this effect to be technique-dependent and either pharmacological (caused by a high rate of drug delivery), or artifactitious (due to contamination of the results by injection-induced nonspecific brain tissue injury), or both.

Acknowledgments

We thank Dr. I. P. Tsvetkova for histological and Mr. L. S. Pavlov for technical assistance. The facilities for preparation of the manuscript were provided by the Department of Physiology and Biophysics, University of Tennessee at Memphis, through Dr. C. M. Blatteis. The authors also appreciate the financial help of the Thermoregulation Research Laboratory, University of California at San Francisco, through Dr. D. I. Sessler. Dr. L. Janský was on leave from the Faculty of Science, Charles University, Prague, Czech Republic. Present address of Dr. A. A. Romanovsky: Department of Physiology and Biophysics, University of Tennessee at Memphis, Memphis, Tennessee, U.S.A.

References

1. Kasting, NW (1989): Criteria for establishing a physiological role for brain peptides. A case in point: the role of vasopressin in thermoregulation during fever and antipyresis. Brain Res Rev 14, 143-153.

2. Cooper, KE, Naylor, AM, Veale, WL (1987): Evidence supporting a role of endogenous vasopressin in fever suppression in the rat. J Physiol (Lond) 387, 163-172.

3. Naylor, AM, Ruwe, WD, Kohut, AF, Veale, WL (1985): Perfusion of vasopressin within the ventral septum of the rabbit suppresses endotoxin fever. Brain Res Bull 15, 209-213.

4. Malkinson, TJ, Bridges, TE, Lederis, K, Veale, WL (1987): Perfusion of the septum of the rabbit with vasopressin antiserum enhances endotoxin fever. Peptides 8, 385-389.

5. Bernardini, GL, Lipton, JM, Clark, WG (1983): Intracereboventricular and septal injections of arginine vasopressin are not antipyretic in the rabbit. Peptides 4, 195-198.

6. Janský, L, Vybíral, S, Romanovsky, AA, Gourine, VN (1990): The mode of neuropeptide action on thermoregulation. In: Neuropeptides and Thermoregulation. Gourine, VN ed. Minsk: Navuka i Tekhnika, 9-31 (in Russian).

7. Tsvetkova, IP (1978): The Rabbit Hypothalamus. Stereotaxic and Cytoarchitectomic Atlas. Leningrad: Nauka (in Russian).

8. Williams, JW, Rudy, TA, Yaksh, TL, Vinwanathen, CT (1977): The extensive exploration of the rat brain for sites mediating prostaglandin-induced hyperthermia. Brain Res 120, 251-262.

9. Cheng, SWT, North, WG (1989): Vasopressin reduces release from vasopressin-neurons and oxytocin-neurons by acting on V_2-like receptors. Brain Res 479:35-39.

10. Zeisberger, E, Merker, G (1992): The lamina terminalis contributes to the pathology of fever. In: Thermoregulation: The Pathophysiological Basis of Clinical Disorders. Lomax, P, Schönbaum, E eds. Basel: Karger, 167-171.

11. Moravec, J, Pierau, Fr-K (1993): Arginine vasopressin excites neurons in slices of anterior hypothalamus of rats and affects temperature sensitivity of thermosensitive neurons. In: International Symposium "Intergrative and Cellular Aspects of Autonomic Functions: Temperature- and Osmoregulation," Bad Nauheim (Germany), July 29-30, 1993; Programme and Abstracts. Bad Nauheim: Max-Planck-Institute, 46.

12. Selye, H (1964): From Dream to Discovery; on Being a Scientist. New York: McGraw Hill.

13. Romanovsky, AA, Shido, O, Ungar, AL, Blatteis, CM (1993): Genesis of biphasic thermal response to intrapreoptically microinjected clonidine. Brain Res Bull 31, 509-513.

Thermal Balance in Health and Disease
Advances in Pharmacological Sciences

CONTRIBUTION OF THE HYPOTHALAMIC SUPRAOPTIC NUCLEUS (SON) TO ENDOGENOUS ANTIPYRESIS IN GUINEA PIGS

M. Bock, J. Roth and E. Zeisberger

Physiologisches Institut, Klinikum der JLU Giessen, Aulweg 129, 35392 Giessen, Germany

The pro-opiomelanocortin derived peptide γ-melanocyte stimulating hormone (γ-MSH) has been discovered and cloned in 1979 (1). Since that time several studies have been performed providing evidence for a role of γ-MSH in the regulation of cardiovascular responses. Thus, it has been shown that intravenous administration of γ-MSH resulted in an activation of the sympathetic nervous system and an increase in blood pressure and heart rate (2). The observed cardiovascular responses of systemic γ-MSH are mediated by an activation of descending vasopressinergic pathways originating in the hypothalamic supraoptic nucleus (SON) (3, 4). The activation of the SON by γ-MSH seems to be mediated by the anteroventral region of the third ventricle (5) including some of the circumventricular organs like the organum vasculosum laminae terminalis (OVLT), where signal transduction of γ-MSH into the brain is possible. In a previous investigation in guinea pigs we observed an attenuation of endotoxin-induced fever by systemic infusions of γ-MSH (6). This suppression of fever by systemically administered γ-MSH was not accompanied by a reduction of circulating endogenous pyrogens (interleukin-6, tumor necrosis factor α), so that rather a central nervous mechanism caused the observed antipyretic effects. In the present study we tried to find out if the excitatory pathway to the hypothalamic SON, which is activated by γ-MSH (2, 3, 4, 5), contributes to the observed antipyretic effects of this peptide after systemic application. Therefore we microinfused γ-MSH into the OVLT or we simulated the activation of the hypothalamic SON by direct electrical stimulation and analyzed the febrile response.

Methods

This study was performed in male guinea pigs with a mean body weight of 352 ± 11 g at the beginning of the experiments. Fever was induced by intramuscular injections of bacterial lipopolysaccharide (LPS from E. coli, 0111:B4, Sigma Co., St. Louis, U.S.A.) into the thigh muscle with a dose of 20 μg/kg. Colonic temperature was measured every 30 min starting at least 90 min before pyrogen application in unrestrained animals. For this purpose a thin plastic coated thermocouple was repeatedly reinserted 6 cm beyond the anus for 2 min at each time point of temperature measurement. Before the experiments the animals were accustomed to this procedure which did not cause excitement or any apparent changes in their body temperature. In animals, anesthetized with 100 mg/kg ketamine hydrochloride and 4 mg/kg xylazin, a bipolar electrode was chronically introduced into the SON or a guide cannula was implanted above the OVLT, through which a microinfusion cannula could be introduced. The electrode tip was located 10 mm in front of the interaural plane, 1mm lateral to midline and 11 mm below the brain surface. The tip of the microinfusion cannula was located 12 mm in front of the interaural plane at midline and 9.5 mm below the brain surface. The implantation of electrodes and cannulae was performed according to a stereotaxic atlas of the guinea pigs' brain (7). Prior to the start of stimulation experiments the implanted electrodes of the non-anesthetized, freely moving animals were connected by cables to a pulse generator (HSE stimulator II, Sachs Electronic, March-Hugstetten, Germany) and stimulated with 10-s trains of current pulses (duration 1 ms, frequency 20 Hz, amplitude 8 V, current 0.205 ± 0.017 mA) applied in intervals of 40 s for 6 h. Microinfusions of γ-MSH into the OVLT were performed for 6 h by connecting the microinfusion cannula to a 5 ml glass syringe fixed to an infusion pump by polyethylene tubing (PE 10) filled with the infusion solution. The infusion speed was 0.1 µl/min. Within 6 h a total amount of 360 pMol γ-MSH was administered. At the end of the experimental series, the animals were deeply anesthetized with pentobarbital (80 mg/kg) and the position of the electrode tip was indicated by iron deposition after electrolysis, which was induced by a 10 fA direct cathodal current delivered for 15 s. The animals were then killed by perfusing their brains transcardially and the brains were put into fixative for at least 48 h. The histological localization of stimulation sites was performed on 10 µm frontal tissue sections stained with Prussian blue, which shows a color reaction with the deposed iron. Microinfusion sites were also localized on 10 µm frontal tissue sections by the trace of the tip of the microinfusion cannula. The mean fever indices, the integrated areas between thermal responses of febrile and normothermic animals, were expressed in °C·h (for 6 h). The mean of all temperatures measured at 30 min intervals before pyrogen injection was calculated as the baseline temperature. Values are expressed as means ± standard error. A one-factor analysis of variance (ANOVA) of data was used to test whether there were differences in temperature responses in response to LPS injection between experimental and control animals followed by the Scheffe post-hoc test. Fever indices were compared by Student's t-tests.

Results

To investigate, if the antipyretic effects of systemically administered γ-MSH (6) is mediated by an excitation of neurons located within the OVLT, the febrile response was tested under the influence of microinfusions of 360 pMol γ-MSH into the OVLT. The microinfusion procedure into the OVLT per se increased the animals' colonic temperature already before injections of LPS. The fever response could therefore not be analyzed appropriately under these experi-mental conditions. Histological examination of brain tissue slices showed that the implantation of guide cannulae partly destroyed the OVLT so that possible effects of γ-MSH on neurons within the OVLT could not be studied in vivo.

In the next experimental series, we tested if a direct excitation of vasopressinergic neurons within the hypothalamic SON might contribute to endogenous antipyresis in guinea pigs. The results of these experiments are summarized in Fig. 1. The upper part of Fig. 1 shows that electrical stimulation of the SON without pyrogen application caused a slight decrease in colonic temperature by 0.2-0.3 °C.

LPS-induced fever was significantly reduced by SON stimulation from 60-360 min after LPS injection ($p < 0.001$, ANOVA), as shown in the lower part of Fig. 1.

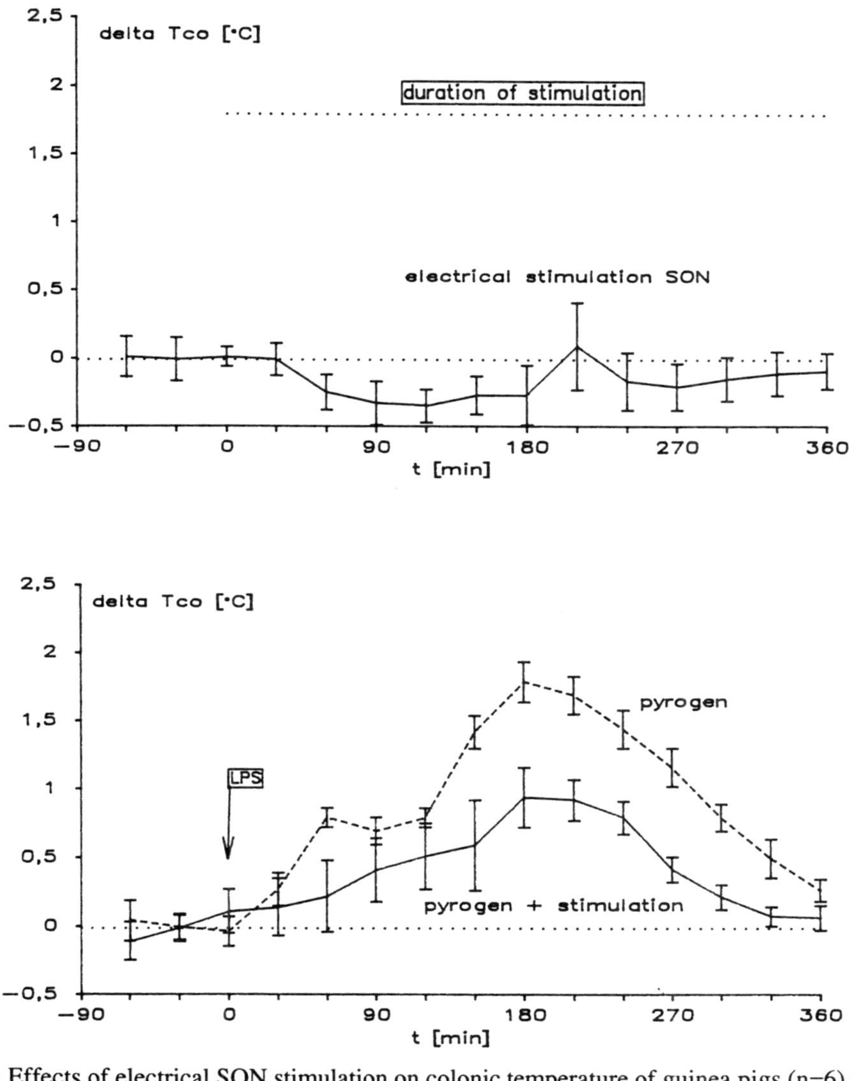

Fig. 1: Effects of electrical SON stimulation on colonic temperature of guinea pigs (n=6) without injection of LPS (upper part) and on LPS induced fever (lower part).

The mean fever index (c.f. Fig. 2) was reduced from 5.7°C·h (for 6 h) in controls to 2.7°C·h in electrically stimulated animals. The integrated thermal response of stimulated animals, which were not injected with LPS, was -1.03°C.h. The sites of the electrode tips within the hypothalamic SON are shown on Fig. 3.

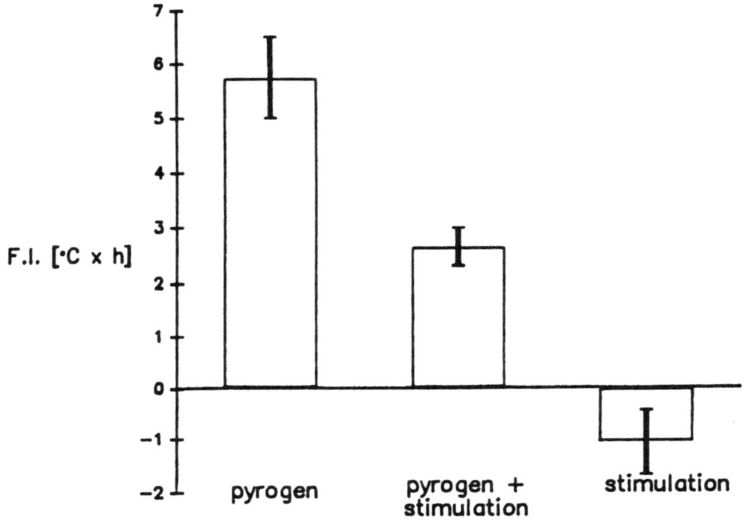

Fig. 2: Mean fever indices in animals receiving LPS without SON stimulation (left column), LPS combined with SON stimulation (middle column) and animals being electrically stimulated but not being injected with LPS (right column).

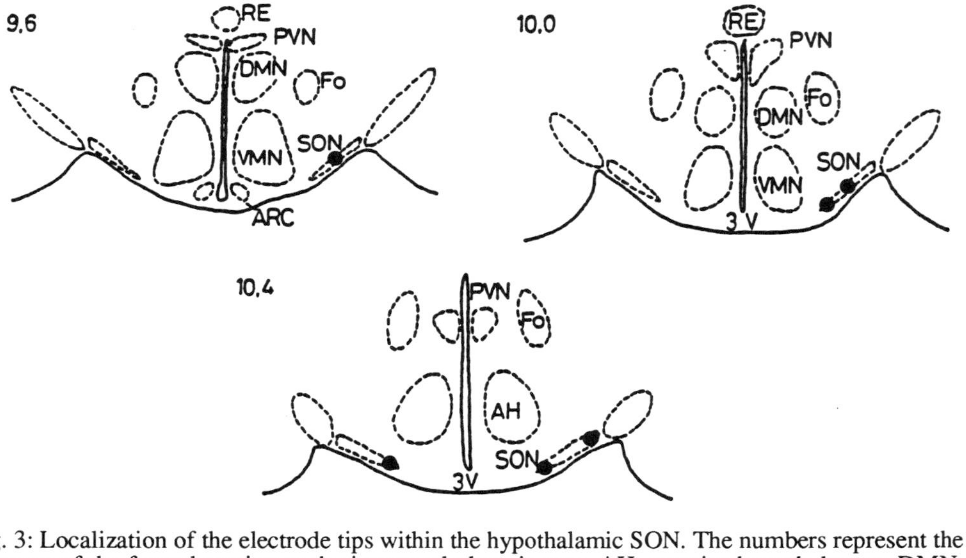

Fig. 3: Localization of the electrode tips within the hypothalamic SON. The numbers represent the distance of the frontal section to the interaural plane in mm. AH, anterior hypothalamus; DMN, dorsomedial nucleus; Fo, fornix; OT, optic tract; PVN, paraventricular nucleus; RE, reunient nucleus; SON, supraoptic nucleus; VMN, ventromedial nucleus; 3 V, third ventricle.

Discussion

The results of this study clearly show that stimulation of the hypothalamic SON, predominantly producing arginine-vasopressin (AVP), results in an attenuation of the febrile response to bacterial pyrogen. The activation of the SON by systemically administered γ-MSH, described by Mitchell et al. (4), has been shown to contribute to cardiovascular responses. Since systemically infused γ-MSH reduced LPS fevers in guinea pigs (6), an activation of the SON could also, in part, be responsible for the observed antipyretic effects. It would be interesting to elucidate the neuronal pathway which finally activates the vasopressinergic neurons within the SON. A possible transmission of γ-MSH-mediated signals into the brain within the OVLT could not be studied in vivo (see above). It would, however, be interesting to investigate, if γ-MSH is able to alter the activity of neurons within the OVLT adjacent area. This could be studied in vitro using brain slices within the OVLT, an experimental approach, which has already been used to analyze neuropeptide effects within circumventricular organs (8).

Finally, the question arises how an activation of vasopressinergic SON neurons exerts antipyretic effects. Most experimental evidence for the role of endogenous AVP in fever regulation and fever suppression supports the hypothesis that AVP-induced antipyresis is mediated in the septal area of the limbic system (9, 10, 11). Indeed, it could be demonstrated that electrical stimulation of vasopressinergic projections to the septum reduced fever in different species (6, 12). According to our knowledge no anatomical evidence for direct neuronal projections from the SON to the septum is available at present. Therefore, the exact mechanism by which SON stimulation contributes to endogenous antipyresis remains still speculative.

If a stimulation of the SON during fever is mediated by a signal transfer into the OVLT, it would mean that this circumventricular organ does interestingly not only contribute to the generation of fever (13, 14), but also to its limitation and finally to its suppression (15).

Acknowledgements

This study was supported by the DFG, project Ze 183 4/1. Excellent technical assistance of Birgit Störr is gratefully acknowledged.

References

1. Nakanishi, S, Inoue, A, Kita, T, Nakamura, M, Chang, ACY, Cohen, SN, Numa, S (1979): Nucleotide sequence of cloned cDNA for bovine corticotropin-beta-lipotropin precursor. Nature 278, 423-427.

2. Gruber, KA, Callahan, MF (1989): ACTH-(4-10) through γ-MSH: evidence for a new class of central autonomic nervous system - regulating peptides. Am J Physiol 257, R681-R694.

3. Gruber, KA, Eskridge, SL (1986): Central vasopressin system mediation of acute pressor effect of γ-MSH. Am J Physiol 251, E134-E137.

4. Mitchell, LD, Callahan, MF, Wilkin, LD, Gruber, KA, Johnson, AK (1989): Activation of supraoptic magnocellular neurons by γ2-melanocyte stimulating hormone (γ2-MSH). Brain Res 480, 388-392.

5. Callahan, MF, Cunningham, JT, Kirby, RF, Johnson, AK, Gruber, KA (1988): Role of the anteroventral third ventricle (AV 3V) region of the rat brain in the pressor response to γ2-melanocyte stimulating hormone (γ2-MSH). Brain Res 444, 177-180.

6. Bock, M, Roth, J, Kluger, MJ, Zeisberger, E (1994): Antipyresis caused by stimulation of vasopressinergic neurons and intraseptal or systemic infusions of γ-MSH. Am J Physiol, R614-R621.

7. Luparello, TJ (1967): Stereotaxic atlas of the forebrain of the guinea pig. Karger, Basel.

8. Schmid, HA, Simon, E (1992): Effects of angiotensin II and atrial natriuretic factor on neurons in the subfornical organ of ducks and rats in vitro. Brain Res 588, 324-328.

9. Cooper, KE (1987): The neurobiology of fever: Thoughts on recent developments. Annu Rev Neurosci 10, 297-324.

10. Kasting, NW (1989): Criteria for establishing a physiological role for brain peptides. A case in point: The role of vasopressin in thermoregulation during fever and antipyresis. Brain Res Rev 14, 143-153.

11. Zeisberger, E (1990): The role of septal peptides in thermoregulation and fever. In: Thermoreception and Temperature Regulation. Bligh, J, Voigt, K eds. Berlin, Heidelberg: Springer, 273-283.

12. Naylor, AM, Pittman, QJ, Veale, WL (1988): Stimulation of vasopressin release in the ventral septum of the rat brain suppresses prostaglandin E_1 fever. J Physiol (London) 399, 177-189.

13. Blatteis, CM, Bealer, SL, Hunter, WS, Llanos, QJ, Ahokas, RA, Mashburn, TA (1983): Suppression of fever after lesions of the anteroventral third ventricle in guinea pigs. Brain Res Bull 11, 519-526.

14. Stitt, JT (1985): Evidence for the involvement of the organum vasculosum laminae terminalis in the febrile response of rabbits and rats. J Physiol (London) 368, 501-511.

15. Zeisberger, E, Merker, G (1992): The role of OVLT in fever and antipyresis. Prog Brain Res 91, 403-408.

Thermal Balance in Health and Disease
Advances in Pharmacological Sciences

PERIPHERAL ADMINISTRATION OF THE OPIATE ANTAGONIST NALOXONE ATTENUATES THE FEBRILE RESPONSE TO POLYINOSINIC: POLYCYTIDYLIC ACID

A. S. Milton, N. C. Eastmond and J. Davidson.

Department of Biomedical Sciences (Pharmacology), Marischal College, University of Aberdeen, Aberdeen. AB9 1AS, Scotland, U.K.

Introduction

A number of agents, when injected peripherally into mammals, have the ability to induce physiological responses similar to those observed during the acute phase response to infection. These agents include lipopolysaccharide (LPS) and muramyl dipeptide which are components of the cell walls of gram negative and gram positive bacteria respectively. In addition to these bacterial agents, the acute phase can also be induced by polyinosinic:polycytidylic acid (poly I:C) which is a synthetic double-stranded polyribonucleotide, similar in structure to viral polyribo- or poly-deoxyribo-nucleotides. Perhaps the most striking and readily measurable manifestation of the acute phase response is fever and poly I:C has been shown to be a potent pyrogen (1). Recently Blatteis, *et al.*, (2) have reported that fever induced in guinea pigs by bacterial LPS and the cytokines interferon-α (IFN-α), tumour necrosis factor-α (TNF-α) and interleukin-6 (IL-6) can be blocked by peripheral administration of the opiate receptor antagonist naloxone, suggesting that endogenous opioids may modulate the fever response. Since poly I:C is not thought to be directly pyrogenic but may act by the induction and subsequent action of cytokines such as interleukin-1 (IL-1) and IFN the aim of the present study was to investigate whether endogenous opioids may modulate the febrile response to the exogenous pyrogen poly I:C. The effect of various doses of naloxone, given by different routes on poly I:C-induced fever was investigated.

Methods

Measurement of body temperature. Male Dutch rabbits (1.8 - 2.3) kg were lightly restrained in conventional stocks throughout all experiments. Experiments were carried out between 10.00 h and 16.00 h at an ambient temperature of 22 ± 1 °C. Rectal temperatures were measured continuously using rectal thermistor probes (Yellow Springs type 401) connected to a Jacquet multi-channel chart recorder which plotted temperature measurements every 20 seconds. Rabbits were left unhandled after insertion of the probes until their body temperatures were stable for approximately 1 hour before any agents were administered. The basal body temperature of each animal was calculated as the average core temperature measurement over 30 minutes prior to the first injection. Magnitudes of the febrile responses were expressed as the change in body temperature from basal (ΔT °C). Data, expressed as the mean of n experiments ± s.e. mean were analysed using Student´s paired t-test. Treatments were randomized, each rabbit acting as it's own control. Differences were considered to be significant when $P < 0.05$.

Cannulation of the third cerebral ventricle. A stainless steel Collison type guide cannula was implanted stereotaxically into the third cerebral ventricle as described previously (3). Rabbits were allowed to recover for at least 2 weeks after surgery before being used in any experiments.

Injections. Polyinosinic : polycytidylic acid, sodium salt and naloxone hydrochloride (Sigma Chemical Co., Dorset, England) were dissolved in saline (sterile non-pyrogenic 0.9 % sodium chloride, Travenol Laboratories, Norfolk, UK). All drugs were sterilized by filtering through a Millex-GS 0.22 µm filter (Millipore S.A. Molsheim, France.) immediately prior to use. Intravenous injections of poly I:C and naloxone were given via the marginal ear vein in a volume not exceeding 0.5 ml. Subcutaneous injections of naloxone were given into the scruff at the nape of the neck in a volume not exceeding 1 ml and intracerebroventricular (i.c.v.) injections of naloxone were given into the third cerebral ventricle via a previously implanted guide cannula in a volume of 50 µl.

Results

Intravenous injection of poly I:C (2.5 µg/kg) produced a biphasic increase in temperature, the first peak occurring approximately 2.5 h and the second peak occurring approximately 4 h after pyrogen administration (Figs. 1 & 2). 5 µg naloxone administered into the third cerebral ventricle had no significant effect on normal body temperature or poly I:C-induced fever when given either 1 h or 2.5 h after i.v. injection of pyrogen (data not shown). The effect of administering larger doses of naloxone on poly I:C-induced fever was not studied as doses of 10 µg - 400 µg given i.c.v. produced dose dependent hypothermias.

When given intravenously, approximately 1-2 minutes prior to Poly I:C, naloxone appeared to attenuate the febrile response in a dose-dependent manner, 12.5 µg/kg having no effect, 25 µg/kg producing a small but not statistically significant attenuation of the fever and 50 µg/kg significantly attenuating the second peak of the febrile response (Fig. 1). Subcutaneous administration of 10 mg/kg naloxone 1 h after Poly I:C had no significant effect on normal body temperature but significantly attenuated the febrile response to Poly I:C (Fig. 2). Subcutaneous and i.v. injection of naloxone produced a small decrease in normal body temperature but this was not significantly different compared to saline controls (Figs. 1 & 2).

Discussion

The exact mechanism by which poly I:C produces fever is not known. Intravenous injection of poly I:C results in a fever which is qualitatively similar to that seen following administration of bacterial LPS. The febrile response to poly I:C is biphasic in time course and has been shown to be prostaglandin dependent (1). It is possible that like LPS-induced fever, poly I:C-induced fever may occur as a result of the production of various intermediates. Immune challenge by foreign antigens, such as poly I:C and LPS, leads to a cascade of cellular events and the release of endogenous cytokines from immunocompetent cells. Some of these cytokines, have been shown to be pyrogenic and to stimulate the synthesis of prostaglandins. Prostaglandins, particularly E type prostaglandins, are widely regarded as the final mediators of fever and are thought to act at the level of the preoptic anterior hypothalamus to initiate the increase in core body temperature (4).

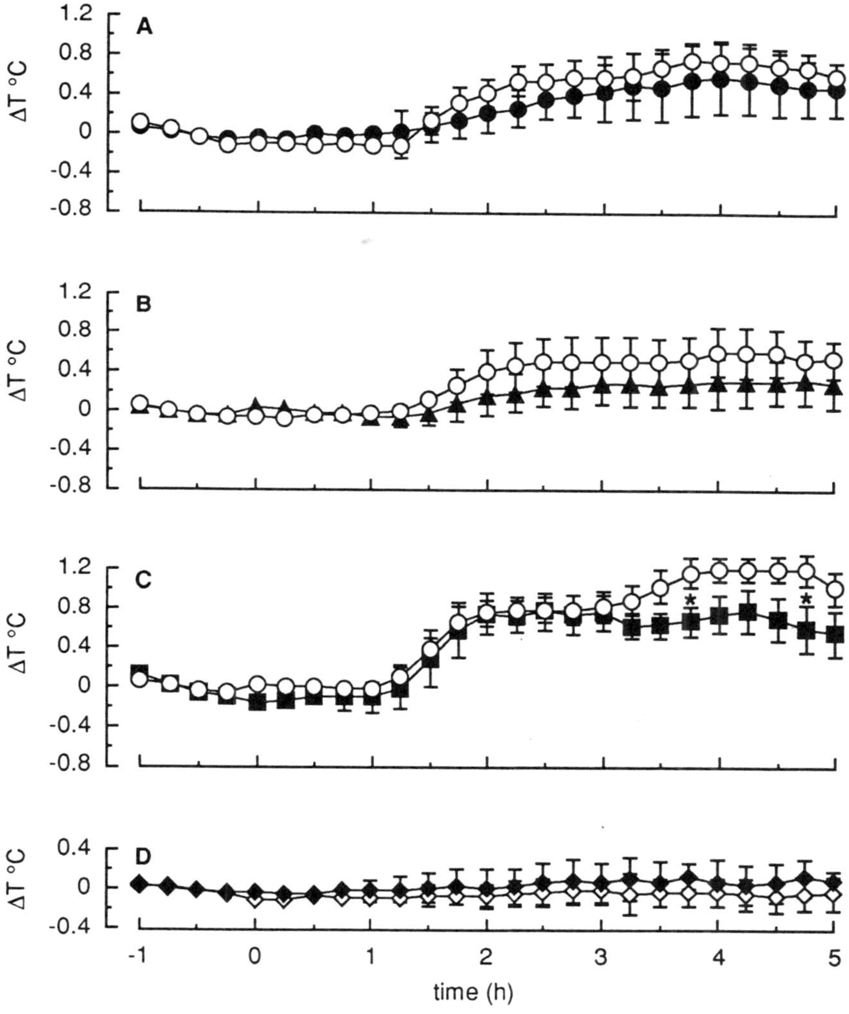

Fig.1: Effect of i.v. administration of various doses of naloxone on poly I:C fever (A, B, C). Rabbits were injected i.v. with either naloxone (12.5 μg/kg,●; 25 μg/kg,▼; 50 μg/kg,■) or saline (○) immediately prior to poly I:C (2.5 μg/kg) which was given i.v. at time zero. (D) Rabbits were injected i.v. with either 50 μg/kg naloxone (◆) or saline (◇) immediately prior to a second i.v. injection of saline which was given at time zero. Each value represents the mean change in body temperature (°C) from basal for n=4; s.e mean shown by vertical bars.

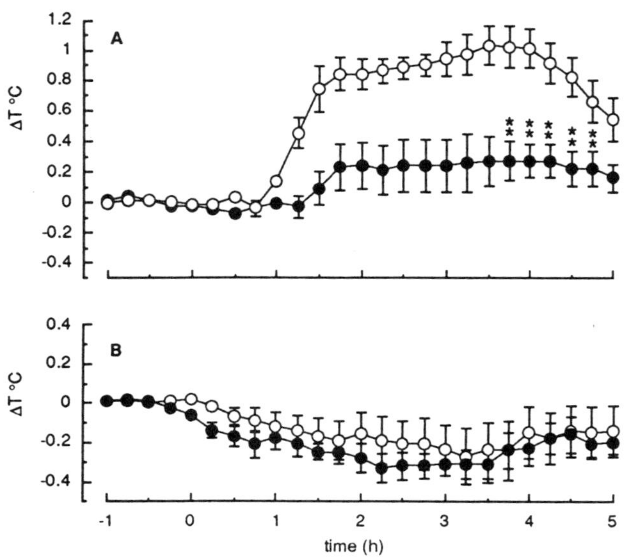

Fig.2: Effect of s.c. administration of naloxone on poly I:C fever. (A) Rabbits were injected s.c with 10 mg/kg naloxone (●) or saline (○) 1 h after poly I:C (2.5 μg/kg) which was given i.v. at time zero. (B) Rabbits were injected s.c. with either 10 mg/kg naloxone (●) or saline (○) 1 h after a second i.v. injection of saline at time zero. Each value represents the mean change in body temperature (°C) from basal for n=6; s.e. mean shown by vertical bars. ** $p < 0.01$, between naloxone and saline pretreatments.

Poly I:C is a potent inducer of the cytokine IFN (5). Both IFN-α (6) and IFN-γ (7) produce fever when injected i.v. into rabbits and these fevers can be reduced by treatment with the cyclo-oxygenase inhibitor indomethacin, suggesting that IFNs are intrinsically pyrogenic by inducing the production of prostaglandins. IFN is also thought to interact with the opioidergic system. Naloxone has been shown to block IFN-α-induced activity of hypothalamic thermosensitive neurones in brain slice preparations (8) and IFN-α has been found to bind to opioid receptors in mouse brain homogenates and to cause, naloxone reversible, endorphin-like behavioural responses when injected i.c.v. (9). Although their exact role remains unclear, it has been suggested that opioids may have a modulatory effect on body temperature during fever. Levels of endogenous opioid peptides have been observed to increase during fever, for example Murphy, *et al.*, (10) report that concentrations of β-endorphin are elevated in the plasma, cerebrospinal fluid and hypothalami of rabbits made febrile with LPS and IL-1 and central injection of low doses of β-endorphin or met-enkephalin cause an increase in body temperature (11) .

The present study investigated the effects of central and peripheral administration of the opiate receptor antagonist naloxone on fevers stimulated by i.v. injection of the exogenous pyrogen poly I:C. When given centrally, into the third cerebral ventricle in a non-hypothermic dose, naloxone was found to have no effect on poly I:C-induced fever, however, when given peripherally, naloxone attenuated the febrile response (Figs. 1 & 2). 10 mg/kg naloxone administered s.c. appeared to attenuate both peaks of the fever (Fig. 2), whereas 50 μg/kg naloxone given i.v. appeared to attenuate only the second peak of the response (Fig. 1). As these two routes of peripheral administration result in different bioavailabilities of the antagonist, further work is required to determine if the effect is purely concentration dependent and whether injecting larger doses of naloxone i.v. would result in attenuation of both peaks of the fever. It is unlikely that poly I:C-induced fever is wholly dependent on the induction of IFN since poly I:C is also a potent inducer of interleukin-1, which itself is a potent stimulus for prostaglandin synthesis (12). This may possibly explain our observation that 10 mg/kg naloxone administered s.c. attenuated, but did not completely abolish the febrile response (Fig. 2). Further work is required to determine if the response can be abolished either by increasing the dose of opiate antagonist or administering naloxone together with a cyclo-oxygenase inhibitor. Although naloxone given alone, both i.v. and s.c., produced a small hypothermic response, the magnitude of this response did not appear to be sufficient to account for the attenuation of fever observed.

Our results are consistent with those of Blatteis, *et al.* (2) who report that, in the guinea pig, fever induced by LPS and the cytokines, IL-6, TNF-α and IFN-α can be blocked by peripheral administration of naloxone. Data presented here indicates that peripheral opioids, in addition to PGE_2, may play an important role in modulating the febrile response to poly I:C.

Acknowledgements

This work was supported by the Wellcome Trust.

References

1. Rotondo, D., Abul, H. T., Milton, A. S., Davidson, J. (1987): The pyrogenic effects of the interferon-inducer, polyinosinic:polycytidylic acid are antagonised by ketoprofen. Eur. J. Pharmacol. 137, 257-260.

2. Blatteis, C. M., Xin, L., Quan, N. (1990): Neuromodulation of fever apparent involvement of opioids. Brain Res. Bull. 26, 219-223.

3. Milton, A. S., Abul, H. T., Davidson, J., Rotondo, D. (1989): Antipyretic mechanism of action of dexamethasone, In: Thermoregulation: research and clinical applications. 7th international symposium on pharmacology of thermoregulation, Lomax, P. and Schönbaum, E. (Ed.) pp 74-77, Karger, Basel.

4. Milton, A. S. (1989): Thermoregulatory actions of eicosanoids in the central nervous system with particular regard to the pathogenesis of fever. Ann. New York. Acad. Sci. 559, 392-410.

5. Lindsay, H. L., Trown, P. W., Brandt, J., Forbes, M. (1969): Pyrogenicity of poly I:poly C in rabbits. Nature, 223, 717-718.

6. Dinarello, C. A., Bernheim, H. A., Duff, G. W., Le, H. V., Nagabhushan, T. L., Hamilton, N., C., Coceani, F. (1984): Mechanisms of fever induced by recombinant human interferon. J. Clin. Invest. 74, 906-913.

7. Morimoto, A., Murakami, N., Takada, M., Teshirogi, S., Wantanabe, T. (1987): Fever and acute phase response induced in rabbits by human recombinant interferon-γ. J. Physiol. (Lond.) 391, 209-218.

8. Nakashima, T., Hori, T., Kuriyama, K., Kiyohara, T. (1987): Naloxone blocks the interferon-α induced changes in hypothalamic neuronal activity. Neurosci. Lett. 82, 332-336.

9. Blalock, J. E., Smith, E. M. (1981): Human leukocyte interferon (HuIFN-α): potent endorphin-like opioid activity. Biochem. Biophys. Res. Commun. 101, 472-478.

10. Murphy, M. T., Koenig, J. L., M., L. J. (1983): Changes in central concentrations of β-endorphin in fever. Fed. Proc. 42, 464.

11. Adler, M. W., Geller, E. B., Roscow, C. E., Cochin, J. (1988): The opioid system and temperature regulation. Ann. Rev. Pharmacol. Toxicol. 28, 429-449.

12. Dinarello, C. A. (1984): Interleukin 1. Rev. Infect. Dis., 6, 51-95.

Section VI:

PHARMACOLOGICAL ASPECTS

Thermal Balance in Health and Disease
Advances in Pharmacological Sciences

ENDOGENOUS OPIOIDS AND FEVER

C. M. Blatteis and A. A. Romanovsky

Department of Physiology and Biophysics, University of Tennessee at Memphis, Memphis, Tennessee 38163, U.S.A.

The possible involvement of μ-opioid receptor agonists in fever production has been suggested by various reports in the literature. Thus, elevations of plasma, cerebrospinal fluid, and hypothalamic levels of ß-endorphin, an endogenous μ-opioid peptide, have been found during bacteremia and both lipopolysaccharide (LPS)- and endogenous pyrogen-induced fevers (1-3). Interleukin (IL)-1, which is generally thought to be the primary endogenous pyrogen, has been shown to induce ß-endorphin secretion by pituitary cells (4) and to modulate opioid receptor binding in the brain (5, 6). Naloxone (Nal), a μ-receptor antagonist, prevents the effect of the pyrogenic cytokines IL-6 (7) and interferon (IFN)-α (8) on the activities of hypothalamic thermosensitive neurons in slice preparations. Furthermore, both the peripheral and central injections of ß-endorphin and the prototypic μ-opioid agonist morphine cause "fever-like" (ascribed to an increase in the thermoregulatory set-point) body temperature rises. This hyperthermia is blocked by μ-receptor antagonists, particularly Nal and naltrexone, which also can produce hypothermia by themselves when injected either centrally or peripherally (9-12). In spite of these lines of evidence, several attempts to block fever by administration of Nal in cats (13), rabbits (11, 14-16), and guinea pigs (12) were, however, unsuccessful, although Nal-induced antipyresis was observed in one study in humans (17) and in one recent study in rats (18). But in different studies carried out in our laboratory (19-21), we repeatedly were able to show that the febrile response of guinea pigs to intravenous (i.v.) *S. enteritidis* LPS was significantly attenuated by the prior subcutaneous (s.c.) injection of Nal-hydrochloride (Nal-HCl) (Fig. 1). This paper is devoted to the analysis of the mechanisms of the s.c. Nal-induced inhibition of febrigenesis.

Attenuation of fever by naloxone: peripheral *versus* central effect?

Since Nal-HCl readily crosses the blood-brain barrier (BBB), it was necessary to differentiate between the peripheral and central thermoregulatory actions of this opioid antagonist in order to answer the question: are opioid peptides involved in fever production acting within or outside the brain? For this, we compared (21) the thermal responses of unanesthetized guinea pigs to s.c. Nal-HCl (which has both peripheral and central actions), intracerebroventricular (i.c.v.) Nal-HCl (which has a central action only), and two s. c. quaternary Nal analogues, Nal-methiodide

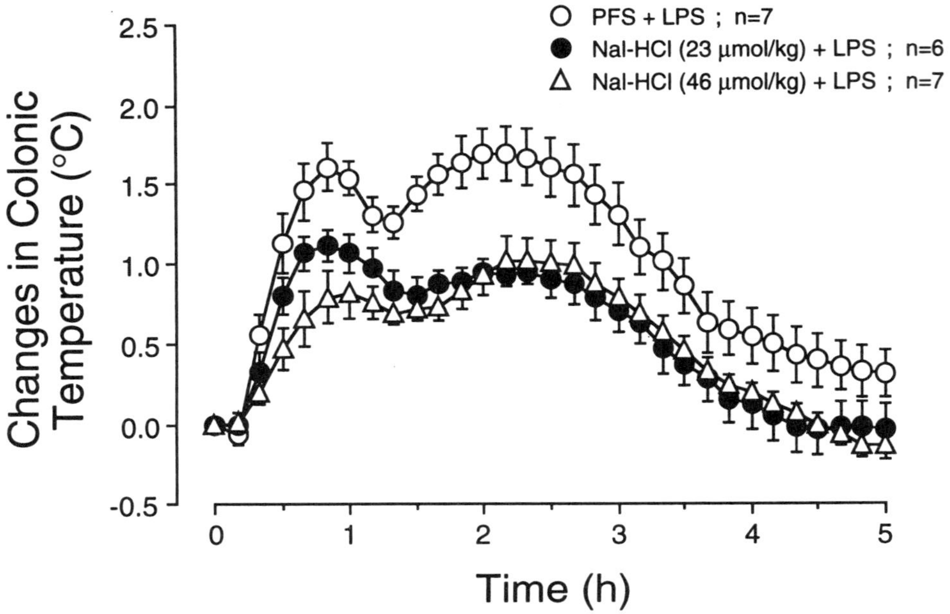

Fig. 1: Changes in colonic temperature induced by subcutaneous injection of naloxonehydrochloride (Nal-HCl; doses indicated) or pyrogen-free saline (PFS; 1 ml/kg) just before intravenous injection of lipopolysaccharide (LPS; 2 µg/kg) at time 0. [Adapted from (21)].

(Nal-mI) and Nal-methobromide (Nal-mBr). Quaternary ammonium derivatives do not cross the BBB and, when administered peripherally, are presumed to exert their actions at peripheral sites only (22-25). Their receptor binding and antagonistic properties are 10^1-10^2 times lower than those of their relatively nonpolar tertiary counterparts (25), but when administered at relatively high doses *in vivo*, under circumstances providing access to either their central targets (peripheral administration in animals with acute spinal transection, central administration in "intact" animals) or their peripheral targets (peripheral administration), their biological effects are similar to those of their parent tertiary compounds (23, 25).

In our study (21), both Nal-mI and Nal-mBr, injected s.c., attenuated both the first and second phases of the characteristically biphasic LPS-induced fever, similarly to the effect of s.c. Nal-HCl. Hence, Nal appeared to attenuate both phases of LPS fever by a peripheral rather than a central action. This conclusion was further supported by the finding that no attenuation of the febrile response occurred when Nal-HCl was injected i.c.v. at two different doses (0.25 µmol and 1.25 µmol) (21). The inability of Nal-HCl to attenuate fever when administered i.c.v. at various doses either before or after administration of a pyrogen was also reported previously in leukocytic pyrogen (i.v.)-induced fever in cats (13) and rabbits (14), IFN-α_2 (i.c.v.)-induced fever in rats (18), and LPS- and prostaglandin E_2 (both i.c.v.)-induced hyperthermia in rabbits (11). Yet even 100 µg (equal to approximately 0.25 µmol) i.c.v. Nal-HCl has been shown to effectively antagonize the body temperature rise produced by a centrally delivered opiate agonist in guinea pigs (26). Therefore, the failure of i.c.v. Nal to induce antipyresis in the present study can probably not be attributed to species differences, insufficient dose, or inadequate time of administration in relation to LPS.

The peripherally mediated attenuation of LPS fever by Nal may *a priori* be attributed to its action(s) on one or several of the following peripheral stages in the genesis of fever: 1) synthesis and release of endogenous pyrogens; 2) activation of thermoregulatory effectors; and/or 3) translation of the pyrogenic signal from the periphery to the brain.

Does naloxone affect production and/or release of endogenous pyrogens?

The mechanism of the antipyretic effect of Nal is unlikely to involve the blockade of the production of endogenous pyrogens. Although μ-opioid agonists can influence the synthesis and/or release of various cytokines (27, 28), and although some processes leading to cytokine production are Nal-sensitive [*e.g.*, systemic IL-6 induction by central IL-1 (22)], we showed that s.c. Nal does not block the acute-phase proteinemic response (as determined by elevations in plasma copper and N-acetyl-neuraminic acid levels) to i.v. LPS (19), implying that endogenous pyrogen production is not impeded by the administered Nal. Furthermore, Nal still exerted an obvious antipyretic effect when fevers were induced by cytokines (IFN-α_2, tumor necrosis factor-α, and IL-6) instead of exogenous pyrogens, thus bypassing the stage of endogenous pyrogen production (20). The inability of Nal to affect the production or release of IL-1 has also been shown by direct measurements (28).

Does naloxone attenuate fever through its direct action upon thermoregulatory effectors?

Under afebrile conditions, both Nal-HCl and its quaternary analogues, administered s.c., induced hypothermic responses associated with cutaneous vasodilation (21); the ear skin vasodilatory response to Nal-mI is shown in Fig. 2. The duration of these responses (*ca.* 1 h) corresponded to the duration of the first phase of the biphasic LPS fever. Not surprisingly, the dose-dependent attenuation of the first phase of fever by both the quaternary and tertiary Nals was accompanied by the commensurate blockade, or even reversal, of the typical, LPS-induced cutaneous vasoconstriction. This represented, therefore, a thermolytic rather than a truly antipyretic effect of this opioid antagonist. Since various vascular beds contain opiate receptors (33), and morphine has recently been shown to possess Nal-sensitive vasoconstrictor action *in vitro* (34), it is likely that the thermolytic action of peripheral Nal was achieved, at least partly, by an alteration in cutaneous vasomotor tone. However, metabolic rate/heat loss experiments with quaternary antagonists have yet to be carried out to clarify the effector pattern of the hypothermic response to peripheral Nal under these conditions.

In contrast to the effect of Nal on the first phase, its effect on the second febrile phase would seem to be separate from its thermolytic action. Indeed, we found recently (35) that intraperitoneal cooling applied at the beginning of LPS fever, which mimicked the thermolytic effect of Nal and was sufficient to completely block the first phase, did not result in any attenuation of the second phase. Hence, it would seem unlikely that the relatively short-lasting thermolytic effect of Nal during the first febrile phase could have accounted for the attenuation of the second phase, which lasted up to 5 h post-LPS administration (21). We suggest, therefore, that in contrast to the thermolytic attenuation by Nal of the first febrile phase, the attenuation of the second phase of fever represents a truly antipyretic effect.

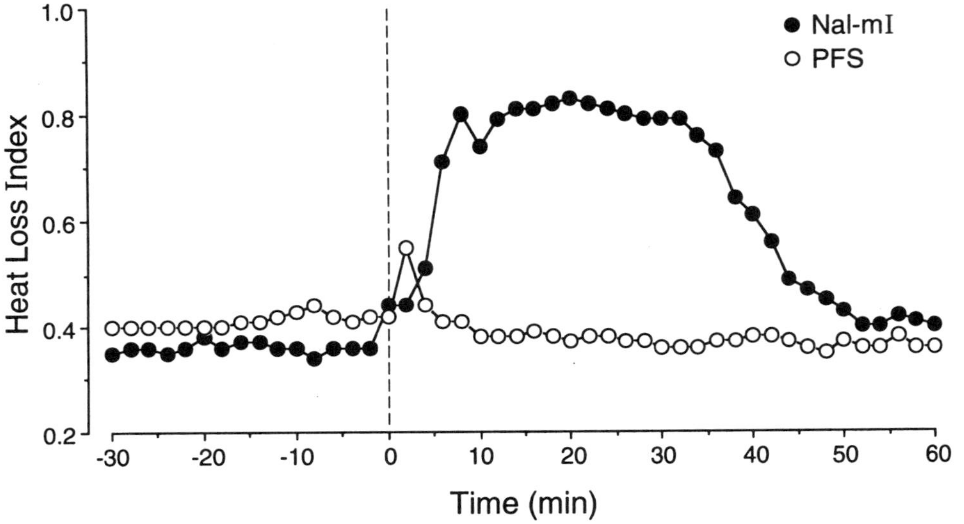

Fig. 2: Changes in the heat loss index (HLI) induced by the subcutaneous injection of naloxone-methiodide (Nal-mI; 92 µmol/kg) or pyrogen-free saline (PFS; 1 ml/kg) (individual curves). The HLI was calculated, according to the formula of Székely (29), as the difference between the skin (ear) temperature and ambient temperature related to the difference between the deep body (colonic) temperature and skin temperature. The HLI has been successfully used in our previous experiments in rats (30) and guinea pigs (31, 32). The low and high theoretical limits of the HLI are, respectively, 0 (full vasoconstriction) and 1 (full vasodilation).

Does peripheral naloxone alter the translation of the pyrogenic signal from the periphery to the brain?

Since the antipyretic effect of Nal seems to be mediated through its action outside the BBB and does not involve the blockade of the production of endogenous pyrogens and the alteration of thermoregulatory effectors, we conjecture that Nal initiates antipyresis by interfering at a peripheral site with the translation of the pyrogenic signal into the brain. The involvement of peripheral opiates in centrally mediated responses has been documented; *e.g.*, both opioid agonists and antagonists modulate luteinizing hormone secretion in the hypophysis by acting outside the BBB (24, 36). Moreover, the circumventricular organs of the brain, which are considered the sites of the translation of circulating pyrogenic stimuli into neural signals (37), have been shown to contain opioid receptors (38). Furthermore, cytokines can influence the specific binding of opioid agonists to their receptors (5, 6). The exact mechanism of the antipyretic effect of Nal remains, however, to be elucidated.

The physiological significance of the antipyretic effect of Nal, obvious during the second phase of LPS fever, cannot be determined from the present data alone. But it is noteworthy that the blood concentrations of endogenous ß-endorphin are elevated during bacteremia (2) and LPS-induced fever (1). Moreover, a dramatic increase in plasma ß-endorphin develops during the second phase rather than the first phase of LPS fever (1).

Multiple actions of endogenous opioids on fever

The fact that Nal appears to attenuate fever through a peripheral rather than a central action nevertheless does not allow excluding the potential existence of opioid-mediated intrabrain mechanisms of fever production. Indeed, Nal has recently been reported to prevent the effects of IFN-α [reviewed in (8)] and IL-6 (7) on the activities of thermosensitive neurons in slice preparations of the rat and guinea pig preoptic area. However, results obtained *in vitro* must be interpreted with caution from the point of view of their *in vivo* implications. Thus, *in vivo*, i.c.v. Nal does not alter the hyperthermic response to i.c.v. IFN-α (18), in spite of its *in vitro* effects already mentioned. By the same token, s.c. Nal attenuates fever induced by both i.v. (20) and i.c.v. (18) IFN-α and sleep changes evoked by i.c.v. IFN-α (39).

Conclusions

Endogenous opioids have been shown to be involved in the mechanisms of fever. Their action is likely to target more than one stage of the febrile process. Our data suggest that the process of the pyrogenic signal transduction across the BBB may be opiate-dependent.

References

1. Carr, DB, Bergland, R, Hamilton, A, Blume, H, Kasting, N, Arnold, M, Martin, JB, Rosenblatt, M (1982): Endotoxin-stimulated opioid secretion: two secretory pools and feedback control *in vivo*. Science 217, 845-848.

2. Leshin, LS, Malven, PV (1984): Bacteremia-induced changes in pituitary hormone release and effect of naloxone. Am J Physiol 247, E585-E591.

3. Murphy, MT, Koenig, JI, Lipton, JM (1983): Changes in central concentration of ß-endorphin in fever. Fed Proc 42, 464.

4. Fagarasan, MO, Aiello, F, Muegge, K, Durum, S, Axelrod, J (1990): Interleukin 1 induces ß-endorphin secretion via *Fos* and *Jun* in AtT-20 pituitary cells. Proc Natl Acad Sci USA 87, 7871-7874.

5. Ahmed, MS, Llanos-Q, J, Dinarello, CA, Blatteis, CM (1985): Interleukin 1 reduces opioid binding in guinea pig brain. Peptides 6, 1149-1154.

6. Wiedermann, CJ (1989): Interleukin-1 interaction with neuroregulatory systems: selective enhancement by recombinant human and mouse interleukin-1 *in vitro* opioid receptor binding in rat brain. J Neurosci Res 22, 172-180.

7. Xin, L, Blatteis, CM (1992): Hypothalamic neuronal responses to interleukin-6 in tissue slices: effects of indomethacin and naloxone. Brain Res Bull 29, 27-35.

8. Hori, T, Nakashima, T, Take, S, Kaizuka,Y, Mori, T, Katafuchi, T (1991): Immune cytokines and regulation of body temperature, food intake and cellular immunity. Brain Res Bull 27, 309-313.

9. Adler, MW, Geller, EB, Rosow, CE, Cochin, J (1988): The opioid system and temperature regulation. Ann Rev Pharmacol Toxicol 28, 429-449.

10. Clark, WG (1979): Influence of opioids on central thermoregulatory mechanisms. Pharmacol Biochem Behav 10, 609-613.

11. Kandasamy, SB, Williams, BA (1983): Central effects of some peptide and non-peptide opioids and naloxone on thermoregulation in the rabbit. In: Environment, Drugs and Thermoregulation. Lomax, P, Schönbaum, E eds. Basel: Karger, 98-100.

12. Kandasamy, SB, Williams, BA (1983): Hyperthermic responses to central injections of some peptide and non-peptide opioids in the guinea-pig. Neuropharmacology 22, 621-628.

13. Clark, WG, Harris, NF (1978): Naloxone does not antagonize leukocytic pyrogen. Eur J Pharmacol 49, 301-304.

14. Davidson, J, Milton, AS, Rotondo, D (1987): A study of the involvement of opioid receptor sub-types in normal thermoregulation and during fever in the rabbit. Br J Pharmacol 91, 494P.

15. Hellon, RF, Townsend, Y, Cranston, WI (1985): Naloxone does not influence a pyrogen fever in rabbits. Pflügers Arch 404, 290-291.

16. Kapás, L, Krueger, JM (1993): Effects of naloxone pretreatment on interleukin 1-induced sleep and fever. Sleep Res 22, 435.

17. Norman, DC, Lin, JH, Morley, JE, Yoshikawa, TT (1986): Aging and fever: naloxone blocks the thermogenic effect of interleukin-1. Clin Res 34, 87A.

18. Zawada, WM, Clarke, J, Ruwe, WD (1992): Systemic and central administration of naloxone (NLX): differential alteration of fevers induced by intracerebroventricular (ICV) endogenous pyrogens (EPs). FASEB J 6, A1199.

19. Ahokas, RA, Seydoux, J, Llanos-Q, J, Mashburn, TA, Jr, Blatteis, CM (1985): Hypothalamic opioids and the acute-phase glycoprotein response in guinea pigs. Brain Res Bull 15, 603-608.

20. Blatteis, CM, Xin, L, Quan, N (1991): Neuromodulation of fever: apparent involvement of opioids. Brain Res Bull 26, 219-223.

21. Romanovsky, AA, Shido, O, Ungar, AL, Blatteis, CM (1994): Peripheral naloxone attenuates lipopolysaccharide fever in guinea pigs by an action outside the blood brain barrier. Am J Physiol (in press).

22. De Simoni, MG, De Luigi, A, Gemma, L, Sironi, M, Manfridi, A, Ghezzi, P (1993): Modulation of systemic interleukin-6 induction by central interleukin-1. Am J Physiol 265, R739-R742.

23. Milne, RJ, Coddington, JM, Gambe, GD (1990): Quaternary naloxone blocks morphine analgesia in spinal but not intact rats. Neurosci Lett 114, 259-264.

24. Panerai, AE, Martini, A, Casanueva, F, Petraglia, F, Di Giulio, AM, Mantegazza, P (1983): Opiates and their antagonists modulate luteinizing hormone acting outside the blood brain barrier. Life Sci 32, 1751-1756.

25. Valentino, RJ, Katz, JL, Medzihradsky, F, Woods, JH (1983): Receptor binding, antagonist, and withdrawal precipitating properties of opiate antagonists. Life Sci 32, 2887-2896.

26. Kandasamy, SB, Williams, BA (1983): Hyperthermic effects of centrally injected (D-ala^2, N-Me-Phe^4, Met-(O)^5-ol)-enkephalin (FK 33-824) in rabbits and guinea pigs. Neuropharmacology 22, 1177-1181.

27. Carr, DJJ (1991): The role of endogenous opioids and their receptors in the immune system. Proc Soc Exp Biol Med 198, 710-720.

28. Yang, S-X, Li, XY (1989): Enhancement of interleukin-1 production in mouse peritoneal macrophages by methionine-enkephalin. Acta Pharmacol Sin 10, 266-270.

29. Székely, M (1986): Skin temperature-skin blood flow: assessment of thermoregulatory changes. Acta Physiol Acad Sci Hung 68, 284.

30. Szelényi, Z, Székely, M, Romanovsky, AA (1992): Central thermoregulatory effects of cholecystokinin-octapeptide and prostaglandin E_1. Sechenov Physiol J 78, 94-101.

31. Romanovsky, AA, Blatteis, CM (1993): Body temperature elevation *per se* induces the late phase syndrome. In: Temperature Regulation. Milton, AS ed. Basel: Birkhäuser, 41-46.

32. Romanovsky, AA, Blatteis, CM (1994): Thermoregulatory vasomotor responses of the guinea pig ear. FASEB J (in press).

33. Altura, BM, Altura, BT, Carella, A, Turlapaty, PDMD, Weinberg, J (1980): Vascular smooth muscle and general anesthetics. Fed Proc 39, 1584-1591.

34. El-Sharkawy, TY, Al-Shireida, MF, Pilcher, CWT (1991): Vascular effects of some opioid receptor agonists. Can J Physiol Pharmacol 69, 846-851.

35. Romanovsky, AA, Blatteis, CM: Biphasic fever: what triggers the second temperature rise? Am J Physiol (submitted for publication).

36. Forman, LJ, Sonntag, WE, Meites, J (1983): Elevation of plasma LH in response to systemic injection of ß-endorphin antiserum in adult male rats. Proc Soc Exp Biol Med 173, 14-16.

37. Blatteis, CM (1992): Role of the OVLT in the febrile response to circulating pyrogens. Prog Brain Res 91, 409-412.

38. Holaday, JW (1983): Cardiovascular effects of endogenous opiate systems. Ann Rev Pharmacol Toxicol 23, 541-594.

39. Birmanns, B, Saphier, D, Abramsky, O (1990): α-Interferon modifies cortical EEG activity: dose-dependence and antagonism by naloxone. J Neurol Sci 100, 22-26.

INTERACTION BETWEEN NEUROTENSIN AND SELECTIVE OPIOID AGONISTS MODULATES BODY TEMPERATURE IN THE RAT

C. M. Handler, D. J. Mondgock, E. A. Bradley, E. B. Geller and M. W. Adler

Dept. of Pharmacology, Temple University School of Medicine, Philadelphia, PA 19140 USA

Biologically active peptides found within the central nervous system affect a variety of physiological processes, including food intake, sleep, hibernation, reproduction, nociception and temperature regulation. The opioid peptides affect temperature regulation via specific opioid receptors (1). At normal ambient temperatures, opioid agonists selective for the μ receptor induce hyperthermia with alteration of set point through an increase in metabolic rate (1-3). Kappa-selective agonists, such as dynorphin $A_{1\text{-}17}$ (DY) or U50,488H (U50), reduce metabolic rate, resulting in hypothermia (1-3). Within normal ambient temperatures, the δ opioid receptor appears to have only a minor role in temperature regulation (1-3). In contrast to opioid-induced hypothermia, neurotensin-induced hypothermia (4, 5) results from activation of heat loss mechanisms (6). Several studies have examined the interaction between neurotensin (NT)- and opioid-induced antinociception (7-10). The purpose of the present study was to examine the interactions of the highly selective opioid agonists, PL-017, DY, U50, and (D-Pen^2-D-Pen^5)-enkephalin (DPDPE) and NT on body temperature (Tb).

Methods

Male, albino Sprague-Dawley rats (Zivic-Miller) were housed 4-5 per cage in an animal room maintained at 22° ± 2°C and 50 ± 10% relative humidity on a 12 hr L/12 hr D cycle. Rats were fed standard rat chow with water available ad libitum. Seven days prior to testing, a polyethylene cannula was implanted into the right lateral ventricle under ketamine anesthesia (150 mg/kg, ip). The cannula was secured to the cranium with dental acrylic. Animals were housed individually after surgery and weighed 250-300 g at time of testing. Cannula placement was confirmed after testing and results from animals with incorrectly placed cannulae were not included in the analysis. All experiments were carried out at 20° ± 2°C and 55 ± 5 % relative humidity.

After a 60-min adaptation period, Tb was recorded by inserting a thermistor probe 7 cm into the rectum. The initial three readings were taken at 30-min intervals; the first reading was discarded to allow for adaptation to the temperature measurement process while the second and third readings were averaged together for a baseline measurement. Data were recorded at 15-min intervals for 3 hr post-injection. Results are reported as change from baseline (ΔTb, °C) ± SEM for each time interval. The changes in Tb were analyzed using a Student's t-test; statistical significance was accepted at $p \leq 0.05$.

NT (Sigma Co, St. Louis, MO), PL-017 and DPDPE (Peninsula Labs, Belmont, CA), DY (Bachem Labs, Torrence, CA), were dissolved in pyrogen-free sterile saline (0.9%) and injected icv in a volume of 5 µl, followed by a 3-µl saline flush. CTAP (Peninsula) and nor-BNI (RBI) were given in a volume of 3 µl followed by a 3-µl saline flush. U50 (a gift from NIDA) and naloxone (a gift from DuPont-Merck) were dissolved in pyrogen-free saline and injected sc. Doses were selected from full dose-response curves, using rectal Tb as the end point.

Results

NT, DY and U50 all induced a dose-dependent hypothermia. NT (0.625 - 5.0 µg, icv) lowered Tb 1.6 - 2.7° C. Hypothermia following administration of DY (10 or 15 µg, icv) was 1.1 - 1.5° C. U50 (10 - 40 mg/kg, sc) reduced Tb 0.45 - 1.5° C. The hypothermia produced by the combination of DY(10 or 15 µg, icv) and NT (0.625 - 5.0 µg, icv) did not differ significantly from that produced by NT alone (Fig. 1).

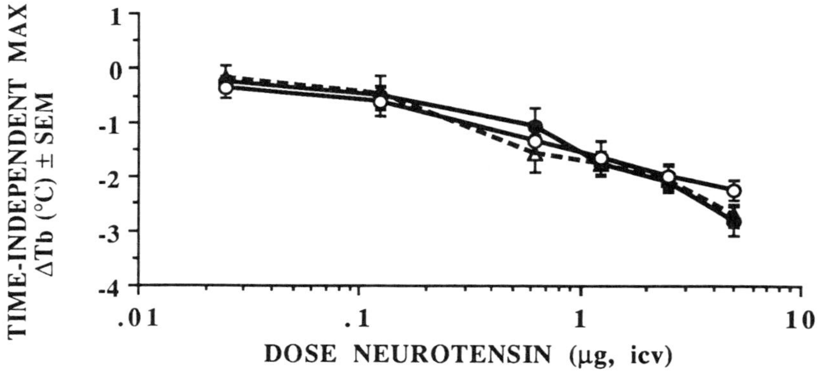

Fig. 1: Effect of neurotensin and dynorphin A_{1-17} ON Tb. ----Δ---- = NT + saline (0.625 - 5.0 µg, icv), n = 6 - 10; ○ = DY (10 µg, icv) + NT, n = 6 - 9 ; ● = DY (15 µg, icv) + NT, n = 5 - 8.

Similarly, U50 (10 mg/kg)/NT in combination did not differ significantly from the NT/saline controls (Fig. 2) However, U50 (20 or 40 mg/kg)/NT in combination produced an additive effect on the reduction in Tb (Fig. 2). In addition to the additive effect on the maximum time-independent change in Tb, U50 and NT in combination altered the time course of recovery from hypothermia. The hypothermia was prolonged in a dose-dependent manner and recovery to baseline was not completed at the end of the 3-hr post-injection period. Pretreatment with the opioid antagonist, naloxone (NLX, 10 mg/kg, t= -15 min) blocked the opioid-mediated portion of the effect of U50/NT, leaving the NT-induced component of the hypothermia intact (data not shown).

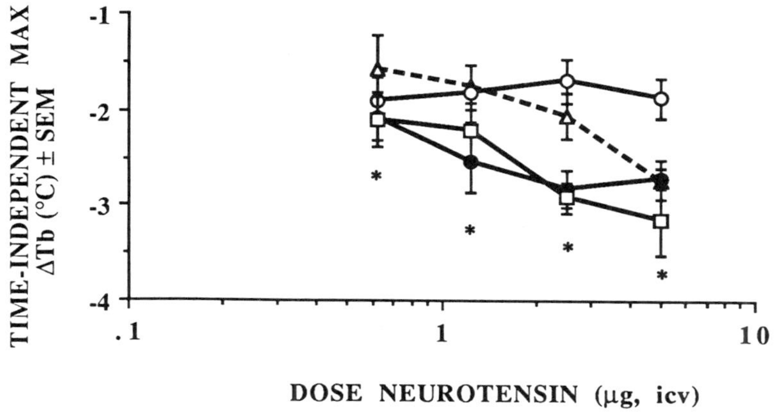

Fig. 2: Effect of U50, 488H in combination with neurotensin on Tb. Δ = NT + saline (sc), n = 6 - 10; ○ = U50 (10 mg/kg, sc) + NT, n = 5 - 8; ● = U50 (20 mg/kg, sc) + NT, n = 6 - 10; □ = U50 (40 mg/kg, sc) + NT, n = 5 - 9; * $p \leq 0.05$.

PL-017-induced hyperthermia is completely blocked when NT and PL-017 are given in combination except at a dose of NT (0.125 μg) which does not cause statistically significant hypothermia (Fig. 3). At the higher NT doses (0.625 - 5.0 μg) used in this study, PL-017 induced a significant, dose-dependent enhancement of the NT-induced hypothermia.

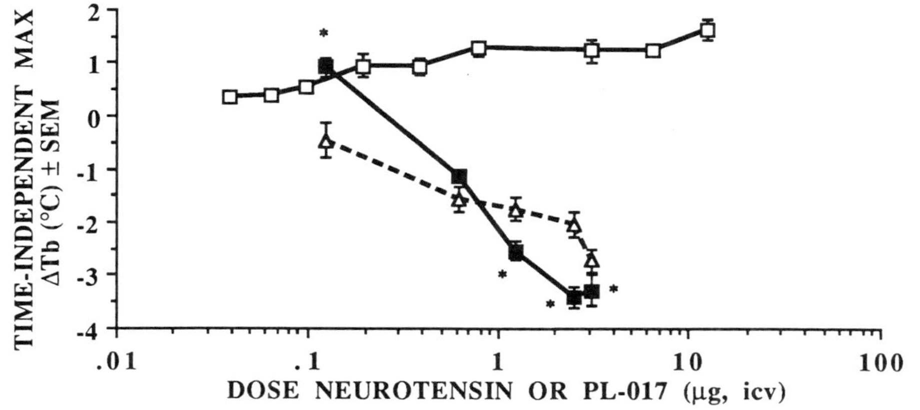

Fig. 3: Effect of PL-017 and neurotensin on Tb. Δ = NT + saline, n = 6 - 10; □ = PL-017 (μg, icv), n = 8 - 11; ■ = PL-017 (1 μg, icv) + NT, n = 6 - 12; * $p \leq 0.05$.

Hypothermia following PL-017/NT in combination lasted over the entire 3-hr post-injection period, contrasting with NT-induced hypothermia which lasts for 60 - 75 min post-injection. Pretreatment with CTAP, a μ-selective antagonist, blocks the PL-017-induced, time-independent change in Tb, leaving the NT-induced hypothermia in place.

DPDPE, the δ-selective agonist, has no significant effect on Tb, alone or in combination with NT (data not shown).

Discussion

The functional role of the opioid receptors and their ligands in the regulation of body temperature involves interaction with neurotransmitter and neuropeptide systems. In the rat, it has been demonstrated that ligands such as morphine, DAMGO and PL-017, selective for the μ receptor, induce hyperthermia (1-3). This hyperthermia is the result of a post-injection increase in metabolic rate. The return of Tb to baseline follows an increase in heat loss and a reduction of the elevated metabolic rate (3). Central or peripheral administration of κ-selective agonists such as DY or U50, respectively, results in a reduction in Tb (1). The hypothermia is caused by a post-injection decrease in metabolic rate (3, 11). Recovery to baseline levels follows an increase in metabolic rate accompanied by a reduction in heat loss (3). Handler et al. (3) have shown that the δ receptor plays a relatively minor role in the regulation of body temperature at thermoneutral ambient temperatures. As with the effects of central administration of opioid ligands, the effects of centrally administered NT are site specific. Administration of NT into the central tegmental area (VTA) increases locomotor activity, while icv administration of NT induces hypothermia, locomotor inhibition and analgesia (12). This hypothermia has been shown to result from a post-injection increase in heat loss (6), probably utilizing vasodilatation through A_1 adenosine receptors (13).

Centrally administered NT has been characterized by Nemeroff and co-workers as an endogenous "neuroleptic" (14), with extensive interactions with ascending dopamine (DA) pathways in the mesolimbic system. Adler et al. (15) have shown that chlorpromazine (CPZ), an exogenous neuroleptic, and κ-selective opioid agonists interact in both rats and guinea pigs, potentiating both the duration and peak effect of the hypothermia. The kappa agonist alone had a greater effect in the guinea pig, probably due to the greater density of central kappa receptors in that species.

NT in combination with U50 had a dose-dependent, additive effect on Tb, with a significant shift of the NT dose-response curve to the left. Not only was the peak effect greater following U50/NT, but also the duration of the hypothermia was prolonged (> than 3 hr post-injection). The changes are qualitatively similar to those reported by Adler et al. (15) with CPZ and were expected since both NT and CPZ produce an increase in heat loss. However, the combination of NT and U50 was additive, whereas the combination of U50 and CPZ was superadditive. We also observed that the sedative effects and reduction in locomotor activity of the combination were greater than that of either drug alone. NLX completely blocked the kappa-mediated component of the hypothermia, while central administration of nor-BNI (20 nmol, t= -30 min) attenuated about 20% of the total combination effect.

DY/NT in combination did not induce hypothermia that differed significantly from that of NT alone, leading to the conclusion that NT blocked DY-induced hypothermia. However, in the studies that used 15 μg of DY in combination with a range of NT doses, stereotypic barrel rolling was observed in approximately one-third of the animals. If DY is given alone, the same proportion of the animals exhibit this stereotypic behavior. Thus, the centers for activity and temperature regulation may be separate. DY/NT in combination should activate central NT and κ receptors. The inability of this combination to affect either the duration or peak effect of the hypothermia may be due to the low density of central κ receptors in the rat. There also may be a feedback loop that is activated when DY and NT are given together, thus preventing a further reduction in Tb. In contrast, the combination of U50/NT would be expected to activate both central and peripheral κ receptors. The data appear to indicate that both central and peripheral effects of the kappa opioid agonist may be necessary for the additive effect on Tb when given in combination with NT in the rat.

The antagonism of PL-017-induced hyperthermia by co-administration of NT was not unexpected. Xin et al. (16) has shown that activation of opioid receptors within the POAH and periaqueductal gray area can increase the release of NT and that NT is involved in the effects of opioids on both analgesia and regulation of body temperature. The shift of the NT dose-response curve to the left, following co-administration of PL-017 may well be due, at least in part, to this stimulated release of endogenous NT. The release of endogenous NT is additive to the administered NT, causing the release of a greater quantity of metabolic heat (6), thus shifting the curve to the left. It is also possible that the post-injection increase in heat loss following PL-017-induced hyperthermia is due to this same stimulated release of endogenous NT, thus returning Tb to baseline (3).

Co-administration of DPDPE and NT did not produce any significant shift of the NT dose-response curve. There appears to be little interaction between δ opioid receptors and NT, and this further confirms the minor role of the δ receptor in the thermoregulatory process at ambient temperatures within the thermoneutral zone.

Thermoregulatory actions do not take place in isolation but are integrated and intricately related with the actions of other regulatory systems. Prange (17) has noted that neuropeptides may be classified into families based on origin, modification, localization and common properties. Thus, the selectivity that we see from various peptides is probably not the result of a wide array of receptors but rather a function of localization, modification and interaction. The data presented in this study support the hypothesis that interaction between the opioid peptides and NT has a modulatory effect on thermoregulation. Further, these thermoregulatory changes do not occur in isolation but are part of the adaptive responses of the organism as a whole.

Acknowledgements

Support provided by grants T32DA07237 and DA00376 from NIDA

References

1. Adler, MW, Geller, EB, Rosow, CE, Cochin, J (1988): The opioid system and temperature regulation. Ann Rev Pharmacol Toxicol 28, 429-449.

2. Spencer, RL, Hruby, VJ, Burks, TF (1990): Alteration of thermoregulatory set point with opioid agonists. J Pharmacol Exp Ther 252, 696-704.

3. Handler, CM, Geller, EB, Adler, MW, (1992): Effect of μ-, κ-, and δ-selective opioid agonists on thermoregulation in the rat. Pharmacol Biochem Behav 43, 1209-1216.

4. Kalivas, PW, Nemeroff, CB, Miller, JS, Prange Jr., AJ (1985): Microinjection of neurotensin into the ventral tegmental area produces hypothermia: Evaluation of dopaminergic mediation. Br Res 326, 219-227.

5. Nemeroff, CB, Bissette, G, Prange Jr., AJ, Loosen, PT, Barlow, TS, Lipton, MA (1977): Neurotensin: central nervous system effects of a hypothalamic peptide. Br Res 128, 485-496.

6. Handler, CM, Bradley, EA, Geller, EB, Adler, MW (1994): A study of the physiological mechanisms contributing to neurotensin-induced hypothermia. Life Sci 54, 95-100.

7. Nemeroff, CB, Osbahr, AJ, Manberg, PJ, Ervin, GN, Prange Jr., AJ (1979): Alterations in nociception and body temperature after intracisternal administration of neurotensin, β-endorphin, other endogenous peptides and morphine. Proc Natl Acad Sci USA 76, 5368-5371.

8. Urban, MO, Smith, DJ (1993): Role of neurotensin in the nucleus raphe magnus in opioid-induced antinociception from the periaqueductal gray. J Pharmacol Exp Ther 265, 580-586.

9. Suarez-Roca, H, Maixner, W (1993): Activation of kappa opioid receptors by U50488H and morphine enhances the release of substance P from rat trigeminal nucleus slices. J Pharmacol Exp Ther 264, 648-653.

10. Rhim, H, Glaum, SR, Miller, RJ (1993): Selective opioid agonists modulate afferent transmission in the rat nucleus tractus solitarius. J Pharmacol Exp Ther 264, 795-800.

11. Lynch, TJ, Martinez, RP, Furman, MB, Geller, EB, Adler, MW (1987): A calorimetric analysis of body temperature changes produced in rats by morphine, methadone, and U50,488H. In Problems of Drug Dependence 1986. NIDA Res Monograph 76. Harris, LSed. Washington, DC, U.S. Dept Health and Human Services, p.82.

12. Kalivas, PW, Jennes, L, Nemeroff, CB, Prange Jr., AJ (1982): Neurotensin: Topographical distribution of brain sites involved in hypothermia and antinociception. J Comp Neuro 210, 225-238.

13. Joilcoeur, FB, Menard, D (1993):Evidence for the involvement of adenosine A_1 receptors in neurotensin-induced hypothermia. In: The Neurobiology of Neurotensin. Kitabgi, P, Nemeroff, CB, eds. Ann New York Acad Sci 668, 353-354.

14. Nemeroff, CB, Levant, B, Myers, B, Bissette, G (1993): Neurotensin, antipsychotic drugs, and schizophrenia: Basic and clinical studies. In: The Neurobiology of Neurotensin. Kitabgi, P, Nemeroff, CB, eds. Ann New York Acad Sci 668, 146-156.

15. Adler, MW, Geller, EB (1987): Hypothermia and poikilothermia induced by a κ-agonist opioid and a neuroleptic. Eur J Pharmacol, 140, 233-237.

16. Xin, L, Geller, EB, Adler, MW (1993): Neurotensin (NT) release from the periaqueductal gray (PAG) and preoptic anterior hypothalamus (POAH) of the rat: Effects of selective μ and κ opioid receptor agonists. Soc Neurosci Abst 19, 1160.

17. Prange Jr., AJ (1993): The manifold actions of neurotensin, a trophotropic agent. In: The Neurobiology of Neurotensin. Kitabgi, P, Nemeroff, CB, eds. Ann New York Acad Sci 668, 298-310.

THERMOREGULATORY AND CARDIOVASCULAR EFFECTS OF ONO-3144 IN THE RABBIT

M. J. Dascombe

Division of Neuroscience, School of Biological Sciences, Stopford Building 1.124, University of Manchester, Manchester M13 9PT (England)

ONO-3144 (2-aminomethyl-4-tert-butyl-6-propionylphenol hydrochloride) is a nonsteroidal anti-inflammatory agent, which does not inhibit cyclo-oxygenase activity (1). The anti-inflammatory action of this compound is thought to be due to its ability to scavenge proinflammatory free radicals (1-3). Prostaglandin (PG) biosynthesis is affected, however, by ONO-3144 stimulating PG hydroperoxidase activity converting PGG_2 to PGH_2, and inhibiting thromboxane synthetase. Despite the facilitated formation of PGH_2 and consequently of PGE_2, an important mediator of fever (4), ONO-3144 has been reported to produce antipyresis in febrile rats (5). In view of these apparently paradoxical effects by ONO-3144 on PGE_2 production and fever in the rat, and because antipyretic drugs can have different effects in models of fever in rats and rabbits (6), the potential antipyretic activity of ONO-3144 in rabbits has been investigated in this study.

Methods

Female half lop rabbits (2-4 kg) were used in this study. Colonic temperature was measured with a thermistor inserted 8-10 cm past the anus while rabbits were restrained in stocks at an ambient temperature of 15°C, 20°C or 26°C. Skin temperature was measured with a surface thermistor over the arteriovenous region on the dorsal surface of the pinna contralateral to the ear used for i.v. injections. A submaximal fever was induced by injection i.v. of *Salmonella abortus equi* lipopolysaccharide (LPS, 1μg/kg) into a marginal ear vein. Afebrile control rabbits received vehicle (0.9% saline 0.2 ml/kg i.v.). ONO-3144 or its vehicle (water for injection, 1 ml/kg) was injected i.p. 1 h after LPS or its vehicle. Procedures minimising contamination by extraneous pyrogens were used for the preparation and injection of solutions (7). Temperature responses were assessed as the change in temperature from preinjection values (Δ°C) and as a temperature response index (TRI) integrating Δ°C against time (h).

The effects of ONO-3144 on blood pressure and heart rate *in vivo* were determined in pentobarbitone-anaesthetised rabbits via a catheter in a femoral artery. Drug solutions were injected through an indwelling catheter in a femoral vein. The effects of ONO-3144 on two *in vitro* preparations, the isolated perfused rabbit heart and the rabbit aortic strip, were determined using methods described elsewhere (8,9), using Ringer-Locke or Krebs' solution respectively, aerated with 95% O_2/5% CO_2 and at 37°C, to maintain the preparations. Cardiovascular responses *in vivo* and *in vitro* were assessed as percentage changes (Δ%) from preinjection values.

Results are presented as the mean ± 1 s.e.m. (unless otherwise indicated) for n experiments. The probability (P) of the significance of the difference between means was evaluated by a 2-tailed Student's t-test. Doses cited relate to the following salts of the drugs or compounds used: acetylcholine chloride, adrenaline hydrochloride, 2-aminomethyl-4-tert-butyl-6-propionylphenol hydrochloride, atropine sulphate, sodium flurbiprofen, indomethacin, pentobarbitone sodium, pentolinium ditartrate, potassium chloride, propranolol hydrochloride.

Results

In rabbits held at a room temperature of 20°C, ONO-3144 (10 mg/kg i.p.) reduced the febrile response to LPS assessed as the TRI ($P < 0.05$, Figure 1); a lower dose of ONO-3144 (2 mg/kg i.p.) had no effect. The reduction in fever was associated with a rise in skin temperature (Figure 1), which was maximal (+7.4 ± 1.7°C; $P < 0.05$) about 1 h after i.p. injection (control skin temperature 26.9 ± 0.7°C). A similar rise in skin temperature, together with a similar fall in colonic temperature, was seen in afebrile rabbits receiving ONO-3144 (Figure 1). Temperature responses to ONO-3144 (10 mg/kg i.p.) were accentuated ($P < 0.05$; n, 5) by exposure to mild cold (room temperature 15°C; control skin temperature 23.2 ±1.0°C) and abolished by warm exposure (room temperature 26°C; control skin temperature 35.9 ± 0.3°C). The hypothermic effect of ONO-3144 in afebrile rabbits at 15°C was not affected by either indomethacin (10 mg/kg i.p.; n, 6) or flurbiprofen (8 mg/kg i.v.; n, 5) injected 15 min before ONO-3144.

In anaesthetised rabbits, ONO-3144 produced transient, dose related falls in systolic blood pressure (Figure 2), diastolic blood pressure and heart rate. Low doses of ONO-3144 (0.01 and 0.1 mg/kg i.v.) had no effect on these parameters. The hypotension and bradycardia induced by ONO-3144 (0.5 and 5 mg/kg i.v.) were not antagonised by i.v. pretreatment with atropine (1 and 5 mg/kg), indomethacin (10 mg/kg), pentolinium (10 mg/kg) or propranolol (1 mg/kg). Propranolol and pentolinium did antagonise the recovery rises in blood pressure and heart rate.

In the isolated rabbit heart, bolus injections of ONO-3144 (dose range 0.064 - 1mg) produced negative chronotropic and inotropic effects (Figure 3). These effects were not antagonised by either atropine (5 µg/ml; Figure 3) or indomethacin (1 µg/ml) in the perfusion fluid. Atropine (5 µg/ml) did antagonise the negative chronotropic and inotropic effects of acetylcholine (dose range 0.1 - 4 µg). ONO-3144 (5 µg/ml) added to the Ringer-Locke solution reduced the basal force and the normal rate of contraction of the isolated heart, as well as the positive inotropic response to adrenaline (dose range 0.25 - 4µg) by about 50%.

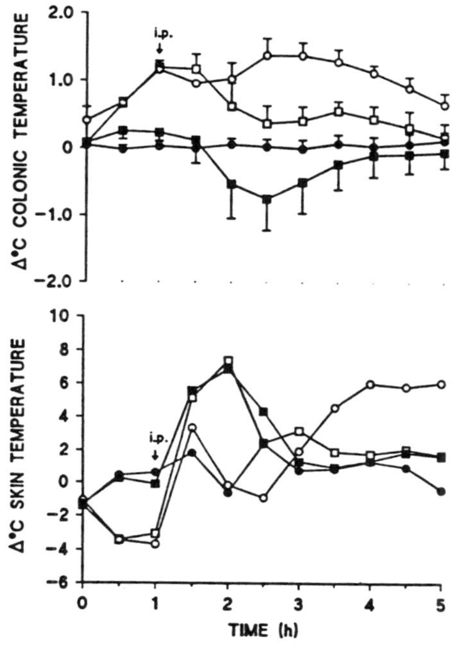

Fig. 1: Changes in colonic and ear skin temperatures in restrained rabbits at a room temperature of 20°C. Points represent the mean change from preinjection control values (Δ°C) and bars the s.e.m., some values omitted for clarity (n,6). LPS (1μg/kg, open symbols) or saline (0.2 ml/kg,closed symbols) was injected i.v. at 0 h. ONO-3144 (10 mg/kg,□) or water for injection (1ml/kg,○) was injected i.p. at 1 h.

Spiral strips of rabbit aorta contracted maximally *in vitro* by either potassium (50 mM) or adrenaline (0.1, 1 or 10 μg/ml final bath concentration) were relaxed by ONO-3144 (50 and 100 μg/ml). In contrast, the contractions produced by these spasmogens were only slightly reduced by acetylcholine (upto 100 μg/ml). The relaxant effect of ONO-3144 on potassium-induced contractions were not antagonised by indomethacin; in paired preparations from the same rabbits, ONO-3144 (50 μg/ml) produced a 50 ± 2% (n, 4) reduction in potassium-induced contraction in the presence of indomethacin (1 μg/ml), compared with a relaxation of 53 ± 4% in concurrent controls in normal Krebs' solution.

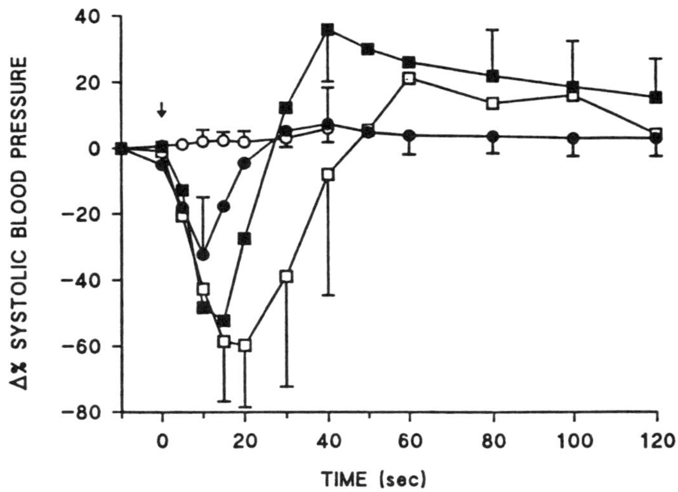

Fig. 2: Effect of ONO-3144 on systolic blood pressure in anaesthetised rabbits. Points represent the mean percentage change from preinjection values (Δ% systolic blood pressure) and bars the s.d. (n, 8-12). ONO-3144 0.1 (○), 0.5 (●), 5 (■) or 10 (□) mg/kg injected i.v. at time 0 (↓). Vehicle (0.9% saline, dose < 0.5 ml/rabbit) had no effect on systolic blood pressure (control value about 75 mm Hg).

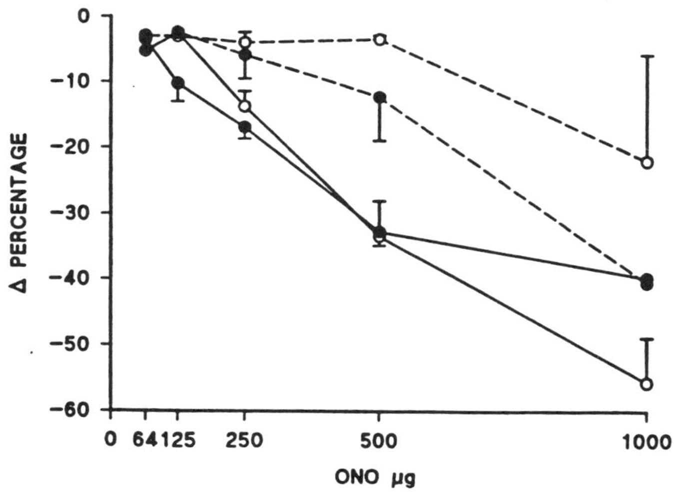

Fig. 3: Effect of ONO-3144 on the force (—) and the rate (- -) of contraction of the isolated rabbit heart perfused with normal physiological salt solution (○) or in the presence of atropine (5 μg/ml, ●). Points represent the mean percentage change from preinjection control values (Δ percentage) and bars the s.e.m. (n, 4 or 5).

Discussion

ONO-3144 has anti-inflammatory activity in various experimental models in mice and rats, its possible mechanism of action involving the scavenging of free radicals rather than the inhibition of cyclo-oxygenase (1-3). In addition, ONO-3144, in doses not affecting normal rectal temperature, has been reported to reduce fever induced in rats by yeast or LPS (5). In contrast with its antipyretic effect in the rat, the findings of this study indicate that ONO-3144 reduces fever in rabbits by a hypothermic action, and not by a specific antipyretic effect. The magnitude of the hypothermia induced by ONO-3144 in febrile and afebrile rabbits was dependent on the ambient temperature, but independent of PG formation. Hypothermic responses in rabbits at room temperatures of 15°C and 20°C were associated with peripheral vasodilatation and increases in skin temperature. Hypothermia was not observed in rabbits at a room temperature of 26°C, when peripheral vasodilatation was already developed. Indications are, therefore, that the fall in deep body temperature caused by ONO-3144 in the rabbits used in this study was passive, and secondary to cardiovascular effects of the drug.

Consistent with this hypothesis, ONO-3144 produced hypotension and bradycardia in anaesthetised rabbits at a site peripheral to the ganglia (pentolinium resistant), and independent of atropine- or propranolol-sensitive receptors, or cyclo-oxygenase activity. ONO-3144 had direct negative chronotropic and inotropic actions on the isolated perfused rabbit heart and a direct relaxant effect on rabbit arterial smooth muscle. The latter effect is unlikely to involve endothelium-derived relaxant factor, because acetylcholine did not have a potent relaxant effect on the spiral aortic strip contracted by adrenaline or potassium. The effects of ONO-3144 on the cardiovascular parameters monitored in these anaesthetised rabbits were transient (< 2 min), whereas the hypothermic effect of ONO-3144 in conscious rabbits was more prolonged, which could indicate a lack of relationship between the cardiovascular and the thermoregulatory effects of the drug. However, the difference in duration of these effects could also be due to the different routes of administration employed, the different parameters monitored and/or the presence or absence of general anaesthesia.

It remains conceivable, therefore, that ONO-3144 does impair thermoregulation in both the febrile and the afebrile rabbit as a consequence, either directly or indirectly, of its cardiovascular effects. The possible mechanisms for the development of hypothermia include heat loss secondary to peripheral vasodilatation, caused perhaps by the direct relaxant effect of ONO-3144 on vascular smooth muscle, especially after i.v. injection, and a reduction in thermogenesis as a result of changes in tissue blood flow.

Acknowledgements

The author thanks Dr. K. Crowshaw (ONO Pharmaceuticals Co Ltd, London) for support, ONO Pharmaceuticals Co Ltd, Osaka for the gift of ONO-3144, and T. Browne and D. Price for technical assistance.

References

1. Aishita, H, Morimura, T, Obata, T, Miura, Y, Miyamoto, T, Tsuboshima, M, Mizushima, Y (1983): ONO-3144, a new anti-inflammatory drug and its possible mechanism of action. Arch Int Pharmacodyn Ther 261, 316-327.

2. Crowshaw, K (1984): Role of inhibition of prostaglandin biosynthesis by antiinflammatory drugs in the pathogenesis of renal damage. Adv Inflammation Res 6, 149-158.

3. Cheeseman, KH, Forni, LG (1988): An investigation of the novel anti-inflammatory agents ONO-3144 and MK-447: studies on their potential antioxidant activity. Biochem Pharmacol 37, 4225-4233.

4. Milton, AS (1982): Prostaglandins in fever and the mode of action of antipyretic drugs. Handbk Exp Pharmacol 60, 257-303.

5. Aishita, H, Morimura, T, Sakai, Y, Sakaguchi H, Tsuboshima, M (1982): Biochemical and pharmacological activities of 2-aminomethyl-4-tert-butyl-6-propionylphenol. Ensho 2, 465-466.

6. Dascombe, MJ (1984): Effects of diflunisal on fever in the rabbit and the rat. J Pharm Pharmacol 36, 437-440.

7. Dascombe, MJ, Milton, AS (1975): The effects of cyclic adenosine 3',5'-monophosphate and other adenine nucleotides on body temperature. J Physiol (Lond) 250, 143-160.

8. Gunn, JA (1913): An apparatus for perfusing the mammalian heart. J Physiol (Lond) 46, 506-508.

9. Furchgott, RF, Bhadrakom, S (1953): Reactions of strips of rabbit aorta to epinephrine, isopropylarterenol, sodium nitrite and other drugs. J Pharmacol Exp Ther 108, 129-143.

Thermal Balance in Health and Disease
Advances in Pharmacological Sciences

EFFECT OF A NON-PEPTIDE ANGIOTENSIN II RECEPTOR ANTAGONIST ON THE ELEVATED BLOOD PRESSURE OF COLD-EXPOSED RATS

M. J. Fregly, Z. Sun and J. R. Cade

Dept. of Physiology, University of Florida, College of Medicine, Gainesville, FL 32610

Previous studies from this laboratory have shown that rats exposed chronically to 5°C develop hypertension, including elevated blood pressure and cardiac hypertrophy (1,2). The renin-angiotensin (RA) system is believed to play a role in initiating the elevation of blood pressure since plasma renin activity (PRA) increases within the first 3 weeks of exposure to cold, but declines to control level thereafter (3). The mechanism(s) that maintain the cold-induced elevation of blood pressure is unknown. The present experiment was designed to assess the role of the RA system in the maintenance of cold-induced hypertension by chronic treatment of rats with the non-peptide angiotensin II (AngII) receptor antagonist, losartan potassium.

Methods

Sixteen male rats weighing initially 480 to 520 g were used. All rats were kept in individual cages and received Purina Laboratory Chow (#5001) and tap water ad libitum. They were maintained in a vivarium kept at 25±2°C and illuminated from 7 AM to 7 PM. At the beginning of the experiment, the rats were divided randomly into 2 groups, one of which was exposed to a thermoregulated cold room (5±2°C) also illuminated from 7 AM to 7 PM. The second group remained at 25°C. Beginning the third week of exposure to cold, systolic blood pressure of each rat was measured weekly from the tail by the method of Fregly (4). After 6 weeks of exposure to cold, both the cold-treated and control groups were administered losartan potassium (0.75 g/l) in their drinking water. Administration continued for 2 weeks during which the cold-treated group ingested 9.6 mg losartan/kg/day while the warm-acclimated control group ingested 5.4. The difference in drug dosage is related to the greater fluid intake of the cold-treated group (12.7 ml/kg/day) compared to the control (7.1 ml/kg/day). At the end of the 2 week treatment, losartan was withdrawn for the next 4 weeks. At this time, the cold-treated and control groups were divided in half and half of each received losartan potassium (0.75 g/l) in their drinking water. The remaining 2 groups were untreated. Treatment continued for an additional

2 weeks. To assure that treatment with the dose of losartan used was effective, all rats were implanted with a carotid and a jugular cannula (PE50) while anesthetized with sodium pentobarbital (45 mg/kg, i.p.). The arterial cannula was coupled via a Statham pressure transducer to a NarcoBio Systems polygraph. AngII was administered via the jugular vein at 0.032, 0.064 and 0.128 µg/kg as single bolus injections. Mean blood pressure response was recorded continuously from the carotid artery.

At the conclusion of this study, the rats were killed with an overdose of pentobarbital and heart, kidneys, thyroid and adrenal glands were removed, cleaned of extraneous tissue and weighed.

Statistical analyses consisted of Student's t test for comparing the blood pressures in the initial 10 weeks of the experiment and a 2 way analysis of variance for data collected thereafter.

Results

Exposure to cold increased systolic blood pressure significantly ($P<0.01$) (Figure 1).

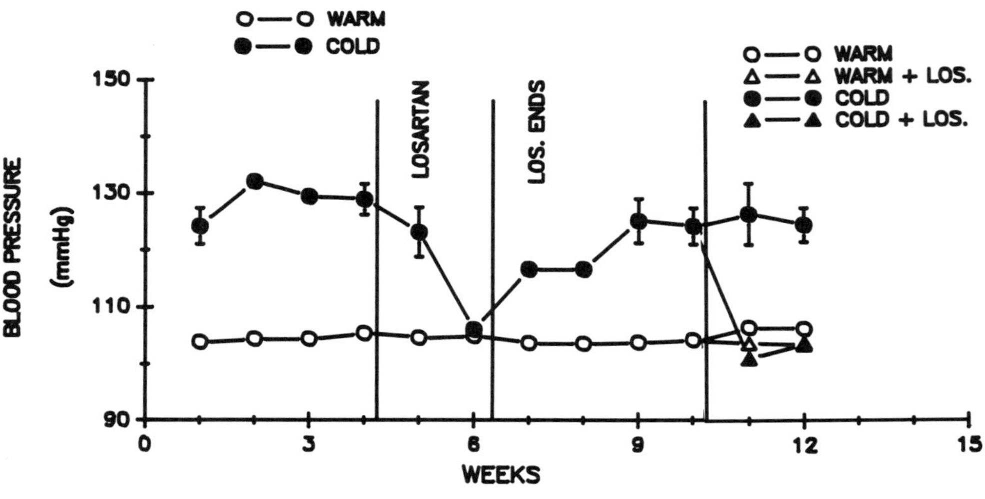

Fig. 1: The effect of administration of losartan potassium on the elevated blood pressure of rats exposed chronically to cold. One SE is set off at each mean. When SE is not shown, it falls within the symbol.

Administration of losartan after 6 weeks of exposure to cold was accompanied by a reduction in blood pressure to the level of the warm-acclimated group within 2 weeks. Blood pressures of the warm-acclimated group were unaffected by treatment with losartan when compared with their blood pressures prior to treatment. Withdrawal of losartan at week 8 resulted in an increase in blood pressure. At the end of week 12, when blood pressures of the cold-treated group had reached previous high levels, both cold-treated and warm-acclimated groups were divided in half and half of each group was treated with losartan. Blood pressure of the treated, cold-adapted group again returned to control level, but this time within 1 week. Blood pressures of the untreated groups remained relatively constant compared with levels prior to treatment.

Body weights of the cold-treated groups increased at a slower rate than warm-acclimated groups. The initial mean body weights of the cold-treated, and warm-acclimated rats at the time cold exposure began were 496 and 493 g, respectively. After 12 weeks of exposure to cold, they were 549 and 662g, respectively.

The effect of administration of AngII to both warm- and cold-acclimated rats treated with losartan on change in mean blood pressure is shown in Table 1. Only data from the 2 higher doses are shown. At these doses (0.064 and 0.128 μg/kg), the changes in mean blood pressure of the warm- and cold-acclimated groups increased to approximately the same extent. Treatment of each group with losartan significantly prevented the rise of pressure to that of untreated counterparts. This suggests that the dose of losartan used blocked the response to AngII.

Tab. 1: Effect of acute i.v. administration of angiotensin II on the change in mean blood pressure.

Experimental group	Change in mean blood pressure (mm Hg) with:	
	0.064	0.128 μg Ang II/kg, i.v.
Warm-acclimated	49.3 ± 8.3+	47.3 ± 9.0
Warm-accl. + Losartan	4.0 ± 1.5**	14.7 ± 8.1*
Cold-acclimated	48.0 ± 8.0	54.0 ± 9.6
Cold-accl. + Losartan	11.0 ± 5.0*	11.5 ± 6.0**

+one SE of mean; * = $p < 0.05$; ** = $p < 0.01$ compared to control group

Exposure to cold was accompanied by an increase in the weight of the heart, kidneys, thyroid, and adrenal glands (Figure 2A-D).

Treatment with losartan reduced the weight of the kidneys of the cold-treated group, but failed to affect the weights of the other 3 organs. This suggests that mechanisms other than those mediated by the renin-angiotensin system must contribute. On the other hand, it is not clear why treatment with losartan prevented the increase in renal weight of cold-exposed rats.

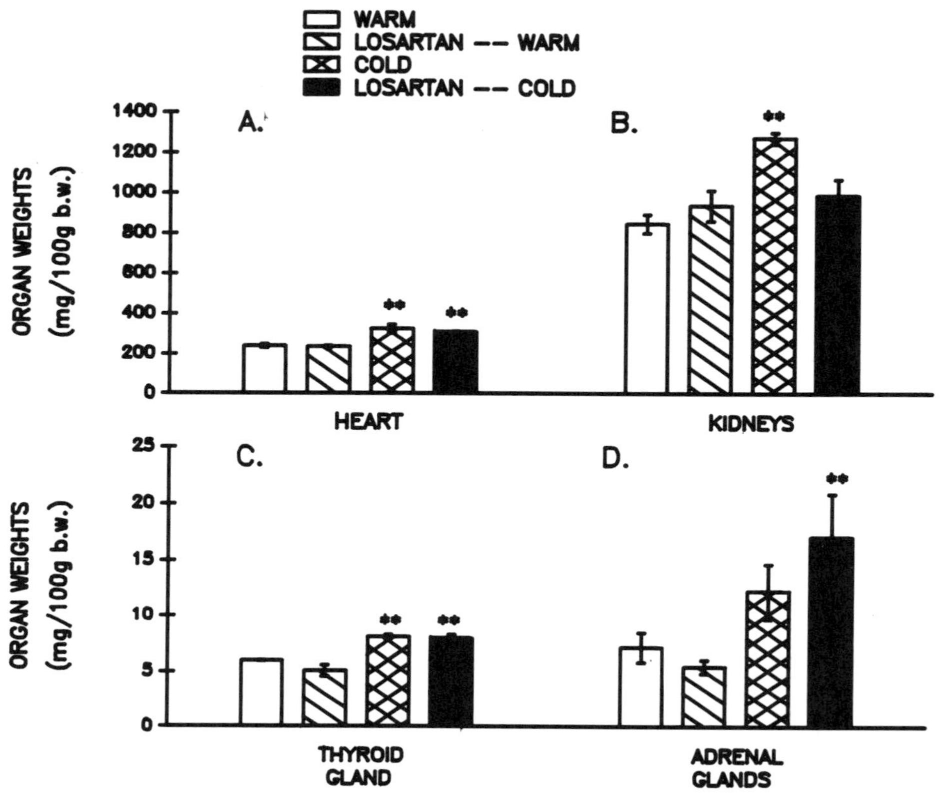

Fig. 2: Effect of treatment with losartan on the weights of heart, kidneys, thyroid and adrenal glands of warm- and cold-treated rats. Mean and SE are shown. ** = P<0.01 compared to warm-acclimated control group.

Discussion

Earlier studies from this laboratory showed that plasma renin activity of cold-exposed rats increased during the first 3 weeks of exposure to cold, but returned thereafter to control level, or slightly below (3). Losartan, administered to rats either prior to or immediately following exposure to cold, prevented the usual elevation of blood pressure (5). These results suggested that the renin-angiotensin system initiated, but did not maintain the elevation of blood pressure during exposure to cold. In the present study, chronic administration of losartan to cold-exposed rats after their blood pressures were significantly elevated reduced blood pressure to the level of controls within 2 weeks. This suggests that the renin-angiotensin system may also play a role in the maintenance of cold-induced hypertension. The mechanism by which this occurs may be related to an up-regulation of AngII receptors. Supporting this possibility is the increased dipsogenic response of chronically cold-treated rats to AngII (3). The vascular responsiveness of cold-treated rats to i.v. administration of graded doses of AngII has also been shown to increase significantly above control level within 1 week of exposure to cold, to return toward the control level at 3 weeks and to reach control level at 5 weeks (6). Thus, vascular responsiveness to AngII is increased at a time during exposure to cold when plasma renin activity is also increased.

The failure of mean blood pressure of the losartan-treated groups to respond to acute administration of graded doses of AngII to the level of controls suggests that the receptors for AngII were blocked by the dose of losartan and route of administration used in this study. There was no significant difference between the responses of the cold-treated and warm-acclimated control groups to administration of AngII. The failure to observe a difference in responsiveness to AngII between these 2 groups after 14 weeks of exposure to cold is consistent with the results of the earlier study mentioned above (6).

Treatment with losartan failed to prevent cardiac hypertrophy in cold-exposed rats although it reduced blood pressure significantly. This observation is consistent with the results of an earlier study in which losartan was administered at the time the rats were exposed to cold and for a much longer period of time. An increase in plasma concentration of catecholamines is well known to induce cardiac hypertrophy (7). Since exposure to cold is accompanied by an increase in the rate of secretion and excretion of norepinephrine, as well as its concentration in plasma, this is most likely the explanation (2). Treatment with losartan has not been reported to affect either the rate of secretion or the response to norepinephrine.

The prevention of renal hypertrophy by treatment of cold-exposed rats with losartan is difficult to explain. The increase in the weight of the kidneys has been attributed to a doubling of food intake by the cold-treated group with a consequent requirement for doubling excretion of electrolytes and metabolites (2,3). Since administration of losartan failed to affect either food intake of cold-exposed rats (6) or their body weight, a decrease in the requirement for excretion of electrolytes and metabolites induced by losartan could not explain these results. Additional study will be required.

Acknowledgements

Supported by grant HL39154-07 from the National Heart, Lung, and Blood Institute, NIH, Bethesda, MD.

References:

1. Fregly, MJ, Kikta, DC, Threatte, RM, Torres, JL and Barney, CC (1989): Development of hypertension in rats during chronic exposure to cold. J Appl Physiol 66:741-749.

2. Papanek, PE, Wood, CE and Fregly, MJ (1991): Role of the sympathetic nervous system in cold-induced hypertension in rats. J Appl Physiol 71:300-306.

3. Fregly, MJ, Shechtman, O, van Bergen, P, Reeber, C and Papanek, PE (1991): Changes in blood pressure and dipsogenic responsiveness to angiotensin II during chronic exposure to cold. Pharmac Biochem Behav 38:837-842.

4. Fregly, MJ (1963): Factors affecting indirect determination of blood pressure of rats. J Lab Clin Med 62:223-230.

5. Fregly, MJ, Rossi, F, van Bergen, P, Brummermann, M and Cade, JR (1993): Effect of chronic treatment with losartan potassium (DuP 753) on the elevation of blood pressure during chronic exposure of rats to cold. Pharmacology 46:198-205.

6. Fregly, MJ and Brummermann, M (1993): Effect of exposure to cold on vascular responsiveness to phenylephrine and angiotensin II. Pharmacology 47:237-243.

7. Rossi, MA and Carillo, SV (1985): Does norepinephrine play a causative role in the process of cardiac hypertrophy? Am Heart J 109:622-624.

Thermal Balance in Health and Disease
Advances in Pharmacological Sciences

NEUROPROTECTIVE EFFECTS OF AN EXTRACT OF GINGKO BILOBA (EGb-761) IN HEAT STRESS INDUCED BRAIN DAMAGE IN THE RAT

H. S. Sharma*[1-4], J. Westman[2], F. Nyberg[3], J. Cervós-Navarro[1] and P. K. Dey[4]

Institute of Neuropathology[1], Free University Berlin, Germany; Departments of Human Anatomy[2] and Pharmaceutical Biosciences[3], Uppsala University, Sweden; Department of Physiology[4], Banaras Hindu University, Varanasi, India

Introduction

Heat stress (HS) is a serious clinical situation in many parts of the tropical world (1). The symptoms of HS include hyperpyrexia, associated with delirium, mental confusion, coma and even death of the victims (1,2). Gross morphological examination of brain showed haemorrhages, edema and neuronal changes (3). The probable mechanisms of such pathological changes in brain following HS are not known in detail.

We showed earlier that increased permeability of the blood-brain barrier (BBB), reduced cerebral blood flow (CBF), brain edema and elevated levels of serotonin in plasma and brain occur in a rat model of HS (4,5). In these animals, marked cell changes are present in many parts of the brain (5,6). A breakdown of the BBB, probably by elevated serotonin, appears to be instrumental in precipitating the pathophysiological symptoms in HS. Thus pharmacotherapy of HS should be directed towards preventing the increased BBB permeability and/or increased serotonin levels.

Recently, an extract on Gingko biloba (EGb-761) has been found to enhance cerebral circulation, promote synaptosomal reuptake of serotonin and reduce ischemic cell damage in animal models (7-9). Since the pathophysiology of HS is associated with ischemia, abnormal 5-HT metabolism and cell changes (4-6, 10-12) we undertook a series of investigations on the probable neuroprotective effects of EGb-761 in HS. To understand the mechanisms of these effects of EGb-761 in HS, the influence of the extract on BBB permeability and brain edema was also examined.

Materials and methods

Experiments were carried out on 36 Sprague Dawley male rats (body weight 80-90 g, age 8-9 weeks) housed at controlled ambient temperature (21±1°C) with 12 h light and dark schedule. The food and water were supplied ad libitum.

Exposure to heat stress

The animals were exposed to HS at 38°C for 4 h in a biological oxygen demand (BOD) incubator (relative humidity 47-50 %; wind velocity 20-25 cm/sec). Animals kept at room temperature served as controls.

Parameters measured

Stress symptoms. The rectal temperature before and after HS was measured using a telethermometer. The behavioural salivation and prostration were qualitatively assessed at every 1 h interval during the 4 h session of HS. At postmortem, the gastric ulceration in the stomach wall was examined as an index of stress (4-6).
Blood-brain barrier (BBB) permeability . The BBB permeability was measured using Evans blue albumin (0.3 ml/100g of a 2 % solution in physiological saline) and ^{131}I-sodium (10 µCi/100 g) as described earlier (4-6).
Brain edema. The brain water content and volume swelling (%) were used to measure edema (5).
Morphology. The cell changes were examined using standard light and electron microscopy. About 3 µm thick multiple paraffin sections were cut and stained with Nissl and glial fibrillary acidic protein (GFAP) and myelin basic protein (MBP) immunostaining (5,6,12). The ultrathins section from epon embedded material were cut using a diamond knife (LKB, Sweden), counterstained with uranyl acetate and lead citrate and examined in a Philips transmission electron microscope (5).

EGb-761 treated groups

Separate groups of animals were treated with EGb 10 mg (n=15), 25 mg (n=15) or 50 mg/kg (n=15) *per os* for 5 days. On the sixth day, ten animals in each group were subjected to 4 h HS . Separate groups of drug-treated normal animals were used as drug-treated controls.

Statistical evaluation

The Student's unpaired *t*-test was used to evaluate statsistical significance. A p value less than 0.05 was considered to be significant.

Results

Untreated animals exposed to HS

Exposure of untreated animals to HS resulted in profound hyperthermia (41.89±0.23°C), behavioural salivation and prostration. The blood-brain barrier (BBB) permeability to Evans blue (550 %) and ^{131}I-sodium (484 %) was markedly increased in these animals as compared to normal animals. The brain water content increased by more than 5 % compared to the normal animals (Table I).

Table I. Influenceof EGb-761 (50 mg/kg) in normal and heat stressed animals. The drug was given orally for 5 days. Experiments were done on the 6th day.

Type of Experiment	Control		4 h HS (38 ° C)	
	Untreated	EGb-761 treated	Untreated	EGb-761 treated
A. Stress symptoms				
1. Δ° C Rect Temp	+ 0.14±0.02	+0.18±0.04	+3.74.±012**	+2.89±0.15**a
2. Salivation	-	-	+++	++
3. Prostration	-	-	+++	++
4. Gastric ulceration^b	-	-	48±12 (microhaemorrhages)	28±8^a
B. BBB permeability				
1. Evans blue mg %	0.24±0.04	0.28±0.06	2.46±0.22**	1.55±0.32**a
2. 131I-sodium %	0.22±0.02	0.24±0.04	2.87±0.33**	1.89±0.28**a
C. Brain edema				
1. Water content %	76.67±0.23	77.54±0.21*	81.56±0.34**	79.56±0.44**a
2. Volume swelling % f	-	+4	+26.5	+14

Value are mean±SD. **=P< 0.01, Dunnet test for multiple group comparison from untreated control group. a=P<0.05, compared from untreated stressed group., b=number of the haemorrhagic spots in the mucosa wall of stomach were counted.

Morphological examination of brain showed marked neuronal, glial and myelin changes. Thus Nissl staining showed profound chromatolysis, dark and distorted neurones in many parts of the brain (Figs. 1 and 2). The immunostaining of glial fibrillary acidic protein (GFAP) showed marked increases in several regions of the brain (Fig. 1). Ultrastructural investigation revealed dark neurones, edema, membrane disruption, synaptic damage, swollen neurones and glial cells as well as myelin disruption (Figs. 1 and 2). The vessels were mainly collapsed and perivascular edema was quite prominent (Fig. 2). Lanthanum, an electron dense tracer was found to be present within the endothelium and in the basement membrane (Fig. 1). Occasionally, the tracer was present in the intracytoplasmic vesicles. However, the tight junctions were mainly intact. These cellular and vascular reactions were most frequent in the brain stem reticular formation, followed by cerebellum, thalamus, hypothalamus, hippocampus and the cerebral cortex.

Effect of EGb-761 on HS

Pretreatment with EGb-761 (50 mg/kg) showed moderate hyperthermia (39.64±0.45°C) and fewer behavioural changes following HS as compared to untreated stressed groups (Table I). This effect was dose dependent. There was a significant reduction in the permeability of the BBB and brain water content with 50 or 100 mg doses. The cell changes and vascular reactions were less apparent in the drug-treated animals (50 mg/kg) as compared to the untreated group (Figs. 1 and 2). EGb-761 alone had little affect on the BBB permeability (Fig. 1). No cell changes were found after EGb-761 treatment of normal animals although the brain water content and immunostaining of GFAP were slightly elevated in drug (50 mg)-treated animals not exposed to HS (Fig. 1).

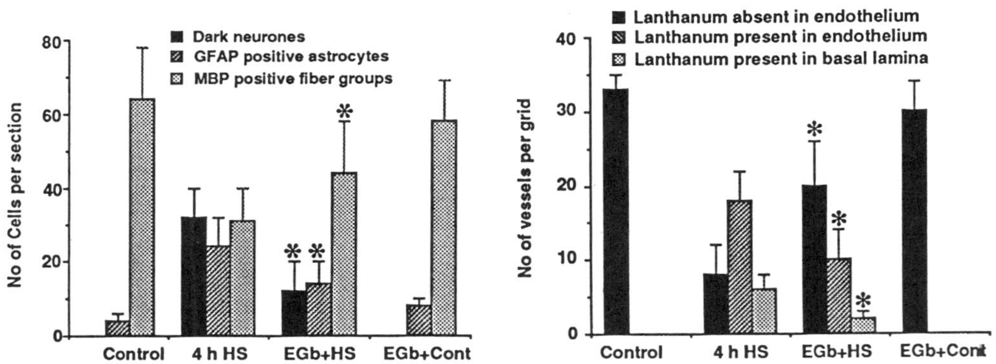

Fig. 1: Semiquantitative data on cell changes at light microscopy (left) and microvascular permeability changes at the ultrastructural level (right) in normal, heat stressed and EGb (50 mg/kg) treated rats. EGb pretreatment reduced the pathology of neurons, glial cells and myelin following HS as compared to the untreated stressed rats (left). Extravasation of lanthanum across the endothelial cells of the microvessels was considerably reduced after HS in the drud treated animals as compared to the untreated rats (right) (* = $p < 0.05$, Student's unpaired t-test, n = 5).

In conclusion, our results for the first time show that EGb-761 is quite effective in reducing brain damage in HS. This neuroprotective effect of the extract appears to be mediated by a reduction in stress response which attenuated BBB permeability and brain edema formation. Obviously, a reduced vascular reaction may partly be responsible for a diminished cellular response.

Discussion

The salient new finding of the present investigation is a marked protective effect of EGb-761 on cell changes in HS. This effect of the drug was dose-dependent. It appears that a reduction in hyperthermia, BBB permeability and brain edema by EGb-761 in HS may contribute to this effect. An increased permeability of the BBB will induce leakage of serum proteins into the cerebral compartment which will induce vasogenic edema (5,6). An increase in tissue pressure

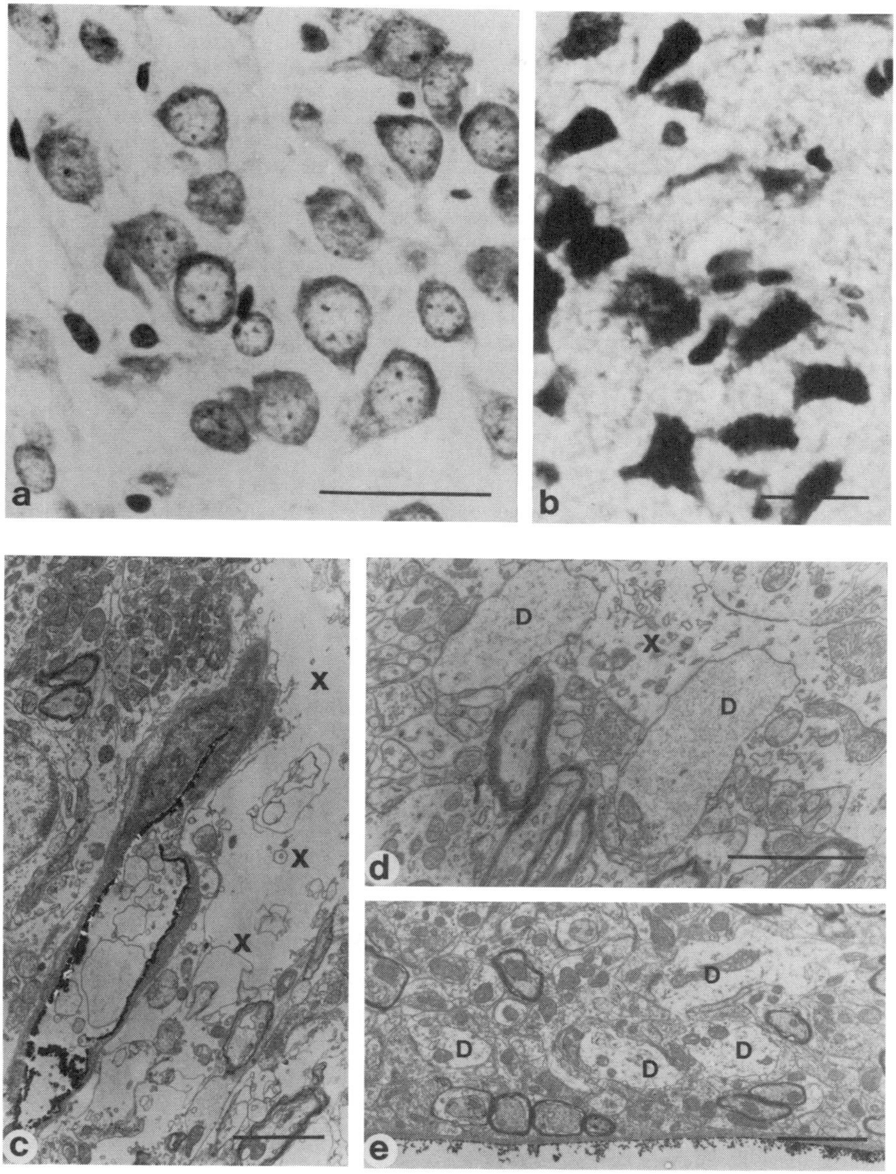

Fig. 2: Dark and distored neurones at light microscopy in the cerebral cortex following 4 h HS in one rat (b) as compared to a normal animal (a) (Bar = 100 μm). Ultrastructural examination showed perivascular edema, membrane disruption (x), swollen dendrites (D), synaptic damage, vesiculated myelin, distorted and partially collapsed vessel containing lanthanum (dark black particles) in untreated animals following HS (c-d). The neuropil appears to be less distorted following HS in EGb (50 mg/kg) pretreated animal (e) (Bar = 5 μm).

due to edema will induce local ischemia. Additionally, severe hyperthermia may alter cellular energy metabolism. All these factors alone or in combination may contribute to the cell damage. The probable mechanisms by which EGb-761 attenuated the hyperthermia, BBB permeability and brain edema following HS are unclear. Less responsiveness of animals to HS appears to be primarily responsible for the neuroprotective effects of EGb-761. A reduction in hyperthermic response and a significant diminution of stress ulcers in the stomach in EGb-761-treated animals following HS are in agreement with the above hypothesis. An attenuation of the stress response will reduce the release and accumulation in brain or plasma of serotonin and may attenuate the BBB leakage, edema formation and subsequently cell damage. A direct effect of EGb-761 on edema formation and membrane damage following HS is less likely because the drug alone induced an increase of brain water content and gliosis in normal animals (7-9). Thus it appears that the neuroprotective effects of EGb-761 in HS is due to its anxiolytic action which might reduce serotonin metabolism (7). A direct effect of EGb-761 on the synaptosomal uptake of serotonin favours this idea (9). To further clarify this point, additional investigations on the measurement of serotonin in EGb-761 treated animals in HS are needed.

Summary

The possibility that an extract of *Gingko biloba* (EGb-761) reduces brain damage following heat stress (HS) was examined in a rat model. Exposure of animals to HS (38°C, 4 h) resulted in hyperthermia (41.68±0.34°C), breakdown of the BBB permeability and brain edema. In these animals, profound neuronal, glial and myelin changes were seen in many parts of the brain. Pretreatment with EGb-761 (50 mg/kg per os for 5 days) attenuated the hyperthermic response (39.68±0.23° C), reduced microvascular permeability, brain edema and cell changes following similar HS. These results show that EGb-761 has a neuroprotective effect in HS.

Acknowledgements

This investigation was supported by Swedish Medical Research Council Project Nos 2710, 9459 and 10357; Göran Gustafssons stiftelse, Sweden; Alexander von Humboldt Foundation, Bonn; Dr William Schwabe Gmbh, Karlsruhe, Germany; and The University Grants Commission, New Delhi, India. The skilful technical assistance of Mrs Kärstin Flink, Kerstin Ryestedt, Madeleina Jarild, Gunilla Tibling, Elisabeth Scherer, Franziska Drum, Hanna Plückhan, Katja Deparade, and secretarial assistance of Mrs Aruna Misra, Angela Ludwig and Katerin Kern are highly appreciated.

References

1. Sterner S (1990) Summer heat illness. Post Grad Med 87: 67-73.
2. Sriramachari S, Patoria N K (1976) Pathology of acute encephalopathy syndrome in children in summer. Indian J Med Res 64: 293-313.
3. Austin M G, Berry J W (1956) Observation on one hundred cases of heatstroke. J Am Med Ass 161: 1525-1529.

4. Sharma H S, Dey P K (1987) Influence of long-term acute heat exposure on regional blood-brain barrier permeability, cerebral blood flow and 5-HT level in conscious normotensive young rats. Brain Res 424: 153-162.

5. Sharma H S, Cervós-Navarro J (1990) Brain oedema and cellular changes induced by acute heat stress in young rats. Acta Neurochir (Wien) Suppl 51: 383-386.

6. Sharma H S, Kretzschmar R, Cervós-Navarro J, Ermisch A, Rühle H -J, Dey P K (1992) Age-related pathophysiology of the blood-brain barrier in heat stress. Prog Brain Res 91: 197-203.

7. Christen Y, Costentin J, Lacour M (1992) Effects of Ginkgo biloba extract (EGb 761) on the Central Nervous System. Advances in Ginkgo biloba Extract Research, Elsevier, Amsterdam, pp. 1-172.

8. DeFeudis F V (1991) Gingko biloba Extract (EGb 761): Pharmacological activities and clinical applications, Elsevier, Amsterdam, pp. 1-188.

9. Dorman D C, Cote L M, Buck W B (1992) Effect of an extract of Gingko biloba on bromethalin-induced cerebral lipid peroxidation and edema in rat. Am J Vet Res 53: 138-142.

10. Sharma H S, Olsson Y, Dey P K (1990) Blood-brain barrier permeability and cerebral blood flow following elevation of circulating serotonin level in the anaesthetized rats. Brain Res 517: 215-223.

11. Dey S, Dey P K, Sharma H S (1993) Regional metabolism of 5-hydroxytryptamine in brain under acute and chronic heat stress. Indian J Physiol Pharmacol 37: 8-12.

12. Sharma H S, Zimmer C, Westman J, Cervós-Navarro J (1992) Acute systemic heat stress increases glial fibrillary acidic protein immunoreactivity in brain. An experimental study in the conscious normotensive young rats. Neuroscience 48: 889-901.

Thermal Balance in Health and Disease
Advances in Pharmacological Sciences

IS THE HYPOTHALAMIC SEROTONERGIC SYSTEM INVOLVED IN SEPTAL 5-HT-INDUCED HYPOTHERMIA IN RATS ?

L. C. H. Wang, Y. Cui and T. F. Lee

Zoology Department, University of Alberta, Edmonton, Alberta, Canada T6G 2E9

It is well documented that stimulation of central serotonergic receptors causes thermoregulatory changes in a variety of species, and the 5-hydroxytryptamine (5-HT) activity within the preoptic anterior hypothalamus (POAH) plays an important role in body temperature (T_b) regulation (for review see 1,2). As compared with other monoamines, few studies have been carried out to examine the local sensitivity to 5-HT in sites other than the hypothalamus. Recently, we have shown that injection of 5-HT into the lateral septum causes a dose-related fall in T_b in rats kept at 20°C (3). This thermoregulatory change appears to be site-specific, since injection of the same doses of 5-HT into the hippocampal area fails to elicit any significant changes in T_b.

Neuroanatomical circuits have been described between the POAH and the lateral septum (4,5) revealing that high densities of 5-HT fibers and terminals are present in both areas (6). This raises the possibility that the serotonergic systems within the septum and POAH may interact with each other in mediating thermoregulation. In order to test this hypothesis and to provide direct evidence, 5-HT and its antagonist were microinjected into these two brain areas of conscious rats and the ensuing thermoregulatory responses documented.

Methods

All experimental protocols used in the present study have received prior approval of the University of Alberta Animal Care Committee following the guidelines of the Canadian Council on Animal Care. Adult male Sprague-Dawley rats weighing approximately 400 g were used. They were housed individually at 20°C in a walk-in environmental chamber set at a 12L:12D photoperiod. Food was rationed daily to maintain body weight around 400 g throughout the experimental period. Water was available at all times. Under halothane anaesthesia, guide cannulae (23-gauge stainless steel tubing) were implanted into the lateral septum, POAH, and the hippocampus of each rat utilizing the following stereotaxic coordinates: AP=8.2 mm, L=0.7 mm, H=5.0 mm; AP=8.0 mm, L=1.0 mm, H=8.3 mm; and AP=3.6 mm, L=2.0 mm, H=3.0 mm below the dura mater for septum, POAH, and hippocampus, respectively (7). The tip of each guide tube was beveled and positioned 1.0 mm above the injection site in order to minimize damage to the actual injection site. After completion of the experiments, the precise anatomical location of the injection site was subsequently verified histologically.

One week post-operatively, each rat was transferred to a circular, plexiglas water-jacketed metabolism chamber in which the ambient temperature could be controlled accurately at 20±1°C. The T_b of the rat was recorded continuously with a pre-calibrated temperature-sensitive radiotransmitter (Model T-M, Mini-Mitter Co.), which was implanted in the peritoneal cavity under halothane anaesthesia at the same time as the guide cannula. Using an IBM personal computer interfaced with an ISAAC 91-I data acquisition system, oxygen consumption and CO_2 production were recorded and integrated simultaneously with the transmitter signal. Heat production (HP) was calculated from oxygen consumption and respiratory quotient using Kleiber's equation, and the T_b was calculated using pre-established calibration curves for each individual transmitter. Both cyproheptadine hydrochloride and 5-HT creatinine sulphate were purchased from Sigma (St. Louis). The doses in the Results section refer to the free base. Each drug was prepared immediately before an experiment in pyrogen-free artificial cerebrospinal fluid (CSF) containing 0.01 mg/ml sodium metabisulfite. All test solutions were passed through a 0.22 μm Swinnex Millipore filter to ensure sterility. Either CSF or the test compound was injected in a dose volume of 1 μl. For antagonism studies, cyproheptadine was injected 15 min. before intraseptal 5-HT injection. Student's t-test (2-tailed) was used for statistical comparisons of the effects of treatments. Significance was set at $p<0.05$, unless otherwise stated.

Results

Control rats exposed to 20°C had an initial mean T_b of 37.2±0.28°C and resting HP of 1.37±0.07 Kcal/rat/h (5.75±0.29 KJ/rat/h; n=7). Unilateral injection of 5-HT (10 and 20 μg) into the lateral septum caused a dose-related fall in T_b (Table 1), which was significantly lower than that of the CSF controls. As shown in Table 1, the decrease in T_b was associated with a reduction in HP, suggesting that the hypothermic response was brought about by a suppression of metabolic activity. Similar to that reported earlier (8), injection of same doses of 5-HT into the POAH produced a dose-related hypothermia (Table 1). However, the magnitude of the decline in T_b and HP was slightly less than that after intraseptal injection. In both cases, the rat spreaded out on the floor of the metabolism chamber for the first 10-20 min. Thereafter, the animal returned to normal posture and displayed a slight increase in activity (grooming and licking) throughout the rest of the experimental period. Simultaneously injecting a submaximal dose of 5-HT (10 μg) into the POAH and lateral septum did not cause any further decrease in either T_b or HP (Table 1).

To investigate the possibility that serotonergic systems within other brain area(s) may be involved in intraseptal 5-HT-elicited hypothermia, the effects of pretreating the animal with cyproheptadine (2.5 μg/site), a 5-HT receptor antagonist, at different brain sites on the thermoregulatory response induced by the optimal dose of 5-HT (20 μg) were examined and the results are summarized in Fig 1. Injecting cyproheptadine into the same site in the lateral septum before 5-HT injection significantly attenuated the intraseptal 5-HT-induced thermoregulatory response (Table 1). After cyproheptadine pretreatment, the changes in behavioral activities typically observed after 5-HT injection were not seen. In contrast, the 5-HT-induced hypothermia was slightly reduced or not affected after bilateral injection of cyproheptadine into POAH and hippocampus, respectively (Fig. 1).

Tab. 1: Thermoregulatory changes after injection of 5-HT into the lateral septum (LS), POAH, or both LS and POAH in rats maintained at 20°C.

Injection Site (n = 7)	Max. Change in T_b (°C)	Max. Change in HP (Kcal/rat/H)
<u>LS</u>		
CSF 1 μl	+0.22±0.15	+0.32±0.06
5-HT 10 μg	- 1.06±0.17*	- 0.26±0.06*
5-HT 20 μg	- 1.42±0.15*	- 0.37±0.13*
<u>POAH</u>		
CSF 1 μl	+0.24±0.11	+0.47±0.11
5-HT 10 μg	- 0.63±0.16*	- 0.02±0.08*†
5-HT 20 μg	- 0.94±0.21*	- 0.21±0.03*
<u>LS + POAH</u>		
CSF 1 μl + 1 μl	+0.19±0.17	+0.30±0.09
5-HT 10 μg + 10 μg	- 1.13±0.16*	- 0.27±0.14*

*Significantly different from concurrent CSF control, $p<0.05$
†Significantly different from intraseptal 5-HT 10 μg group, $p<0.05$

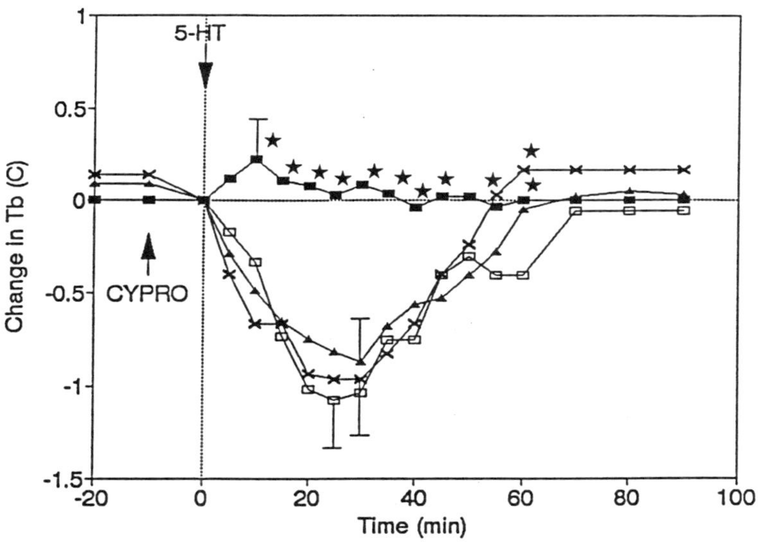

Fig. 1: Effects of intraseptal (■), intrahippocampal (X) or intrahypothalamic (▲) pretreatment with cyproheptadine (2.5 μg/site) on core temperature changes induced by intraseptal injection of 5-HT 20 μg (□) in rats kept at 20°C. Each point represents the mean change from 7 rats. The s.e. mean is represented by a vertical bar and, for clarity, is shown only at the peak response value of each dose. Arrow indicates time of injection.
*Significantly different from intraseptal 5-HT-treated group, $p<0.05$.

Discussion

Similar to our previous observation (4), injection of 5-HT into the lateral septum caused a dose-dependent decline in T_b. Since this decline in T_b was associated with a dose-related decrease in HP, the hypothermia observed after intraseptal injection of 5-HT may involve suppression of metabolic rate. Further, an increase in heat loss appears to be, at least, partly responsible for the 5-HT induced hypothermia as the behavioral changes (increased activity, prone body extension, and grooming) observed after intraseptal 5-HT injection were similar to the heat dissipation responses elicited during exposure to warm environment (9). Other than a slightly reduced hypothermic response, the pattern of intrahypothalamic 5-HT-induced thermoregulatory changes was virtually identical to those observed after intraseptal injection. The similar thermoregulatory responses elicited after injection of 5-HT into the POAH and lateral septum may thus suggest a possible functional relationship between these two brain areas. To test this possibility, a submaximal dose of 5-HT (10 μg) was injected simultaneously into lateral septum and POAH. As shown in Table 1, simultaneous injections of 5-HT into both these brain areas failed to produce an additive response. These data infer that these two structures may function independently to modify the animal's T_b.

To further examine the possibility that the serotonergic systems within the lateral septum may interact with other brain structure(s) to elicit the observed hypothermic response, a specific 5-HT receptor antagonist, cyproheptadine, was injected into different brain structures prior to intraseptal 5-HT injection. Intraseptal pretreatment with cyproheptadine completely abolished the 5-HT induced hypothermia, suggesting that the thermoregulatory effects of 5-HT were elicited by activation of septal 5-HT receptors. However, pretreating the animal with the same dose of cyproheptadine, even injected bilaterally, into either POAH or hippocampus had no effect on thermoregulatory responses consequent to intraseptal injection of 5-HT. These observations indicate that the septal serotonergic system may function independently from the hypothalamic and hippocampal systems. Analogous to this, functionally independent dopaminergic systems in regulating T_b have been demonstrated to exist within the POAH and substantia nigra (10).

In summary, the present pharmacological study indicates that the septal serotonergic system functions independently from the hypothalamic 5-HT system in regulating T_b. However, the physiological role of this thermoregulatory location remains to be elucidated. As the central serotonergic system has been implicated in sleep (10), during which heat loss increases and T_b declines, and the septohippocampal complex has been recognized as a key area in sleep regulation (11), it is plausible that the septal serotonergic system could mediate the thermoregulatory alterations unique to the physiological state of sleep.

Acknowledgements

This research was supported by a NSERC Operating Grant to L. Wang.

References

1. Myers, RD, Lee, TF (1989): Neurochemical aspects of thermoregulation. In: Advances in Environmental and Comparative Physiology, Vol. 4. Wang, LCH ed. Berlin: Springer-Verlag, 161-203.

2. Wang, LCH, Lee, TF (1989): Temperature regulation. In: Psychoendocrinology, Vol. 2, Behaviors and Regulation of Species Adaptation. Levine, S, Brush, R eds. New York: Academic Press, 437-539.

3. Cui, Y, Lee, TF, Wang, LCH (1993): Thermoregulatory responses following injections of 5-hydroxytryptamine into the septohippocampal complex in rats. Pharmacol. Biochem. Behav. 45, 935-939.

4. Kiyohara, T, Hori, T, Shibata, M, Nakashima, T, Osaka, T (1984): Neural inputs to preoptic thermosensitive neurons - histological and electrophysiological mapping of central connections. J. Therm. Biol. 9, 21-26.

5. Staiger, JF, Nürnberger, F (1991): The efferent connections of the lateral septal nucleus in the guinea pig: projections to the diencephalon and brainstem. Cell Tissue Res. 264, 391-413.

6. Steinbusch, HWM (1981): Distribution of serotonin-immunoreactivity in central nervous system of the rat - cell bodies and terminals. Neuroscience 4, 557-618.

7. Pellegrino, LJ, Pellegrino, AS, Cushman, AJ (1979): A stereotaxic atlas of the rat brain. New York: Plenum Press.

8. Cox, B, Davis, A, Juxon, V, Lee, TF, Martin, D (1983): A role for an indoleamine other than 5-hydroxytryptamine in the hypothalamic thermoregulatory pathways of the rat. J. Physiol. (Lond.) 337, 441-450.

9. Roberts, WW, Mooney, RD, Martin, JR (1974): Thermoregulatory behaviors of laboratory rodents. J. Comp. Physiol. Psychol. 86, 693-699.

10. Lee, TF, Mora, F, Myers, RD (1985): Dopamine and thermoregulation: an evaluation with special reference to dopaminergic pathways. Neurosci. Biobehav. Rev. 9, 589-598.

11. Borbely, AA, Tobler, I (1989): Endogenous sleep-promoting substances and sleep regulation. Physiol. Rev. 69, 605-670.

12. Heller, HC (1979): Hibernation: neural aspects. Ann. Rev. Physiol. 41, 305-321.

Thermal Balance in Health and Disease
Advances in Pharmacological Sciences

HYPOTHALAMIC COOLING INDUCED BY CHRONIC ANTIDEPRESSANT DRUGS IN SYRIAN HAMSTERS

W. C. Duncan, Jr., K. A. Johnson and T. A. Wehr

Clinical Psychobiology Branch, NIMH, NIH, Bethesda, Md., USA 20892

Affective disorder is often accompanied by abnormalities in the circadian pattern of body temperature (T_b), which are sometimes reversed in remission (1). These observations suggest that impaired regulation of the circadian variation in T_b is part of the pathophysiology of depression. Both pharmacological and non-pharmacological treatments of depression restore a more normal daily pattern of T_b in patients with these abnormalities. Tricyclic antidepressant drugs (1), lithium (2), electroconvulsive therapy (3), and bright light increase the circadian amplitude of T_b, and/or decrease average T_b during sleep. The fact that a variety of pharmacological and non-pharmacological treatments tend to restore a more normal circadian pattern of T_b suggests that daily variation in T_b is an important correlate of clinical antidepressant response.

Relatively few animal studies have examined chronic effects of antidepressant treatments on T_b. Here we describe effects of chronic treatment with antidepressant and neuroleptic drugs on brain temperature within the hypothalamus. We used radiotelemetry techniques to measure temperature in freely moving animals to reduce the effects of stress on thermoregulatory physiology. Antidepressant and neuroleptic drug treatment effects on the twenty-four hour, the diurnal and the nocturnal averages of hypothalamic temperature (T_{hy}) were examined.

Methods

Male golden hamsters (Mesocricetus auratus, LGV: lak, Charles River), 12-16 weeks old and weighing approximately 140 grams, were used in all studies. Prior to surgery, animals were group-housed in LD 14:10 with access to food and water ad libitum. Transmitters (Model XM-FH, Mini-Mitter Inc.) for recording hypothalamic temperature (resolution = 0.02°C) were implanted using dental cement. Temperature-sensitive probes were positioned with their tips in the hypothalamus. Surgical procedures for transmitter implants were conducted using sodium pentobarbital anaesthesia (90mg/kg, ip). After surgery, hamsters were individually housed in ventilated and sound-attenuated chambers (Napco Model 3810) with free access to food and water.

Animals were maintained under LD 14:10 (L: 01:00-15:00,10-20 μW cm^{-2}; D: 15:00-01:00) in an ambient temperature of 22±1°C . Surgical recovery was judged to have ocurred when the rhythm stabilized (about 7 days). Hamsters received clorgyline (CLG; 2 mg/kg/d; Research Biochemicals, Inc.), administered s.c via mini osmotic pumps (Model 2002, Alzet), fluoxetine (FLX; 2, 10, or 20 mg/kg/d; Eli Lilly and Company), administered in diet, lithium (.075-.45% administered in chow), or haloperidol (administered i.m.; Haldol Deconoate 100, McNeil Pharmaceutical). Different routes of drug delivery were used in order to achieve appropriate clinical doses while minimizing stress to the animals. In the course of the experiments, blood samples were taken from anesthetized hamsters to obtain measurements of plasma drug levels. At the end of the experiment animals were euthanized and their brains rapidly frozen on dry ice, cut into 30 μm sections, examined and verified for placements of transmitter probes. Following the experiment, transmitters were recalibrated and found to be consistent with pre-surgery calibration levels.
T_{hy} (°C) was recorded telemetrically every 5 minutes, and data were stored on a laboratory computer. Between five and seven days of continuously monitored temperature were selected to represent baseline (BL), subchronic (WEEK 1), and chronic (WEEK 2-7) treatment phases. During BL and during each of the treatment weeks, data were averaged over the 14 hour light period (T_{hy}-L), the 10 hour dark period (T_{hy}-D), and the total 24 hour period (T_{hy}-24), to estimate drug effects during each lighting condition and during the entire day. ANOVA and paired t-tests were used to assess differences between baseline and drug treatments. Independent group t-tests were used to assess differences between drug and control treatment.

Results

Experiment One: Clorgyline's effects on hypothalamic temperature

In the first experiment, we examined the effects of four weeks of CLG or saline treatment on T_{hy} in hamsters housed in LD14:10. The level of T_{hy} exhibited a twenty-four hour variation over the course of five weeks. Peak levels occurred during the dark phase and minimum levels occurred during the light phase, as we have previously described (Fig. 1).

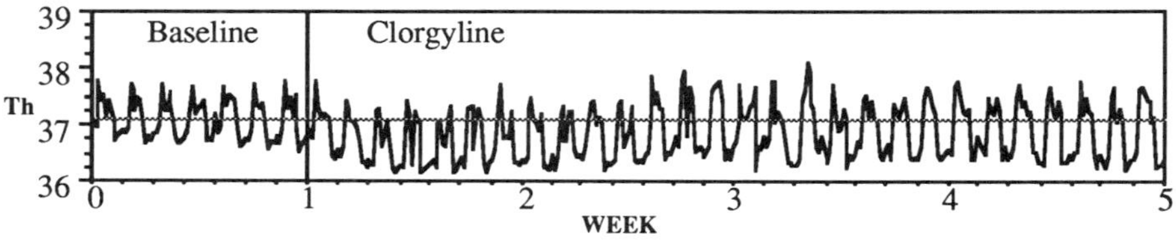

Fig. 1: The mean hypothalamic temperature of five clorgyline-treated hamsters during baseline, and during four weeks of treatment. The horizontal line represents the average Th during the baseline period.

During the first week of treatment, clorgyline decreased T_{hy}. During the second week, peak (nocturnal) T_{hy} gradually increased toward the baseline and then remained near baseline levels during the third and fourth weeks. In contrast, during the second, third and fourth weeks of treatment, minimum (diurnal) T_{hy} showed little recovery toward the baseline .
Two factor (CLG, SAL) ANOVA with repeated measures (BL,WEEK 1,WEEK 3) was used to determine the drug effects on T_{hy} during the twenty-four hour period (T_{hy}-24), and during the light (T_{hy}-L) and dark periods (T_{hy}-D). In saline-treated hamsters, T_{hy} did not change during the

four weeks of treatment. Significant interactions were found between drug treatment (CLG, SAL) and week of treatment (BL, WEEK 1, WEEK 3) for T_{hy}-24 (F=35.2, p<.001), T_{hy}-L (F=45.9, p<.001), and T_{hy}-D (F=14.2, p<.01). Compared with control, CLG decreased T_{hy}-L during WEEK 3 (p<.05).

Experiment Two: Lithium's effects on hypothalamic temperature

Increasing the dose of lithium periodically over an eight week period produced a gradual drop in T_{hy} compared with control animals. Low doses of lithium diet (0.075% to 0.15%), introduced during weeks one and two, had little effect on T_{hy}. Increasing lithium to 0.30%, decreased the level of T_{hy} after 1.5 weeks at this dose. Increasing lithium to 0.45% resulted in a further decline in the level of T_{hy}. This decline was evident in both the peak and trough of T_{hy} (Fig. 2). A gradual decline in the level of T_{hy} was observed in the control-diet hamsters, but this decline was smaller compared with the lithium-treated hamsters.

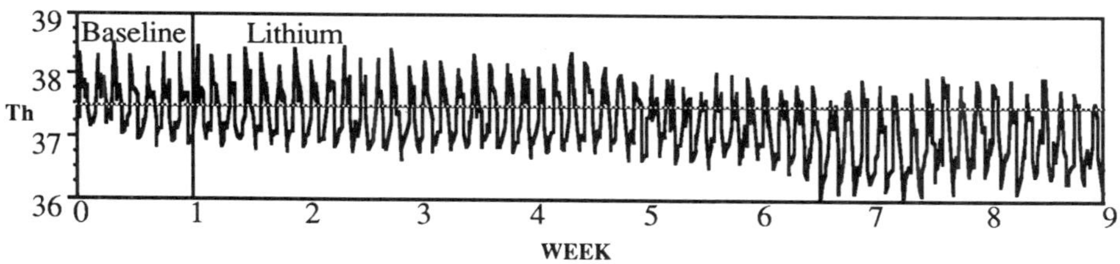

Fig. 2: The mean hypothalamic temperature of four hamsters during baseline and during nine weeks of lithium treament. Lithium diet (0.075%) was provided during week 1; dose was increased to 0.15%, 0.30% and 0.45% on weeks 2, 3, and 6, respectively. The horizontal line represents the average Th during the baseline interval.

In orbital blood samples taken from three of the four hamsters hamsters after four weeks of lithium treatment, plasma lithium levels were 0.61, 0.89 and 1.04 mEq. Plasma lithium levels measured in a separate group of hamsters ranged between 1.22-1.31 mEq/L at week eight. There were significant interactions between drug treatment (TRT: lithium vs control) and the week of treatment (BL, WEEK 4-5, WEEK 6-7) on T_{hy}-24 (F=6.48, p<.02) and T_{hy}-L (7.52, p<.01). Compared with controls, lithium significantly decreased T_{hy}-L (p<.004) and T_{hy}-24 (p<.03) during WEEK 6-7.

Experiment Three: Fluoxetine's effects on hypothalamic temperature

The 2 mg/kg/day dose of fluoxetine produced a slight decrease in T_{hy}, most evident during the trough of the daily temperature curve. Increasing the dose of fluoxetine to 20 mg/kg/day (between weeks 2 and 4) produced a further decline. During fluoxetine withdrawal, T_{hy} began to increase toward the levels observed during baseline.

Repeated measures ANOVA (BL,WEEK 1,WEEK 3, WEEK 4, WD) was used to determine the chronic effects of FLX on T_{hy} over twenty-four hours (T_{hy}-24) and during the light (T_{hy}-L) or dark phases (T_{hy}-D). There were significant measures effects on T_{hy}-L (F=5.691, p<0.034). Compared with baseline, fluoxetine decreased T_{hy}-L during WEEK 3 (p<.05) and WEEK 4 (p<.04). During week 3, 1.5 ml of blood was extracted from the orbital sinus of each hamster. This procedure was associated with a transient 2-3 day rise in the level of T_{hy}. Averaged fluoxetine and norfluoxetine plasma levels at this time were 41 ± 5.5 and 326 ± 63.5 ng/ml, respectively.

Experiment Four: Haloperidol effects on hypothalamic temperature

Injection with haloperidol decanoate 100 produced a 1.5 week elevation of T_{hy}, but two factor ANOVA with repeated measures indicated a non-significant interaction between TRT (haloperidol, vehicle) and measures (BL, WEEK 2, WEEK 3) for T_{hy}-L. Plasma drug levels measured in separate groups of hamsters after 4, 16 and 21 days of treatment averaged 20.25, 15.5 and 6.0 ng/ml respectively. Injection of sesame oil vehicle was not associated with an increase in T_{hy}.

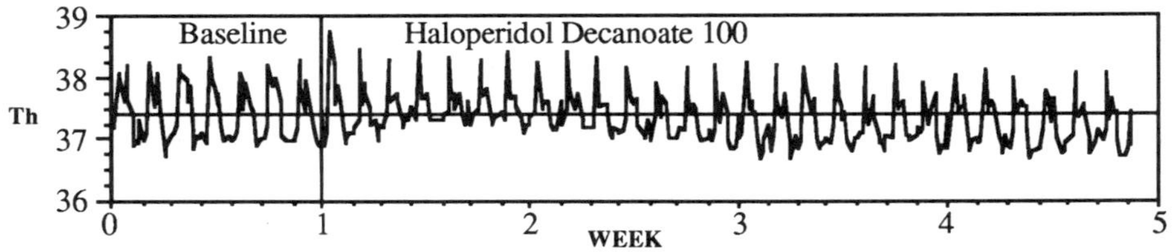

Fig. 3: The mean hypothalamic temperature of three hamsters treated with haloperidol decanoate 100. The horizontal line represents the average Th during the baseline period.

Discussion

The major finding of these experiments was that the antidepressant drugs clorgyline, fluoxetine and lithium decreased hypothalamic temperature. The decrease was most pronounced during the light-(rest-) phase of the twenty-four hour cycle. Clorgyline also decreased T_{hy} over 24 hours and during the circadian activity phase, but these effects were only present during the first week of treatment. In contrast, neither subchronic or chronic treatment with the neuroleptic drug haloperidol, decreased T_{hy} during the light phase.

A novel aspect of these experiments was the use of chronic, non-invasive radio-telemetry techniques to monitor hypothalamic temperature. Handling-stress has been shown to strongly influence body temperature and to obscure the course of fever in rodents (4). Just as telemetry has been shown to be useful in monitoring fever (4), it also appears to be a valuable method for detecting consistent, but subtle decrements in T_b induced by chronic treatment with antidepressant drugs. Since the current studies indicate that chronic antidepressant drug treatment decreases T_b, an effect of stress on T_b should be considered as a possible explanation for the failure of other investigators to detect antidepressant drug effects on T_b. In the future, non-invasive monitoring techniques will probably be essential to identify possible relationships between T_{hy}-cooling and mechanisms of antidepressant drugs.

Antidepressant Drug Effects on T_{hy} Level

The T_{hy}-lowering effects of antidepressant drugs reported here are consistent with body temperature-lowering effects of desipramine (DMI) and clomipramine (5), lithium, and electroconvulsive stimulation reported in rodents. The results of the fluoxetine experiments confirm our earlier findings that fluoxetine decreases brain temperature (6). They further indicate that fluoxetine's effects are not an evoked response to i.p. drug administration, since fluoxetine was delivered ad lib in the diet. The fact that the antidepressant drugs used in the current experiments share the common property of altering 5HT metabolism implicates this neurotransmitter in the brain-cooling effects of the treatments, although other transmitter systems may be involved. The current experiments extend previous observations of antidepressant-induced lowering of T_b to the hypothalamus, the location of a central thermostat that plays a major role in body temperature regulation.

Neuroleptic Drug Effects on T_{hy} Level

As reported by other investigators, we found that haloperidol decanoate 100 had no acute effect on T_{hy}. However, T_{hy} subsequently *increased* during the first ten days of haloperidol treatment, then returned to baseline levels. These changes were most evident during the diurnal (rest) phase of the twenty-four hour interval. Whether repeated injections or implants of these drugs would maintain the elevated T_{hy} is a question for future research. The rise and fall in T_{hy} that we observed may have been secondary to corresponding changes that we detected in plasma drug levels. It is also possible that long-term adaptive processes (desensitization) could have been responsible for the time-course on T_{hy}. A future experiment could examine whether repeated treatments that maintained elevated plasma drug levels would prolong the elevation in T_{hy} that we saw in the acute phase of the experiment.

Brain Cooling Mechanisms

Temperature of the mammalian brain is regulated by a combination of factors including regional brain metabolic rate, the temperature of arterial blood supply, the rate of brain blood flow, and heat exchange through the surface and base of the skull. With regard to metabolism and vascular factors, antidepressant drugs (DMI, and the monoamine oxidase inhibitor [MAOI] phenelzine) decrease glucose utilization (GU) in rats (7), and MAOI (8) and fluoxetine (9) treatments alter 5HT levels and reuptake within cerebral vessels that regulate brain temperature. In addition, the combination of MAOI and noradrenaline alters cerebral blood flow. Observations that a) pharmacological and non-pharmacological manipulations of 5HT decrease brain GU, and b) 5HT is present in brain arterioles (8) and modulates vasomotor function, suggest that this neurotransmitter may play a role in the mechanism of drug-induced lowering of T_{hy}.

Since, on average, the tips of the T_{hy} probes in the current studies were located about 2000μ from the ventral surface of the anterior hypothalamus, they may have been affected by venous drainage flowing from the nasal vein into the opthalamic and pterygoid sinuses. Since antidepressant drugs alter the density of central alpha and beta adrenergic receptors, and since these receptors regulate blood flow to pools at base of the brain (10), it is possible that increased venous drainage to the ventral brain surface may have been partly responsible for hypothalamic cooling induced by drugs.

Clinical Relevance

These are the first experiments to explore chronic effects of psychoactive drugs on brain temperature in mammals. The facts that clinically relevant drug doses were used in these studies, and that therapeutic plasma neuroleptic and antidepressant drug levels were associated with temperature changes suggest that similar effects may occur in humans. Recent experiments suggest that the clinical response to sleep deprivation can be modulated by warm and cold ambient temperatures (11), and by the use of antidepressant or neuroleptic drugs (12). It is possible that the modulating influence of these two treatments on the antidepressant response is related to their common effects on thermoregulatory mechanisms. Finally, during depression, body temperature is often reported to be elevated especially during the rest or sleep phase. The observation that drugs decrease T_{hy} during the rest phase indicates their effect may be partly sleep-dependent, and that these drugs may correct a pathological factor (i.e. elevated nocturnal temperature) that is associated with depression.

References

1. Souetre, E., E. Salvati, T.A. Wehr, D.A. Sack, B. Krebs, and G. Darcourt (1988): Twenty-four hour profiles of body temperature and plasma TSH in bipolar patients during depression and during remission and in normal control subjects. Am J Psychiatry 145, 1133-1137.

2. Mellerup, E.T., A. Widding, G. Wildschiodtz, and O.J. Rafaelson (1978): Lithium effect on temperature rhythm in psychiatric patients. Acta Pharmacol Toxicol 42, 125-129.

3. Avery, D.H., G. Wildschiodtz, R.G. Smallwood, D. Martin, and O.J. Rafaelson (1986): REM latency and core temperature relationships in primary depression. Acta Psychiatr Scand 74, 269-280.

4. Kluger, M.J. (1991): Fever: role of pyrogens and cryogens. Physiol Rev 71, 93-127.

5. Brown, G.M. and J. Seggie (1988): Effects of antidepressants on entrainment of circadian rhythms. Progress Neuropsychopharmacol Biol Psychiatry 12, 299-306.

6. Gao, B., W.C. Duncan Jr., and T.A. Wehr (1992): Fluoxetine decreases brain temperature and REM sleep in Syrian hamsters. Psychopharmacology 106, 321-329.

7. Gerber, J.C., J. Choki, D.J. Brunswick, M. Reivich, and A. Frazer (1983): The effect of antidepressant drugs on regional cerebral glucose utilization in the rat. Brain Res 269, 319-325.

8. Reinhard, J.F., Jr., J.E. Liebmann, A.J. Schlosberg, and M.A. Moskowitz (1979): Serotonin neurons project to small blood vessels in the brain. Science 206, 85-87.

9. Amenta, F., M. De Rossi, M.C. Mione, and P. Geppetti (1985): Characterization of (3H)5-hydroxytryptamine uptake within rat cerebrovascular tree. European J Pharmacol 112, 181-186.

10. Winquist, R.J. and J.A. Bevan (1980): Temperature sensitivity of tone in the rabbit facial vein: myogenic mechanism for cranial thermoregulation. Science 207, 1001-1002.

11. Wehr, T.A. (1989): Sleep loss: A preventable cause of mania and other excited states. J Clin Psychiatry 50, 8-16.

12. Van den Burg, W., A.L. Bouhuys, R.H. Van den Hoofdakker, and D.G.M. Beersma (1990): Sleep deprivation in bright and dim light: antidepressant effects on major depressive disorder. J Affective Disord 19, 109-117.

Thermal Balance in Health and Disease
Advances in Pharmacological Sciences

THERMOREGULATORY EFFECTS AFTER INTRACEREBROVENTRICULAR INJECTIONS OF WU-CHU-YU EXTRACTS IN RATS

T. F. Lee, T. H. Tsai, C. F. Chen and L. C. H. Wang

Zoology Department, University of Alberta, Edmonton, Alberta, Canada T6G 2E9

Wu-chu-Yu, the dry unripe fruits of *Evodiae rutaecarpa* have been used as a Chinese folk medicine for the treatment of gastro-intestinal disorders, headache, or as a cardiotonic and an analgesic (1). The major bioactive components that have been isolated from *Evodiae* are dehydroevodiamine (DeHE), evodiamine (EVO) and rutaecarpine (RUT)(1). Previously, some of these alkaloids have been shown to exhibit calcium channel antagonistic activity: DeHE (2) and EVO (3) have been shown to reduce the vascular contractile responses to $CaCl_2$ or phenylephrine (via Ca^{2+} influx), respectively. Further studies also showed that YC-08, a derivative of DeHE, inhibited intracellular Ca^{2+} rise of Fura-2 loaded platelets stimulated by collagen (4).

There is accumulating pharmacological and physiological evidence that endogenous Ca^{2+} ions may play a functional role in determining the "set-point" in thermoregulation (5). Since altering the availability of endogenous Ca^{2+} ions has been demonstrated to cause a shift in the thermoregulatory "set-point", administration of alkaloids extracted from Wu-chu-yu is thus expected to induce various thermoregulatory changes. Even though the dose and the route of administration were not described, the alcohol extract of Wu-chu-yu has been shown to elevate the T_b of the rabbit (1). It is, therefore, of interest to examine whether the alkaloids, DeHE, EVO and RUT, could account for this thermoregulatory effect.

Methods

All experimental protocols have been approved by the University of Alberta Animal Care Committee, following the guidelines of the Canadian Council on Animal Care. Adult Sprague-Dawley rats (3-6 months) were used in the present study. The rats were kept individually at 20°C and 12L:12D photoperiod. Food was rationed (about 20 g/day) to maintain body weight around 400 g throughout the whole experimental period to maintain constant weight and minimize variation in results due to differences in body size. Water was available at all time. Under halothane anaesthesia, a guide cannula (23-gauge stainless steel tubing) was implanted unilaterally into the lateral ventricle of each rat utilizing the following stereotaxic coordinates: AP=5.6 mm, L=1.8 mm, H=3.5 mm below the dura mater (6). After completion of the experiments, the precise anatomical location of the injection site was subsequently verified histologically.

One week post-operatively, each rat was transferred to a circular, plexiglas water-jacketed metabolism chamber in which the ambient temperature could be controlled accurately at 20±1°C. The T_b's of the rats were recorded continuously with a pre-calibrated temperature-sensitive radiotransmitter (Model T-M, Mini-Mitter Co.), which was implanted in the peritoneal cavity under halothane anaesthesia at the same time as the guide cannula. Using an IBM personal computer interfaced with an ISAAC 91-I data acquisition system, oxygen consumption and CO_2 production were recorded and integrated simultaneously with the T_b transmitter signal. Heat production (HP) was calculated from oxygen consumption and respiratory quotient (7), and the T_b was calculated using pre-established calibration curves for each individual transmitter.

Nimodipine (NIMOD) was purchased from RBI (Natick, MA). Both DeHE, EVO and RUT were extracted from the dried fruit of *Evodiae rutaecarpa* using the method described elsewhere (2) and the purity of these compounds was found to be greater than 99%. All the compounds were prepared immediately before an experiment in dimethyl sulphoxide (DMSO) and all test solutions were passed through a 0.22 μm Swinnex Millipore filter to ensure sterility. Either DMSO or test compounds was infused into the ventricle by gravity in a volume of 4 μl. Statistical comparisons on the thermoregulatory responses to alkaloids and NIMOD of rats were performed by Wilcoxon's Signed Ranks test. Significance was set at $p<0.05$, unless otherwise stated.

Results

Control rats exposed to 20°C had an initial mean T_b of 36.8±0.13°C and a resting HP of 1.33±0.05 kcal/h/animal (5.53±0.21 kJ/h/animal; n=7). As shown in Table 1, intracerebroventricular (ICV) administration of DMSO caused slight increases in T_b and HP, which were not significantly different from the pre-injection values. In contrast, injection of DeHE (66-264 nmole) caused a dose-related increase in T_b of the rat maintained at 20°C. As an increase in HP always preceded the elevation of T_b after DeHE injection, the hyperthermic effect of DeHE may be brought about through an increased metabolic rate. To examine whether DeHE is the active component in the alcohol extract of Wu-Chu-Yu which elicits the hyperthermic effect as reported previously (1), the thermoregulatory effects of two other alkaloids, EVO and RUT, isolated from Wu-Chu-Yu, were also studied. ICV injection of low dose of EVO (66 nmole) did not lead to any change in T_b (Table 1). However, increases in both T_b and HP were observed after administration of higher doses (132 and 264 nmole) of EVO (Table 1). In comparison with DeHE, the thermogenic effect of EVO was less prominent at the higher dose. In contrast to those observed after DeHE and EVO injections, administration of RUT in the same dose range did not cause any significant changes in either T_b or HP of the animal (Table 1). To further elucidate the underlying mechanism by which DeHE and evodiamine elicit the hyperthermic response in rats, the thermoregulatory effect of NIMOD, a Ca^{2+} channel blocker, was also examined. ICV injection of NIMOD (5 and 10 nmole) caused a similar thermogenic effect as those observed after administration of either DeHE or EVO. That is, a rise in T_b which was associated with an increase in HP (Table 1). However, the dose of NIMOD (10 nmole) required to elicit a significant hyperthermic effect was about 13-26 times less than that of DeHE and EVO.

Tab. 1: Thermoregulatory changes after ICV injection of dehydroevodiamine (DeHE), evodiamine((EVO), rutaecarpine (RUT), and nimodipine (NIMOD) in rats maintained at 20°C.

Treatment (n = 7)	Max. Change in T_b (°C)	Max. Change in HP (Kcal/Hr/rat)
DMSO 4 µl	+0.22±0.15	+0.32±0.10
DeHE 66 nmole	+0.48±0.12	+0.56±0.17
DeHE 132 nmole	+0.71±0.10*	+0.79±0.16*
DeHE 264 nmole	+1.09±0.22*	+1.17±0.07*
EVO 66 nmole	+0.16±0.08	+0.27±0.11
EVO 132 nmole	+0.83±0.22*	+1.01±0.18*
EVO 264 nmole	+0.77±0.19*	+0.69±0.13*
RUT 70 nmole	+0.19±0.11	+0.30±0.09
RUT 140 nmole	+0.34±0.16	+0.41±0.14
RUT 280 nmole	+0.27±0.11	+0.39±0.12
NIMOD 5 nmole	+0.56±0.09	+0.58±0.11
NIMOD 10 nmole	+0.88±0.08*	+0.97±0.14*

*Significantly different from DMSO control, p<0.05

Discussion

In an attempt to determine the active component(s) involved in the previously reported hyperthermic effect of Wu-chu-yu, a Chinese medicinal herb (1), the thermoregulatory effect of purified alkaloids from Wu-chu-yu was examined. ICV injection of DeHE and EVO caused a dose-related rise in T_b in rats maintained at 20°C. As the rise in T_b was always associated with an increase in HP, it is most likely that the hyperthermic effect of DeHE and EVO is brought about by increasing the metabolic rate. In contrast to that observed with DeHE and EVO, injection of the same dose range of RUT failed to elicit any thermoregulatory effect. Therefore, it may be concluded that both DeHE and EVO are the active components of Wu-chu-yu in eliciting the hyperthermic effect. Even though both DeHE and EVO have been shown to exhibit calcium channel antagonistic activity in vitro (2,3,4), it is not certain whether their thermoregulatory effects were through the blocking the Ca^{2+} channels. To further elucidate the underlying mechanism, the thermoregulatory effect of NIMOD, a Ca^{2+} channel blocker, was also examined. Similar to those observed with DeHE and EVO, ICV injection of NIMOD also elicited a long-lasting thermogenic response. Further, systemic injection of DeHE and EVO caused a profound hypothermic effect (Tsai et al., submitted), which is similar to those observed after systemic injection of classic Ca^{2+} channel blockers (8,9). These results suggest that DeHE and EVO may elicit their thermoregulatory effect through Ca^{2+} channel blocking. This suggestion is further supported by two other observations: a) the inability of RUT to elicit hyperthermia corresponds to its lack of Ca^{2+} channel blocking activity (10); and b) the relative potency of DeHE and EVO in eliciting the hyperthermic response in vivo parallels with their relative potency in Ca^{2+} channel blocking activity in vitro (11).In summary, the present study demonstrated that both DeHE and EVO are the active components in Wu-chu-yu in eliciting its thermoregulatory response. It is generally believed that Ca^{2+} channel antagonists have profound effects on Ca^{2+}-mediated functions in the hypothalamus, including alteration of the ionic "set-point" and its consequent thermoregulatory responses (5). It is therefore likely that the observed hyperthermic effect of DeHE and EVO are consequent to alteration of neuronal Ca^{2+} influx due to the Ca^{2+} channel blocking property of these alkaloids.

Acknowledgement

The present study was supported by Natural Sciences and Engineering Research Council (Canada) and National Research Institute of Chinese Medicine (Taiwan).

References

1. Chang, HM, But, PH (1986): Pharmacology and applications of Chinese materia medica, Vol. 1. Singapore: World Scientific, 605-609.

2. Chiou, WF, Chou, CJ, Shum, AYC, Chen, CF (1992): The vasorelaxant effect of evodiamine in rat isolated mesenteric arteries: mode of action. Eur. J. Pharmacol. 215, 277-283.

3. Yang, CM, Wu, SL, Kuo, CH, Chen, CF (1990): The hypotensive and negative chronotropic effect of dehydroevodiamine. Eur. J. Pharmacol. 182, 537-542.

4. Yen, MH, Ding, YA, Chen, CF (1991): Effects of dehydroevodiamine and its derivative on human platelet aggregation and cerebral arteries. Abstracts of "Symposium of natural herbal medicines: Team work research on Chinese herbal drug", Taiwan, 18.

5. Myers, RD, Lee, TF (1989): Neurochemical aspects of thermoregulation. In: Advances in comparative and environmental physiology, Vol. 4. Wang, LCH, ed. Berlin: Springer-Verlag, 161-203.

6. Pellegrino, LJ, Pellegrino, AS, Cushman, AJ (1979): A stereotaxic atlas of the rat brain. New York: Plenum Press.

7. Wang, LCH (1980): Modulation of maximum thermogenesis by feeding in the white rat. J. Appl. Physiol. 49, 975-978.

8. Palmi, M, Sgaragli, G (1989): Hyperthermia induced in rabbits by organic calcium antagonists. Pharmac. Biochem. Behav. 34, 325-330.

9. Jourdan, ML, Lee, TF, Wang, LCH (1992): Thermoregulatory effects of nimodipine on cold exposed and hypothermic rats. In: Thermoregulation: the pathophysiological basis of clinical disorders. Lomax, P, Schonbaum, E, eds. Basel: Karger, 88-92.

10. Liao, JF, Yu, L, Chen, CF (1991): Effects of Wu-chu-yu crude extract and its four bioactive principles on the gastrointestinal transit in mice. Abstracts of "Symposium of natural herbal medicines: Team work research on Chinese herbal drug", Taiwan, 20.

11. Yang, HY, Li, SY, Chen, CF (1988): Hypotensive effects of dehydroevodiamine, a quinazolinocarboline alkaloid isolated from *Evodiae rutaecarpa*. Asia Pac. J. Pharmacol. 3, 191-196.

Thermal Balance in Health and Disease
Advances in Pharmacological Sciences

NITRIC OXIDE MODULATES THERMOREGULATORY EFFECTOR MECHANISMS IN THE CONSCIOUS RABBIT

R. Gerstberger, H. Hjelmqvist and R. Keil

Max-Planck-Institut für physiologische und klinische Forschung, W.G. Kerckhoff-Institut, Parkstr. 1, 61231 Bad Nauheim, Germany

Introduction

Body temperature homeostasis in higher vertebrates is achieved by modulation of thermoregulatory effector mechanisms such as shivering, panting or changes in local vascular tone. Regional cardiovascular control by differentiation of the sympathetic outflow represents a well characterized concept to maintain body temperature homeostasis under conditions of physiological disturbances within the thermoregulatory control circuit (1). Peripheral warming leads to a fall in cutaneous sympathetic nerve activity in conscious rabbits, thus reducing vasomotor tone, whereas visceral and cardiac sympathetic activity increases (1,2). Respiratory frequency and evaporative water loss (REWL) are augmented to allow sufficient heat dissipation at unaltered mean arterial pressure and heart rate, and low rates of metabolic heat production (3). Peripheral cold stimuli produce the reverse pattern of autonomic effector activities. Cutaneous vasodilation occurring during heat exposure requires active neurogenic vasodilation in addition to the release of vasoconstrictor tone. Besides vasoactive intestinal polypeptide (VIP) and ATP, gaseous nitric oxide (NO) might represent one of the mediators involved in nonadrenergic, noncholinergic (NANC) transmission processes (4). NO endogenously derived from the amino acid L-arginine through enzymatic activity of nitric oxide synthase (NOS) isoforms represents a rather novel concept of intercellular signalling (5). Synthesized by vascular endothelial cells, smooth muscle cells and neurons, NO appears to activate soluble intracellular guanylate cyclase activity in a paracrine fashion, finally leading to NANC vaso- or bronchodilation and alteration of neuronal activity, respectively. With regard to its possible role within autonomic regulatory circuits to maintain homeostasis, only scarce information is available (5). Employing the conscious rabbit as well-characterized model system, the question was addressed in the present study, whether NO might participate in the control of thermoregulatory effector mechanisms such as cutaneous or respiratory heat dissipation.

Methods

Female Chinchilla-Hybrid rabbits (Ivanovas, FRG) with body weights (BW) ranging from 4.0 to 5.5 kg were kept in large individual cages under a 12:12 h day-night cycle with artificial illumination from 06.00 to 18.00 h at 20°C ambient temperature. Drinking water and conventional laboratory rabbit food were available *ad libitum*.
The experiments were performed between 08.30 and 14.30 h with the rabbits sitting unrestrained in a temperature-controlled (24°C) rabbit box. Back skin and ears were exposed to room temperature (20-22°C). Preliminary experiments were performed with the same animals exposed to an ambient temperature of 5°C or 31.5°C (body warming or cooling). For continuous monitoring of mean arterial blood pressure (MABP) and heart rate (HR), a central ear artery was cannulated under local anesthesia using 1.0% Xylocain^R (Astra Chem., FRG). The catheter (Stericath 0.8x1.2 mm; Vygon, FRG) was continuously perfused with sterile, pyrogen-free isotonic saline (Fresenius, FRG) at 0.1 ml/min. For systemic drug application, a peripheral vein of the contralateral ear was cannulated under local anesthesia by percutaneous venipuncture (Vasocan^R- 20G Braunula; Braun, FRG), and the infusion line protected against light by aluminium wrapping. Drugs infused were the competitive NO synthase blocker N^G-nitro-L-arginine methyl ester (L-NAME; Sigma, FRG) and the NO donors 3-morpholino-sydnonimin-hydrochlorid (SIN-1; gift of Cassella-Hoechst, FRG) and S-nitroso-N-acetylpenicillamine (SNAP; Biomol, FRG) or L-arginine (L-Arg; Sigma, FRG). All substances were dissolved in isotonic saline just before application and applied as bolus injections (L-NAME and L-Arg) at 10 and 40 mg/kg BW, respectively, or infused for 15 min (SIN-1 or SNAP) at 5-15 µg/min/kg BW with a rate of 0.2 ml/min.
Ear skin (T_{ear}) and rectal (T_{rec}) temperatures were measured to an accuracy of 0.1°C, using thermocouples attached to the ear or inserted 10 cm deep into the rectum (3). To monitor MABP and HR, the arterial catheter was connected to a Statham P23XL pressure transducer (Hellige, FRG) fixed at the level of the animal's heart. The output signals were amplified and processed using a pressure computer and Biotach coupler (Gould Inc., USA), and displayed on a chart recorder (Linseis, FRG). Respiratory rate was counted visually. Oxygen consumption and respiratory evaporative water loss (REWL) were determined using open flow respirometric techniques in air continuously withdrawn from the head cage through a gas meter. From the air flow rate (STPD corrected) and oxygen extraction as measured by means of an Ametek oxygen analyzer (Model S-3A/II; Thermox Instr. Div., USA), oxygen consumption was calculated. REWL values were obtained by measuring the water vapour contents of the air entering and leaving the head cage utilizing a Binos^R IR-VIS/UV Gas Analyzer (Leybold-Hereaus GmbH, FRG). All analogous output signals of the measuring equipment were transferred to a hybrid recorder (Modell 3081; Yokogawa Electric Corp., Japan), which was connected to a desk top computer. The data were collected at 6 sec intervals, integrated over periods of 5 min and stored for processing. The experimental protocol was started after recordings of deep body temperature and cardio-respiratory parameters had been stable for at least 60 min.
The results are presented as means with standard errors of the mean (SE). ANOVA was used to test for significant differences using the Sigmastat statistical analysis program. For multiple comparisons the probability coefficients were adjusted according to Bonferroni-Holm. The null hypothesis was rejected when $P \leq 0.05$, with statistical differences presented in the text.

Results

In order to elucidate the possible role of NO in the control of cardiovascular and thermoregulatory effector mechanisms in conscious rabbits acclimatized to a thermoneutral-warm environment (24°C), alterations in the endogenous NO system were mimicked by systemic application of NO donors or NO synthase antagonists. During control conditions, the rabbits were in a state of

incipient panting with respiratory frequencies between 90 and 150 breaths/min and REWL of 6-8 mg/min/kg BW. Oxygen consumption as the indicator of metabolic heat production amounted to basal values of 6.2-6.8 ml/min/kg BW and ear skin temperature representing cutaneous vasomotor tone fluctuated between 28 and 32°C depending on the individual animal (Fig.2). The status of the cardiovascular system revealed values for MABP and heart rate of 75-85 mmHg and 180-220 beats/min, respectively (Fig.1).
The systemic infusion of the NO donors SIN-1 and SNAP led to a rapid and transient fall in MABP with concomitant tachycardia ($P \leq 0.05$) (Fig.1), with SNAP eliciting a more rapid action possibly due to faster generation of NO radicals from the mother molecule. Employing the NO donors at two different concentrations (5 and 15 µg/min/kg BW) indicates dose-dependency with regard to the induced cardiovascular responses. Ear skin temperature increased suggestive of cutaneous vasodilation ($P \leq 0.05$) (Fig.2). Both respiratory frequency and REWL were transiently augmented ($P \leq 0.05$) from 93±14 to 156±18 and 6.7±0.9 to 11.3±1.9 mg/min/kg BW, whereas metabolic heat production remained unaltered (Fig.2). In consequence of stimulated cutaneous and respiratory heat dissipation, core body temperature significantly decreased by 0.1°C ($P \leq 0.05$). In time control experiments rectal temperature rather tended to in- than to decrease.
Blockade of endogenous NO formation via bolus application of L-NAME at 10 mg/kg BW induced a fast and plateau-like rise in MABP of about 15 mmHg ($P \leq 0.05$). Heart rate was concomitantly and markedly reduced by 80 beats/min on average, thus possibly reflecting direct cardiac effects of NO synthase inhibition rather than mere reflex bradycardia (Fig.1) (9). Both cardiovascular parameters proved equally affected irrespective of the ambient temperature chosen. Administration of L-Arg at the dose of 40 mg/kg BW only partially reversed the L-NAME induced cardiovascular effects, and L-Arg at higher doses was found to reveal intrinsic cardiovascular and

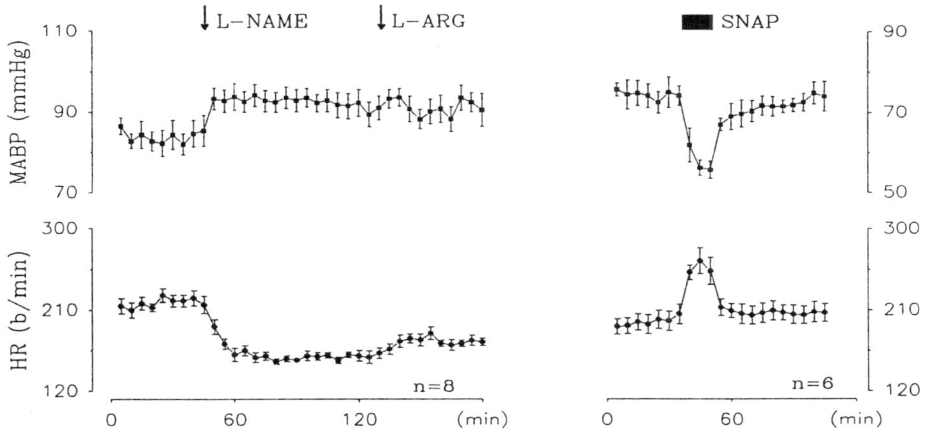

Fig. 1: Modulatory role of the NO donor SNAP (15 µg/min/kg bw for 20 min) (right) and the NO synthase blocker L-NAME (10 mg/kg bw) (left) on mean arterial pressure (MABP) and heart rate (HR) in conscious rabbits exposed to 24°C environmental temperature. L-ARG (40 mg/kg bw) was applied to competitively displace L-NAME from NO synthase.

endocrine side effects in the conscious rabbit (6). With regard to cutaneous vascular tone, NO synthase blockade did not significantly alter ear skin temperature under conditions of 24°C (Fig.2) ambient temperature, while leading to mild vasoconstriction during whole body warming ($P \leq 0.05$) with values for T_{ear} of 38.4±0.5 and 37.1±0.8 °C, respectively. In contrast to the stimulatory actions of exogenous NO donors on both respiration frequency and REWL, competitive blockade of NO synthase activity caused L-Arg reversible reduction of respiratory frequency ($P \leq 0.05$) (Fig.2). REWL decreased simultaneously by 2.47±0.64 mg/min/kg BW. Diminuation of respiratory heat loss was even more pronounced under conditions of body warming. At 31.5°C ambient temperature, RF proved to be diminished from 170±30 to 85±18 breaths/min, and REWL was reduced from 10.9±1.6 to 7.1±1.0 mg/min/kg BW ($P \leq 0.05$). L-NAME treatment led to a slight but significant ($P \leq 0.05$) decrease in oxygen consumption as a measure of metabolic heat production with values of 6.77±0.29 and 6.02±0.32 ml/min/kg BW, respectively, during exposure of the animals to an ambient temperature of 24°C (Fig.2), and this response was markedly enhanced during cold exposure with values of 8.03±0.17 and 6.79±0.20 ml/min/kg BW, respectively.

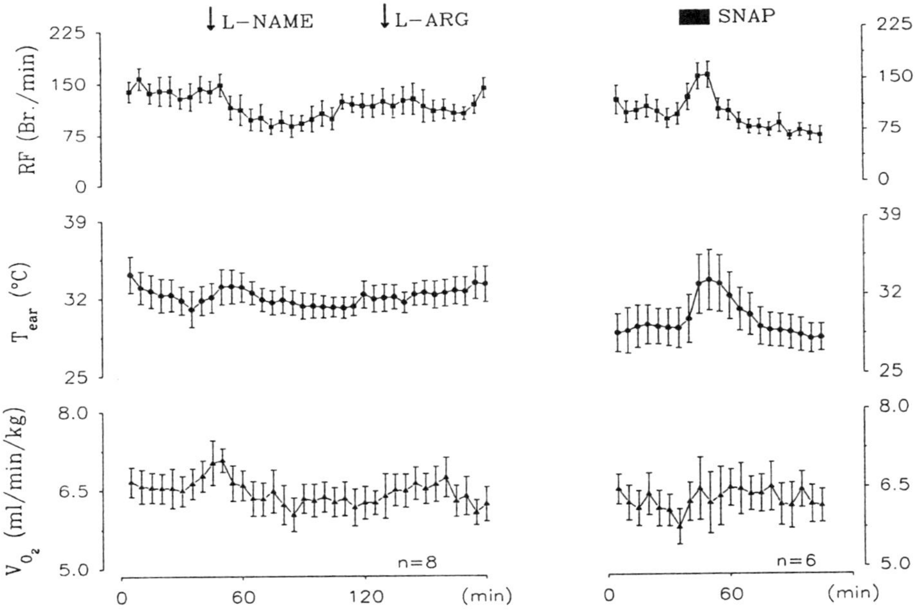

Fig. 2: Modulatory role of the NO donor SNAP (15 µg/min/kg bw for 20 min) (right) and the NO synthase blocker L-NAME (10 mg/kg bw) (left) on respiratory frequency (RF), ear skin temperature (T_{ear}) and oxygen consumption (V_{O2}) in conscious rabbits exposed to 24°C environmental temperature. L-ARG (40 mg/kg bw) was applied to competitively displace L-NAME from NO synthase.

Discussion

Alterations in thermoregulatory effector mechanisms such as cutaneous and respiratory heat dissipation or metabolic heat production during the systemic application of either NO donors or competitive inhibitors of endogenous NO production have been investigated very sparcely. The observed combination of ear vasodilation, increased respiratory heat loss and decrease in core temperature during the application of NO donors would suggest that NO is capable of eliciting a coordinated heat defence response. However, with regard to the involvement of NO in the thermoregulatory control of cutaneous vasodilation in the rabbit, contradictory results have been reported, at least partially depending on the status of anesthesia, method of blood flow determination or differences in the thermoregulatory status chosen. In the present study performed in conscious rabbits, the application of NO donors under thermoneutral conditions clearly induced ear skin vasodilation, whereas the inhibition of endogenous NO synthase activity proved to be ineffective. Only during whole body warming, ear skin temperature was transiently reduced by L-NAME treatment. Total suppression of neurogenic ear vasodilation recorded with Doppler ultrasonic flow probes in conscious rabbits during whole body warming occurred after bolus injections of the NO synthase antagonist N^G-nitro-L-arginine, however, with a marked time delay (4). As in the present study, specificity of the reaction was proven by treatment with the endogenous NOS substrate L-arginine. Employing laser Doppler flowmetry under thermoneutral conditions in restrained rabbits, ear skin blood flow was shown to be significantly reduced in rabbits chronically treated with the NOS antagonist L-NAME. Cutaneous vasodilation caused by body warming, however, could not be inhibited by L-NAME suggesting that NO did not participate in the thermally induced peripheral vasodilation in the study of Khan and coworkers (7). On the other hand it could be shown in the anesthetized pig, that endogenous NO release might indeed provide a local, rapid mechanism to adapt to warm ambient temperature. Tissue blood flow measurements with radioactive microspheres of two different sizes allowed determination of blood flowing through skin and pulmonary arteriovenous anastomoses to be markedly reduced during NOS blockade by L-NAME treatment (8). A general increase in pulmonary resistance was also determined in artficially ventilated rabbits during NOS blockade (see 5).

Without doubt systemic inhibition of NO synthesis leads to augmented peripheral vascular resistance as indicated by appropriate changes in MABP (9), however, with the alterations occurring not evenly distributed to all vascular beds. This could be clearly demonstrated in the conscious rat, rabbit and sheep (6,10), where, e.g., femoral blood perfusion remained either unchanged or even slightly increased at markedly reduced renal blood flow. Taking the study of Van Gelderen and coworkers into consideration (8), local control of organ-specific blood flow by the endogenous NO system might be possible through differentiated activation or inhibition similar to the model of differentiated sympathetic activity during body warming or cooling. The present study, using systemic NO synthase inhibition during cold, thermoneutral and warm ambient temperatures as experimental tool, could verify cutaneous and respiratory heat dissipation as well as metabolic heat production as major targets for the endogenous NO system(s). However, as expressed by the subsequent reduction in both heat and cold defence mechanisms in a non-coordinated manner, general systemic NO synthase manipulation did not allow to differentiate various target-specific local NO systems controlling thermoregulatory effector mechanisms. Immunocytochemical and very recent physiological reports describe the presence of NOS in various functional populations of sympathetic preganglionic neurons and spinal cord neurons mediating sympathetic outflow to various peripheral organs (11,12). The question therefore arises whether NO might affect neural regulation of the cardiovascular system via modulation of sympathetic vasoconstrictor neurons.

References

1. Iriki, M, Simon, E (1978): Regional differentiation of sympathetic efferents. In: Integrative control functions of the brain, Vol. 1. Ito, M, Tsukahara, N, Kubota, K, Yagi, K eds. Tokyo, Kodansha Ltd., Elsevier Biomedical Press, 221-238.

2. Riedel, W, Kozawa, E, Iriki, M (1982): Renal and cutaneous vasomotor and respiratory rate adjustments to peripheral cold and warm stimuli and to bacterial endotoxin in conscious rabbits. J Aut Nerv sys 5, 177-194.

3. Keil, R, Gerstberger, R, Simon, E (1994): Hypothalamic thermal stimulation modulates vasopressin release in hyperosmotically stimulated rabbits. Am J Physiol (submitted).

4. Taylor, F, Bishop, VS (1993): A role for nitric oxide in active thermoregulatory vasodilation. Am J Physiol 264, H1355-H1359.

5. Moncada, S, Higgs, A (1993): The L-arginine-nitric oxide pathway. New Engl J Med 329, 2002-2012.

6. Gerstberger, R, Keil, R, Jurzak, M, Hjelmqvist, H, Eriksson, S (1994): Hormonal and hemodynamic effects of NOS inhibition in the conscious sheep and rabbit. In: Biology of nitric oxide. Moncada, S, Feelisch, M, Busse, R, Higgs, EA eds. Colchester, Portland Press (in press).

7. Khan, F, Palacino, JJ, Coffman, JD, Cohen, RA (1993): Chronic inhibition of nitric oxide production augments skin vasoconstriction in the rabbit ear. J Cardiovasc Pharmacol 22, 280-286.

8. Van Gelderen,EM, Den Boer, MO, Saxena, PR (1993): NG-nitro-L-arginine methyl ester: systemic and pulmonary haemodynamic, tissue blood flow and arteriovenous shunting in the pig. NS Arch Pharmacol 348, 417-423.

9. Du, Z-Y, Dusting, GJ, Woodman, OL (1992): Baroreceptor reflexes and vascular reactivity during inhibition of nitric oxide synthesis in conscious rabbits. Eur J Pharmacol 214, 21-26.

10. Sigmon, DH, Carretero, OA, Beierwaltes WH (1993): Renal versus femoral hemodynamic response to endothelium-derived relaxing factor synthesis inhibition. J Vasc Res 30, 218-223.

11. Vizzard, MA, Erdman, SL, Erickson, VL, Stewart, RJ, Roppolo, JR, Degroat, WC (1994): Localization of NADPH diaphorase in the lumbosacral spinal cord and dorsal root ganglia of the cat. J Comp Neurol 339, 62-75.

12. Kumagai, H., Averill, DB, Khosla, MC, Ferrario, CM (1993): Role of nitric oxide and angiotensin II in the regulation of sympathetic nerve activity in spontaneously hypertensive rats. Hypertension 21, 476-484.

DOES CENTRAL NITRIC OXIDE PLAY A ROLE IN THERMOREGULATION?

A. V. Gourine

Institute of Physiology, Academy of Sciences of Belarus, Skorina Street 28, Minsk 220725, Republic of Belarus, CIS.

The diffusible stable free radical gas nitric oxide (NO) has been shown to function as a messenger molecule in the cardiovascular, immune, central and peripheral nervous systems (1,2). Concerning the nervous system, most of the principal criteria for considering NO to be a new neurotransmitter have been satisfied: the presence of a specific synthesizing enzyme, NO-synthase (NOS), specific Ca^{++}-mediated activation of NOS in presynaptic neurones, and the presence of specific intracellular receptors: soluble guanylyl cyclase or ADP-ribosyltransferase in postsynaptic neurones (2,3). There is evidence that putative NO-synthesizing neurones are widespread over most parts of mammalian brain (4,5). However, the data related to the functional role of NO in the central nervous system are limited. Recent evidence favours a possible role for NO in central autonomic regulation (6). It was shown that endogenous formation of NO is required in the mechanisms associated with central respiratory rhythm generation (7). NO is also involved in the control of vasomotor tone within the ventrolateral medulla (8), and in cardiovascular regulation within the nucleus tractus solitarii (9,10).

In this report a possible involvement of nitric oxide in the central mechanisms of thermoregulation was studied by determining the effects of centrally administered NO-related drugs (i.e. NO-donor and NOS-inhibitor) on three types of elevation of body temperature: (i) during rewarming after deep experimental hypothermia; (ii) during overheating; and (iii) lipopolysaccharide(LPS)-induced fever.

METHODS

Adult male rats weighing 180-200g were used in all experiments. Aqueous drug solutions were injected into the lateral cerebral ventricles in volumes not exceeding 20 µl. The stereotaxic coordinates were 0.8 mm posterior to bregma, 1.5 mm right to the midline, and 3.5 mm below the skull surface.

Measurement of body temperature: Rectal temperature was measured at 30-min intervals by a thermistor probe inserted 3cm in the rectum. On the day of the experiment, basal temperature of each animal was determined twice, 60 and 30 min before administration of any drug.

Experimental hypothermia: Deep hypothermia was induced by immersing the animals in a water bath of 21°C for 30 min. During this period of time a fall in rectal temperature by 10.3-10.9°C was observed. The animals were withdrawn, dried, and relevant solutions were immediately injected into the lateral cerebral ventricles. These experiments were performed at an ambient temperature of 20°C.

Overheating: The rats were treated intracerebroventricularly (i.c.v.) with the required solutions and then placed in a temperature controlled chamber at 37±0.5°C.

Fever: Fever was induced by acute intraperitoneal administration of *E.coli* lipopolysaccharide (50 μg/kg). The NOS-inhibitor was injected 30 min and the NO-donor - 180 min after the endotoxin treatment (Ta=18°C).

Drugs: LPS from *Escherichia coli* (SIGMA Chemical, product no. L2630), the NO-donor - molsidomine (N-(ethoxycarbonyl)-3-(4-morpholinyl)-sydnoneimine, SIGMA Chemical, product no. M2901), the NO-synthase inhibitor N-nitro-L-arginine (SIGMA Chemical, product no. N5501) were used. LPS was dissolved in isotonic saline. Molsidomine and N-nitro-L-arginine were dissolved in pyrogen-free bidistilled water.

Statistical analysis: Data were expressed as means ±SE. Comparisons between the experimental and control groups were analysed for statistical significance using Student's t-test. The significance level was set at $P<0.05$.

RESULTS

(i) Intracerebroventricular (i.c.v.) injections of the NO donor molsidomine (50μg) or NO-synthase inhibitor N-nitro-L-arginine (75μg) had no significant effect on the rectal temperature of rats at an ambient temperature of 20°C.

TABLE I: Effect of i.c.v. injections of NO-related drugs on the rise in rectal temperature during rewarming after deep hypothermia in rats

Drug	n	dose,μg	Tb,°C	Tf,°C	ΔTre30min,°C	ΔTre60min,°C
Solvent	7		36.54±0.15	25.83±0.07	4.70±0.24	10.72±0.44
M	7	50	36.83±0.12	25.94±0.08	4.17±0.26	8.98±0.27**
L-NA	8	75	36.41±0.13	26.11±0.06	4.78±0.35	10.78±0.60

Abbreviations and symbols: Tb: basal rectal temperature. Tf: final rectal temperature after cooling. ΔTre30min(60min),°C: rectal temperature changes during rewarming, 30min (60min). M: molsidomine. L-NA: N-nitro-L-arginine. **Significantly different from controls ($P<0.01$).

TABLE II: Effect of i.c.v. injections of NO-related drugs on rectal temperature changes during overheating in rats

				Change in rectal temperature °C Mean ±SE		
				Time (min)		
Drug	n	dose,μg	Tb,°C	30	60	90
Solvent	8		36.87±0.12	1.75±0.12	2.39±0.13	2.60±0.13
M	6	50	36.69±0.15	1.35±0.16	2.43±0.20	2.62±0.20
L-NA	7	75	37.00±0.13	1.50±0.21	3.23±0.10***	3.40±0.11***

Abbreviations and symbols: Tb: basal rectal temperature. M: molsidomine. L-NA: N-nitro-L-arginine. ***Significantly different from controls (P<0.001).

TABLE III: Effect of acute treatment with NO-related drugs on febrile response induced by intraperitoneal injection of E.coli endotoxin (50 μg\kg) in rats

			Change in rectal temperature °C Mean ±SE			
			Time (min)			
Drugs	dose, μg	n	120	180	240	300
LPS+Solvent (30 min)		7	0.97±0.14	0.77±0.19	0.71±0.21	0.80±0.17
LPS+L-NA (30 min)	75	7	1.46±0.23	1.59±0.21*	1.81±0.13***	1.60±0.13**
LPS+Solvent (180 min)		8	0.81±0.20	0.77±0.14	0.83±0.18	1.25±0.20
LPS+M (180 min)	50	8	0.53±0.18	0.51±0.17	0.51±0.25	0.87±0.19

Abbreviations and symbols: LPS: lipopolysaccharide. M: molsidomine. L-NA: N-nitro-L-arginine. *,**,***Significantly different from controls (P<0.05, P<0.01, P<0.001, respectively).

(ii) As shown in Table I, the rise in rectal temperature during rewarming after deep experimental hypothermia was significantly reduced by i.c.v. injection of molsidomine (50μg). N-Nitro-L-arginine (75μg) had no significant effect on rectal temperature changes during rewarming.

(iii) Rats treated i.c.v. with NOS-inhibitor N-nitro-L-arginine (75 μg) and placed at an ambient temperature of 37°C had a greater rise in body temperature as compared with controls. The results are presented in Table II. As seen, molsidomine (50 μg) had no effect on changes in rectal temperature during overheating.

(iv) I.c.v. administration of N-nitro-L-arginine (75 μg) 30 min after the intraperitoneal injection of *E.coli* endotoxin (50 μg/kg) significantly enhanced the febrile response. The data are presented in Table III. As shown, i.c.v. application of the NO-donor had no significant effect on the magnitude of the febrile response.

DISCUSSION

There is evidence that NO-synthesizing neurones exist in several areas and nuclei of the rat hypothalamus and brain stem (5). The results of the present investigation support the hypothesis that endogenous nitric oxide is involved in the central mechanisms of thermoregulation. The results demonstrate that the thermoregulatory responses to high and low ambient temperatures and pyrogens can be affected by the central administration of the NO-related drugs. Furthermore, the results indicate that the central N0-synthesis is almost completely inhibited during rewarming after deep hypothermia and almost completely activated during overheating and fever. The peripheral mechanisms underlying the effects of centrally administered NO-related drugs during rewarming, overheating and fever still remain unclear and await additional study.

It is concluded that central nitric oxide is involved in the mechanisms of thermoregulatory responses to high ambient temperature and pyrogens, thus playing a certain role in thermoregulation.

ACKNOWLEDGEMENTS

This study was partly supported by ASB grant B1-120.

REFERENCES

1. Moncada, S., Palmer, R.M.J. & Higgs, E.A. (1991): Nitric oxide: physiology, pathophysiology, and pharmacology. Pharmacol Rev 43, 109-142.

2. Nathan, C. (1992): Nitric oxide as a secretory product of mammalian cells. FASEB J 6, 3051-3064.

3. Snyder, S.H. (1992): Nitric oxide and neurons. Curr Opin Neurobiol 2, 323-327.

4. Bredt, D.S., Hwang, P.M. & Snyder, S.H. (1990): Localization of nitric oxide synthase indicating a neural role for nitric oxide. Nature 347, 768-770.

5. Vincent, S.R. & Kimura, H. (1992): Histochemical mapping of nitric oxide synthase in the rat brain. Neuroscience 46, 755-784.

6. Iadecola, C., Faris, P.L., Hartman, B.K. & Xu, X.-H. (1993): Localization of NADPH-diaphorase in neurons of the rostral ventral medulla - possible role of nitric oxide in central autonomic regulation and oxigen chemoreception. Brain Res 603, 173-179.

7. Ling, L., Karius, D.R., Fiscus, R.R. & Speck, D.F. (1992): Endogenous nitric oxide required for an integrative respiratory function in the cat brain. J Neurophysiol 68, 1910-1912.

8. Shapoval, L.N., Sagach, V.F. & Pobegailo, L.S. (1991): Nitric oxide influences ventrolateral medullary mechanisms of vasomotor contol in the cat. Neurosci Lett 132. 47-50.

9. Lewis, S.J., Ohta, H., Machado, B., Bates, J.N. & Talman, W.T. (1991): Microinjection of S-nitrosocysteine into the nucleus tractus solitarii decreases arterial pressure and heart rate via activation of soluble guanylate cyclase. Eur J Pharmacol 202, 135-136.

10. Machado, B.H. & Bonagamba, L.G.H. (1992): Microinjection of S-nitrosocysteine into the nucleus tractus solitarii of conscious rats decreases arterial pressure but L-glutamate does not. Eur J Pharmacol 221, 179-182.

HYPOTHERMIA AND SOCIAL BEHAVIOUR IN MICE

J. W. G. M. Thuring and E.Schönbaum*

Department of Drug Safety, N.V. Organon, P.O. Box 20, Oss, and (*) Department of Medical Pharmacology, Leiden - Amsterdam Institute for Drug Research, Leiden State University, The Netherlands

Introduction

One of the standard protocols used in murine drug safety studies requires daily oral dosing for a period of fourteen days. In such studies a characteristic syndrome was often observed: hypothermia, together with "isolated" behaviour and other clinical phenomena such as bradycardia, prostration, lethargy, depressed food and water intake and weight loss. In spite of daily dosing, the severity of these symptoms decreased with time and often had almost or entirely disappeared after the first week of dosing.

To explore this syndrome further, we selected a small number of substances reported to produce hypothermia, viz. clozapine, chlorpromazine, propranolol and oxotremorine. This paper describes the results.

Methods

Animals

Male NMRI mice, spf bred, Charles River, Sulzfeld, FRG, body weight (BW) 35-40 g on the initial day of the experiment, were maintained at 20°C ± 1°C in Makrolon (clear plastic), type III, cages with a sawdust bedding. Each treatment group of four animals was acclimatized for approximately ten days; the dosing period lasted seven days.
Relative humidity in the air-controlled animal quarters was approximately 55%, air was exchanged approximately 10 times per hour and light-dark cycles adjusted 12:12 with lights on between 07:00 h and 19:00 h.
Animals had free access to pelleted food (RMH-B, Hope Farms B.V., Woerden, The Netherlands) and tap water from drinking bottles. For purposes of cage enrichment a stainless

steel baby shoebox-like shelter (measurements 14.5 x 7.5 x 5.5 cm, with the smallest side as the entrance and the opposite side provided with perforations) was placed in a corner of the cage (because of animal welfare considerations).

Drug treatment

Drugs were supplied by the Dispensing Section, Department of Pharmaceutics, N.V. Organon, The Netherlands. The drugs were suspended/dissolved in aqueous gelatin (0.5% w/v) and mannitol (5% w/v) at concentrations 0.3; 3.0; 5.0 and 30 mg/ml. This range of concentrations enabled us to administer a daily volume of 0.1 ml/10 g BW with an accuracy of 0.01 ml. The vehicle was given to a control group as placebo. An iron needle with a ball tip attached to a syringe was used for oral dosing. Treatments were given in the morning between 08:00 and 09:30 h.

Clinical measurements

Food and water consumption and general appearance were recorded daily, BW twice weekly. Colonic temperature was measured using an Ellab Copenhagen type ctd 85 apparatus with a probe type PRF-A with a blunt thermocouple (diameter 2.3 mm). Mice were manually restrained and the thermistor (lubricated with glycerine) was inserted into the rectum for a length of approx. 4.5 cm and kept in situ for approx. 20 seconds, until the temperature measured was more or less stable. Tail temperature was recorded using a YSI Tele-thermometer. A probe was developed for this purpose using a silicone cuff, length 18 mm, outer diamter 8 mm, inner diameter 5 mm. The thermocouple was pierced through the silicone wall, and was glued to the cuff in such a way that the tip protruded into the lumen and was not covered by glue. The cuff with the thermocouple at the ventral side was moved along the tail until resistance was encountered, generally some 1.5 cm from the tail base. Both thermometers were calibrated daily.

Electrocardiograms were made using a monochannel Scharzer electrocardiograph. Needles were inplanted subcutaneously (neck, left hind leg, sacrum; modified Nehb-Spörri lead D). The animals were manually restrained for this procedure. The recordings were made over a period of approx. 5 seconds per mouse per interval. Heart rate was calculated from R-R intervals.

This experiment was performed in compliance with GLP (Good Laboratory Practice) regulations except that the study was not inspected. The internal Animal Welfare Officer's consent was obtained prior to the start of the experiment.

Results

Clozapine (50 mg/kg/day) (Fig. 1)

About 30 minutes after the initial dose, mice began to show the characteristic syndrome. It reached a maximum after about two hours. Thereafter it slowly disappeared. After subsequent dosing, the syndrome returned, but with reduced intensity. The syndrome consisted of torpor; i.e. prostration, lethargy and "isolated" behaviour. Mice no longer showed the usual huddling in the baby shoebox but positioned themselves throughout the cage, leaving several centimeters distance between them. There was hypothermia and bradycardia, BW loss, less intake of food and water. On the first two days colonic temperature fell from 38.5°C to 29.4°C; this drop lasted for over eight hours. Thereafter hypothermia became less prominent. On days 1 and 2, but not on experimental day 0 (after first dosing) tail skin temperature was slightly higher in the clozapine-treated animals. The course of bradycardia was similar to that of hypothermia.

Chlorpromazine (30 mg/kg/day) (Fig. 1)

The extent and degree of the cageside observation and clinical measurements were of the same order of magnitude as those described for clozapine. The tail skin temperature was slightly higher on day 1 (on day 2 it was not measured). On day 5, when mice were challenged with a high dose of chlorpromazine (100 mg/kg), symptoms reappeared.

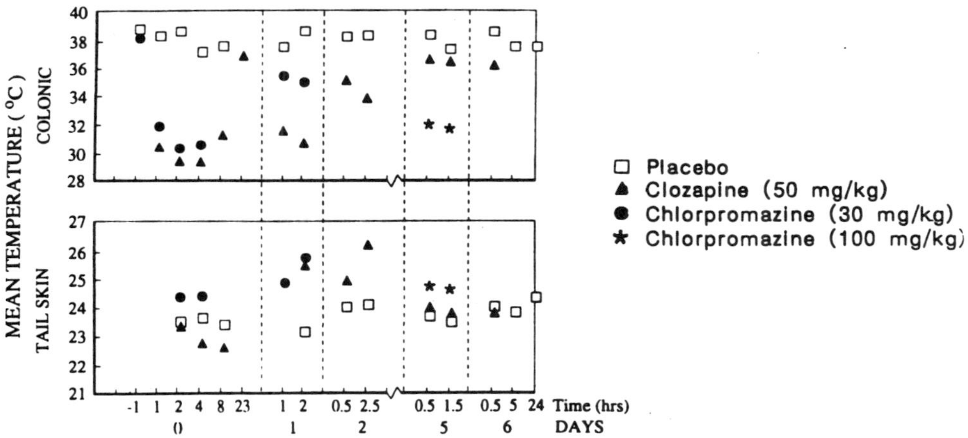

Fig. 1: Effects of daily oral administration of clozapine and chlorpromazine on tail skin and colonic temperature at various intervals after dosing. Each point represents the mean of four mice.

Propranolol (300 mg/kg/day) (Fig. 2)

Results were similar to those described for clozapine and chlorpromazine, with the following exceptions: sporadically, mice displayed loss of righting reflex, that could last up to two hours. Bradycardia was more prominent in the beginning of the study.

Oxotremorine (3 mg/kg/day) (Fig. 2)

This group showed daily transient signs of generalised tremors, hypersalivation, soft defaecation, wet belly due to abnormal micturition and isolated behaviour. The onset was within 30 min of dosing and the most prominent signs were seen approximately 1 hour thereafter. The character and degree of these cageside observations were more or less similar throughout the treatment period. The mean daily food and water consumptions were depressed initially but reached normal levels on experimental day 3, i.e. after the fourth treatment. There was evidence of hypothermia (a fall in body temperature of approx. 20%) between 1-2 h after dosing throughout the study. It was not accompanied by consistently higher tail skin temperatures when compared with values obtained in placebo treated mice. Heart rate was also depressed during the hypothermic episodes, more or less in proportion to the degree of hypothermia. On day 5 when the animals were challenged with the high dose of 10 mg/kg of oxotremorine, three of four mice died within 30 min of dosing.

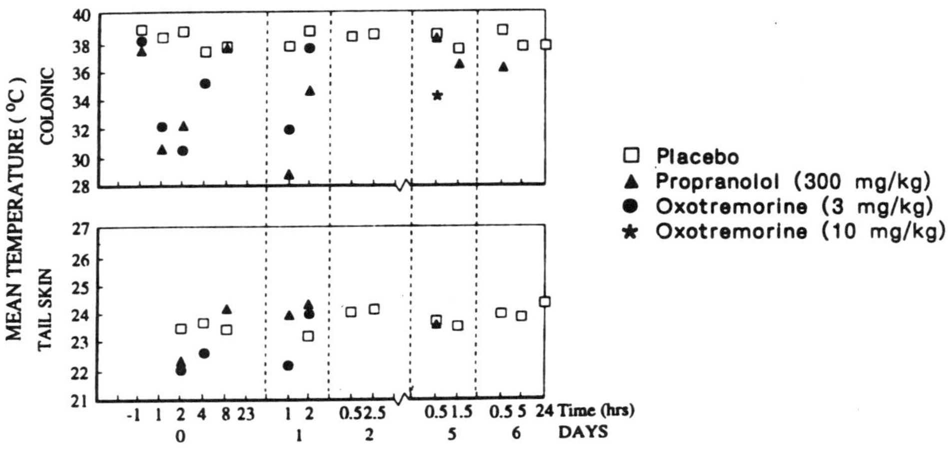

Fig. 2: Effects of daily administration of propranolol and oxotremorine on tail skin and colonic temperature at various intervals after dosing. Each point represents the mean of four mice unless otherwise indicated (* - one mouse only)

DISCUSSION

Choice of a lower ambient temperature by mice after treatment with high doses of drugs seems to be an aspecific response.

Normal mice exposed to an ambient temperature below that of thermoneutrality, try to maintain homeothermia by huddling and sleeping inside the shelter (inside temperature 25.5 - 27.5°C). In contrast, drug-treated mice move to the open space where the temperature (at the sawdust level) is that of the room temperature (21°C). Morever, the choice to lie separately enhances the exposure of the total body surface to the lower environmental temperature. This behaviour increases heat loss.

Activities which enhance thermogenesis, e.g., increased muscular activity are absent. Oxotremorine-treated mice became hypothermic despite prolonged episodes of autonomic muscular movements. Other autonomic reactions to conserve heat such as a hunched posture and piloerection were not observed. It appeared as if the animals seemed to avoid heat gain and promote heat loss by thermoregulatory mechanisms, both autonomic and behavioural. The set point mechanism no longer seems to function to maintain homeothermia.

Hypothermia in rodents can be induced by sublethal doses of various chemical entities (1-8). Hypothermia and isolated behaviour could contribute to survival by modifying the animal's response to the toxic insult (9-12).

References

1. Clement, J.G. (1993): Pharmacological nature of soman-induced hypothermia in mice. Pharmacol Biochem Behav 44, 689-702.

2. Coudray-Lucas, C., Prioux-Guyonneau, M., Tassel, A., Coq, H.M., Cohen, Y., Wepierre, J. (1981): Influence of intoxication by anticholinesterase agents on core temperature in rats: relationship between hypothermia and acetylcholinesterase inhibition in different brain areas. Acta Pharmacol et Toxicol 49, 215-222.

3. Moser, P.C. (1991): The effect of the putative 5-HT1A receptor antagonist in 8-OH-DPAT-induced hypothermia in rats and mice: Eur J Pharmacol, 193, 165-172.

4. Tang, A.H., Code, R.A., Himes, C.S. (1993): Antagonism of hypothermia produced by benzodiazepine-related compounds by U-78875 in mice. Eur J Pharmacol, 326; 1-5.

5. Wang, L.C.H., Peter, R.E. (1975): Metabolic and respiratory responses during helox-induced hypothermia in the white rat. Am J Physiol, 229, 890-895.

6. Gearhart, J.M., Seckel, C., Vinegar, A. (1993): In vivo metabolism of chloroform in B6C3F1T mice determined by the method of gas uptake. The effects on body temperature on tissue partition coefficients and metabolism. Toxicol Appl Pharmacol 119, 258-266.

7. Katzman, G.M., Penney, D.G. (1993): Electro-cardiographic responses to carbon monoxide and cyanide in the conscious rat. Toxicol Letters, 69, 139-153.

8. Sabouraud, A., Urtizberea, M., Grandgeorge, M., Gattel, P., Makula, M.E., Scherrmann, J.M. (1991): Dose dependent reversal of acute murine cholchicine poisoning by goat cholchicine-specific Fab fragments. Toxicol, 68, 121-132.

9. Hultborn, R., Lundgren-Eriksson, L., Ottosson Lohn, S., Rye, W., Weiss, L. (1990): Chlorpromazine induced hypothermia in tumor-bearing mice, acute cytotoxic drug lethality and long-term survival. Acta Oncol, 29, 941-944.

10. Klein, M.S., Conn, C.A., Kluger, M.J. (1992): Behavioural Thermoregulation in mice inoculated with influenza virus. Physiology & Behaviour, 52, 1133-1139.

11. Gordon, C.J., Mohler, F.S., Watkinson, W.P., Rezvani, A.H. (1988): Temperature regulation in laboratory mammals following acute toxic insult.Toxicol, 53, 161-178.

12. Watkinson, W.P., Gordon, C.J. (1993): Caveats regarding the use of the laboratory rat as a model for acute toxicological studies: modulation of the toxic response via physiological and behavioral mechanisms. Toxicol, 81, 15-31.

Thermal Balance in Health and Disease
Advances in Pharmacological Sciences

PHARMACOLOGICAL ANALYSIS OF ANTICHOLINESTERASE-INDUCED HYPOTHERMIA AND HYPERTHERMIA

C. J. Gordon

Neurotoxicology Divison, Health Effects Research Laboratory, U.S. Environmental Protection Agency Research Triangle Park, North Carolina, U.S.A. 27711

Organophosphate (OP) compounds such as diisopropyl fluorophosphate (DFP) cause irreversible inhibition of acetylcholinesterase (AChE) activity in the central nervous system (CNS) and peripheral tissues. Following acute exposure to DFP and many other OP's, rodents exhibit signs of cholinergic stimulation characterized by salivation, lacrimation, tremor, and hypothermia (1-4). The acute signs of cholinesterase inhibition, including the hypothermia, generally recover within 24 hours after exposure in spite of the fact that AChE activity in the CNS remains inhibited (1).

Humans acutely exposed to various OP agents are reported to exhibit a fever lasting in some cases for several days (5). Until recently it was thought that cholinesterase inhibition led exclusively to hypothermia in rodents. However, using radiotelemetric techniques, this laboratory has found that an elevation in core temperature lasting several days is observed in the rat during recovery from DFP-induced hypothermia (1). Hence, with proper techniques to monitor core temperature in the unstressed rodent, it is possible to study the mechanisms of hypothermia and hyperthermia following cholinesterase inhibition. The purpose of this study is to use cholinergic blocking agents to study the mechanisms of DFP-induced hypothermia and hyperthermia in the rat while measuring body temperature, motor activity, and heart rate with radiotelemetric procedures.

Materials and methods

Animals used in these studies were male rats of the Long-Evans strain at an approximate age of 3 months. The rats were maintained at an ambient temperature (Ta) of 22 RC, 50% relative humidity, and a 12:12 L:D photoperiod. Details on the methods for surgical implantation of the transmitters have been published (1). Briefly, the transmitters (Data Sciences, St. Paul, MN) are implanted into the abdominal cavity under nembutal anesthesia. The rats are permitted at least one week to recover before used in an experiment. The transmitter provided continuous monitoring of core temperature, heart rate, and motor activity 24 hours per day. All data are recorded at 10 min intervals and stored on a hard disk for later analysis. In general, the telemetric variables were recorded for several days prior to administration of chemical agents. DFP (Sigma, St. Louis, MO) was dissolved in anhydrous peanut oil and injected subcutaneously (sc) at a dose of 1.5 mg/kg. Immediately prior to the administration of DFP the rats were treated with a cholinergic blocker or other pharmacological agent. In the experiment presented in this paper, the rats were treated with 1.0 mg/kg methylscopolamine or physiological saline. The telemetry variables were usually monitored for at least 72 hours after the administration of DFP. The data recorded at 10 min intervals were averaged into one hour bins and analyzed statistically using a two-way repeated measures analysis of variance.

Results

An example of the time course of core temperature over a four day period following the administration of a relatively low and high dose of DFP illustrates the hypothermic and hyperthermic effects of this OP (Figure 1). 1.0 mg/kg of DFP, a relatively high dose leading to ~80% inhibition in AChE activity, causes a marked decrease in core temperature reaching a nadir approximately 5 hour after injection. This is followed by a gradual recovery and then an increase in core temperature above the preinjection day time level, an effect which persisted for 4 days. The low 0.1 mg/kg dose of DFP causes a 47% reduction in AChE activity and does not have the hypothermic effects seen with the higher dose.

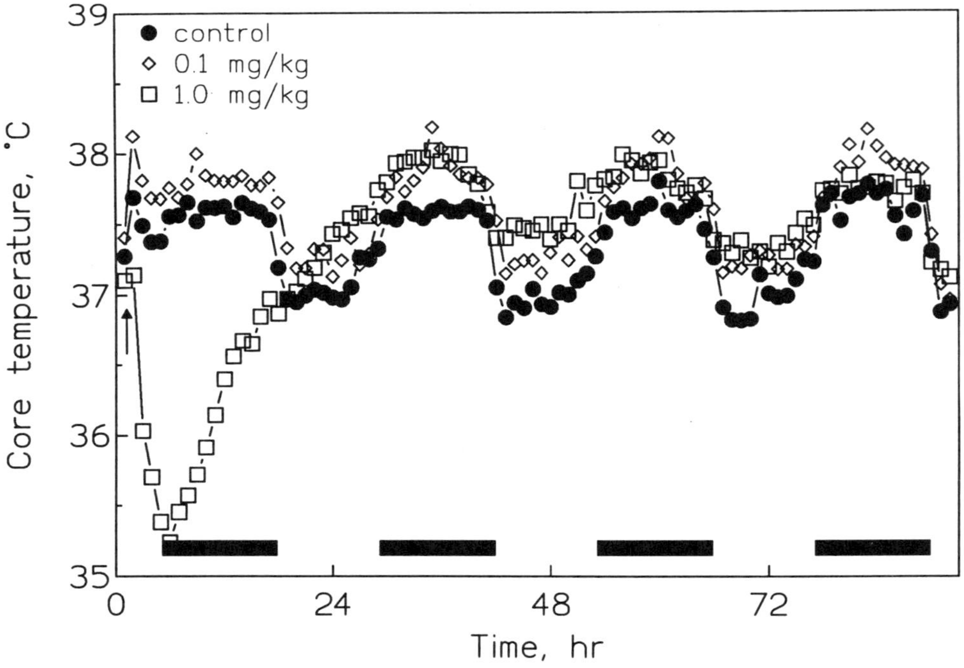

Fig. 1: Time-course of core temperature following administration of 0.01, 0.1 or 1.0 mg/kg DFP (sc) in the Long-Evans rat. Modified from (1). Bars indicate periods of lights out.

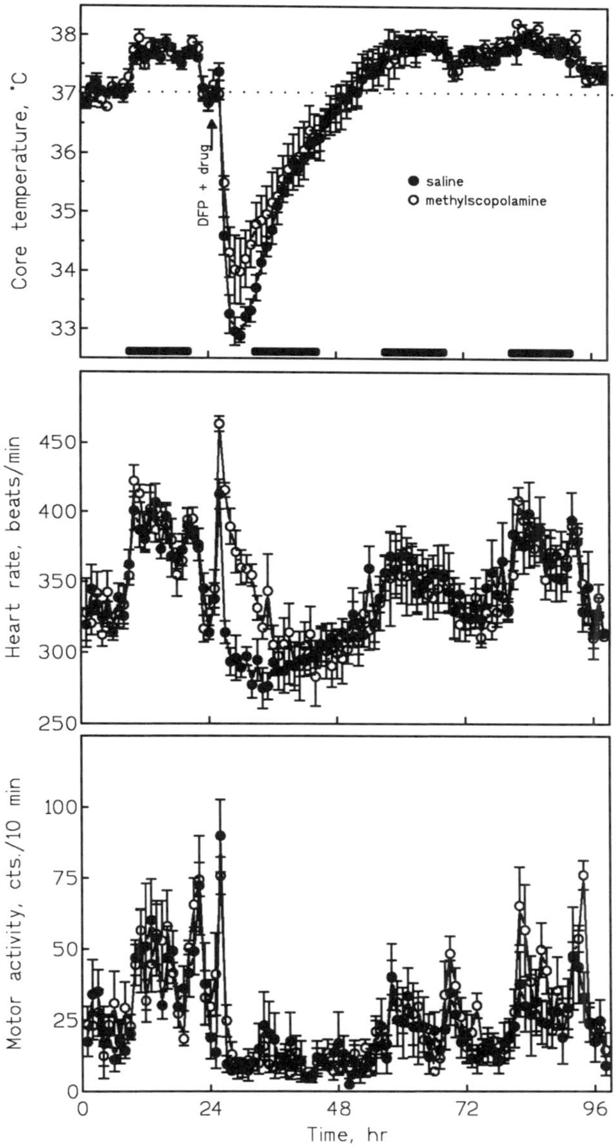

Fig. 2: Effect of methylscopolamine on thermoregulatory response to 1.5 mg/kg DFP in the Long-Evans rat.

However, there is nonetheless a subtle elevation in core temperature which persisted for several days after DFP treatment. The hypothermic and hyperthermic effects of cholinesterase inhibition were also seen following administration of 1.5 mg/kg DFP immediately after treatment with either saline or 1.0 mg/kg methylscopolamine (Figure 2). Methylscopolamine alone stimulates heart rate but has little effect on body temperature and motor activity in the rat (6). Treatment with methylscopolamine attenuated the maximum DFP-induced hypothermia by approximately 1.0 RC. The methylscopolamine also temporarily reversed DFP's depressant effects on heart rate but it had no discernible effect on motor activity. The recovery period, characterized by a 0.7 RC elevation in the day time core temperature above basal levels, was virtually identical in the control and methylscopolamine treatment groups 48 hr after DFP. Heart rate and motor activity were also similar between the two groups during the recovery period.

Discussion

Using radiotelemetric procedures it is possible to observe key behavioral and physiological responses following cholinesterase inhibition in the unrestrained rodent. The hypothermic response to DFP has been noted in many other studies using conventional colonic probe techniques to monitor body temperature (2-4). However, the hyperthermic response during the recovery to DFP has only recently been observed using radiotelemetry.

Methylscopolamine is a quartenary muscarinic antagonist and does not penetrate the blood brain barrier in appreciable quantities. The fact that methylscopolamine administration elicits a marked acceleration on heart rate but has little effect on motor activity or core temperature is clear indication of a predominant peripheral effect. In this study, methylscopolamine reversed the depressant effects of DFP on heart rate, attenuated the maximum drop in core temperature by 1.0 RC, but had little effect on motor activity. In another study, chronic administration of methylscopolamine via osmotic minipump has similar effects on heart rate and core temperature following 1.5 mg/kg DFP (Gordon, unpublished). From these data it is apparent that the hypothermic response to DFP is driven predominantly from changes in central neural function, whereas the effects on heart rate are controlled mainly from the peripheral effect of DFP.

The DFP-induced hyperthermia during the recovery period was similar between the saline and methylscopolamine groups. During this period it is likely that methylscopolamine has been cleared from the blood. However, this experiment nonetheless demonstrates that the hyperthermic effects of DFP are apparently not influenced if DFP's acute effects on core temperature and heart rate are attenuated. That is, the methylscopolamine-treated rats had a higher heart rate and smaller drop in core temperature immediately after DFP but exhibited an increase in core temperature during the recovery period that was similar to that of the controls.

DFP-induced hypothermia is thought to be mediated via the cholinergic activation of heat loss pathways in the preoptic/anterior hypothalamic area and other thermoregulatory centers in the CNS. (1,4,7). During the initial stages of hypothermia there is an elevation in tail skin temperature, reduction in metabolic heat production, and preference for cooler ambient temperatures, all of which suggest a stimulation in heat loss processes. On the other hand, the mechanism responsible for the rise in core temperature following low doses of OP's in rodents, recovery from high doses of OP's in rodents, and following acute exposure to OP's in humans is not understood. Combined use of pharmacological techniques with radiotelemetric procedures to collect data in the unstressed rodent will be of paramount importance for understanding the mechanism of OPinduced hyperthermia.

References

1. Gordon, C.J. (1993): Acute and delayed effects of diisopropyl fluorophosphate on body temperature, heart rate, and motor activity in the awake, unrestrained rat. J. Toxicol. Environm. Hlth. 39, 247-260.

2. Kokka, N., Clemons, G.K., and Lomax, P. (1987): Relationship between the temperature and endocrine changes induced by cholinesterase inhibitors. Pharmacology 34, 74-79.

3. Clement, J.G. (1991): Variability of sarin-induced hypothermia in mice: Investigation into incidence and mechanism. Biochem. Pharmacol. 42, 1316-1318.

4. Meeter, E. (1969): The mode of action of cholinesterase inhibitors on temperature regulation of the rat. Arch. Int. Pharmacodyn. 182, 416-419.

5. Namba, T., Nolte, C.T., Jackrel, J., and Grob, D. (1971): Poisoning due to organophosphate insecticides. Am. J. Med. 50, 475-492.

6. Gordon, C.J. (1993): Thermoregulatory effects of chlorpyrifos in the rat: Long-term changes in cholinergic and noradrenergic sensitivity. Neurotoxicol. Terat. (in press).

7. Gordon, C.J., Fogelson, L., Lee, L., and Highfill, J. (1991): Acute effects of diisopropyl fluorophosphate (DFP) on autonomic and behavioral thermoregulatory responses in the Long-Evans rat. Toxicol. 67, 1-14.

Thermal Balance in Health and Disease
Advances in Pharmacological Sciences

THERMOREGULATORY CONSIDERATIONS AFFECTING BOTH ACUTE AND PROLONGED EXPOSURES TO OZONE IN RODENTS

W. P. Watkinson, M. J. Wiester, J. W. Highfill, A. A. Aileru[1], M. J. Campen[2], J. S. Tepper[3] and D. L. Costa

U.S. Environmental Protection Agency, Health Effects Research Laboratory, Pulmonary Toxicology Branch, MD-82, Research Triangle Park, North Carolina, 27711 U.S.A.
[1]Howard University, Washington, D.C.; [2]Virginia Polytechnic and State University, Blacksburg, Virginia; [3]Genentech Corporation, South San Francisco, California, U.S.A.

Introduction

Our laboratory has been investigating the thermoregulatory effects of xenobiotic agents in rodents for a number of years [1, 2]. The general hypothesis which has evolved as a result of these studies can be stated as follows: *In response to exposure to a wide variety of toxic agents, rodents physiologically and behaviorally lower their body core temperature; the resulting moderate hypothermia confers specific advantages with respect to decreased toxic response and increased survival. If measures are employed to block or exacerbate the moderate hypothermic response, toxicity and lethality may be potentiated.* More recent work which examined pulmonary and extrapulmonary effects of air pollutants has provided additional support for the above hypothesis. In these studies [3, 4], exposure to near environmental levels of ozone (O_3) triggered a profound hypothermia in the rat which was accompanied by significant decreases in heart rate (HR) and activity, as well as increases in selected biochemical indices of pulmonary damage. The hypothermic response may precede the other observed responses and appears to play an important role in determining the extent of O_3-induced toxicity. The present studies investigated the effects of both acute and prolonged exposures to O_3 in rats maintained at normal ambient temperatures (T_a), as well as the modulatory influence on the induced toxicity exerted by altering the T_a at which the exposures were conducted.

Methods

General. Radiotelemetry methodology was used in these studies to permit continuous monitoring of HR, body core temperature (T_{co}), and activity in unanesthetized, unrestrained adult Fischer 344 rats. Animals were anesthetized with sodium pentobarbital and implanted with radiotelemetry transmitters (Data Sciences, International, St. Paul, MN, USA; Model TA11CTA-F40). They were allowed 1 week to recover from the surgery. In some experiments, respiratory frequency and oxygen consumption were also monitored using conventional techniques. Ozone was generated by passing pure oxygen across an ultraviolet light. The output of the O_3 generator was connected to the exposure chamber via a T-fitting that allowed the chamber to be supplied with either filtered air (Control Exposures) or a variable quantity of O_3 mixed and diluted with filtered air (Ozone Exposures) to achieve the desired concentrations. A number of experimental protocols were employed to permit examination of both acute and prolonged exposures (2 h-5 day) to near environmental levels of O_3 (0.25-1.0 ppm) over a range of T_a's (10-34°C). Following O_3 exposure, rats were anesthetized, their lungs were lavaged with saline, and the lavage fluid was examined for recognized biomarkers of O_3 toxicity.

Statistics. Portions of the T_{co} data from these studies are presented as actual values and the remainder are presented as relative values in order to facilitate comparisons among groups. The relative data were normalized with respect to an average T_{co} control value obtained during the preexposure period and are presented as the magnitude of change from the control value. Specific experimental protocols and conditions are given in the figure legends. Normalized data were analyzed using a repeated measures multivariate analysis of variance, with significance levels ($p \leq 0.05$) adjusted using a Bonferroni-type correction for multiple comparisons.

Results and Discussion

Acute Exposures. Acute exposure of rats to varying concentrations of O_3 while maintained at near normal room T_a's (0.0-1.0 ppm; 2 h; 18-20°C) yielded a modified dose-concentration relationship with respect to both T_{co} and HR [3]. While only the temperature data are presented here, it should be noted that the shapes of the response curves of these two parameters were essentially identical. There were no T_{co} effects observed at 0.25 ppm (Figure 1) while moderate but statistically significant decreases were seen during exposure to 0.37 ppm O_3. The effects observed at 0.5 and 1.0 ppm O_3 were more severe than those seen at the lower concentrations and were statistically indistinguishable from each other. The magnitude of the decrease in T_{co} induced with acute exposure to intermediate levels of O_3 (0.5 ppm; 2 h; 18-20°C) averaged 3.6 °C in these studies, and was accompanied by a 220 bpm decrease in HR and a 145% increase in respiratory frequency. A second study which repeated the protocol of the previous study but utilized a higher O_3 concentration and exposure T_a (1.0 ppm; 2 h; 30-32°C), resulted in an induced decrease in T_{co} of less than 1.0 °C (Figure 2), in conjunction with a HR decrease of 115 bpm. A third study employing similar exposure conditions but at a lower T_a (0.5 ppm; 6 h; 10°C) yielded comparable results, although of different magnitudes [5]. In the third study, T_{co} decreased an average of 2.9 °C while HR showed a 160 bpm decrement.

Prolonged exposures. Prolonged exposure of rats to a moderate concentration of O_3 maintained at various T_a's (0.5 ppm; 23 h/day x 5 day; 10-34°C) demonstrated both T_a and length of exposure dependency [5]. The data obtained during the initial portion of a prolonged exposure conducted at normal room T_a (22°C) were similar to previous results observed during equivalent acute exposure conditions. With continued exposure to O_3, however, both T_{co} and HR continued to decrease, reaching nadirs averaging 1.8 °C and 150 bpm, respectively, by the end of the first exposure day. Despite maintenance of the O_3 exposure, these functional parameters gradually returned to preexposure levels and, by the third exposure day, were not significantly different from those of air-exposed animals. In a second study in which identical exposures were conducted at a lower T_a (10°C), the magnitude and duration of the response were greatly potentiated (Figure 3).

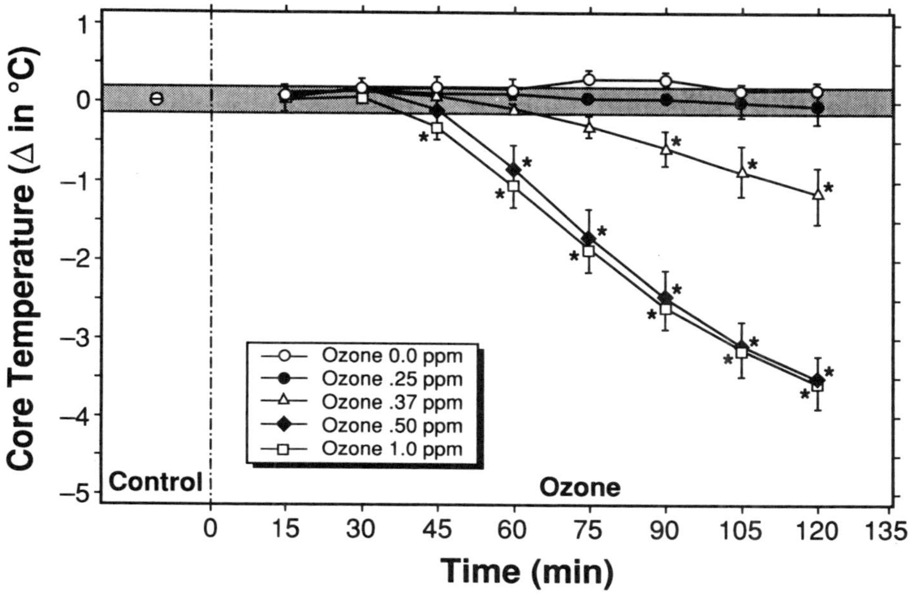

Fig. 1: Relative change in core temperature of unanesthetized, adult Fischer 344 rats (n=4-6 /group) following exposure to different concentrations of ozone. Data have been normalized with respect to preexposure values and are presented as magnitude of change in °C from initial control value. All experiments followed the same protocol: 0.25 h control period, 2 h exposure period, and 3 h recovery period (not shown). Ambient temperature was maintained at 18-20°C. Circle with bar indicates initial control value for all groups and shaded band represents ± 2 SEM of initial control value for 0.0 ppm group animals. Data points shown are means ± 1 SEM.
* Indicates significant difference ($p \leq 0.05$) from time-paired control value.

Under these conditions, T_{co} decreased 3.5 °C and HR dropped 210 bpm, with both parameters requiring nearly three days to return to preexposure levels. The responses observed in a third study that employed the same exposure regimen at a higher T_a (34°C) were significantly attenuated. There was only a slight decrease in T_{co}, averaging 0.5 °C, while HR effects were characterized primarily by a shift in circadian rhythm. In a final study in which both concentration and T_a were changed simultaneously (1.0 ppm; 10°C), the results were especially dramatic, demonstrating 75% lethality in the exposure group after approximately 24 h of continuous exposure [6].

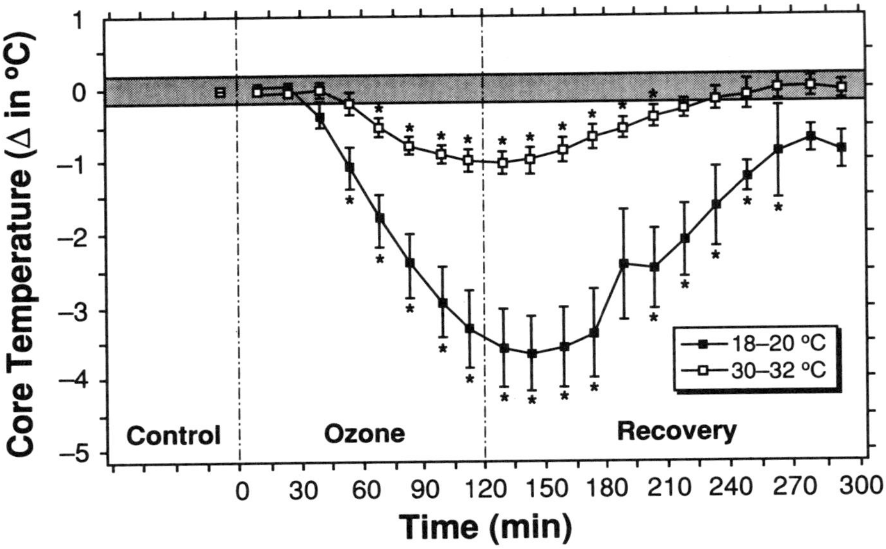

Fig. 2: Comparison of relative change in core temperature of unanesthetized, adult Fischer 344 rats (n=6/group) following exposure to 1.0 ppm ozone at different ambient temperatures. Data have been normalized with respect to preexposure values and are presented as magnitude of change in °C from initial control value. All experiments followed the same protocol: 0.25 h control period, 2 h exposure period, and 3 h recovery period. Ambient temperature range was maintained at either 18-20°C or 30-32°C. Square with bar indicates initial control value for both groups and shaded band represents ± 2 SEM of initial control value. Data points shown are means ± 1 SEM.
* Indicates significant difference ($p \leq 0.05$) from initial control value. Note: SEM values for data points that do not appear to have SEM bars fall within the value range covered by the graphic symbol.

Extrapolation Concerns. Ozone exposure in rats is capable of inducing significant changes in T_{co} (and other functional parameters) and it appears that the magnitude of the decrease in T_{co} may play a role in determining the extent of the induced toxic response. Furthermore, it has been amply demonstrated that the resultant T_{co} in the rodent exposed to O_3 is highly dependent upon the T_a at which the exposure is conducted. Other experimental factors known to influence the magnitude and duration of the hypothermic response include O_3 concentration, animal mass, stress, level of activity, and housing conditions [4, 7]. Unless such factors are clearly defined and closely regulated, the ability to extrapolate and compare results obtained from one rat study to another may be seriously compromised. Comparisons of data from studies using different species of varied sizes further complicates the extrapolation process. For example, hypothermic responses in mice following O_3 exposure (2.0 ppm; 2 h; 22°C) were even more dramatic than those observed in rats, averaging 7.5 °C [7]. Comparisons between rat and human studies present the greatest difficulty, however, as humans employ different thermoregulatory effectors and do not appear to exhibit a hypothermic response to O_3 exposure. It is clear that the thermoregulatory response in the rat invokes complex changes in a number of parameters and that it must be considered when evaluating O_3 toxicity and extrapolating these results to humans.

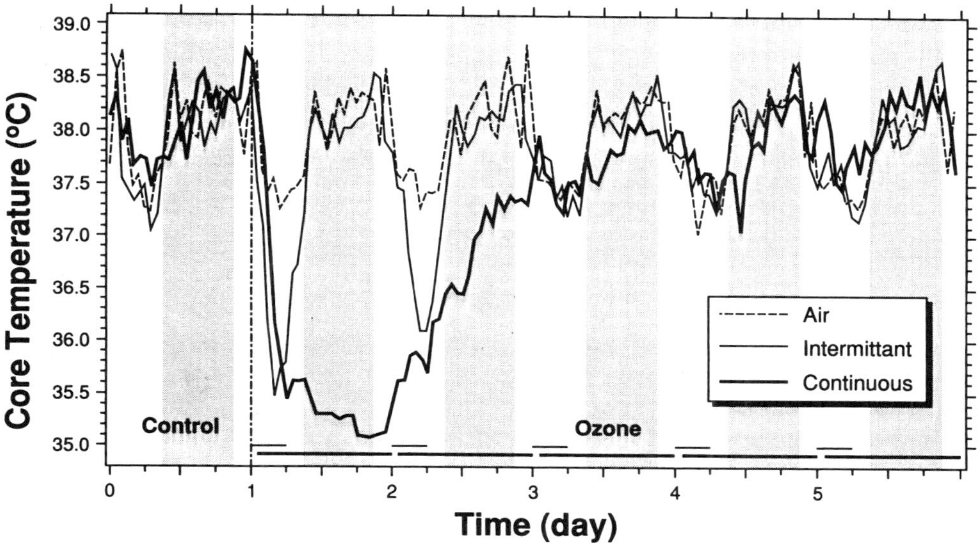

Fig. 3: Effects of prolonged ozone exposure on core temperature of unanesthetized, adult Fischer 344 rats (n=4-6/group) maintained at 10°C. All animals were exposed to filtered air for the first day (Day 0). Non-treated animals (Air) were exposed to filtered air throughout the experiment while treated animals were exposed to 0.5 ppm ozone for either 6 h/day (Intermittent) or 23 h/day (Continuous) for 5 days (Days 1, 2, 3, 4, 5). Data were obtained at 10 min intervals but for presentation purposes have been averaged across 1 h periods, with each data point representing the combined mean of 24-36 individual values. All time periods (Days) began at 0900 h. Ozone exposure regimens for Intermittent and Continuous animals are represented as light and heavy bands, respectively, shown at the bottom of the figure. Periods of light and dark are indicated as light and dark vertical bands, respectively.

References

1. Gordon, C.J., Mohler, F.S., Watkinson, W.P., Rezvani, A.H. (1988): Temperature regulation in laboratory mammals following acute toxic insult. *Toxicol* 53, 161-178.

2. Watkinson, W.P., Highfill, J.W., Gordon, C.J. (1989): Modulating effect of body temperature on the toxic response produced by the pesticide chlordimeform in rats. *J Toxicol Environ Health* 28, 483-500.

3. Watkinson, W.P., Aileru, A.A., Dowd, S.M., Doerfler, D.L., Tepper, J.S., Costa, D.L. (1993): Acute effects of ozone on heart rate and body temperature in the unanesthetized, unrestrained rat maintained at different ambient temperatures. *Inhal Toxicol* 5, 129-147.

4. Watkinson, W.P., Gordon, C.J. (1993): Caveats regarding the use of the laboratory rat as a model for acute toxicological studies: Modulation of the toxic response via physiological and behavioral mechanisms. *Toxicol* 81, 15-31.

5. Watkinson, W.P., Wiester, M.J., Campen, M.J., Richardson, V.M. (1992): Ozone toxicity in the unanesthetized, unrestrained rat: Effect of changes in ambient temperature on physiological parameters. *Toxicologist* 12, 230.

6. Watkinson, W.P., Wiester, M.J. (1991): Modulation of ozone toxicity with changes in ambient temperature in the unanesthetized rat. *FASEB Journal* 5, A1402.

7. Watkinson, W.P., Highfill, J.W., Slade, R., Hatch, G.E. (1994): Effects of acute exposure to ozone on body temperature and bronchoalveolar lavage fluid constituents in C57BL/6J and C3H/HEJ mice. *Am Rev Respir Dis* (in press).

Keyword Index

Author Index